John S. Scott, who was born in 1915, is a chartered structural
and mining engineer, a certified colliery manager, and a qualified
linguist. He has worked in Austria, France, Germany, Saudi
Arabia and the UK and has spent ten years in civil engineering,
but now writes books when he can afford it. At other times he
translates texts from French, German or Russian. He has six
children and lives in Germany.

THE PENGUIN DICTIONARY OF
CIVIL ENGINEERING

JOHN S. SCOTT

FOURTH EDITION

PENGUIN BOOKS

PENGUIN BOOKS

Published by the Penguin Group
Penguin Books Ltd, 27 Wrights Lane, London w8 5TZ, England
Penguin Books USA Inc., 375 Hudson Street, New York, New York 10014, USA
Penguin Books Australia Ltd, Ringwood, Victoria, Australia
Penguin Books Canada Ltd, 10 Alcorn Avenue, Toronto, Ontario, Canada M4V 3B2
Penguin Books (NZ) Ltd, 182–190 Wairau Road, Auckland 10, New Zealand

Penguin Books Ltd, Registered Offices: Harmondsworth, Middlesex, England

First published 1958
Second edition 1965
Third edition 1980
Fourth edition 1991
3 5 7 9 10 8 6 4

Printed in England by Clays Ltd, St Ives plc
Set in Monophoto 8½/9½ pt Imprint

CONTENTS

PREFACE TO FOURTH EDITION

At the time of this revision in 1989, metrication, which had started with the building industry and its corresponding government departments in 1970, was still only half completed in the UK. The BSI had metricated, as had several large nationalized industries including the British Coal Corporation and British Steel, but others had not. Feet, inches, yards, pounds, acres and miles have therefore occasionally been left in, but new emphasis has been given to kilograms, kilometres, metres, centimetres and millimetres. Newtons (N), newtons per square metre (N/m^2), pascals (Pa) and bars have been introduced to the text.

There is a bewildering variety of choice and little authoritative guidance, even from the continent of Europe where the Système International (SI) originated. In practice, for stress in metals the N/mm^2 is widely used, since it is conveniently large ($1 \ N/mm^2 = 145$ psi) though the meganewton per square metre or megapascal might serve equally well ($1 \ N/mm^2 = 1 \ MN/m^2 = 1 \ MPa$). For modulus of elasticity, a unit 1000 times larger is convenient and the GN/m^2 has been used, though the kN/mm^2 might be more logical. For soil stresses, which are much smaller than metal stresses, common units are kN/m^2 ($1 \ kN/m^2 = 0.145$ psi).

The bar had legal status on the continent of Europe and is still widely used there (10 bars $= 1 \ N/mm^2 = 145$ psi), being roughly one atmosphere. The standard atmosphere is slightly different: 760 mm mercury $= 0.101 \ N/mm^2$.

Another unit widely used on the European continent is the 'technical' atmosphere of $1 \ kg/cm^2$ or (in Germany) the kilopond/cm^2 (kp/cm^2). Roughly equivalent to 1 bar and to 10 metres of water, it is almost exactly $0.098 \ N/mm^2$.

When in doubt refer to BS 350 'Conversion Factors'.

J.S.S.

ACKNOWLEDGEMENTS

Enormous help was given by Anita Witten of the Structurals Library, Mike Chrimes of the Civils, Janet Flynn of the British Cement Association; Peter Ridd of Pilkington Glassfibre; Barry Theobald of Holliday Concrete Testing; Denys Parsons, consulting chemist; John M. Dransfield of Fosroc Technology; Dr Peter Bosworth of the Concrete Admixtures Association; Valentine H. Lewin, IWEM, water and waste water consultant; Jan Karalus, FRICS, of Clyde Surveys; P. R. Head of Maunsell Structural Plastics; Bernard Gambrill, Permanent Way Maintenance Engineer, British Rail; Ivan Jameson, Chief Plant Engineer, Opencast Executive, British Coal Corporation; Alan Maddison, freelance site engineer; George Reason, freelance site engineer; Jack Dixon, concrete technologist, British Rail research; James MacLean, builder, Tasmania; Ronald Birse, chartered civil engineer and civil engineer biographer, Edinburgh; Kiran Grimm, illustrator; Isobel Barrill of the Institute of Measurement and Control. Last but not least, from my wife I have had constant inspiration.

The publications of the BSI, the ASTM, and the ASCE have been helpful; so have CIRIA and the ACI, publishers of many useful guides. The Mechanical Handling Engineers' Association sells useful glossaries. The following books particularly have helped:

Application of Artificial Intelligence Techniques to Civil and Structural Engineering, ed. B. H. V. Topping, Civil-Comp Press, 1986.

Architects' Standard Catalogue, annually, 4 vols.

Biographical Dictionary of Railway Engineers, J. Marshall; David & Charles, 1978.

Concrete Technology, A. M. Neville and J. J. Brook, 1987.

Concrete Year Book, Palladian Publications, 1988.

Construction and Design of Prestressed Concrete Segmental Bridges, W. Podolny, Jun., and J. M. Mullen, John Wiley, 1982.

Design of Small Dams, US Dept of the Interior, Bureau of Reclamation, 3rd edn 1987. (Some US small dams are like large ones in the UK.)

Dictionary of Geotechnics, S. H. Somerville and M. A. Paul, Butterworth, 1983.

Dictionary of Soil Mechanics & Foundation Engineering, J. A. Barker, Construction Press, 1981.

Dictionary of Waste & Water Treatment, J. S. Scott and Paul Smith, Butterworth, 1981.

Earthmoving and Heavy Equipment, ASCE, 1986.

Earthquake-resistant Design for Engineers and Architects, D. J. Dowrick, Wiley, 1987.

Ferrocement, S. Abercrombie, Schocken, 1977.

GRC and Buildings, M. W. Fordyce and R. G. Wodehouse, Butterworth, 1983.

Great Engineers and Pioneers in Technology, ed. R. Turner and S. Goulden, St Martin's Press, 1981.

Guinness Book of Structures, 1976.

Handbook of Water Purification, ed. W. Lorch, Ellis Horwood, 1987.

History of Civil Engineering, H. Straub, Leonard Hill, 1952.

McGraw-Hill Yearbook of Science and Technology, 1989 (annually).

Modern Construction Equipment and Methods, Frank C. Harris, Longman, 1989.

Oxford English Dictionary, supplements of 1982 and later.

Planning the Rehabilitation of Water Mains, Water Research Centre, 1986.

Properties of Concrete, A. M. Neville, Pitman, 3rd edn, 1981.

Samuel Smiles: Several books on Victorian and earlier engineers.

Sell's Building Index 1989, Sells Publications Ltd.

Sewerage Rehabilitation Manual, Water Research Centre, 1984.

Short History of Technology, T. K. Derry and Trevor Williams, Oxford, 1979.

Specification, Architectural Press, 1987 (6 vols.).

Surveyor's Guide to Civil Engineering Plant, ed. Stephen Booth, FCES, Institution of Civil Engineering Surveyors, 1984.

Testing of Concrete in Structures, J. H. Bungey, Surrey University Press, 1982.

Transitions in Engineering: Guillaume Henri Dufour and the Early Nineteenth Century Cable Suspension Bridges, Tom F. Peters, Birkhäuser Verlag, 1987.

Wastewater Engineering: Treatment, Disposal, Re-use, Metcalf and Eddy Inc., McGraw-Hill, 1972.

ABBREVIATIONS

See also Units and Conversion Factors, p. xiv

(1) *Abbreviations indicating the branch of engineering for each entry (given in square brackets at the beginning of the entry)*

air sur.	photogrammetry	r.m.	rock mechanics
d.o.	drawing office practice	sewage	sewage treatment and
elec.	electrical engineering		sewerage
hyd.	hydraulics and	s.m.	soil mechanics
	hydrology	stat.	statistics
mech.	mechanical engineering	stru.	structural design
min.	mining	sur.	topographical or mine
rly	railways		surveying

Most of these headings are briefly described in the text.

(2) *Abbreviations of units (in singular or plural)*

AC	alternating current	kg	kilogram
bhp	brake horsepower	kJ	kilojoule
BThU or Btu	British thermal unit	km	kilometre
cc, cm³	cubic centimetre	kN	kilonewton
	(millilitre)	kN/m²	kilonewtons per square
cm	centimetre		metre
cu. ft, ft³	cubic foot	kph	kilometres per hour
cu. m, m³	cubic metre	lb	pound
cu. yd, yd³	cubic yard	m	metre
DC	direct current	m³	cubic metre
dia.	diameter	mg	milligram
E	modulus of elasticity	MN/m²	meganewtons per
fpm	feet per minute		square metre
ft	feet	MPa	megapascal
g	gram	mm	millimetre
g	gravity (acceleration)	Pa	pascal
gpm	gallons per minute	psi	pounds per square
I	moment of inertia	rpm	inch
in.	inch	s	revolutions per minute
J	joule		second

(3) *Abbreviations of organizations*

Below are a few mainly UK institutes, institutions and organizations for research, trade or parliamentary lobbying. Many more are listed in books by CIB and CIRIA. Many trade associations provide free publications on their subject.

AASHTO	American Association of State Highway and Transportation Officials
ACC	Association of County Councils
ACE	Association of Consulting Engineers
ACI	American Concrete Institute
AFNOR	Association française de normalisation (French standards body)
AMA	Association of Metropolitan Authorities
ASCE	American Society of Civil Engineers
ASHRAE	American Society of Heating, Refrigeration and Air Conditioning Engineers
ASME	American Society of Mechanical Engineers
ASTM	American Society for Testing and Materials
BACMI	British Aggregate Construction Materials Industries
BNCOLD	British National Committee on Large Dams (see also **ICOLD**)
BRE	Building Research Establishment
BRF	British Road Federation
BSI	British Standards Institution
Bureau Veritas	The French *classification* society
CEBTP	Centre Expérimental de Recherches et Études du Bâtiment et des Travaux Publics
CIB	International Council for Building Studies, Research and Documentation (The Hague, Netherlands)
CIBSE	Chartered Institution of Building Service Engineers
CIOB	Chartered Institute of Building
CIRIA	Construction Industry Research and Information Association
CS	Concrete Society
DIN	Deutsches Institut für Normung e.V. (German standards body)
DNV	Det Norske Veritas (Norwegian *classification* society)
EPA	Environmental Protection Agency of the US Federal Government
FCEC	Federation of Civil Engineering Contractors
FIDIC	Fédération Internationale des Ingénieurs Conseils (consultants)
GOST	USSR standards organization
IABSE	International Association for Bridge and Structural Engineering
ICE	Institution of Civil Engineers
ICOLD	International Commission on Large Dams (World Power Conference)

ISSMFE	International Society for Soil Mechanics and Foundation Engineering
IStructE	Institution of Structural Engineers
ITBTP	Institut Technique du Bâtiment et des Travaux Publics (Paris)
IWEM	Institution of Water and Environmental Management
Lloyd's Register of Shipping	The British *classification* society
Lloyd's of London	An organization of insurers
NAMAS	National Measurement Accreditation Service (c/o NPL)
NPL	National Physical Laboratory
NRA	National Rivers Authority
SECED	Society for Earthquake and Civil Engineering Dynamics (c/o ICE)
SETRA	Société d'Études Techniques des Routes et Autoroutes
TRADA	Timber Research and Development Association
TRRL	Transport and Road Research Laboratory
UBC	USA's Uniform Building Code
WI	Welding Institute

CROSS-REFERENCES

Cross-references in text are indicated by italic or bold type, those to the *Dictionary of Building* by the letter (*B*).

The great bulk of the terms used in building and civil engineering has forced the publisher to divide the material into two volumes, a *Dictionary of Building* and a *Dictionary of Civil Engineering*. Since many terms have two or more senses, these senses are likely to be listed separately in the two volumes. It will help the reader if he or she will first study the list of subject abbreviations at the beginning of each book, which is a rough guide to its contents. Both volumes contain a few essential terms dealing with contracts, bills of quantities and quantity surveying. Some other subjects, less easily allocated to one of the books, are in the list below.

Dictionary of Civil Engineering

concrete, reinforced concrete, concrete formwork, lightweight concretes, prestressed concrete, cast iron, wrought iron and steels

welding

underpinning, bridges, space frames, timber preservation

land surveying, photogrammetry, sewerage, sewage treatment and disposal, water purification, contract terms, bills of quantities

scientific terms such as dewpoint, electro-osmosis, geomorphology

Dictionary of Building

building trades, their tools and materials including plastics

brazing, soldering, capillary joints, silver brazing

carpentry, timber houses, shoring of buildings, air houses

heating and ventilation, insulating materials and techniques, degree-days

UNITS AND CONVERSION FACTORS

See also BS 350, Conversion Factors

Lengths

1 μm = 0.001 millimetre = 1 micron (officially micrometre)
25.4 millimetres = 1 inch
39.37 inches = 3.281 ft = 1 metre
5280 ft = 1760 yd = 1 mile = 1.609 kilometre = 1609 metres
1 international sea mile = 1852 metres
1 fathom = 6 ft = 1.829 metres

Areas

645 sq. mm = 6.45 sq. cm = 1 sq. inch
10.76 sq. ft = 1 sq. metre = 1.196 sq. yd
100 sq. metres = 1 are = 119.6 sq. yd
10,000 sq. metres = 100 ares = 1 hectare = 2.47 acres
1,000,000 sq. m = 100 hectares = 1 sq. kilometre (km)
4840 sq. yd = 1 acre = 0.4047 hectare
1 sq. (Gunter's) chain = 484 sq. yd = 0.1 acre
640 acres = 1 sq. mile = 2.59 sq. km

Volumes

There are three different gallons in use in the English-speaking world. The imperial gallon of Britain is so called to distinguish it from the two others, one wet and one dry, in use in the USA. Generally the wet US gallon, 0.833 of the imperial gallon, is the only one that concerns engineers.

62.4 lb of water occupy 1 cubic ft = 6.24 imperial gallons
1 US wet gallon = 0.833 imperial gallon = 3.785 litres
1 US dry gallon = 0.967 imperial gallon = 4.404 litres
1 imperial gallon of water weighs 10 lb (4.546 kg) and occupies 4.546 litres. (A pint of water weighs $1\frac{1}{4}$ lb.)
1 litre of water weighs 1 kg and occupies 0.22 of an imperial gallon
1 US barrel (bbl.) contains 5.62 cu. ft
35.28 cu. ft = 1 cu. metre
1 acre foot = 43,560 cu. ft = 1235 cu. metres

Weights

16 ounces = 1 pound (lb) = 454 grams (g)
112 lb = 1 hundredweight (cwt)
20 cwt = 2240 lb = 1 long ton = 1016 kilograms (kg)
1 US cental = 100 lb
20 centals = 1 short ton = 2000 lb = 2 kilopounds (kips)
50 kg = 1 $\left\{ \begin{array}{l} \text{Zentner} \\ \text{quintal} \end{array} \right\}$ = 110 lb = 0.98 cwt

But 100 kg = 1 quintal métrique
1000 kg = 1 tonne = 2200 lb
1 US sack (or bag) of cement weighs 94 lb = $\frac{1}{4}$ barrel
112 lb of cement is taken as occupying 1.25 cu. ft
1 Canadian bag of cement weighs 87.5 lb
1 UK bag of cement weighs 50 kg
1 tonne = 1000 kilograms (kg)

Forces

1 newton (N) = 0.2248 lb force
1 kilonewton (kN) = 224.8 lb force
10 kN = roughly 1 ton
10 newtons = roughly 1 kg
The German unit: 1 kilopond (kp) = 1 kg force = 9.81
  newton

Pressure, stress,
force per
unit area

1 pascal (Pa) = 1 newton per sq. metre (N/m²) = 0.000,145
  psi
1,000,000 Pa = 1 megapascal (MPa) = 1 N/mm² = 1,000,000
  N/m² = 1 MN/m² = 10 bar = 10,000 millibar = 145 psi =
  0.065 ton/in²
1 N/sq. mm = 1 newton/sq. mm ⎫
1 MN/sq. m = 1 meganewton/sq. m ⎬ = 145 psi
1 MPa = 1 megapascal ⎭
1 'technical' atmosphere = 1 kg/cm² = 14.2 psi = about 1
  bar (and 10 m water pressure)
1 'standard' atmosphere = 760 mm mercury = 101.325 kPa
1 torr = 1 mm mercury = 133.3 Pa (within 1 in 7,000,000)
1 psi = 6895 Pa = 6895 N/m² = 0.06895 bar

For engineers, almost all pressures are gauge pressures, i.e.
pressure above atmospheric, and gauge pressures therefore can
be negative. Absolute pressures are never negative.

Atü is the German abbreviation for the English 'gauge
pressure or psig'. Ata is the German abbreviation for the
English 'absolute pressure or psi abs'.

Energy, work
and power

SI units: 1 joule = 1 newton-metre = 1 watt-second = 1
  pascal-cubic metre
1 kilogram-metre = 7.233 foot-pounds = 9.81 N-m = 9.81
  joules
1 horsepower USA (and British) = 746 watts = 76 kg-m/
  s = 550 ft-lb/s
But 1 metric horsepower = 75 kg-m/s
Thus 1 USA (and British) horsepower = 1.014 metric horsepower

Heat and
insulation

273 Kelvin = 0 degree Celsius (formerly 0 degree Centigrade)
  but the interval between degrees Kelvin and degrees C is the
  same, so 373 K = 100°C

1 British thermal unit = 0.252 kilogram-calorie =
   1055 joules = 1.055 kJ
1 Btu/lb = 2.326 kilojoule/kilogram

Thermal conductivity k-value
1 Btu/in. sq. ft hour degree F = 0.144 watt/metre degree C
Therefore to translate k imperial to k metric, divide by 7

Thermal transmittance, U-value
(for a particular wall or roof of known thickness)
1 Btu/sq. ft hour degree F = 5.68 watt/m^2 degree C
To translate U-value imperial to U-value metric, multiply by
   5.7

Metric prefixes

		Prefix	*Symbol*	*Example*
million million	10^{12}	tera-	T	
thousand million	10^9	giga-	G	gigahertz (GHz)
million	10^6	mega-	M	megawatt (MW)
thousand	10^3	kilo-	k	kilogram (kg)
hundred	10^2	hecto-	h	
ten	10	deca-	da	decagram (dag)
tenth	10^{-1}	deci-	d	decimetre (dm)
hundredth	10^{-2}	centi-	c	centimetre (cm)
thousandth	10^{-3}	milli-	m	millimetre (mm)
millionth	10^{-6}	micro-	μ	microsecond (μs)
thousand millionth	10^{-9}	nano-	n	nanosecond (ns)
million millionth	10^{-12}	pico-	p	picofarad (pF)
thousand million millionth	10^{-15}	femto-	f	
million million millionth	10^{-18}	atto-	a	

Because they are not multiples of powers of three (10^3), hecto-,
deca-, deci- and centi- are not approved and should be avoided
in formal texts.

SELECTED BRITISH STANDARDS

There is not space to print the titles of the thousands of British standards concerned with building and civil engineering. The searcher for information may find what is needed by scanning the index of the British Standards Institution's annual catalogue. If you look up 'sand', you may also find 'sandlime bricks' and 'sandwich panel'; 'material' can lead to 'materials in contact with potable water'; 'reinforce' can lead to 'reinforced concrete, reinforced materials, reinforcing materials, reinforcing steels', etc. Below is a small selection of useful standards. Standards are often updated or superseded; check the BSI catalogue.

Aggregates for concrete and roads	blast-furnace slag, air-cooled 1047 blast-furnace slag, foamed lightweight aggregate 877 clinker 1165 glossary 6100 (30 sections) gravel for road surfaces 1984 guide to use of industrial waste materials 6543 lightweight aggregates 3797, 3681 natural aggregates 882 single-sized roadstone 63 testing aggregates 812, 4550 (4), 5835
Concrete and mortar standards	admixtures 5075, 4887; ASTM C-494 concrete code (design, construction, design charts) 8110 concrete linings for steel flues 4207 concrete mixers 1305, 3693 dry prepacked concrete 5838 post-tensioned prestressing anchorages 4447 precast units 5642, 5911, 6073; BSCP 297 prestressed concrete pressure vessels 4975 prestressing wire and strand 5896 sand: see **aggregates** (above) sewer pipes and fittings, incl. manholes of precast concrete 5911 specifying concrete, including ready-mix 5328 strength assessment of existing buildings 6089 structural design of concrete 8110 (3 parts in 1988) truck mixers 4251 vibrators 2769
Glossaries	Although the 150 or so BS glossaries are useful, they are not always immediately helpful to the uninitiated, since they aim to guide the technical people who write standards.

building and civil engineering 6100 (about 30 sections)
steels 2094
welding 499

Non-Portland   high-alumina c. 915, 4550
cements   supersulphated c. 4248, 4550

Portland cements   Portland blast-furnace c. 146
low-heat c. 1370
low-heat Portland blast-furnace c. 4246
masonry c. 5224, 4550
powder cement paints 4764, 4550
ordinary and rapid-hardening Portland c. 12
pigments for c. 1014
PFA: see **pozzolans**
sulphate-resisting Portland c. 4027
testing c. 4550

Pozzolans   blast-furnace slag, ground or granulated 6699
fly-ash 3892 (2 parts), 6588, 6610

Steels for use in   austenitic stainless-steel bars 6744
concrete   bending schedules 4466
cold-reduced steel wire 4482
cold-worked weldable bars, deformed high-yield steel
   (grade 460) 4461
high-tensile steel bar 4486
hot-rolled alloy steel bars for prestressing 5896
hot-rolled bars, mild steel and high-yield steel 4449
mesh reinforcement 4483
mild-steel wire 1052
prestressing wire and strand 5896

Structural steels   bolts and nuts 3692, 4190, 4933, 4395
building design 449, 5950
code for design and for hot-rolled steels (5 parts
   in 1988) 5950
cold-formed sections up to 8 mm thickness 5950 (5)
high-strength friction-grip bolts, nuts and washers 4395
hot-dip galvanizing 729
plate, sheet and strip of carbon and carbon–manganese
   steel 1449
hot-rolled steel sections 4848
profiled steel sheet floors 5950, part 4
sprayed aluminium or zinc protective coatings 2569
weldable structural steels 4360

Testing   cement testing 4550 (14 parts in 1988)
concrete testing and sampling 1881 (30 parts in 1988)

non-destructive testing 4408, glossary 3683
testing of resins used in building 6319 (10 parts in 1988)

General

adhesives glossary 6138
air compressors and pneumatic tools glossary 5791
bridges, steel, concrete or composite, incl. bearings 5400 (13
    sections)
builders hardware glossary 3827
building drainage 8301
building joints glossary 4643
building maintenance management 8210
cathodic protection code CP1021
cladding of buildings 8200
ductile iron pipes 8010
earthmoving equipment 6295; glossaries 5718, 6296, 6685
elastomeric joint rings for pipes 2494
electrically conductive or antistatic earthed rubber
    flooring 2050, 3187
energy efficiency of buildings 8207
external cavity walls 8208
fire protection including sprayed mineral coatings 8202
fixed offshore structures 6235
flooring of timber 8201
foundations 8004
glassfibre-reinforced-plastic water or sewer pipes 5480
loading of structures 8100
project network techniques 6046; glossary 4335
quality assurance glossary 4778; systems 5750 (many parts)
road tars 76, 5273
sealants 3712, 5889, 6213
security against crime 8220
setting out 6953
sound insulation code 8233
structural timber 5268 (several parts)
tubular steel scaffolding 1139
water-retaining structures 8007
water-well casing 879

Water

flow measurement glossary 5875
water quality 6068 (many parts including glossary)

Welding

argon arc w. 4365, 5135
carbon and carbon–manganese steels with carbon equivalent
    below 0.54%; procedures, consumables, preparation,
    inspection, approval of welders; manual, semi-automatic,
    automatic and mechanized arc w. 5135
Class 1 welding 2633, 1821
Class 2 welding 2640, 2971

NORTH AMERICAN STANDARDS

Most US standards are published by ASTM, the American Society for Testing and Materials, 1916 Race St, Philadelphia, PA 19103, though other standards organizations do important research. All the very numerous ASTM standards are republished annually. The 1987 text included 66 volumes classified into 16 sections of which No. 4, Construction, is the one which concerns us. Of the other sections, Section 1 is on iron and steel, Section 2 is on nonferrous metals, and Section 3 is on metal analysis and testing, each with five or six volumes. On average, 30% of each volume is new or revised. Each volume is indexed to show the code numbers of the standards in it.

ASTM Section 4, *Volume*
Construction
    04.01 Cement, lime, gypsum, 118 standards
    04.02 Concrete and aggregates, 176 standards
    04.03 Road and paving materials, traveled surface
        characteristics
    04.04 Roofing, waterproofing, bituminous materials, 126
        standards
    04.05 Chemical-resistant materials, vitrified clay, concrete,
        fiber-cement products, mortars, masonry
    04.06 Thermal insulation, environmental acoustics
    04.07 Building seals and sealants, fire, building
        constructions
    04.08 Soil and rock, building stones, geotextiles
    04.09 Wood

NOTE TO READERS

For abbreviated organizations, such as AASHTO, ACI, ASCE, ASME, CIRIA, ICE, RIBA, etc. please see p. xi, 'Abbreviations of organizations'.

A

Abney level [sur.] A *hand level* for measuring vertical angles, used for taking levels up steep slopes and as a *clinometer*.

Abrams' law Other things being equal, the *water/cement (W/C) ratio* determines the strength of a concrete. Provided that the mix is workable, the drier the mix, the stronger will the concrete be. Abrams' law was also discovered in 1892 by Féret in France. The least W/C ratio at which *hydration* of the cement is possible is about 0.25 but more than this is needed to wet the sand and stones. In practice precast concrete makers can use a W/C ratio of only 0.32 with good mix design and vibration, but see **Seattle**.

ABS *Acrylonitrile butadiene styrene*.

abscissa See **graph**.

abseil survey Examination of the surface of a structure by an engineer or technician who lowers himself or herself down it on a rope. Abseil surveys have been made of the *Humber Bridge* towers, and of Birmingham City housing blocks 32 storeys high. Cf. **access platform**.

absolute humidity See **humidity of air**.

absorbent liner Blotting paper, textile, etc., fixed inside concrete *formwork* to improve the surface finish.

absorbing well, w. drain A *well* to drain away water. See also **vertical sand drain**.

absorption loss [hyd.] Water lost by the wetting of the ground during the first filling of a canal, *reservoir*, etc. Cf. **seepage**.

absorption pit A *soakaway*. Cf. **disposal well**.

abutment A support of an arch or bridge, etc., which may carry a horizontal force as well as weight. See **arch dam**, **prestressed concrete**; also (*B*).

abutment wall See **wing wall**.

Abyssinian well A pointed, perforated tube driven into the ground by sledge hammer or by ramming with a light *pile hammer*. Water can be extracted from it by pumping. It is the ancestor of the *wellpoint*.

accelerated curing The use of a warm, moist environment for concrete *curing* to increase its early strength. *Steam curing* is common. Accelerated curing should never be used with *high-alumina cement*, nor with *calcium chloride*, but it is often essential for *precast concrete*. Many methods that achieve high early strength do lower the final strength. But if, for the first three hours, the concrete does not heat up by more than $15°C$ and thereafter does not heat up or cool down by more than $35°C$, and if its temperature never exceeds $85°C$, the effect may be minimized. Accelerated curing test methods that provide cube or cylinder test results on concrete matured in only one or two days are described by ASTM C 684–81 and BS 1881 part 112:1983. See also **jet cement**.

accelerator An *admixture* which hastens the hardening rate and/or *initial setting time* of concrete. Calcium chloride ($CaCl_2$) was widely used, but because it can corrode embedded steel it is now banned in the UK except in unreinforced concrete. Chloride-free accelerators that are safe with steel are based on inorganic chemicals including formates, nitrates and thiocyanates. Sodium carbonate (washing soda) can be used to make a flash set for quick repairs but it weakens the concrete.

accelerometer A *sensor* that measures acceleration. A simple one has a mass fixed to the end of a cantilevered spring-steel bar, which flexes when-accelerated. The amount of flexing is measured. See **inertial surveying system**.

access platform, aerial p., cage lift, sky l., cherry picker (manbucket USA) One type has a hydraulically operated crane boom that lifts a work platform, e.g. for maintenance of street lamps. One self-propelled vehicle having a maximum lift of 30 m (99 ft) has a maximum outreach of 9 m and a travelling speed of 20 mph, but some can lift to 60 m (200 ft). With all types, the controls are in the platform. Some platforms are towed. A towed, Danish access platform can lift 200 kg to a height of 25 m (82 ft). It has an outreach of 11 m (35 ft) and slews 360°. All this demands only four 6 volt batteries having a capacity of 240 amp-hours. A built-in mains charger with 15 m of cable enables the batteries to be charged, but no fewer than 75 lifts to 25 m are possible between charges.

Accidents with access platforms are serious and some training in their use is needed. No ladder should ever rest on one. Guniting, grit blasting, paint spraying or simply hosing from one produces a thrust which may overturn the platform. For such jobs the platform should be tied to the building or solidly strutted, or both. See **out-rigger**; cf. **rack-and-pinion mast platform**, **scissors lift**.

accidental error, random e. [sur.] *Compensating error*.

accumulator, hydro-pneumatic a. (see diagram) A pressure tank containing *hydraulic fluid*, sometimes separated by a rubber *diaphragm* from inert gas such as nitrogen. Another type has no inert gas or diaphragm but a spring and piston instead. It is commonly used in hydraulic systems, often with an oil pump (PM) used also as a motor. It is an energy store when the PM works as a pump to accumulate energy. If a PM is fitted to each wheel of a four-wheeled vehicle, differential gears and transmission shafts can be eliminated. With a hydraulic accumulator, the electric accumulator does not need to be so large, since it may be not needed for starting the engine. The pressure in a hydraulic system can be maintained for months. Consequently it is dangerous to start working on such a system without releasing the pressure.

accuracy [sur.] Accuracy is nearness to truth. Because absolute accuracy is unobtainable, it is approached by taking as many measurements as are practicable and averaging them. Practical limits of accuracy in civil engineering are about 3 mm for the placing of walls and floors and about 20 mm (1 in.) for long tunnels meeting in the middle of a mountain. Cf. **precision**.

acid Most commercial metals and all *Portland cement* concretes are attacked by acids. Some *polymers* in concrete may reduce acid attack.

acidity See **pH scale**.

acid rock An *igneous* rock containing much *silica*. Cf. **basic rock**.

acoustic strain gauge A *strain gauge* for measuring the extension or shortening of parts of a structure by tight wires fixed into them. If plucked, these

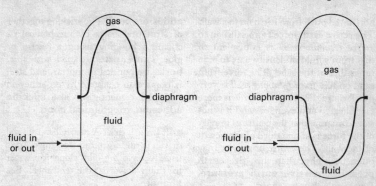

Pressure vessel for hydro-pneumatic accumulator. The vessel is storing more energy at left than at right. The diaphragm moves up under high pressure as fluid is pumped in and moves down as the fluid is released. The quantity of gas remains the same but is compressed as more fluid is pumped in, raising its pressure energy.

wires emit a note which becomes higher when they are stretched and lower when they are shortened. One instrument was developed by the *Building Research Establishment* (*B*).

acre foot [hyd.] A unit of measure for the capacity of a reservoir: the volume contained in 1 acre of water 1 ft deep. See 'Conversion factors', p. xiv.

Acrow prop A proprietary adjustable telescopic prop which supports deck *formwork* and enables it to be accurately levelled.

acrylic resin A *thermoplastic* resin used as a bonding agent or surface sealer (ACI). Perspex (Plexiglas, polymethyl methacrylate) is a common acrylic.

acrylonitrile butadiene styrene (ABS) A *thermoplastic* with a high softening point and good impact strength. ABS pipes are accepted by most authorities for cold water supply, soil and waste systems.

Ac system See **airfield soil classification**.

activated carbon See **GAC**.

activated-sludge process [sewage] Treatment of the *effluent* from *primary treatment* (sedimentation) by blowing air through it in an aeration tank for several hours, followed by settlement in a *secondary sedimentation* tank. Of the *sludge* that settles out, a proportion (activated sludge) is returned to the inlet of the aeration tank, the remainder (waste sludge) flows away for *de-watering* and disposal. The returned sludge, that has been activated by several hours' aeration and has separated in the sedimentation tank, inoculates the incoming effluent (or influent) with its flourishing microbial population. The process is much more efficient than a *trickling filter*, in that it needs only one seventh of the land area for the same volume of sewage but it also needs expensive, skilled, careful supervision to operate the compressors and other mechanical plant. Consequently it is almost universal in city treatment plants.

active earth pressure [s.m.] The

horizontal push from earth on to a wall. The active earth force from sand on to a *free retaining wall* is equivalent to that from a fluid of density 0.25 to 0.30 times that of the sand. The force from sand on to a *fixed retaining wall* is very much more. For clays the force is more complicated and even for sand it is not safe to assume that the centre of gravity of the force is at one third of the height of the wall. The centre of gravity may often be higher than this. See **earth pressure**, **passive earth pressure**, **Rankine's theory**, **yield**.

active layer [s.m.] The layer at the surface of the ground, which moves seasonally as the soil-volume changes, expanding when frozen in winter, and shrinking when it thaws and dries in summer. *Foundation* bases should be below the active layer. See **permafrost**.

actuator An electrical, pneumatic, hydraulic or mechanical device that provides a push, pull, rotation or other controlled motion when it receives a signal, usually electrical. *Solenoids* are often the main part of an actuator. A rotary actuator is formed by the motor and controls of a motor-controlled valve. A linear actuator provides straight-line movement only. An output *transducer* also is often an actuator.

addition, additive A substance blended with hydraulic cement during manufacture. Cf. **admixture**.

additive constant [sur.] In *stadia work*, a length added to the product of the *intercept* on the staff and the *multiplying constant* to give the true distance between telescope centre and staff. It is often less than 0.3 m. In *electronic distance measurement* it is a length added algebraically to give the true distance between instrument and reflector. See **anallactic telescope**.

adhesion, bond The sticking together of structural parts by mechanical or chemical bonding using a *cement* or glue. Timber parts are stuck with glue, bricks are bonded in mortar, and steel is bonded to concrete by its adhesion with the cement. See also **specific adhesion**, **mechanical bond**.

adit [min.], **drift** A nearly level tunnel driven to underground workings. Adits usually have a slight slope for water to drain towards the entrance. See **khanat**.

adjusting screw [sur.] A finely threaded screw on a surveying instrument to give it the final adjustment for level, focus or position.

adjustment [sur.] (1) When a series of survey observations is inconsistent with itself, each observation is adjusted to bring it into agreement with the others. For instance if the three angles of a triangle do not add up to 180° but to 180° + 3 minutes, one minute of arc is subtracted from each angle. See **Bowditch's rule**.

(2) An operation carried out on the bubbles, foot screws, centring device, etc., of a surveying instrument to make the instrument truly level, the bubble truly parallel with the line of sight, etc. An adjustment may be permanent or temporary. Adjustment of *automatic levels* and of *total-station instruments* must however be left to experts.

admeasurement See **measurement contract**.

admixture In concrete or mortar, a substance other than *aggregate*, *cement* or water added in small quantity, normally less than 5% of the weight of the cement, to alter the properties of the mix or the hardened solid. Some 80% of the concrete made in North America, Australia, Japan and most of

Europe contains an admixture, and more than half contains *air-entraining agent*. Other admixtures are *accelerators, bonding adixtures, superplasticizers, water reducers, retarders, anti-freeze, corrosion inhibitors, pore fillers* and *thickening agents. Shrinkage* preventers, colouring (pigments), damp-proofing, expanding, fungicidal, gas-forming, *grouting* and *flocculating* agents also exist.

For concretes with high cement content, water-reducing admixtures may reduce the cement needed to such an extent that the admixture pays for itself without harm to the physical properties. But this saving probably never occurs with very *lean concretes*.

adobe See (*B*).

adopted street See **dedicated street**.

adsorption Condensation of a film of gas or liquid on the surface of a solid. See **held water**.

advanced composites See **composites**.

advanced wastewater treatment, (AWT) [sewage] See **tertiary treatment**.

aeolian [s.m.] Wind-blown, a description often applied to *loess*.

aerated concrete Aerated concrete is in reality foamed calcium silicate, but the names aerated concrete, gas concrete, and foamed concrete are probably too well established to be changed. It has only three essential constituents, since it does not need the sand and stone which are essential to concrete. It needs only binder (cement or lime or a mixture of the two), water, and gas bubbles (which later fill with air). The bubbles are usually made by aluminium powder reacting with the lime of the binder. Air bubbles may be made by whisking with an air-entrain-

ing agent or similar admixture. A number of other admixtures improve foaming or reduce the cost of binder (fly-ash instead of cement). The two main sorts of aerated concrete are the autoclaved factory-made variety, subjected to curing under high steam-pressure for some 10–20 h, and the site-cast variety, which is much weaker than the autoclaved concrete but is nevertheless useful for insulating steam pipes, etc. In countries where high insulation in walls and roofs is essential (Sweden, Canada) the lightweight autoclaved reinforced floor beams and wall panels, about 0·4 m wide and up to 6 m long, have become extremely popular. In fact 70% of new factory roofing is made with them in Sweden, where they originated. Aerated concretes are made at any density from 0.4 to 1 g/cm^3 (occasionally up to 1.44). At the lowest density the k-value is about 0·1 W/m per degree C and the wet cube strength about 1.4 N/mm^2, whereas at 0.8 g/cm^3, k = 0.23 W/m per degree C and the cube strength is 4.8 N/mm^2. Trade names for this material in the UK are Celcon, Durox, Siporex, Thermalite and *Ytong*. See also **lightweight concrete** and **sand-lime brick** (*B*).

aeration of raw water *Raw water* is aerated to improve the taste and remove undesirable gases such as carbon dioxide and hydrogen sulphide, as well as dissolved iron and manganese, which are precipitated. Several aeration methods exist: splashing over trays, *stripping*, etc.

aeration tank [sewage] A concrete, steel or fibreglass tank in which sewage *effluent* is brought into contact with dissolved oxygen and *flocs* in the *activated-sludge process*. Oxygen enters the mixed liquor either by air blown in below the water surface as coarse or fine bubbles or by agitation with rotating paddles. See **deep-shaft process**.

aerial ropeway (aerial tramway USA) A line of towers carrying steel ropes which serve as tracks for buckets (called carriers) for coal, ore or building material in mountainous regions or into the sea where a road, railway, bridge or tunnel would be cumbersome or costly or slow to build. It has more than two towers so is not a *cableway*. Loads vary up to 1 ton and speeds may be up to 150 m/min. For different types see **bi-cable r., continuous r., jig back r., monocable r., twin-cable r**.

aerial surveying *Photogrammetry*.

aerial tramway American term for an *aerial ropeway*.

aerobic [sewage] Description of any treatment or activity that needs air or oxygen. Cf. **anaerobic**.

aerodynamic instability [stru.] Flutter during wind that is so large as to endanger a structure, a term used to describe the failure of the *Tacoma Narrows bridge*. See **Golden Gate bridge**.

A-frame [stru.] A *frame* which may or may not be symmetrical but consists generally of two sloping legs joined at the top and braced farther down by a horizontal or diagonal brace. An A-pole is a wooden A-frame.

aftercooler [mech.] A cooler inserted in the compressed-air circuit between the compressors and the mine (or other consumer of air) to cool and dehumidify the air and reduce its volume, and thus effectively increase the capacity of the pipeline. See **intercooler**.

ageing (aging USA) An alteration in the strength properties of a material with time, usually a strengthening or hardening, such as with *duralumin* after heat treatment, mild steel after cold working, concrete throughout its life, or plywood a few days after it has been glued. Age-hardening implies embrittlement because of age.

agent Formerly the *contractor*'s most responsible representative on a site; but see **project manager** for the current term.

aggregate *Gravel, sand, slag*, crushed rock or similar inert materials which form a large part of *concretes, asphalts* or *roads* including *macadam*. See also **angularity number, coarse aggregate, elongation index, fine aggregate, lightweight aggregate**.

aggregate/cement ratio The weight of aggregate divided by the weight of cement in a concrete.

aggressive See **corrosion**.

aging See **ageing**.

agitating truck A mobile concrete-mixer, carrying ready-mixed concrete, and turning slowly (2–6 rpm). Cf. **truck-mixer**.

Aglite A UK *lightweight aggregate* made from *expanded clay* (*B*).

agonic line [sur.] A line on a map of the earth along which the magnetic declination is zero.

agricultural drain *A field drain*.

AGVS *Automated guided vehicle system*.

A-horizon [s.m.] The uppermost of the three layers (A-, B-, C-) of soil science. Its upper part accumulates organic matter from plant roots while its lower part is leached of its soluble constituents and fine grains into the *B-horizon*. It may be 2 m (6½ ft) deep. See **podzol**.

air base [air sur.] The distance between the exposure stations of adjacent, overlapping aerial photographs. See **timing**.

air-bubble curtain See **bubble barrier**.

air compressor See **compressor**.

air content of fresh concrete The air content of fresh concrete is measured, under BS 1881, by applying air pressure to an enclosed sample of known volume and measuring the reduction in volume.

air embolism *Caisson disease*.

air-entrained concrete A concrete used for making roads. It has about 5% air and is therefore less dense than ordinary good concrete, but it has excellent frost resistance. The strength loss is roughly 5% for each 1 % of air entrained.

air-entraining agent An *admixture* to concrete or grout that drags tiny air bubbles, from 0.01 to 0.1 mm into the mix, amounting to about a quarter million per cm³. The bubbles increase *workability*, allowing both sand and water contents to be minimized. One agent is vinsol resin, made by distilling pine tree stumps. *Bleeding* and *segregation* are reduced, and the frost resistance of the *cement paste* is improved both during setting and after hardening, but the improved durability is gained at the expense of strength so the amount of air must be strictly limited. *ACI*, for frost resistance, normally requires 7.5% air but with large aggregate of 50 mm (2 in.) maximum size, only 5% because such concrete has less cement paste. BSI is in general agreement.

Some specialists think that all concrete should be air entrained except where the very highest strength is needed. See also **ASTM Portland cements, superplasticizer**.

airfield soil classification, Ac system [s.m.] A soil *classification* published by Arthur Casagrande in the USA in 1948. It is based on sieve analyses and on the *consistency limits*. *Cohesive* soils are divided into those with a *liquid limit* below or above 50%, the former being generally *silts* and the latter *clays*.

air-flush drilling Exploratory drilling using compressed air instead of water to cool the drilling bit and to clear the chippings. Developed in deserts where water was scarce, it is used also to drill water wells, etc., in the UK.

Airform construction Unreinforced 9 m (30 ft) dia. domes built of 50 mm (2 in.) thick *shotcrete* sprayed on to 'balloons' of inflated plastic, used by Eliot Noyes Associates in Florida in 1954 for houses and school buildings, and more recently in Malvern, UK. To increase the volume of a building two or more domes may intersect, since the domes have openings in opposite faces, about 7.5 m (25 ft) wide which can also be used for windows and doors. See **Binishell**.

air knife See **supersonic air knife**.

air-lift pump A pump consisting of two pipes hanging vertically in a well or sump (one may be within the other). Compressed air is injected into the larger pipe at its base from the smaller one. This air, as it expands and rises, reduces the density of the water in the larger pipe. The pressure of the water around the large pipe forces more water into it. The large pipe must be submerged to a considerable depth, but if this can be done, large volumes of very dirty water or mud can be pumped, at a high expense, in compressed air. See **compound air lift, hydraulic ejector, mammoth pump**.

air lock [min.] A compartment with two airtight doors, one to a tunnel, caisson, etc., under excess air pressure, the other to an area at lower pressure.

The air lock can thus give access between these two areas, but the two doors must never be open simultaneously. Similar air doors exist in mines to ensure that the ventilation is not short-circuited. See **man-lock**, **materials lock**.

airplane mapping *Photogrammetry*.

air pollution See **cancer cluster**, **pollutants**.

air pump See **vacuum pump**.

air rights The right, sold or leased by a local authority, to put a building above an open space such as a street. This may profit both the developer of the building and the local authority.

air scouring A method of cleaning out a water main by controlled injection of short, high-speed air blasts which remove loose deposits. Three skilled workers can clean up to 8 km/day, sometimes including a length of 1 km in one operation. But experienced people are needed, possibly also extra hydrants and valves, and the method is unsuitable for pipes larger than 200 mm (8 in.).

air skates An 80 mm (3 in.) thick metal platform which acts as a miniature hovercraft and is supplied with compressed air to lift it fractionally off the floor. Very heavy weights are easily moved. Roller skates are not supplied with air but are otherwise similar, with hard rubber rollers.

air stripping See **stripping**.

air-supported structure Balloon structures, a special class of *inflatable structure*, have a single skin (membrane) and are anchored to the ground. Though the internal air pressure to hold up the skin is small, the anchorage force needed is large. It should be regarded (for safety) as equal to double the weight of the skin and any supporting ropes or cables on it. The skin may be of nylon, polypropylene, polyester, glassfibre or other textile which does not support combustion. Snow loading can be disregarded if the internal air temperature is at least 12°C, but wind must be considered. All lights must stand on the ground; no loads may be hung from the skin.

To prevent air losses from a balloon structure, access to it is through air locks or revolving doors. The emergency fan and power supply should start within 30 s. Emergency doors should be at least 800 mm (32 in.) wide, and for over 100 people at least 1100 mm (3 ft 7 in.), or 530 mm for every 100 people. Air-supported boat hulls also exist – two-skinned inflatables, rescue dinghies like the picnic tents that are carried in a car. See also **air house** (*B*), and BS 6661.

air survey A survey made by *photogrammetry*.

air valve [hyd.] A valve in a water pipe at a summit. It allows air but not water to pass out automatically.

air vessel, receiver [mech.] In a pipeline on the delivery side of a *reciprocating pump*, a small air container which smooths out the pulsations of the pump.

air voids In the *cement paste* of mortar or concrete, there are at least two types of void caused by air bubbles: the irregular, unwanted, *entrapped-air* voids of 1 mm or larger and the much smaller, rounded, intentional *entrained-air* bubbles of 0.1–1 mm which make the concrete workable, protect the paste from frost and enable the sand and water contents to be minimized.

Akashi–Kaikyo Bridge, Japan A suspension bridge with many changes of plan and decades of research behind it for the earthquake area of the Akashi

Strait. The bridge towers are in waters 35 m (115 ft) deep with heavy scour. The 1986 proposal, being built at the time of writing (1988), was for the world's longest span of 1990 m (6500 ft), 960 m side-spans, and towers originally designed for a height of 333 m (1090 ft). They may be reduced to about 300 m. The railway originally proposed will not now be carried.

Alaska pipeline One of the largest civil engineering jobs of all time, begun in 1969 and completed in 1977. This 1.22 m (48 in.) pipeline, 1290 km (800 miles) long, runs from Prudhoe Bay on the north coast of Alaska (its North Slope) to Valdez, a relatively ice-free port on the Alaskan south coast. Construction was first planned to meet the expected world oil shortage of the early 1970s, but work was stopped for three years because of lawsuits by Alaskans for their land and by environmentalists for the *ecology* until the US Secretary of the Interior gave permission in January 1974 in the wake of the 1973 oil crisis. The National Environmental Policy Act had taken effect in June 1970, forcing the oil companies to give guarantees of their cooperation in an *ecosystem development stategy* to protect wild life of every sort. The oil companies wholeheartedly cooperated, but this most far-reaching of US laws on the environment was the first to require a statement of *environmental impact*, and the Alaska pipeline was probably the first to conform to this law.

At its peak over 20,000 men were working on the pipeline, but before it could begin to be built a road 360 miles long was needed from Fairbanks to the Arctic coastline. In the Arctic, where no roads exist, winter is the best season for transporting materials. Ice bridges are built over rivers by laying logs on the ice wider than the road width needed, then spraying these with water to make more ice and so on until the ice is 2 m (6 ft) thick.

No American or European steelmaker had a plant able to make 48 in. pipe in the time available, though several were willing to build a plant for the purpose, which would have taken a year. Japanese pipe was therefore bought, of high quality, weldable at temperatures well below freezing ($-30°$C), in three grades with *yield points* of 410, 450 and 480 MPa (60,000, 65,000 and 70,000 psi).

The oil comes out of the wells at $71°$C ($160°$F) and cannot be allowed to cool because it would then cease to flow. Wax would precipitate out of the oil and the pipe would freeze solid. All pipes are therefore jacketed with 95 mm ($3\frac{3}{4}$ in.) of foam which prevents the oil cooling below $60°$C ($140°$F). Around the insulation is a galvanized steel sheet casing. The environmentalists had emphasized the danger to the *permafrost* of the heat from the pipe. Therefore, where the pipe is buried in permafrost likely to thaw (near the southern end), cooling pipes are buried alongside the insulated oil pipe to make sure that the ground cannot thaw out. Near by a large refrigerator provides the cooling power and the pumps to circulate the coolant.

Where the pipe is above ground, and this is for about 580 km (360 miles), the pipe is carried on steel pipe-piles of 46 cm (18 in.) dia. If the permafrost were alternatively to freeze and thaw it would force these piles out of the ground. To prevent this happening, a *heat pipe* is inserted into each pipe to make sure that its base is kept cold. More than 60,000 heat pipes are installed. See **commissioning**.

Alclad Trade name for aluminium alloy coated with pure aluminium to give high corrosion resistance. See **clad steel**.

algae (one **alga**, two or more **algae**)
Primitive tiny plants which are undesirable in drinking water or filters but are helpful in natural waters and *stabilization* ponds because of their *photosynthesis*. Seaweeds are some of the most sophisticated algae, but most algae are invisible except in millions floating on ponds. In sunlight they convert carbon dioxide into oxygen by photosynthesis. They are larger than bacteria and viruses but smaller than all other *microorganisms*.

alidade, sight rule [sur.] An instrument used in *plane-tabling*, placed on the table, used as a straight edge. It carries, above it, sights by which the straight edge can be aligned with the object. In the USA the term is also applied to the telescope and its attachments on a *theodolite*.

align [sur.] To arrange in line; also spelt 'aline', mainly in the USA.

alignment [sur.] (1) The fixing of points on the ground in the correct lines for setting out a road, railway, wall, transmission line, canal, etc.

(2) A ground plan showing a route, as opposed to a profile or section, which shows levels and elevations.

alignment chart A *nomogram*.

aline See **align**.

alkali–aggregate reaction, concrete cancer Bursting of concrete caused by internal swelling resulting from a chemical reaction between some types of *aggregate* and the soluble *alkali compounds* in *Portland cement* or mixing water (*ACI*). A. M. Neville states that as little as 0.5% of vulnerable aggregate can cause damage. This breakdown of concrete became common after 1970 when cements acquired higher contents of alkalis and specially when they were used for making high-strength concretes rich in cement. Low-alkali cements are those with less than 0.6% Na_2O equivalent, but most concretes have less than 1%. See below.

alkali–carbonate–rock reaction
Alkali–aggregate reaction with some carbonate rocks, especially dolomite (calcium–magnesium carbonate). By 1989 it had not been noticed in the UK.

alkali compounds For the purposes of *alkali–aggregate reaction*, only the compounds of sodium (Na) and potassium (K) matter.

alkalinity See **pH scale**.

alkali-resistant fibre See **glassfibre-reinforced concrete**.

alkali–silica reaction (ASR) The commonest *alkali–aggregate reaction* in the UK. It occurs with highly reactive siliceous minerals such as opaline *chert* and acid volcanic glass. *BRE* Digest 330 of 1988 considers that *fly-ash* and ground granulated *blast-furnace slag* (ggbfs) are not reactive but that one sixth of the total *alkali* in fly-ash and half that in ggbfs should be included in the calculation to keep the equivalent alkali below the accepted maximum value of 3 kg/m³ of concrete.

alkali–silicate–rock reaction *Alkali–aggregate reaction* with greywackes, argillites and phyllites containing *vermiculite* (*B*), which expands.

alkali threshold *Alkali–aggregate reaction* can be prevented or reduced in several ways: by keeping the concrete dry, or by ensuring that it is impermeable or contains no *alkali compounds*. Cement is often the main source of alkali and this source can be diluted by substituting a non-alkaline *pozzolan* for some of the cement. The salt spread on concrete roads is blamed in part for the trouble, though it has been claimed that some corrosion is caused by sulphate rather than chloride. The total alkali content should be kept below 3 kg/m^3 of concrete.

all-in contract See **design and build**.

allowable bearing capacity, permissible ground pressure, etc. The figure, expressed in $tons/m^2$, k/m^2 or other unit, from which foundation engineers calculates the area needed for their foundations, knowing the weight they must carry. It is based on information from *soil mechanics* engineers who can estimate the ground strength and how much it may sink, as well as on the *consultant*'s feelings about how much *differential settlement* may be alowed before failure.

alloy A mixture of two or more metals made usually with the intention of combining their qualities.

alloy steel, specials. A *steel* which contains elements not in *carbon steel* or present in smaller amounts. Carbon steels generally contain less than 5% – alloy steels contain much more – all added together of the following metals: nickel, chromium, manganese, molybdenum, vanadium.

all-terrain vehicle (ATV) At least three types of ATV exist: three-wheelers, four-wheelers, and four-axled or five-axled vehicles with either tracks or flotation tyres. They are ridden like a motor-bike, with two wheels at the back. 'All-terrain' also describes wheeled *mobile cranes* which can travel on any rough surface.

alluvium Any geologically recent deposit of silt, sand, gravel, etc. from an ancient river. Many cities, including London and much of Paris, are built on alluvium.

altar A step back in the wall of a dry dock, used for holding the feet of the wooden shores which steady the vessel when the dock is empty.

alternate-bay construction, hit and miss *Chequer-board construction.*

alternator [elec.] A machine which generates alternating current by the rotation of its rotor, driven usually by a steam or water *turbine* or other *prime mover.*

altitude [sur.] (1) The height of a place above sea level.

(2) The angular height of an object such as a star above the horizon.

altitude effects Because the density of air falls with rising altitude, internal combustion engines lose power as their height above sea level increases. From 0 to 300 m (1000 ft) altitude the loss can be ignored. Above 300 m the loss can be taken as 3% for each 300 m (1000 ft). From sea level to 2000 m (6000 ft) the air becomes cooler by about 2°C per 300 m increase of altitude.

altitude level A precise *level tube* on the *vertical circle* of a *theodolite*. It establishes the horizontal datum from which vertical angles are measured.

alum Aluminium sulphate $(Al_2(SO_4)_3 \cdot 18H_2O)$, the commonest *coagulant*, used at *Camelford*.

alumina, aluminium oxide Al_2O_3. An important constituent of ordinary *clays* (in chemical combination), as well as of *corundum*.

aluminium (aluminum USA) Chemical symbol Al. The most important element in *light alloys* (except magnesium alloys). A metal which has the low *relative density* of 2.70 and the low *ultimate tensile strength* (for a metal) of 93 N/mm^2. For this reason when used to carry loads it is always *alloyed* to form stronger metals called light alloys. Its *reduction* is performed by *electrolysis* of *bauxite*.

Its good corrosion resistance compared with steel enables it to be used in *Alclad* and other *clad steels*. Its high electrical conductivity is only worse than that of copper. Aluminium alloys can reach *proof stresses* of 500 MPa (72,000 psi) but more usually 130 MPa. Aluminium alloy wheels can reduce the weight of a lorry by 46 kg per wheel.

aluminium oxide Alumina.

aluminized steel pipe Steel pipe made of helically wound corrugated sheet hot-dipped in aluminium, claimed to have high corrosion resistance, and used for building *culverts*.

alumino-thermic reaction The chemical reaction which takes place when powdered aluminium is ignited with the oxides of other metals. The aluminium takes the oxygen from the other metals, burning fiercely, usually melting them. It is used for welding steel in the *thermit* process.

aluminous cement See **high-alumina cement**.

aluminum See **aluminium**.

American caisson See **box caisson**.

American Ephemeris and Nautical Almanac [sur.] An annual publication containing astronomical data for several years ahead, published by the Government Printing Office, Washington. See **Nautical Almanac**.

American Society for Testing and Materials (ASTM) The main organization in the USA that corresponds to the *British Standards Institution* in publishing standards.

American wire gage, Brown and Sharp w. g. Thicknesses of non-ferrous wire and sheet, some 25% different from *Standard wire gauge* (*B*).

amino plastics The main amino plastics are three slightly different resins, all of them thermosetting, derived from the reaction of formaldehyde with urea – urea-formaldehyde, urea and melamine formaldehyde resins. Urea-formaldehyde resins are excellent glues for plywood. Melamine formaldehyde resins make hard, decorative, chemical-resistant table surfaces. One disadvantage of phenolic and amino plastic resins is the need for high pressures during moulding. *Glassfibre-reinforced polyesters* and *epoxides* are easier.

amplitude The depth of a wave measured from the level of calm water. The word is used for every sort of wave. The double amplitude is twice the amplitude.

anaerobic, anoxic Description of any process, activity or treatment that needs no air or oxygen. Cf **aerobic**.

anallactic lens [sur.] The additional lens used in a telescope so that for *stadia work* the *additive constant* is zero.

anallactic telescope [sur.] One in which for *stadia work* the *additive constant* is zero.

analogue (analog USA) In computing, analogue is the opposite of digital. Analogue values vary in the same way as every natural value does: rivers flowing, time passing, children growing. A digital watch shows figures (digits) only but the old-fashioned watch with two hands is analogue. All wave forms are analogue but can be digitized for the purpose of sending clear messages. Digital signals can be amplified without adding electrical noise, and are much more single-valued than analogue.

The first analogue computer was probably the slide rule, invented by Oughtred in 1622. Analogue readings on a thermometer scale or slide rule are less precise than digital readings, and perhaps slower to take, but possibly no less accurate than digital. A water engineer can devise an electrical network to simulate (copy) a water network and this will be an analogue computer if its results are read from a scale, not as digits.

analogy A comparison between two effects, of which one is the better known. This comparison makes easier the understanding of the less well-known effect. The alternating current analogy is used in investigating tides, other electrical analogies in the *seepage* of water through sands, the *column analogy* and the *membrane analogy* in *structural analysis*. They often correspond to a similarity between the mathematical relationships.

analysis See **dimensional a., mechanical a., structural a**.

anchor Something restraining movement, such as a bolt screwed into concrete or, on a larger scale, a *tie rod* holding back a *sheet pile* retaining wall or *diaphragm wall* to a buried block, or the whole assembly of tie and block. See also **anchor bolt** and below.

anchorage (1) **anchor wall** A *dead man* (see diagram, p. 60).
(2) [stru.] An essential piece of *post-tensioning* equipment cast into the *anchorage zone* at the end of a *tendon*. It grips the tendon and transfers the load from it to the concrete with the minimum of slip or other *loss of prestress*.

anchorage distance The distance behind a quay wall at which the *dead man* must be placed so as to ensure that it will not slip with the quay wall, and that it anchors it effectively.

anchorage zone The concrete around the *anchorage* in *post-tensioning*. It has to be strengthened with **links** or hoop steel to prevent it bursting.

anchor and collar A heavy metal hinge for lock gates. It is built into the lock masonry and carries a projecting plate with a hole into which the *pintle* (*B*) of the gate drops.

anchor block A *dead man*.

anchor bolt, foundation b., holding-down b. A bolt with its threaded part projecting from masonry or concrete and secured into it so as to hold down a steel building-frame against wind loads, or machinery against its own vibration. See **Lewis bolt, expansion bolt, Rawlbolt**.

anchor gate A gate like a canal lock gate, held in position at the top by the pintle of an *anchor and collar* or similar hinge.

anchor ice [hyd.] Ice on the bed of a stream.

anchoring spud See **spud**.

anchor pile A *pile* to take tension or sideways pull.

anchor plate A plate at the foot of an *anchor bolt*, buried in the concrete.

anchor tower When a Scotch *derrick* has to be built on towers to achieve the height it needs, the anchor towers are the two that anchor its legs. The *crane tower* carries the mast.

anchor wall See **anchorage**.

anemometer An instrument for measuring wind speed. See **flow meter, ultrasonic flow meter**.

aneroid barometer [sur.] A portable barometer working on the principle of a sealed box which expands with a drop in air pressure and contracts with an increase in pressure. It is used in altimeters for aircraft and by surveyors in approximate measurements of altitude and may be from pocket watch size to 18 cm cube. See **barometric pressure**.

ANFO (ammonium nitrate + fuel oil) A cheap, safe explosive but with relatively little explosive energy. It also easily absorbs water from the air. FGAN (fertilizer grade ammonium nitrate) is an allied explosive.

angle [stru.], **a. bar** See **angle section**.

angle cleat A small bracket of *angle section* fixed in a horizontal position, normally to a wall or stanchion, to support or to locate a structural member. It may be a *shelf angle*.

angledozer A *bulldozer* with the mouldboard set at an angle so that it pushes earth partly sideways and partly ahead.

angle iron An *angle section*.

angle-iron smith [mech.] A skilled man who shapes and welds *angle sections*, directing the *strikers* who hammer the hot metal to shape round the pegs of a *template* (*B*).

angle of friction [mech.] In the study of bodies sliding on plane surfaces, the angle between the perpendicular to the surface and the resultant force (between the body and surface) when the body begins to slide.

angle of internal friction [s.m.] For quite dry or quite submerged soils without *cohesion* such as clean sands, the angle of internal friction is approximately the *angle of repose*. More precisely it is the angle ϕ in the equation: Shearing resistance of soil = Normal force on surface of sliding × tan ϕ where tan ϕ is determined by experiment. ϕ is not related to absolute grain size but to the roughness and relative sizes of the particles, and is larger for angular, dense, well-graded sands than for loose, uniform, rounded sands. See **angle of shearing resistance**.

angle of repose [s.m.] For any given granular material the steepest angle to the horizontal at which a heaped surface will stand in stated conditions. See **angle of internal friction**.

angle of shearing resistance [s.m.] The value of ϕ in *Coulomb's equation* for cohesive soils, $s = c + p$ tan ϕ. ϕ is determined by experiment. It is 0 for a saturated clay sheared without change of water content, but for most silts or clays in other conditions ϕ is not 0. See **angle of internal friction**.

angle section, angle, a. bar A *rolled steel* or other metal section shaped like an L, made in sizes up to about 0·2 m leg length. It may be equal or unequal in leg lengths.

angularity number A number between 0 and 12 which roughly indicates the roundness of the pebbles in an *aggregate*. The more rounded the stone, the better the concrete *workability* for a given water content. The angularity number of an aggregate is 67 minus its solid volume determined in a standard way. Spheres of the same size have 67% solid volume and their angularity number is 0. Good concreting aggregates have a number less than 11. Thus the lower the angularity number, the better the aggregate is for making workable concrete. Cf. **elongation index**.

anion An *ion* which moves to the *anode* in *electrolysis*. See **cation**.

anisotropic [stru.] Not *isotropic*.

annealed wire, binding w., iron w., tying w. Soft steel wire for tying *reinforcement*.

annealing Softening a metal (originally steel, now copper and alloys or metals, as well as glass) by heating it to a suitable temperature and, for steel, holding it there for several hours, ending this treatment with gradual cooling. Annealing may remove stresses and weakness caused during casting, or brittleness, or it may produce machinability, softness, or cold-working properties. For copper, see **dead-soft temper** (*B*). See **normalizing**.

annual variation [sur.] The change each year in the magnetic *declination* of a place.

annulus grouting In *pipe renewal*, grouting the ring-shaped gap (annulus) between the new lining and the old pipe. One advantage of *swaging, rolldown* etc. is that *polythene* pipe does not need annulus grouting.

anode [elec.] The opposite *pole* from the *cathode* in *electrolysis*. The pole at which the basic chemical part of the salt (radical) collects or at which oxidation occurs. DC enters at the anode: to remember this, the *mnemonic* is: the anode is the in-ode. See **cathodic protection**.

anodizing, anodic oxidation A protective, durable film of oxide formed on the surface of aluminium and its alloys, as well as magnesium or titanium, by *electrolysis*. The metal is made the anode in a bath of acid. Many brilliant colours can be obtained.

anthracite filter A *multi-layer filter*.

anticlastic shell, saddleback s. [stru.] A horse's saddle is curved up in front and behind to hold the rider in place. A saddleback shell is similarly curved, convex upwards in one direction and concave upwards at right angles to it. The shape occurs naturally in awnings or in tents with two or more tentpoles, in soap bubbles or in rubber sheet or other pure-tension materials, and is useful to structural engineers building shells in compressible materials, because of its stiffness and strength.

anti-crack reinforcement A close mesh of light steel rods or chicken-wire placed just below the surface of concrete to reduce surface cracking.

antidune [hyd.] A sandhill like a *dune* but formed at higher flow rates and with the steep face upstream.

antifer [hyd.] *Armour* consisting of precast blocks of a modified cubical shape, with a central hole and indentations in four of its six faces. The holes and indentations reduce the central mass and make for uniform heating during *hydration*. The Hydraulics Research Station has listed 21 types of blockwork armour, many of them adapted cubes. Antifer is the oil port of Le Havre, where antifer blocks were used. See **SHED**.

anti-flood and tidal valve A *check valve* in a drain which is below flood level or high-tide level.

antifreeze Any *admixture* for concrete, claimed to be an antifreeze, is likely to be an *accelerator* of set and hardening. It may not protect concrete from freezing or frost damage, but in accelerating the set and speeding the hydration of the cement, it does warm up the concrete for the first 24 hours or so after placing. In the USSR concreting is regularly done at $-30°$C. Soviet engineers use sodium nitrite, also calcium nitrate–nitrate with urea and other substances. At relatively warm temperatures from 0 to $-5°$C the proportion of antifreeze is about 6% of the mixing water, increasing as the temperature drops, to about 25% of the mixing water at $-25°$C.

Crystalline sodium nitrite burns in contact with wood, cotton, straw, etc., and should be kept in alkaline solution (pH above 8). In acid solution below pH 7 it decomposes, emitting the poisonous gases NO and NO_2.

The salt spread on roads in winter to melt snow and ice corrodes steel, but a substance made from surplus cereals, calcium magnesium acetate, will melt snow without corroding metal, according to Michael Dukakis, US Democratic presidential candidate in 1988.

anti-friction metal See **white metal**.

antinoise, antisound Sound made deliberately out of phase with nuisance noise so as to cancel it, resulting in silence. Either of these separately would be noisy. The first antinoise patent was in 1934 but recent high-speed electronics have made it easier to realize, for example in an aircraft cockpit.

anti-sag bar, sag b. [stru.] A vertical *tie rod* from the ridge of a truss down to the horizontal tie beam. It may also connect horizontal steel rails in wall framing or purlins in roofing, for the same reason, to reduce their deflection.

apparent horizon See **sensible horizon**.

apron (1) A *hard standing* in an airfield.
(2) [hyd.] A hard surface to the sea bed or to the bed or banks of a stream or canal to prevent *scour*. An apron may be of *bagwork, mass concrete, reinforced concrete*, timber or *rip-rap*, or a *mattress*. See also (*B*).

aquaplaning Skidding of motor vehicles on wet roads, or of aircraft on wet runways, a defect that can be avoided on concrete surfaces by roughening the surface texture during casting or by grooving the slab transversely afterwards.

aqueduct A conduit (which may include tunnel and bridge) for carrying water over long distances. The term is usually confined to water-carrying bridges. Cf. **viaduct**.

aquiclude [hyd.] An *aquitard*.

aquifer [hyd.] Buried rock or soil that not only contains water but also yields it up easily to a pipe or well driven into it. It is therefore porous and of high permeability, like a uniformly sized gravel or a fissured limestone or sandstone. It is normally sealed below as well as above by impervious strata like clay, though water flowing into it may provide the lower seal. An aquifer treated by *artificial recharge* may reduce the need for an *impounding reservoir*. Underground oil or natural gas reservoirs are similar and would be aquifers if they were filled with water instead of oil and gas.

Apart from water, aquifers are also used for storing natural gas and, in *compressed-air energy storage*, air. The gas or air injected through a borehole at the top of the aquifer may make space for itself by forcing the water level down if its injection pressure is high enough. See **artesian well**.

aquifuge [hyd.] Ground that neither contains nor transmits water in useful quantities. Cf. **aquifer**.

aquitard, aquiclude [hyd.] A geological formation of low *permeability*, that delays the flow of water from an *aquifer*. It may itself contain a large quantity of water, but gives it up too slowly to be considered an aquifer.

Aral Sea A frightening example of efficient civil engineering that is ecologically disastrous. What was the world's fourth largest inland sea and the finest fishing area of the USSR is now too saline for the fish that lived in it, and too shallow for navigation. Its level has dropped 13 m (43 ft) since 1960. The surrounding land is becoming a saline desert. The drinking-water wells are becoming more and more saline, the harvests are failing and the health of the Karakalpak people is seriously affected by the salinity. The waters of the rivers Amu Darya and Syr Darya are irrigating the deserts of Kazakhstan and Uzbekistan to grow cotton and little of them now flows into the Aral Sea.

arbitration *Civil engineering contracts* usually have a clause naming an arbitrator who is an engineer and will help to settle any dispute before it reaches the law courts. ICE's *Arbitration Procedure* is published in English and Scottish versions for use with the ICE *Conditions of Contract*. See **mediation**.

arbor [mech.] In the USA a *mandrel* or a drive shaft.

arch [stru.] A *beam* curved usually in a vertical plane, for carrying heavy loads such as bridges or long-span roofs. Arches are made of any solid material but usually of *light alloy, prestressed concrete, steel* or *reinforced concrete*, stone, *mass concrete*, timber or brick.

arch dam A *dam* which is held up by horizontal thrust from the sides of the valley (*abutments*). It must therefore be built on rock, as yielding ground would cause the dam to fail. It may be of concrete, stone or brick, concrete being the commonest. In dams the *arch* is curved in a horizontal plane. See **gravity-arch dam**.

Archimedean screw [hyd.] An ancient water-lifting device. An inclined spirally threaded pipe (or screw in a pipe) which turns and lifts water from its submerged lower end. It is efficient in lifting large volumes of dirty water against a low head. The screw shaft may slope at 27–40° to the horizontal but one preferred angle is 38°. The steeper the angle, the smaller is the flow rate. The output can vary from 7 litres/s to 10 m³/s.

Archimedes, principle of See **principle of Archimedes**.

arch rib [stru.] A main load-bearing member of a ribbed *arch*, a local deepening of an arch.

arch ring (1) [stru.] The load-bearing part of an *arch*.
(2) See **ring**.

arc welding *Fusion welding* in which the heat comes from an electric arc.

are [sur.] The metric unit of area, 100 m². One hectare is 100 ares.

argon-arc welding *Shielded-arc welding* with the inert gas argon.

arm, lever a. [stru.] (1) See **bending moment**.
(2) The arm of an *eccentricity* is the distance by which a force is out of centre, eccentric.

armour A *revetment* of large, heavy rocks (*rip-rap*) or precast concrete blocks (*antifers, SHEDs, tetrapods,* etc.) on top of a *breakwater*, protecting it from *scour*. Since very large blocks are difficult not only to quarry but also to move, they are often precast. The top layer (main armour) has the largest blocks, often weighing 25 tons or more, while those below it are progressively smaller. Reinforcement of precast blocks with steel or plastic fibres improves their resilience and ductility considerably. See also **geotextile**.

armoured cable [elec.] Electrical power cable used in mines, or buried under streets, with two layers of hard-steel wire wound in opposite directions round the cable (double wire armoured). See also (*B*).

armoured pipe [hyd.] Pipe made of plastics protected by steel wire. The first underwater armoured water supply pipe in Britain was laid in October 1973 between the Ayrshire coast and the Isle of Bute, 150 mm bore and 2 km long, in eight hours.

arrest point

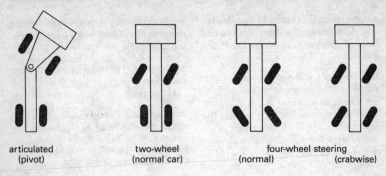

articulated
(pivot)

two-wheel
(normal car)

four-wheel steering
(normal)   (crabwise)

Right turn with wheeled steering systems. (J. Singh, *Art of Earthmoving: Equipment and Methods*, A. A. Balkema, 1986.)

arrest point [mech.] A *critical point*.

arrissing tool A tool like a *float* used in road-making to round off the edge of a concrete slab.

arterial road A main *road*, *street* or *avenue*, with tributary roads, etc., joining it. Cf. **freeway, parkway**.

artesian well [hyd.] A well, nowadays usually a borehole, from which water flows without pumping. Because too much water is drawn from them, many artesian wells have become pumped wells. The first known to history were dug in Flanders, near Béthune, after 1100 A.D., into a fractured chalk bed that extended underground up into the hills of Artois and so provided ample water under pressure. Early North American artesian wells gushed to a height of 45 m or so. Modern hydrologists have shown that some artesian pressure may be caused by sinking of the ground surface. Artesian wells draw their water from *confined aquifers*.

articulated floor slab, caterpillar f.s. A concrete ground floor slab with steel only at the mid-depth. Frequent joints allow each area of floor to settle in relation to its neighbours, so that the slab can be used in areas of mining *subsidence*.

articulated steering, articulation (see diagram) Instead of wheels that turn relative to the frame, a whole axle is pivoted. Wheeled *loaders* and large *dump trucks* can thus turn in a relatively small space.

articulated vehicle, a. lorry, artic, semi-trailer A heavy goods vehicle in two parts: the front, driving, powered part, with the driver's cab, and the rear part towed, with one, two or three axles. The two parts can be separated, enabling the driver to leave one loaded part at a customer and to tow away another one. Instead of a tow bar, there is a massive pivot (articulation) over the rear axle of the powered section.

artificial aggregate Usually a *lightweight aggregate* or a very dense, *neutron-absorbing aggregate*, but not one from a quarry.

artificial cementing, grouting The permanent or temporary strengthening or waterproofing of rocks or loose soils through holes drilled or *jetted* into them. Stiff-fissured clays can be cement grouted, other clays can be *electro-chemically* hardened. Where the range of void sizes is wide, it is usually economical to seal the large voids first with coarse grout, then to inject thin grout to seal the small voids. The four traditional procedures for grouting are: open-ended pipe, stage grouting, sleeve grouting and claquage grouting. The open-ended pipe is the simplest technique, and suitable for coarse material like an open gravel. Stage grouting can be used in rock, drilling and grouting by stages of 1.5–3 m of hole at a time, re-drilling through each grouted length as the hole is deepened. Sleeve grouting involves the use of the *tube à manchette*. In claquage grouting, fine-grained soils can be penetrated by tongues of grout injected at high pressure, that compact the soil and reduce its permeability. The high pressure can lift the ground and may affect nearby structures. The finest solutions can be injected with solutions of water-soluble acrylamide or phenoplasts or polysaccharides, sometimes with a metal salt to give an insoluble precipitate. See **base exchange**, **ground improvement**, **jet grouting**.

artificial harbour A harbour formed by building one or more *breakwaters* round an area of sea.

artificial horizon [sur.] A saucer of mercury used for giving by reflection a truly horizontal direction when measuring *altitudes* with a *sextant*. See **horizon**.

artificial intelligence Long before computers existed, float-operated, electrically driven pumps in mines started to pump when the water rose and stopped when they had lowered it enough. This is intelligent but comput-

ers properly programmed can do much more, playing chess better than most people or acting as consultants in many engineering or scientific applications of *expert systems*. For civil engineering, probably the whole of artificial intelligence is contained in *expert systems*, their *shells, knowledge manipulators* or inference engines.

artificial islands In the Arctic conditions of the Beaufort Sea, north of Canada, with ice floes endangering ships even in the summer, exploration wells for oil or gas are best drilled from land. The first artificial island there was built by *suction dredger* in August 1975 in 7 m of water behind a breakwater, built first 180 m long and 60–90 m wide. In less than 800 hours' dredging, 1.75 million m^3 of sand were placed using a 500 m long *floating pipeline*, achieving an island 97 m across and 5 m above sea level. Many others have been built north of Alaska. See ACI's book *Offshore Arctic Structures*, 1986; **Chesapeake Bay bridgetunnel, Kansai International Airport, Tokyo Bay**.

artificial pozzolan The commonest artificial pozzolan is *fly-ash* but many others exist, including the ash of rice hulls and *microsilica*.

artificial recharge [hyd.] Replenishment of *groundwater* artificially, through shafts, wells, boreholes, pits or trenches, with water from a lake or river or used water that has been treated. A great improvement in water quality can be achieved underground because the water may filter many hundreds of metres through air-filled ground. A beach that has been scoured can be artificially recharged by dumping on it sand or rock – often called beach replenishment.

The Great Salt Lake, Utah, has no outlet to the sea although it is some 1280 m (4200 ft) above sea level. Its

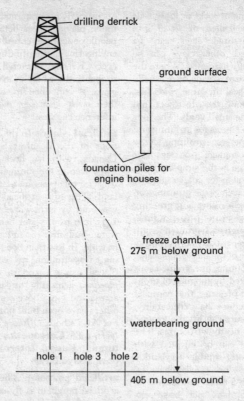

drilling derrick

ground surface

foundation piles for
engine houses

freeze chamber
275 m below ground

waterbearing ground

hole 1   hole 3   hole 2

405 m below ground

Directional drilling at Asfordby shafts of British Coal 1985–8, enabling engine houses to be built without waiting several years for completion of freezing and sinking. Figures for the accuracy of these holes are stated on p. 182 (diagram). Directional drilling is usual also in micro-tunnelling.

floods have damaged railways, farms, waterworks and housing, so the recharging of aquifers below it has been considered, but in 1987 this was rejected in favour of pumping to a desert area some 14 miles (22 km) to the west. The State of Utah paid for the pumping and for the canal at a gradient of 0.15 m per 1000 m. The level of the Great Salt Lake was lowered thus by 33 cm (13 in.) in 1987–8.

asbestos [min.] A mineral silicate, consisting of fibrous crystals found in veins in rocks. The crystals are thin, tough fibres like textile, which can withstand very high temperatures without change when pure. The use of asbestos is discouraged for health reasons so it may not now be sprayed. See (B).

as-drawn wire *Prestressing* wire supplied coiled remains curved when it comes off the coil unless it is the pre-straightened grade that has had *heat treatment* with this in mind. Any wire which bonds with concrete must be fully degreased.

aseismic design, seisonic d. [stru.] *Structural design* to withstand *earthquakes*.

Asfordby mine shafts (see diagram) Two shafts for the British Coal Corporation to a depth of about 600 m. Sinking of the first was begun by Cementation Ltd in 1986. These very large shafts (7.32 m. (24 ft) dia.) had to go through heavily waterlogged ground between 275 and 405 m below ground. This length only was frozen, and a freeze chamber was dug at 275 m to cut all the previously drilled and cased holes. The upper parts were grouted and the lengths below the freeze chamber used for freezing. The land above the freeze holes was thus freed for building the winding engine-house foundations.

asphalt Black mineral hydrocarbons containing bituminous substances, which are decomposition products of petroleum, found at the surface of the ground near petroleum deposits. This waterproof surfacing material can also be made by distilling petroleum. It is applied in two main ways, first as *mastic asphalt* in *tanking* (*B*) or roofing, secondly as *rolled asphalt*. The distinction between asphalt and bitumen is mainly chemical. To the layperson they are similar except that asphalt is reserved for solid surfacings, the term bitumen being used for liquids suitable for coating aggregates.

asphalt cement US term for asphalt or bitumen being used as a binder.

asphaltic concrete US term for a road surfacing of *rolled asphalt*. Cf. **asphalt cement**.

ASTM Portland cements Eight Portland cements standardized by the American Society for Testing and Materials are listed below. They are numbered I, IA, II, IIA, III, IIIA, IV and V. 'A' implies an *air-entrained* cement. Type I is ordinary; Type II has moderate sulphate resistance; Type III is *rapid-hardening* – it must achieve 12.4 MPa (1800 psi) after one day and nearly double this at three days, but the air-entrained version does not do so well. Type IV is a *low-heat cement* and Type V has high sulphate resistance. All types except II (for which the figure is not stated) have a minimum *specific surface* of 160 m^2/kg by turbidimeter test and 280 by the air permeability test. (ASTM volume 04–01, standard C 150–86.)

astronomy [sur.] The study of the sun and other stars and their movements. Observations of stars are made to obtain precisely the latitude, longitude, time or a true geographical *bearing*.

attenuation Reduction, e.g. of sound. But attenuation of the flow in a sewer implies reducing the peak flow by spreading it over a longer time. This is done by storage in *stormwater tanks* or *detention tanks*, or occasionally and less easily by throttling the flow either in the sewer or before it reaches the sewer.

Atterberg limits [s.m.] The *consistency limits* of a clay.

ATV *All-terrain vehicle*.

auger [s.m.] A soil auger is like a corkscrew that pulls soil up from a hole it makes in the ground. Small ones, from 4 cm dia., are hand-turned by one or more workers in *post-hole augering* or *shell-and-auger boring*. *Truck-mounted drilling rigs* can drill large holes of up to 1.8 m (6 ft) dia. which may be enlarged at the base by *under-reaming*. Horizontally augers are used for mining coal from an outcrop or for removing the soil from the face of a *pipe-jacking* tunnel. But they are unsuitable for working below the *standing water level*. It is claimed that a 500 mm (20 in.) dia. auger can remove boulders 250 mm (10 in.) in size. See **continuous-flight auger**.

austenitic steel Many non-magnetic, corrosion-resistant, hard-wearing, tough, *alloy steels* with high resistance to shock; often alloyed with manganese, chromium, nickel or all three. A common one is stainless steel with 18%Cr and 8%Ni. Alloy steels began with Hadfield's manganese steel in Sheffield in 1882 (12–14%Mn, 1%C), still used for *rock drills*, *breakers*, *dredges*, *turnout* rails, etc. It is hard to machine but can be forged or hot rolled.

Austrian tunnelling method See **NATM**, **shotcrete**.

autoclaving, high-pressure steam curing Treatment of freshly precast concrete, *aerated concrete* or *sand-lime bricks* (*B*) at about 10 bar of steam pressure (190°C) – a way of *accelerated curing* using dry steam which therefore must be very carefully done. Steam below 100°C ensures wet curing, which should be easier for the non-specialist.

autocollimation [sur.] In alignment surveys, a technique for checking squareness. The cross hairs of a telescope focused at infinity are illuminated and their image reflected back in a mirror. When the reflected image coincides with the cross hairs the mirror is square to the line of sight. See **plumb box**.

autogenous healing The closing up and disappearance of breaks or cracks in concrete when the concrete parts are kept damp and in contact. In *prestressed concrete*, cracks heal without damping provided that the overload is released sufficiently for the cracks to close.

automated guided vehicle system (AGVS) Automatic vehicles began about 1974 in the Volvo factory. Later with Fiat they became standard units in modern warehousing, and then were developed for moving people, as at Gatwick airport.

automatic compensator See **autocollimation**.

automatic level, autoset l., self-aligning l., self-levelling l. [sur.] Different names of different makers for levelling instruments that, unlike the *dumpy level*, maintain a horizontal line of sight regardless of a slight displacement of the vertical axis. They use *autocollimation*.

automatic siphon spillway [hyd.] See **siphon spillway**.

automatic welding Many highly productive, efficient ways of making welds of first-class quality by *arc welding* or *resistance welding* in the shop but rarely if ever on building sites. One exception is British Rail, which uses *semi-automatic welding* of rails in the open at 200 A with 1.6 mm dia. flux-cored wire. Multiple welding heads may be used and welding currents may be three times what is possible with manual welding. The methods include: *electroslag, electrogas, MIG, resistance-seam, submerged-arc* and *TIG welding*.

automation The automatic operation of production processes, made possible by such devices as the *bi-metal strip*, the *float switch*, the *photo-electric cell*, *remote control*, the steam governor and microprocessors or *microcomputers*.

Automation includes feedback, returning information about output to the controller. At Grimsby main pumping station, commissioned in 1987, the feedback is from some 50 sensors throughout the plant, monitoring all activities including tide level, rainfall and flow rates in and out of the station.

Pumping is mainly on the ebb tide. The controlling computer, and accessories costing £225,000, displaced 25 of the 30 workers formerly needed. The station has three of the country's largest sewage pumps, each capable of moving 3.5 m³/s (one working, two on standby) to the 2 m (6½ ft) dia. *outfall* pipe.

autopatrol, motor grader A *grader*.

avenue A road which in the USA usually runs north–south, the word *street* being reserved mainly for east–west directions. Both streets and avenues are numbered in a logical way from which the position of each number can be closely estimated.

average [stat.] The average of *n* values is the sum of the values divided by *n*. This idea is used in surveying as well as in sampling and the calculation of batting averages in cricket. In a series of *n* survey measurements, the average of the *n* values is likely to have $\sqrt{n}$ less error than any one measurement, other things being equal. See **mean**, **personal equation**.

axial-flow fan [min.] A *propeller fan*.

axonometric projection [d.o.] A way of showing a plan and a part elevation on the same drawing. The plan is turned through 30° or 45° and vertical lines are drawn from the corners to show the part elevations. See **projection**.

azimuth [sur.] The direction, from 0° to 360° of a line measured clockwise from true north. Cf. **bearing**

azimuthal projection [sur.] A map on which from one point, usually the centre, all other points are on their true *bearings*.

B

Bacillus coli [sewage] An old name for *Escherichia coli*.

backacter, backhoe, drag shovel, trench hoe A *face shovel* acting in reverse, that is digging towards the machine. It can easily dig down to a depth of about 4 m below the tracks. It does not make such a clean-walled trench as the trencher. See **loader**.

back cutting Additional excavation required to make up an embankment (or railway, road or canal) where the original amount of cut was insufficient.

back drain See **weephole**.

back gauge, b. mark [stru.] The distance from the back edge of an *angle* or other *rolled-steel section* to the centre line of the rivet or bolt hole through it.

backhoe loader A **backacter** attached to the rear of a *loader*; small versions are used in tunnelling, hung from a monorail or the *shield*. A backhoe can dig below its wheels economically, including short trenches. If the backhoe is on the machine centre-line it is a 'centre-post backhoe', others are 'offset backhoes'.

back-inlet gulley A water-sealed branch entry to a drain, covered by a grating but otherwise open to the air. Rainwater or waste pipes discharge into it under the grating but above the water seal. It is of cast iron, PVC or *vitrified clayware* (*B*).

back mark See **back gauge**.

back observation, b. sight [sur.] Any sight taken towards the last station passed. Cf. **fore sight**.

back prop A raking strut which transfers the weight of the timbering of deep trenches to the ground. Back props are inserted under every second or third *frame*.

back propping Providing props under recently cast slabs or beams to prevent early deflection and thus reduce *creep*. Props under successive floors transfer loads through a multi-storey building to the ground.

back sight [sur.] A back observation. Back sight refers particularly to readings of a levelling staff, for which the ASCE prefers the newer term *plus sight*.

backwashing Cleaning a *rapid sand filter* by reversing the flow and rejecting the wash water.

back water [hyd.] Water held back by an obstruction.

backwater curve [hyd.] The curve of the surface of the water measured along an open channel after inserting a weir or dam in the channel and thus raising the water level. The curve is concave upwards.

bacteria bed [sewage] A *trickling filter*.

baffle, b. plate, baffler [hyd.] A plate used for deflecting the flow of any fluid such as air, water or flue gas.

baffle pier [hyd.] A *groyne*.

bag of cement In the USA 94 lb (43 kg), in Canada 87.5 lb (40 kg), in Europe 50 kg (110 lb).

bag wash, bagging See **sack rub**.

bagwork [hyd.] A *revetment* to protect sea walls or river banks from *scour*, consisting of dry concrete sewn in bags and tamped against the river bank. Occasionally the material in the bags is gravel. Sea-wall bagwork is usually held together by steel dowel rods driven through the bags like skewers.

bail [min.] A steel half hoop over a sinking *kibble* by which it hangs from the hoisting rope.

bailer (1) [s.m.] A *sand pump*.
  (2) A length of 3–12 m of pipe with a foot valve. Lowered down a well *casing* it can raise oil or water which contains sand or for any other reason will not flow or cannot be pumped.

Bailey bridge [stru.] A military bridge developed in Britain about 1942, the first British welded lattice bridge. It is built in panels, which are connected at the four corners by steel pins to the next panels. To build bridges of higher strength, panels can be connected together in two or three storeys with one, two or three panels in each storey. No *lifting tackle* is needed; each part can be lifted by one or at most two people, so it is a most useful temporary bridge.

Baker bell dolphin See **bell dolphin**.

balance bar, beam [hyd.] A large, usually wooden beam projecting from a lock gate. By pushing on this beam when the water level is the same on each side, the lock gate is opened.

balance box A loaded box at the far side of a crane from the *jib* and the load, to counterbalance them.

balance bridge A *bascule bridge*.

balanced earthworks An excavation scheme designed so that the cuts equal the fills. There will be theoretically no earth left over and no *back cutting* to complete the last fill.

balanced steel, semi-killed s. Steel that has been partly deoxidized, allowing some bubbles to form in the ingot. Cf. **killed steel**.

balance point An intersection between a *mass-haul curve* and the datum line. At this point all excavated material has been used up in fill.

balancing (1) [sur.] *Adjustment. Fore sights* and *back sights* are equalized (balanced) as much as possible to eliminate errors such as those due to atmospheric refraction.
  (2) In *sewerage*, the *attenuation* of *run-off* sometimes by oversizing a *sewer* or by flooding an area of land or using a length of a stream as a storage pond. See also **balancing reservoir**.

balancing reservoir In a *waterworks*, a buffering reservoir or water tower which allows steady flow to or from plant upstream or downstream. See also **balancing** (2).

balata Rubber latex from the South American bullet tree for making conveyor belts and belts for power transmission.

balk, baulk Earth between excavations, a *dumpling*. See also (*B*).

ballast [rly] Coarse stone, slag or clinker about 50 mm (2 in.) in size laid as a bed for *sleepers* in the *permanent way*. Cf. (*B*)

ballast cleaner [rly] A self-propelled railborne machine which uses an endless cutting chain to remove *ballast*.

ball mill [min.] A cylindrical or conical mill for grinding mineral, cement, *clinker*, etc. It is charged with steel balls which are of about 13 cm dia. when new and gradually wear down to nothing.

band, b. chain, steel b. An accurate, strong steel or *invar* tape that may be either etched metal or coated with plastic protecting the steel and carrying the digits and graduations. A band is of thicker steel than a tape. Bands are 30, 50 or 100 m long; tapes are only 10, 20 or 30 m in length.

banderolle [sur.] A *range pole*

band screen [hyd.] An endless moving belt of wire mesh for removing solids from water at the intake to a power station or water works.

bank (1) An *embankment*.
(2) Inclined rail track.

banking *Superelevation*.

Bank of China office block, Hong Kong A 70-storey office block, completed in 1989, 315 m (1060 ft) to the roof and 366 m (1200 ft) to the top of the steel masts, designed by a New York structural engineer Leslie E. Robertson, for wind and live loadings twice those of New York. The main contractor was Kumagai Gumi of Japan and the architect I. M. Pei of New York. The Bank of China, together with capitalist banks, financed the Hong Kong harbour *immersed tube*.

bank of transformers [elec.] A set of power *transformers* installed together.

bank protection [hyd.] Devices for reducing *scour*, including *gabions*, *groynes*, *mattresses*, *turfing* of banks, covering banks with pegged-down brushwood, or other *revetments* such as grass, willows or the use of *geosynthetics*.

bank seating A bridge *abutment*, usually a shallow end support to a bridge.

banksman (signalman, spotter USA) A crane-driver's helper. He or she signals to the driver when to raise, lower, swing the jib, etc., and the crane driver obeys no one else.

bank storage [hyd.] Water absorbed by the banks of a stream and returned to it as the stream level falls. It helps to reduce flood peaks.

banquette (1) A *berm*.
(2) A bridge footway above road level.

bar (1) A hot-rolled piece of steel or iron, usually round, rectangular or hexagonal in cross section. Steel bars may also be forged. Light alloy bars are usually *extruded*.
(2) Silt, sand, or gravel dropped at the mouth of a river where its stream velocity falls and its carrying capacity also falls. Such bars are always moving. See **ebb channel**, **flood channel**.

bar bender, iron fighter, steel bender, steel fixer (1) A *skilled man* who cuts and bends steel *reinforcement* and binds it in position ready for the concrete to be poured round it. A bar bender usually also attends while the concrete is being poured, to correct any bars which fall out of place.
(2) A machine for bending *reinforcement*.

bar chair A specially made plastic clip or a bent steel bar or piece of concrete that maintains the concrete *cover*. If it supports the steel in a slab, it rests on the bottom steel or *formwork* or *blinding concrete*. In a beam it supports the links round the bottom steel. Cf. **spacer**.

barge A craft of 4.3 m width or more, carrying in Britain 70–150 tons, on the continent of Europe previously up to 300 tons and now 500 tons, but on parts of the *Grand Canal* even 2000 tons.

barge bed (see diagram) A mud bottom near the bank of a river where barges can moor and sit on the mud at low tide. The bank is often protected by a *double-wall cofferdam*.

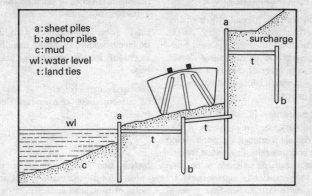

a: sheet piles
b: anchor piles
c: mud
wl: water level
t: land ties

surcharge

Barge bed.

Barnes's formula for flow in slimy sewers A more accurate formula than *Crimp and Bruges*. It states: velocity in m/s equals $46.5m^{0.7} \sqrt{i}$ where m is the *hydraulic mean depth* in metres and i the slope.

barometer An instrument which shows the atmospheric pressure. It may be an *aneroid* or a more cumbersome but accurate instrument, the mercury barometer.

barometric pressure [sur.] The pressure of the atmosphere. As measured by the *aneroid barometer*, air pressure can be used by surveyors to estimate altitude to an accuracy of 1 m, provided that calibrated precision aneroids are used at points of known altitude, and another precision aneroid is taken up to the point whose altitude is needed. If the atmospheric pressure is stable at the time of the survey, this also helps accuracy.

barrage A low dam, gated across its entire width, placed across a river to raise its level, normally for irrigation or navigation. There are several across the Nile as well as other large rivers.

barrel vault, cylindrical v., tunnel v., wagon v. A continuous plain arch or vault of semi-circular shape, usually considerably longer than its diameter; often of brickwork or masonry. The *reinforced* or *prestressed-concrete* barrel-vault *shell* is now being used more and more to roof factories or other areas which must not be obstructed with closely spaced columns.

barrette The short lengths that join to make a *diaphragm wall*.

BART system The San Francisco Bay Area Rapid Transit System, 114 km long.

barytes (BaSO$_4$) (barite USA) A dense mineral used in *biological shields*, oilwell *drilling fluids*, etc.

bascule bridge, balance b., counterpoise b. A bridge which is hinged at the bank to allow ships to pass under it by raising the part over the river and lowering the part over the bank behind the hinge. Modern bascule bridges are of *light alloy*. Tower Bridge, London, is an old bascule bridge. See **drawbridge**.

base The *base course* of a road.

base course In a road the *surfacing* layers other than the *wearing course*, but particularly in the *California bearing ratio* method of designing *flexible pavements*, a layer of chosen and compacted soil which is covered with a thin layer of asphalt.

base exchange [s.m.] A reversible chemical process, *ion exchange* of dissolved metals (*cations*), by introducing soluble sodium salts to replace insoluble magnesium or calcium salts. The device is periodically regenerated by washing with brine. Another type of base exchange is used in the *electrochemical hardening* of clays.

base flow [hyd.] That part of the flow of a stream that comes from the *groundwater* or long-term lake storage or snow melt.

base line (1) [sur.] A line measured very accurately so as to form the starting length for the calculations of a survey, usually a *triangulation*. The base line forms one side of a triangle (or several triangles) whose other sides are calculated from the base line and the measured angles of the triangle. But now the commonest use of base lines is for calibrating *EDM instruments*.
(2) [air sur.] An *air base*.
(3) A line from which a structure is set out.

base plate [sur.] The part of a *theodolite* which carries the lower ends of the three *foot screws* whose upper ends are fixed to the *tribrach*. The base plate is screwed to the tripod head whenever the theodolite is *set up* on the tripod.

basic refractory A *refractory lining* material low in silica content, used for metallurgical furnace linings. It contains metal oxides like lime (CaO),

magnesia (MgO), or calcined dolomite, a mixture of the two.

basic rock Igneous rock with little or no free *silica* and therefore usually dark in colour: basalts, most lavas and gabbros.

bastard cut [mech.] A file type of intermediate coarseness. See **cut** [mech.].

batch, mix The quantity of concrete, mortar, etc. mixed at one time.

batch box A container of known volume for measuring the right proportions of sand and aggregate in a *batch* of concrete or mortar (*ACI*).

batcher, weigh-b., batching plant Equipment for proportioning the parts of a concrete mix, whether with manual control of the switches, or automatically or semi-automatically. Measuring by volume is possible with *batch boxes*, but because of *bulking* (2) is usually too inaccurate.

batch mixer Almost the only type of mixer on medium or small jobs, one which mixes batches of concrete or mortar. Cf. **continuous mixer**.

batch mode, batch system of computing Large computers, being expensive, have to work as economically as possible. Batch mode is a way of doing many jobs in quick succession without costly interruption. For example, many different companies' weekly pay sheets can be quickly produced using the same *program*. Cf. **real-time computing**.

bathotonic reagent See **depressant**.

bathymetric surveying Surveying ocean depths and floors.

batten plate [stru.] In composite stanchions built up from pairs of rolled sections such as *channels*, *angles* or *joists*, a batten plate is a horizontal rectangular plate connected square across the pairs of sections by riveting or welding. Cf. **lacing**.

batter, rake (1) An artificial, uniform steep slope.

(2) Its inclination, expressed as 1 m horizontally for so many metres vertically. See **retaining wall** (diagram).

batter level A *clinometer* for measuring the slope of earth cuts and fills.

batter pile, raking p. A *pile* (usually driven not bored) at an angle to the vertical.

batter rail A sloping *profile* to show the vertical angle at which to cut a slope or fill an embankment.

battery See **blasting machine**.

battledeck [stru.] Steel plates stiffened by flats, angles or other *rolled sections* welded under them. Originally used on ships, they were adapted for bridge decks after 1945.

baulk See **balk**.

bauxite [min.] The most important ore of *aluminium*, $Al_2O_3.2H_2O$, named after Les Baux in Provence. It does not fuse below 1600°C. It is used as a *refractory*, and as the raw material for *high-alumina cement*.

beaching [hyd.] Loose-graded stones from 7 to 20 cm in size used in a layer 0.3 to 0.6 m thick for revetting reservoirs and embankments below the level of the *pitching*.

beach replenishment [hyd.] See **artificial recharge**.

beacon [sur.] See **monument**.

beacon laser [sur.] A laser rotating in a horizontal plane and projecting a horizontal beam of light, used for checking *earthwork* and other levels. It pro-

vides a horizontal datum all over the field of view, enabling an *excavator* driver or *bulldozer* operator to keep to the required earthwork level within 25 mm (1 in.) without the help of a surveyor, merely by using the *laser sensor* on his or her machine. Power is from a 12-volt battery.

beaded section [stru.] A *light-alloy* **angle section** or **channel section** *extruded* with beads or bulbs at the extremities like a *bulb angle*. These beads are an advantage of *extrusion*, since they cannot be easily formed by rolling and they increase the bending strength of the section in a metal-saving way.

beam [stru.] A structural member designed to resist loads which bend it. The bending effect at any point in a beam is found by calculating the *bending moment*. Beams are usually of wood, steel, light alloy, or reinforced or prestressed concrete.

beam and block, beam and pot floor See **pretensioned floor beams**.

beam and slab floor A reinforced-concrete floor in which the *slab* is carried on reinforced-concrete beams. This is the oldest reinforced-concrete floor for large spans but is superseded for offices by the *plate floor* and *hollow-tile floor*, because of its inflexible plan, with beams projecting beneath. It is still used for heavy construction, bridge decks, factories, etc.

beam bender [mech.] A machine for straightening or bending rolled-steel joists.

beam compasses [d.o.] A light wooden beam fitted with adjustable heads to take a point and a pen or a pencil and thus to draw arcs of larger radius than is possible with a pair of compasses.

beam engine [mech.] An early *steam engine* with vertical cylinder. It operated the Cornish pump.

beamless floor [stru.] A *plate floor*.

beam test Measuring the *modulus of rupture* of a concrete or mortar after casting from it a standard beam without reinforcement, supported and loaded in a standard way. The *bending moment* at failure is recorded and from it the maximum tensile stress (assumed equal to the maximum compressive stress) is calculated. In this way an estimate of the strength of the concrete or mortar or cement can be obtained without the expense of a laboratory testing machine.

bearing (1) [stru.] The support of a *beam*, or the length (or area) of the beam which rests on its support.

(2) [stru.] The compressive stress between a beam and its support (bearing pressure), particularly on foundations.

(3) [sur.] A horizontal angle between a stated direction, usually north, but occasionally south (true or magnetic), measured clockwise or counterclockwise, unlike an *azimuth*.

bearing capacity *Allowable bearing capacity*.

bearing pile A *pile* which carries weight, as opposed to a *sheet pile*, which takes earth pressure, or a *raker*, which takes thrust. It may be either an *end-bearing pile* or a *friction pile*.

bearing pressure, b. stress [stru.] The load on a *bearing* surface divided by its area.

bearing stratum [stru.] The stratum (or formation or bed) which has been chosen as the most economical or suitable to carry the load in question.

bearing test See **plate bearing test**.

Beaufort scale An ancient scale of wind speeds used by sailors that varies from o for a mirror-smooth calm (less than 1 kph) to 12 for a hurricane of 120 kph or more. The wind force in newtons, according to BSCP 3, ch. 5, part 2, 'Wind loads', is $0.6\ cv^2 A$: v = wind speed, m/s, A = projected area, m², and the coefficient, c, is 1.5 for flat surfaces and 1 for cylinders. Many other values of c exist. It can even be negative for wind suction on the lee side of a roof.

bed load [hyd.] The weight or volume of silt, sand, gravel or other material rolled along a stream bed in unit time. See **sediment**.

bed plate, bedplate [mech.] A cast-iron plate or steel frame on which a machine sits. It is usually held down to the concrete floor by *anchor bolts*.

bedrock (ledge, ledge rock USA) Hard, cemented rock below gravel or other loose soil. In alluvial mining in gravel, the valuable mineral if dense (e.g gold) is often concentrated at the bedrock surface.

beetle head A *drop hammer*.

Belanger's critical velocity See **critical velocity**.

Belgian truss A *Fink truss*.

bell dolphin, Baker b. d. (see diagram) A large bell-shaped steel or concrete fender, suspended on a cluster of piles in the open sea for the mooring of vessels, first used at Heysham jetty.

belled pile A *bored pile* enlarged by *under-reaming*.

bellmouth overflow [hyd.] An overflow from a reservoir through a tower built up from the bed to the overflow level. The water is led out usually through a tunnel.

belly rod See **camber rod**.

belt conveyor [min.] (see diagram,

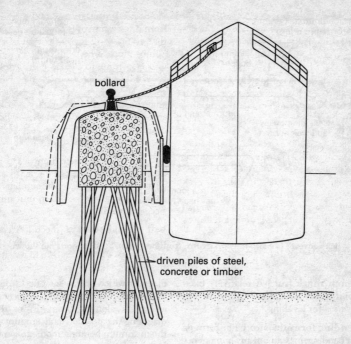

bollard

driven piles of steel, concrete or timber

Baker bell dolphin for mooring at sea. The dotted line shows allowable movement of bell.

p. 32) An endless belt of rubber-covered textile from 0.45 to 1.5 m wide used for carrying coal, gravel or similar loose material. It should be straight in plan but can pass through limited *vertical curves* which are concave above and unlimited curves which are convex above. Coal cannot be carried at an angle steeper than 18° on a smooth belt because it slips back. Belt speeds vary from 15 m/min. in a movable to 150 m/min. in a permanent conveyor. If the conveyor is accurately laid out, curves of large radius are possible. For transporting large volumes, a belt is often cheaper than a *haul road*, especially where a bridge is needed. See **cable belt**, **conveyor loop**, **curved conveyors**, **steel-cored conveyor**, **tension drum**.

bench See also (*B*). (1) A *berm*.

(2) [min.] An artificial long horizontal step from which mineral or stone is quarried, usually by blasting from vertical holes.

benched foundation, stepped f. A foundation on a sloping *bearing stratum*, cut in steps to ensure that it shall not slide when concreted and loaded up, as a sloping foundation could.

benching (1) A *berm* above a ditch.

(2) [min.] Quarrying with benches about 3 m deep or more.

benching iron, change plate [sur.] A triangular steel plate with points at the corners. The points are driven into the ground and the plate is used as a temporary *bench mark* or *change point* in levelling.

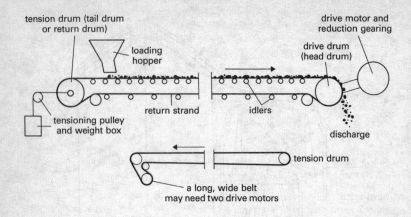

tension drum (tail drum or return drum)

loading hopper

drive motor and reduction gearing

drive drum (head drum)

return strand

idlers

tensioning pulley and weight box

discharge

tension drum

a long, wide belt may need two drive motors

Belt conveyor (see also **tension drum**, **troughed belt**). Steel supports not shown.

bench mark [sur.] A relatively fixed point whose level is known and used as a *datum* for levelling.

bending formula [stru.] The formula for bending, in beams of any homogeneous material, is

Bending moment = stress × *modulus of section*

usually written

$$M = fz \text{ or } M = \frac{fI}{y}.$$

bending moment [stru.] The total bending effect at any section of a beam is called the bending moment. It is equal to the algebraic sum of all the moments to the right of the section (or to the left of the section, which amounts to the same thing) and is called M for short. Every bending moment can be expressed as a force times a distance called the *arm*. The units are pound-inches, ton-inches, kg-m, N-m, tonne-m, etc. See **Navier's hypothesis**.

bending-moment diagram [stru.] A

diagram which shows for one loading the amount of the *bending moment* at any point along a beam. From this diagram, the position and amount of the maximum bending moment can be immediately seen.

bending-moment envelope [stru.] Several bending-moment diagrams (one for each loading carried by the beam), laid on top of each other to show the worst bending moment at any point for all possible loadings.

bending schedule A list of reinforcement which accompanies a reinforcement detail drawing, prepared by a reinforced-concrete designer. It shows the dimensioned shapes of all the bars and the number of bars required. The *bar bender* bends the reinforcement according to the schedule and places it according to the drawing. See **cutting list**, BS 4466.

bends Slang for *caisson disease*.

bend test A test of a weld or of the steel in a flat bar, in which the bar is bent cold through 180° to verify its *ductility*. If there is no cracking the piece is considered *ductile*.

bent [stru.] A two-dimensional *frame* which is stable, but only within these dimensions. It has at least two legs and is usually placed at right angles to the length of the structure which it carries such as a bridge, pipeline, aqueduct, etc. See **trestle**.

benthic deposits Mud, sand or gravel on the bed of a river, lake or sea.

bentonite A *clay* composed, like fuller's earth, mainly of the same clay mineral *montmorillonite*. Found in Wyoming, USA, where it is known for its remarkable accordion-like expansion when its water content increases. It is used for making *refractories* and rubber compounds, as a filler for *synthetic resins* (B) and for making oilwell *drilling fluids*, because it is *thixotropic* in small concentrations.

bentonite–cement pellets Dry pellets of 1:3 cement:bentonite, dumped in to backfill and seal a borehole quickly.

bentonite mud A thixotropic suspension of *bentonite* in water, used for holding up the sides of deep trenches excavated by machine. Bentonite slurry is often used as a lubricant to reduce skin friction in *pipe-jacking* or in *pile-driving* or in the sinking of a *caisson*. See **diaphragm wall**.

bentonite slurry See **bentonite mud**, **diaphragm wall**, **slurry shield**.

bent-up bar, cranked b. (slant b. USA) Near supports, where *shear* force is high, *main steel* in the bottom of a beam is often bent up at about 45° to the horizontal to join the top bars. The bent-up length acts as shear *reinforcement* and reduces the need for overcrowded *links* near the support, thus making it easier to place the concrete.

Berlin wall A retaining wall made by driving H-beams vertically down; during excavation horizontal timbers or precast concrete slabs are fixed between the flanges to hold back the earth.

berm A horizontal ledge in the sloping surface of an *embankment* or *cutting*.

Bernoulli's assumption [stru.] In any bent beam, sections which were plane before bending are plane after bending. See **Navier's hypothesis**.

Bernoulli's theorem [hyd.] The *energy* per unit mass of a stream at one point is equal to that at another point upstream (or downstream) plus (or minus) the friction losses. Stated differently: Pressure energy + *potential energy* + *kinetic energy* + losses = a constant.

berth A *jetty*, *quay*, *wharf* or other place where a ship can tie up.

berthing impact The forces on piers, jetties, etc., during the berthing of vessels. The forces are usually estimated from the *kinetic energy* of a large vessel berthing at about 15 cm/s.

beton Concrete, in French, German, Hungarian, Romanian, Russian, etc.

Betws ventilation shaft A shaft 210 m deep and 3 m (10 ft) finished dia., drilled from the surface for the British Coal Corporation. It is an unusual shaft because nobody has ever travelled through it.

It was drilled with a 3.75 m dia. (12 ft) bit with the shaft full of *bentonite* mud, and lined by welding together lengths of two steel tubes over the finished hole, the inner one being of 3 m dia. The gap between them was filled with concrete and the lining was eventually lowered to its final position with the bottom sealed to allow it to be floated down. Water was added to sink it as required. The space between the steel lining and the rock was grouted.

bevelled washer, tapered w. A *washer* made from a wedge-shaped piece of steel plate. It is thinner at one edge than the other so as to fit under a nut on the flange of a *rolled-steel joist* or other section with tapered flanges.

B-horizon [s.m.] The lowest part of the *topsoil*, between the *A-horizon* and the *C-horizon*. It contains metal oxides and other soluble materials leached from the A-horizon above it.

bi-cable ropeway An *aerial ropeway* in which one or more *track cables* carry the loads and a *traction rope* moves them along. See **monocable**.

bid-rigging, tender fixing *Collusion*.

bid shopping (USA but not exclusively) On being awarded a *contract*, *main contractors* do not automatically award subcontracts to those *subcontractors* who helped them with their successful *tenders* but instead choose any lower-priced subcontractors. This unfair practice can be prevented if the *consultants* insist on main contractors naming their subcontractors when they tender.

billet An intermediate product made by *cogging* a steel *ingot*. It is usually less than 160 cm² in cross-section. See **bloom**.

bill of quantities A list of numbered *items*, each describing the quantity, measurement unit, and sometimes also the price of work to be done in a building or *civil engineering project*. See **priced bill**.

bi-metal strip (see diagrams) A strip of two metals with different *coefficients of expansion*, one metal forming each side of the strip. As the temperature changes, the strip curves one way or the other by a calculated amount. Bi-metal strips are used in *thermostats* (*B*), gas heaters for water, thermal *relays*, etc.

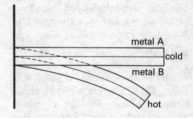

Bimetal actuator. Metal A has a larger coefficient of thermal expansion than metal B.

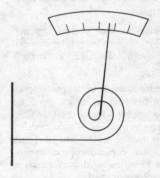

Bimetal thermometer. (E. A. Parr, *Industrial Control Handbook, Collins, 1986.*)

binder (1) *Hydraulic cement, bitumen, gypsum plaster* (*B*), *lime* (*B*) or other substance used for joining masonry or concrete.
(2) The clay, silt, etc. in *hoggin* or the *matrix* of rock.

(3) A *link* in a *reinforced-concrete* column, beam or pile.

binding wire *Annealed wire*.

Binishell A dome of either 18 or 36 m (59 or 118 ft) dia. erected by inflating a plastic 'balloon'. The Pittwater High School Binishell dome in north Sydney collapsed on 4 August 1986, probably because of lightning. According to Dr Bini, 1500 such shells had been erected by 1986 without trouble. The collapse may have been caused by lack of grounding and electrical continuity in the *reinforcement*. Each dome is reinforced and concreted before inflation lifts the plastic balloon carrying 300 tons of wet concrete and steel to create a roof varying in thickness from 40 mm ($1\frac{1}{2}$ in.) to 110 mm ($4\frac{1}{2}$ in.). Any doors or windows are cut through the hardened concrete.

biochemical oxygen demand (BOD) [sewage] The oxygen absorbed by a sewage sample in five days at 20°C. Most rivers used for drinking-water have a BOD of 4 mg/l or less. See also **Royal Commission recommendation**.

biodegradable Description of wood, paper, leaves, etc. that rot.

bioengineering Use of grass or other plants as *revetment*.

biofiltration Removal of smell from air or gas by passing it up through a bed of compost or peat containing bacteria, in a process developed by the Water Research Centre, Stevenage. The compost and heather mixture with 50% water is nearly 1 m (3 ft) deep. It works best on smells other than hydrogen sulphide. Cf. **bioscrubber**.

biogas digester See **digestion**.

biological filter [sewage] A trickling filter.

biological shield A thick wall surrounding a nuclear reactor, which protects workers from radiation. The shield may be of concrete, sheet lead or other heavy material, and the concrete can be made denser than usual by replacing the stone with iron or lead shot or barytes (specific gravity 4.5) or other heavy mineral.

biological treatment [sewage] Any treatments such as *trickling filters, rotating biological contactors*, or the *deep shaft* or *activated-sludge processes* which make use of *micro-organisms* to lower the *biochemical oxygen demand* of *sewage* or *effluent* or *sludge*. It may be *aerobic* or *anaerobic*.

bioscrubber A wet tower packed with plastic that removes smell from gas passing up through it. The plastic medium provides a large surface area for bacteria to make contact with the gas. Sewage *effluent* flowing down removes hydrogen sulphide in the form of very dilute sulphuric acid. It can reduce 300 ppm hydrogen sulphide to 2 ppm or less.

Birmingham (or **Stub's**) **wire gauge (BWG)** Numbers that describe thicknesses of wire and sheet steel, close to but different from the *Standard Wire Gauge* (*B*). A further complication is that another slightly different gauge is the Birmingham gauge (BG).

bit See (*B*), also **detachable bit**.

bitumen A black sticky mixture of hydrocarbons completely soluble in carbon disulphide. It can be obtained from natural deposits (usually of oxidized petroleum) or by distilling petroleum. See **asphalt**.

bitumen-coated steel A way of preventing rusting. See **epoxy-coated steel**.

bituminous carpet A road wearing

course containing a tar or bitumen binder and that is not thicker than 38 mm.

bituminous emulsion, bitumen road e. A liquid mixture of water and tar or *bitumen* which has the advantage that it can be applied to roads in cold damp weather. See **emulsion**.

black [mech.] Steel as it comes from the forge or rolling mill, covered with *mill scale*, which is not removed except when the steel is prepared for a protective paint or *metal coating* (*B*). See **clearing hole**.

black diamond, carbon, carbonado [min.] *Diamond* which is not used as a gem because of its grey or black colour but forms a valuable cutting point for many tools. It comes from Brazil mainly and is preferred to *bort* because it does not have the crystalline cleavage of bort.

black top *Tarmacadam.*

bladed shield A tunnelling *shield* in which the protective steel cylinder consists of separate steel plates (blades). A 10 m (33 ft) dia. shield in Hamburg (1988) for a tunnel to contain two rail tracks had 32 blades, each about 1 m (3 ft) wide and 6.5 m (21 ft) long. The shield passes through soft (treated) ground but the permanent concrete lining has to be cast close behind the shield, and the tunnelling equipment cannot jack itself forward on concrete which is *green*. Instead, the blades are used to pull it forward by retracting them all simultaneously. As soon as the shield has moved forward, the blades are again jacked forward one by one.

blade grader A grader.

blading back, back-blading Smoothing a surface made by a bulldozer or loader by dragging the mouldboard or shovel backwards.

Blake breaker See **jaw breaker.**

blank carburizing [mech.] A *carburizing* carried out without the carbon. It is a *heat treatment.*

blank flange [mech.] An undrilled *flange*.

blank nitriding [mech.] A *nitriding* without ammonia or nitrogen.

blast furnace A smelting furnace in which heated air is blown in at the bottom, for reducing iron, copper or other metals from their ores. The charge is fed in at the top, the metal and slag tapped off at convenient holes at the foot.

blast-furnace cement See **Portland blast-furnace cement.**

blast-furnace slag Waste from iron smelting. The compounds in it are the same as those in *Portland cement* (silicates of calcium, magnesium and aluminium) but their proportions are different. Properly used it can make more durable concretes than those made with pure Portland cement. Probably all the UK and Continental output is used. See also **ground granulated blast-furnace slag, Portland blast-furnace cement**, and (*B*).

blasting (1) **shot firing** (see diagram) [min.] The breaking of rock by boring in it a hole which is filled with *explosives* and detonated.
(2) See **sand blast.**

blasting fuse [min.] *Safety fuse.*

blasting machine (battery, exploder UK) [min.] US term for a portable electric generator worked by a *rack* and pinion or by a turning handle, the whole weighing about 5 kg. The

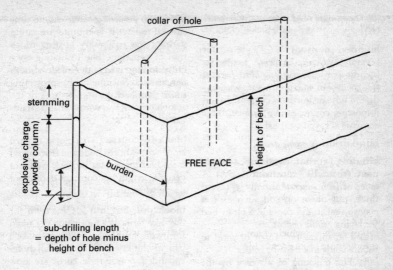

Blasting terms.

most powerful exploder available in Britain generates 1200 volts and will fire 500 shots simultaneously, but most exploders are designed for only a small number of shots.

blast mixer A truck-mounted concrete mixer which can mix an almost dry concrete or mortar for *shotcreting* or *guniting* and place some 50 tons/h through a pipe 50 m long.

bleeding (also called **water gain** in the USA) Separation of clear water from the cement paste of mortar or concrete. Two types are known, the first beneficial, the second harmful to concrete strength, but they may coexist. During compaction, water can flow out of concrete, lie on its surface, and thus encourage good curing for the first few hours during hot weather. Some *admixtures* are said to encourage this sort of bleeding. In the other sort

of bleeding, water segregates beside or under the steel or larger stones, weakening the bond between them and the body of the concrete. A *water reducer* should enable the *water/cement ratio* to be lowered, reducing this sort of bleeding. A second vibration, some hours after the first one, is sometimes used to expel this excess water.

Occasionally, when the bleed water on a slab has evaporated, *plastic settlement* cracks may be seen over the top steel. These cracks should not be confused with those due to drying *shrinkage* in concrete that has been allowed to dry too fast. Bleeding of an asphalt road in hot weather may seriously reduce its skid resistance.

blemishes *ACI* lists several types of blemish in a concrete surface: *blow holes, honeycombing, laitance, lift* *joint, pop-out, sand streak, efflorescence* (*B*).

blended cement *ACI*'s term for several *hydraulic cements* made by blending or intergrinding *blast-furnace slag*, *pozzolan*, *microsilica*, etc. with hydrated lime or Portland cement. In oil-importing countries like France, these cements are well established.

blind drain A *rubble drain*.

blinding (1) **mat, mattress, sealing coat (mudsill, mudmat** USA) A layer of *lean concrete* usually 5–10 cm thick, put down on soil such as wet clay to seal it and provide a clean bed for laying *reinforcement*.

(2) The spreading of grit or sand to fill the voids in a *wearing course*.

(3) The blocking of a *screen* by the material being screened.

blind rivet A *pop rivet*.

blind shield, full-face diaphragm s. A tunnelling *shield* used in very soft uniform soils near Tokyo and elsewhere in Japan. As the shield is jacked forward, soil is extruded into the tunnel through a hole, in the *bulkhead*, of about 800 mm (2½ ft) for a 6 m (20 ft) dia. shield. No cutting head and no moving parts are needed except the jacks and the muck transport equipment. See **silt displacement**.

blister lock An *air lock* for a *pneumatic caisson* – a conspicuous swelling on top of a vertical steel tube projecting some way above the caisson. The tube contains the ladderway into the caisson.

bloated clay See **lightweight aggregate.**

block (1) [mech.] The frame holding the pulley or pulleys of *lifting tackle*.

(2) See below and (*B*).

blocked-in stanchion A stanchion in which the web-flange areas have been filled with *aerated concrete* can be considered to have a 30-minute fire rating without the customary coating of 50 mm of concrete. The insulating concrete infilling covers the inside of both flanges and all the web, protecting more than half the surface area of the stanchion. See also **water-filled structure**.

block-in-course Large blocks of hard stone with worked beds and hammered faces laid in courses in dock walls, of variable length but constant depth, not deeper than 0.3 m.

block out, box out *ACI*'s term for box-like *formwork* put inside the main formwork for a slab or beam to make a gap in the concrete. For small holes expanded polystyrene cores are now usual. They are easily removed by a hand torch.

block pavement A road *wearing course* made of rectangular blocks of stone or wood, as opposed to *sheet pavement*.

blockwork Masonry in *breakwaters* and other marine structures built of precast concrete (or stone) blocks weighing from 10 to 50 tons to resist movement by waves. See **coursed blockwork, sliced blockwork** and (*B*).

blockyard, casting yard A space where precast concrete pieces are poured and allowed to harden before use.

Blondin After Blondin the tight-rope-walker; the French word for a *cableway*.

bloom A half-finished rolled or forged piece of steel or wrought iron with a cross-sectional area greater than 160 cm² (cf. **billet**). If used as a stanchion base, the top surface is machined.

blow See **boil**.

blow down (1) **blowing d.** (**blow off** USA) [mech.] Opening a valve in a steam *boiler* mud drum or other place where boiler sediment collects, so as to eject it. This may be done periodically by hand or continuously by an automatic arrangement.

(2) A *pneumatic caisson* has to be blown down when its downward movement stops and its internal air pressure is reduced to persuade it to move. The men are first withdrawn and the air pressure is lowered but by not more than a quarter of the gauge pressure. If this does not move it, more *kentledge* must be added.

blow down lance, b. pipe A tube connected to an air compressor, and inserted into *formwork* just before it is filled with concrete, so as to clean it out.

blow hole (**bug h.** USA) Surface cavities in concrete, usually not more than 15 mm across, from air trapped during casting. See **blemishes**.

blow off (1) An outlet on a pipeline for discharging sediment or water or for emptying a low sewer.

(2) See **blow down** (1).

blow out (1) In compressed air work a sudden loss of compressed air from the tunnel or *caisson*, which increases rapidly and may become disastrous for the work and dangerous for the workers.

(2) [min.] An oilwell from which oil and gas may flow out uncontrollably because the *drilling fluid* was not dense enough.

blow pipe See **blow down lance**.

blueprint [d.o.] A *contact print* on *ferro-prussiate paper* of a drawing made on transparent paper or linen. It is developed either in water or in a special solution. It has white lines on a blue ground. Because of the dark ground, blueprints are now obsolescent, although the print from a faint pencil drawing is much better than from a *dyeline*.

Board of Engineers' Registration The section of the *Engineering Council* which establishes engineers' qualifications – whether they are entitled to be called *chartered engineers, incorporated engineers* or *engineering technicians*.

Board of Trade unit See **kilowatt hour**.

BOD See **biochemical oxygen demand**.

bog blasting See **peat blasting**.

boil, blow A flow of soil, usually fine sand or *silt*, into the bottom of an excavation, forced in by water or water and air under pressure. It starts as a small spring and like *piping* may increase rapidly. Therefore boils should be stopped immediately, before they bring disaster. They can be prevented by reducing the pressure difference. This can be done in one of several ways. (i) By *wellpoints* or other relatively slow processes of *groundwater lowering* outside the excavation. (ii) The sinking of *relief wells* inside the excavation, deep below the final excavation level, if filled with gravel (or drain pipes) provides a safe path for water under pressure to flow up. (iii) The sinking of pumped wells below final excavation level inside the cofferdam may be possible only combined with *groundwater lowering* outside. (iv) The pressure inside the excavation can be increased by flooding it or filling it with gravel. Flooding or refilling are the only possible courses in an emergency. A spring is a boil with very little sand in the flow. See **graded filter, quicksand**.

boiler, steam b. [mech.] A plant for

raising steam. Large boilers evaporate about 400 kg of water per second and are usually coal or oil fired. Such boilers are used in power stations, the power plants of mines, rolling mills and similar heavy industry. See **pulverized coal**; also (*B*).

boiler-house foundations See **refrigerator foundations**.

boiler rating [mech.] The heating capacity of a boiler expressed in kilowatts, *British thermal units* (see p. xvi) per hour, kg of steam/second, etc.

bollard (1) A cast-iron post anchored deeply into the masonry of a quay wall, used for mooring vessels. See **bell dolphin**.
(2) A post anchored in a road to protect a kerb, wall or street *refuge*, or to divert traffic.

bolster (1) A support for a bridge truss on an abutment.
(2) Padding or lagging generally; in particular that used round the edge of a *limpet dam* to make a watertight joint with a dock wall.

bolt [mech.] A cylindrical bar which is screwed at one end for a nut and forged with a square or hexagonal head at the other end. The oldest (except forge welding) and commonest way of fixing steel parts together. Part of the bar next to the head has no screw thread. Cf. **screw**.

bolt sleeve A cardboard, plastic, steel or asbestos cement tube round a bolt in a concrete wall. It prevents the concrete sticking to the bolt and also acts as a *distance piece* to keep the shutters at the correct distance apart.

bond (1) **grip, interface strength** [stru.] *Adhesion*.
(2) **whip** [mech.] A short length of wire rope by which loads are fixed to a crane hook. See **sling**.

(3) [elec.] A short conductor between the ends of rails at a rail joint, or between the parts of an earthing system or lightning protective system or other metal work. The ends of rails even when welded together are also electrically bonded to reduce the electrical resistance of the joint. See also (*B*).

bond breaker A *release agent*, or for *lift-slabs* a *sealant* (3).

bonder See (*B*).

bonding admixture A latex (polymer emulsion) mixed into a concrete or mortar to improve its tensile strength, bending strength, durability and bond strength.

bonding capacity US term for the financial strength of e.g. a *contractor*.

bonding grout, b. layer About 2 mm thickness of grout containing special *admixtures*, occasionally epoxy resin, to make sure that a *topping* sticks to the concrete below it. Some admixtures may prevent rusting of steel. See **monolithic topping**.

bond length The *grip length* of a reinforcing bar.

bond prevention *Bond* between steel and concrete can be prevented, perhaps unintentionally, by grease on the steel or on purpose by enclosing *prestressing* strand in a plastic or steel sheath.

bond stress A shear stress at the surface of a reinforcing bar, which prevents relative movement between bar and concrete. It is helped by *mechanical bond*. See **adhesion**. The allowable bond stress is about one tenth of the concrete compressive stress.

bone necrosis Damaged hip or shoulder joints, appearing sometimes years after a worker in *compressed air* has left that work, even if he spent only 18 months in it. About ten men who had worked in the Dartford tunnels between 1974 and 1976 had received more than £1 million compensation by spring 1988. See **decanting**.

boning [sur.], **boning in** Setting out a slope by *boning rods*. Two rods are held on pegs which have been previously set on the slope at correct levels and a third rod is lined in as it is moved about between them. A laser instrument is more expensive than boning rods but slopes can be set out easily with it.

boning rod [sur.] A T-shaped staff about 1.2 m long made of two pieces of 50 × 12 mm board nailed together. The short T-piece, which is held uppermost, is used for sighting and lining up with the other boning rods. See **boning, sight rail, traveller**.

boogie box See **boojee pump**.

boojee pump, boogie box, grouting machine, grout pan A container for pressure *grouting* with cement *slurry* behind tunnel linings or into fissures in rock. It is usually stirred by a compressed-air engine and forced out by air pressure, but for mining work in which very high pressures may be needed pumps often provide the high pressure.

booking [sur.] The recording of field observations in a legible and understandable way for later use in the field or drawing office, often by another person.

boom (1) Any moving beam, such as the *jib* of a crane, or in a *boomheader* the excavating arm.

(2) A *chord*. The horizontal member, top or bottom, of any built-up *girder* or *truss*.

(3) A barrier across a stretch of water that may be designed to stop submarines, or be a *floating boom* to hold oil slicks, etc.

boomcat See **sideboom cat**.

boomer A *drill carriage*.

boomheader, roadheader, cutter boom tunneller Machines of many different types for *breaking ground* in rock that does not have to be blasted. They have existed since the 1930s for driving headings in mines and until about 1970 they were relatively small, with up to 25 kW power. (The largest, of 500 kW, now weigh more than 100 tons.) They move on caterpillar tracks or wheels, or on a *gantry* or *monorail* in a *shield*. The 'pineapple' rotating head at the end of the massive boom projecting ahead can be directed upwards, sideways or downwards. In some machines the boom is telescopic. Like *fullface machines* they make no *overbreak* as explosives do, nor do they shatter the rock, and they can choose their tunnel profile, unlike fullface machines. The many makers in Europe and the USA have machines that cut 6 m (20 ft) wide and 6 m high from one position. The cutting-head motors use half the power. See *water-jet-assisted boomheader*.

booster [mech.] A pump or compressor inserted into a water or compressed-air pipeline near the consumer, so as to increase his pressure.

boot, b. lintel (1) The lower end of a bucket elevator, from which the buckets lift material.

(2) A projection from a concrete beam to carry facing brick, stone or other *cladding* (B).

boot man A labourer wearing the company's boots, for standing in fresh concrete or mud, and who is paid extra while wearing the boots (boot money).

Bordeaux connection [mech.] A *thimble* for the end of a steel wire rope with a link fitted permanently into it. It is a convenient way of joining wire rope to short-link crane chain.

border stone A *kerb* stone.

bore (1) [mech.] The internal diameter of a pipe or other cylinder.

(2) [min.] A borehole, or to make a borehole.

(3) [hyd.] A wave advancing with a nearly vertical front upstream during the flowing tide in an estuary, as in the Severn estuary. Occasionally, in a flood, a bore may travel downstream.

bored pile, b. cast-in-place/cast-in-situ pile A *pile* formed by pouring concrete into a hole formed in the ground, usually containing some light reinforcement. They can now be made 1.25 m dia., 60 m deep, and 12° off vertical, apart from *under-reaming*. These piles were first made in about 1905 and are now a favourite method of taking foundations through soft surface soil on to a *bearing stratum* in London. They may be bored by tripod (*shell-and-auger*), *continuous-flight auger* or *rotary drilling*. In water-bearing gravels, uncased bored piles may be inadvisable: the water may remove the cement and ruin the concrete. See **diaphragm wall**, **secant piling**; cf. **driven cast-in-place pile**.

borehole A hole driven into the ground to get information about the strata, or to release water pressure by *vertical sand drains*, or to obtain water, oil, gas, salt, sulphur, etc. Occasionally boreholes are sunk from the surface to shallow mines to admit water pipes, hydraulic fill, power cables or fresh air. See **bored pile**, **cable drill**, **diamond drilling**, **pipe-jacking**, **rotary drilling**.

borehole log [s.m.] A record of the findings at a borehole, including details of the groundwater level (and gain of water in *air-flush drilling*) with descriptions of the cores, percentage core recovery, etc. *Well logging* is a specialized activity developed for the deep holes of oilwell drilling but can also be applied in shallow holes.

borehole pump [min.] A centrifugal pump, electrically driven through rods from a motor on the surface, or (for deep well *submersible pumps*) by an electric motor at the foot of the pump casing. These pumps are used for dewatering mine shafts or for pumping drinking water and are therefore obtainable in capacities from 50 to 50,000 l/min.

borehole samples [min.] Samples obtained from boreholes, namely *diamond drill* or shot-drill cores, or sludge or chippings from other drilling methods. For an investigation into the strength of a *clay*, an *undisturbed sample* must be obtained.

borehole surveying, directional s. Measuring the *azimuth* and slope of a borehole at selected depths, often to help with *directional drilling*, Cf. **well logging**.

borescope See **fibrescope**.

boring [min.] (1) Making a hole in rock for blasting, using a rotative or percussive drill.

(2) Driving a borehole. See above; also **auger, bort, jetting, wash boring**.

borrow Material dug to provide *fill* elsewhere.

borrow pit An excavation dug to provide *fill*.

bort, boart [min.] A crystalline *diamond* which lacks the brilliance and purity of colour that would make it a gem but is nevertheless very valuable as a cutting agent in *diamond drills*.

Bosporus Bridge When built (1973) it was the longest (1074 m) *suspension bridge* in Europe. It cost $36 million, but there was so much traffic that this cost was recovered from tolls by mid-1976.

Built like the *Severn Bridge* to Freeman Fox's design, it aroused interest with its economical aerofoil deck design and the off-vertical, continuous suspender ropes each side of the deck, carrying its weight up to the main cables above them. It was built in $3\frac{1}{2}$ years instead of the normal $5\frac{1}{2}$.

The second Bosporus bridge with a main span of 1090 m and two four-lane carriageways (two lanes wider than the first bridge) was started in May 1985 and completed in June 1988, six months ahead of schedule, but it could not then be opened because the 28 km of approach roads were not ready. Designed by Freeman Fox it was built by a three-country joint venture, including Impregilo of Italy; three Japanese contractors – Ishikawajima Heavy Industries, Mitsubishi and Nippon Kokan; and, from Turkey, Turkes-Feyzi Akkaya Insaat. One difference from the first bridge is that its suspenders are vertical.

Each main cable has 18,648 galvanized 5 mm dia. hard-drawn steel wires in 37 strands. The cables are coated with red-lead paste, wrapped with soft-steel galvanized wire and then painted.

Boston caisson A *Gow caisson*.

BOT (build–operate–transfer) A type of *franchise*.

bottom cut, draw c. [min.] In small tunnels in rock, a *cut* consisting of two converging rows of holes. One row is horizontal and near the floor, the other is placed about the middle of the tunnel. The two rows are fired together and a wedge of rock is blasted out. Where more space is available, the *wedge cut* is preferred.

bottoming (1) Large stones in a road laid on the *formation*.

(2) [rly] The ballast in a *permanent way*.

(3) The last few centimetres of excavation usually removed by spade to ensure that the bottom is to the correct level and smooth.

bottom-opening skip A *drop bottom bucket*.

bottom sampler [hyd.] A *sounding lead* with adhesive on the underside, or a coring device for picking material up from the sea bed.

boulder clay (till, tillite USA) [min.] Material consisting of rocks crushed by moving glaciers and containing scratched stones of boulder size down to sand, with some clay, all unstratified.

boulevard In the USA a wide road which in cities is usually planted with shade trees, sometimes also on the central strip. See **street**.

Bourdon pressure gauge [mech.] (see diagram, p. 44) A tube of oval cross-section which tends to straighten as the pressure inside it increases. It is a very simple, robust and useful gauge for boilers, *pore-water pressure* measurements, and so on. Perhaps the most sophisticated Bourdon pressure gauge is a helical tube made entirely of quartz.

Boussinesq equation [s.m.] In 1885 Boussinesq published his analysis of the stresses in the soil beneath a loaded foundation assuming the soil to be a semi-infinite elastic solid. Although most soil is not strictly elastic, Boussinesq's rigorous mathematical work was proved broadly correct 50 years later by workers in *soil mechanics* who showed that the lines of equal vertical stress under a loaded point are roughly circular. See **bulb of pressure**.

Bowditch's rule [sur.] An *adjustment* for a *closed traverse* between points whose coordinates are accepted as

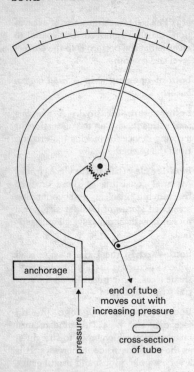

anchorage

end of tube
moves out with
increasing pressure

pressure

cross-section
of tube

Bourdon tube. (E. A. Parr, *Industrial Control Handbook*, vol. 1, *Transducers*, Collins, 1986.)

fixed. Angles and sides are assumed equally liable to error. The correction to apply to any line for an error in *latitude* is

$$\frac{\text{length of line}}{\text{perimeter of traverse}} \times \begin{matrix}\text{total error in}\\\text{latitude}\end{matrix}$$

The correction for errors in departure is made in the same way.

bowk [min.] A *kibble*.

bowl scraper, wheeled s., tractor s. (see diagram) A massive self-propelled steel box on four rubber-tyred wheels. As it moves forward the bottom edge digs into the ground, filling the bowl. At the dumping-point the soil is ejected by a 'tail gate' or ejector which sweeps the soil out. In stiff clay a pusher tractor (bulldozer) is often essential during loading.

Scrapers vary in capacity from 7 to 40 m³ (9 to 50 yd³), from 150 to 615 hp per engine and over 1000 hp for twin engines. Their top speeds are from 40 to 64 kph (25–40 mph). All have automatic or semi-automatic transmission and high-pressure hydraulics. About half have self-loading elevators at the cutting edge of the bowl, reducing the need for a pusher tractor in heavy soil. But open-bowl scrapers are not thereby made obsolete. Scrapers not only transport soil, but also spread and level it. Their highest outputs are for short hauls of less than 100 m but their most profitable haul is from 150 to 500 m.

The large rubber tyres of scrapers are so expensive that they amount to 30% of total scraper costs.

bowstring girder A *girder* shaped like a bow with the string downwards. It may be of concrete, steel or timber. Modern laminated timber girders are made up to 50 m span, and can be delivered ready made in spans up to 24 m.

box beam A *box girder*.

box caisson, American c., stranded c. A large open-topped box (closed at the bottom) built on shore, floated out, and sunk over the ground chosen for a foundation. The box forms part of the final structure. It is commonly used for bridge piers, as it enables construction to be done in the dry. It is usually built of reinforced concrete. See **caisson**.

box culvert A culvert of rectangular or square cross-section.

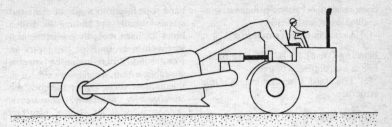

Bowl scraper.

box dam A *cofferdam* completely surrounding an area.

box drain A small rectangular *drain* of brick or concrete.

box frame construction [stru.] A method of building a long thin block of flats, offices, etc., with concrete slab floors carried on load-bearing walls across the thickness of the building. The walls may be of brick or concrete but for flats of up to six storeys are usually of brick. When both walls and slabs are seen on the elevation, it looks like an egg box. This construction is of simple design and economical in steel, particularly with the relatively deep slabs of *hollow-tile* construction.

box girder, b. beam [stru.] A hollow, square or rectangular girder made of steel or light alloy, or reinforced or prestressed concrete, or timber. Steel box girders caused alarm in the early 1970s because two bridges which were being built then collapsed, causing 40 deaths – one bridge was over the river Cleddau at Milford Haven, Wales, the other over the river Yarra at Melbourne, Australia. The Merrison committee reported optimistically on box girders in 1973, emphasizing that *responsibility for construction* and design must

be clearly allocated. For a concrete box girder see **incremental launching**.

box heading A *heading* close-timbered both in roof and sides.

boxing [rly] A bed of *ballast* between rail sleepers.

boxing in [rly] Packing ballast under sleepers to raise sagging track.

box out See **block out**.

box pile A steel *pile* made from two *steel sheet piles* or *channels* or *angle sections* or joists, welded along their contact lines.

box sextant [sur.] A compact *sextant* suitable for rapid land surveys. Like the nautical sextant it has a small telescope but can read angles with a vernier only to the nearest minute of arc. The telescope is removable and may be not always needed. The size may be of up to 10 cm dia.

box shear test [s.m.] A simple standard method of measuring the *shear strength* of soil in a box split in two, to which pressure is applied at the same time as a shearing force.

brace, bracing (1) [stru.] A usually diagonal member which takes direct compression, occasionally tension, generally against wind force.

(2) A strut in trench timbering.

bracing [stru.] (1) A stiffening member in a structure; a brace.

(2) The act of inserting braces into a structure.

bracket See (*B*).

Braithwaite piles See **screw pile**.

branding iron An *indenting roller*.

brass [mech.] An alloy generally of copper and zinc, mainly copper. See (*B*).

braze welding *Welding* with a *filler rod* melting at a temperature below that of the metal or metals to be joined, without relying on the *capillary* action of *brazing*.

brazing High-temperature *soldering* (*B*). Hot molten metal from the *filler rod* is drawn by *capillary* action into the space between the two parts to be joined. The melting point of the filler rod is usually above 450°C but always below that of the metals to be joined. See also (*B*).

brazing spelter See **hard solder** (*B*).

breakdown, breaking of an emulsion The separation of an *emulsion* into two or more of its constituents; for *bituminous emulsions*, their separation into bitumen and water.

breaker, crusher [min.] A *gyratory breaker*, *jaw breaker* or other rock breaker. Between breakers and crushers the distinction is not sharp but crushing implies a smaller product than breaking. The jaws of some hydraulic excavator buckets are powerful enough to act as breakers. Cf. **hydraulic hammer**.

breaking ground [min.] Changing a

hard rock face into a pile of shattered stones, usually by *blasting* in drilled holes for hard rock, by *pneumatic pick* in medium ground, or hand pick in soft ground. Where hydraulic power is available, a *hydraulic hammer* can break the hardest rock. Other methods are *boomheaders*, *broaching*, *channellers*, driving in *gads*, *jet drilling*, *plug and feathers*.

breaking point, break piece Like a boiler *safety valve* or domestic electrical fuse, every rock *breaker* is fitted with an easily replaceable part that breaks whenever an unbreakable lump enters its jaws. Without this arrangement serious damage could be done to the breaker or its motor. In *jaw breakers* it is a *toggle* that breaks.

breaking strength See **ultimate strength** and below.

breaking stress [stru.] The *crushing strength* of a concrete, brick or stone, or the *ultimate strength* of a *test piece*.

break-out force In a *face shovel* or *loader*, the greatest forward force available at the teeth of the bucket. Cf. **crowd force**.

breakpoint chlorination [sewage] Addition of a small calculated amount of chlorine to *effluent* in *tertiary treatment*. It converts the ammonia to nitrogen and avoids the air pollution of *stripping* (4).

break-pressure tank [hyd.] Small open tanks placed at the level of the *hydraulic gradient* of a *gravity main* in hilly country to reduce the maximum pressure on the main. The main discharges into each tank in turn, and a new section of main takes off from each tank. Large savings in the cost of the main can thus be made, since it is under very much smaller pressure than is a continuous main.

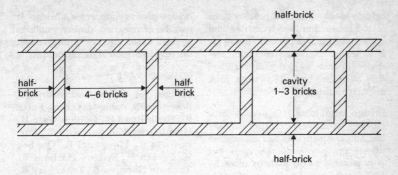

Brick or block diaphragm wall, suitable for walls without windows. (W. G. Curtin, *Structural Engineer*, February 1980.)

breakwater, mole A wall built out into the sea to protect a harbour (natural or *artificial*) from the waves. The two main types of construction are *rubble mound* and *blockwork*. **See tetrapod, floating breakwater**.

breast The *mouldboard* of a plough or *dozer*.

breathing apparatus See **respirator**.

breeze See (*B*).

brick diaphragm wall, blockwork d. w. (see diagram) A tall wall without windows for a shed, warehouse, gymnasium, swimming-pool or other single-storey structure. Its wide cavity provides thermal and sound insulation and fire resistance. It is quick to build and less expensive than the conventional steel-framed structure. The walls, outer leaf, inner leaf and crosswalls across the cavity are half a brick thick. The cavity is from one to three bricks across and the crosswalls are 1–1.5 m (3–5 ft) apart.

bridge A structure which covers a gap. Generally bridges carry a road or railway across a river, canal or another railway. For the longest spans (600 m or more) only steel *suspension bridges* are used. From 300 to 600 m the suspension bridge, the steel arch bridge and the cantilever bridge are equally suitable. For spans under 240 m the concrete arch can be used as well as the above three. See **long span**.

bridge bearings The supports on a bridge *pier*, which carry the weight of the bridge and control the movements at the bridge supports, including the temperature expansion and contraction. They may be metal rockers, rollers, knuckles or slides, or merely rubber or laminated rubber (rubber with steel plates glued into it).

bridge cap, bridge pier c.. The highest part of a bridge pier, on which the *bridge bearings* or rollers are seated. It may be of stone, brick, or plain or reinforced concrete, usually the last for heavy loads.

bridge deck The load-bearing floor of a bridge, that which carries and spreads the loads to the main beams. It is either of reinforced concrete, *prestressed concrete*, welded steel or (rarely) *light alloy*. See **bascule bridge, jack arch, troughing**.

bridge launching It is usually difficult and expensive, sometimes impossible, to place supports for a bridge in midstream, even temporary ones. But it is possible to do without them by building the complete bridge on one or both banks, provided it is a beam bridge and not an arch. The structure is pushed out when it is complete, after making sure there is enough counterweight on the landward end. Often a lightweight launching nose is connected to the front end of the bridge and discarded on completion. In civil engineering a launching may take place from both banks but in military bridging from only one bank. See **incremental launching**.

bridge loading [stru.] Loadings on bridges for the purpose of the design of the structure are laid down by authorities such as AFNOR, ASTM, BSI, DIN, often by agreement, Railway bridge loadings (*RU loading, RL loading*) include allowances for *centrifugal force, lurching, nosing*, etc.

bridge overlays A layer of silica sand mixed with synthetic resin can repair a road or bridge more quickly than any other method, and some can be laid only 6 mm ($\frac{1}{4}$ in.) thick. Three resins are in common use: epoxies (since 1976), polyesters and polymethyl methacrylate (PMMA) more recently, PMMA has an unpleasant smell but costs less than epoxy though more than polyesters. PMMA cures in two hours even at 10°C of frost. Epoxies have the advantage that they are unaffected by spilt oil.

bridge pier A support for a bridge. It may be of masonry, timber, concrete or steel, but in any case is founded on firm ground below the river mud.

bridge pier cap A *bridge cap*.

bridges See **Akashi-Kaikyo Bridge, Bosporus B., cable-stayed b., Forth B., Forth Road B., Golden Gate B., Humber B., Kill Van Kull B., Parramatta B., Plougastel B., Quebec B., Severn B., Sydney Harbour B., Tacoma Narrows B., Träneberg B., Verrazano Narrows B.**

bridge thrust A horizontal force in an arch bridge caused by the arch shape. It is resisted at the abutment by a horizontal reaction from the ground, or, as in the *bowstring* girder, by a pull in a tie beam (the bowstring).

bridge truss A *truss* suitable for carrying bridge loads, such as a *Warren* or *Vierendeel girder* or *Pratt truss*.

bridging piece [rly] An inverted U-section placed over a broken rail to allow trains to travel over it at walking pace.

bright steel Steel without *mill scale* and usually not rusty.

Brinell hardness test [mech.] A hard steel ball of 1, 2, 5 or 10 mm dia. is loaded with a weight for 15 s, and the diameter of the indentation is then measured. The Brinell hardness number (BHN) is worked out from

$$\frac{\text{load in kilograms}}{\text{spherical area of impression in mm}^2}$$

From this figure can be obtained corresponding numbers for the Vickers test and the *Rockwell hardness test*, which are more suitable for case-hardened material. The Brinell method is suitable for soft metal and mild steel, and even for tough steel which is not case-hardened. For a steel that has not been hardened by cold work, the ultimate tensile strength in tons/cm^2 is roughly 0.034 times the (BHN). Thus the BHN for mild steel is between 120 and 140; both hardened tool-steel and white cast-iron have a BHN between 400 and 600. See BS 240.

briquette (1) [min.] A lump of fuel or ore (green pellet) made from dust by compression or sintering at high temperature, enabling dust to be used that would otherwise pollute the air.

(2) A specimen of cement cast in the shape of a thick-waisted hour glass, used in the tensile testing of cement.

Britannia Bridge (1849) A railway bridge built of rectangular wrought-iron tube over the Menai Strait by Robert Stephenson with a 31 m (105 ft) clearance above high water for sailing ships to pass. Until this bridge was built the longest wrought iron girder had been only 9.6 m (31.5 ft) long. Its two equal main spans were about 200 m (600 ft) long. The ironmaster, William *Fairbairn*, and the materials expert, *Hodgkinson*, in 1845 tested 33 model tubes of different cross-sections and proportions including elliptical, and eventually built a one-sixth scale model of the proposed bridge. Showing that it could carry 86 tons they eventually persuaded Stephenson that suspension chains were not needed. The bridge was destroyed by fire in 1970 and replaced by a concrete one. It had been a model for the steel plate girders and box girders that came later.

British Standard, British Standard Specification, BS A numbered publication of the *British Standards Institution* describing the quality of a material or the dimensions of a manufacture such as pipes or bricks. Frequently the dimensions and the quality are described in two different standards. The use by architects or engineers of British Standards in their *specifications* can reduce the bulk of the description to a reference to the number of the standard.

British Standards Institution (BSI) The British organization for standardizing, by agreement between maker and user, the methods of test and dimensions of materials as well as *codes of practice* and nomenclature. Corresponding abbreviations in other countries are: France, AFNOR (Association Française de Normalisation); Germany, DIN (Deutsches Institut für Normung e.V.) USA, ASTM (American Society for Testing and Materials) and ASA (American Standards Association); Soviet Union, GOST.

British thermal unit See p. xvi.

brittle fracture [stru.] *Cleavage fracture*, usually at a low stress and low temperature. Such failures can be reduced by avoiding *notch effects*, *residual stresses* and high *strains*. Thick plates are more likely to suffer brittle fracture than thin plates. The *Charpy test* is relied on to find suitable steels or safe stresses for the temperature of use. Brittle fracture is usually not considered except for members under direct tension. See **impact test**.

broaching, broach channelling, line drilling A rock excavation method used where the rock left in place must not be shattered by explosive. A line of holes is drilled close together along the break line. The rock between them is knocked out with a chisel called a broach and the block is finally moved with wedges. If the holes are as far as 10 cm apart, light charges may be fired or the *plug and feathers* used. See **channeller**.

broad gauge [rly] Any width of track wider than the 1.435 m of *standard gauge*. Brunel originally used 2.13 m ft) for the Great Western Railway. Finland, Ireland, the USSR, Spain, Portugal, and parts of Australia and India have broad gauge track.

broad irrigation An old name for *land treatment* of sewage *effluent* or occasionally of *sludge*, using *contour ploughing*.

bronze welding A form of *braze welding* using a *filler rod* of copper-rich alloy containing zinc and sometimes manganese, nickel or other elements. It can be used to join many metals.

brooming (1) The crushing and spreading of a wooden pile head with no *driving band* when driven into hard ground:

(2) **broomed finish, brushed surface** A surface texture obtained by brushing fresh concrete with a stiff brush, either for architectural effect or to improve the skid resistance of a road.

brothers A sling of chain or rope, the term may mean either a *two-leg* or a *four-leg* sling.

Brown and Sharp wire gauge [mech.] Another name for *American wire gauge*.

BS *British Standard*.

BSCP British Standard *code of practice*.

BSI *British Standards Institution*.

BThU (British thermal unit) See **Conversion factors**, p. xvi.

bubble [sur.] The air bubble within a *level tube*, or the level tube itself.

bubble barrier, air bubble curtain (see diagram) Releasing air bubbles from a perforated pipe laid on the bed of a sea or lake. Such a 'curtain' can hold back an oil slick, reduce the power of incoming waves, block an intrusion of salt water, or melt surface ice. Ice is melted because the densest water is on the bed at 4°C and this water is brought up by the air bubbles. Finally, air bubbles reduce any pollution by domestic sewage in a body of water.

The compressed air used may be costly but for most of these applications, except the last, it is not a year-round expense. Air bubble barriers have been tried in Holland and Japan, and are reported by Atlas-Copco, the Swedish compressed-air equipment builders. The number of atmospheres of air pressure required from the air compressors used is equal at least to the water depth in metres divided by 10.

bubble trier [sur.] A *level trier*.

bubble tube [sur.] A *level tube*.

buck, door b. A *door frame* (*B*).

bucket [hyd.] (1) A cup on the perimeter of a *Pelton wheel*.

(2) A reversed curve in the profile of a *spillway*, designed to deflect the water horizontally at its foot from the steep overflow face on to the *apron* below the dam (USA).

(3) The digging cup on an *excavator*, *loader*, *bucket elevator* or *bucket-ladder dredger*. Many types of bucket exist; see **four-in-one bucket, side-dump bucket**.

(4) A *kibble*.

bucket elevator An endless chain (or two linked) with buckets attached for raising loose material such as *slurry*, coal or stone at slopes varying from 45° to vertical.

bucket-ladder dredger A *dredger* whose main equipment is a *bucket elevator* reaching below its keel into the mud to be dredged. The buckets dig mud as well as lift it. They discharge at the top into a chute which leads to a *dumb* barge or to a hold for mud in the dredger.

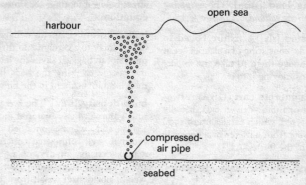

open sea

harbour

compressed-air pipe

seabed

PNEUMATIC BREAKWATER AT HARBOUR ENTRANCE

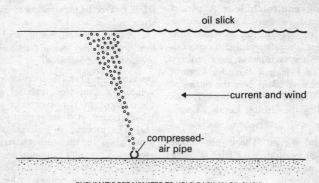

oil slick

current and wind

compressed-air pipe

PNEUMATIC BREAKWATER TO HOLD BACK AN OIL SLICK

Bubble barriers. (R. H. Waring, *Pneumatic Handbook*, Trade and Technical Press, 1982.)

bucket-ladder excavator An endless chain of digging buckets resembling a *bucket elevator*, often used in a *trench excavator*.

bucket-wheel excavator [min.] See **rotary excavator**.

buckle [stru.] To load a *column* too heavily or with *eccentric loading* until it bends sharply and crumples or wrinkles at its *crippling load*. This happens on the inside of the curve of a pipe bent too sharply, if it does not become oval. The enormously expensive *offshore platforms* have aroused interest in pipe buckling because many of them are carried on steel tubes 100 m or more long and of up to 4 m dia., with very thin walls a few centimetres thick that crumple easily. Rail track also can buckle in hot weather. See **de-stressing**.

buckling load [stru.] The *crippling load* of a *long column*.

buffer stop [rly] A fixture bolted to track-rails, consisting of a sleeper set at 1.05 m above them to take the *impact* of wagons moving into it.

buggy, concrete cart (USA) A two-wheeled or motor-driven cart, usually rubber-tyred, which carries up to 170 l of concrete from a mixer or a concrete hopper to the forms (ACI).

builders' plant See **contractors' plant**.

building code In the USA, local building laws, which correspond to by-laws in Britain.

building control officer A person authorized to inspect work during design or construction to make sure both that it is safe and that it complies with the law. In London, since January 1986, 14 Principal Building Control Officers have taken over the duties of the 27 district surveyors formerly employed by the Greater London Council.

building owner The owner of the works to be built on a site: an individual, an organization, a government, etc. See **client**.

building paper See (*B*).

built-up [stru.] Description of a steel, light alloy, or wooden beam, stanchion, etc. built of different sections firmly joined by bolting, gluing, riveting, welding, etc.

bulb angle A steel *angle section* enlarged to a bulb at one end. See **beaded section**.

bulb of pressure [s.m.] The mass of compressed soil below a loaded foundation. The term is also used to describe the bulb-shaped lines of equal vertical stress below a footing; see **Boussinesq equation**.

bulk cement Cement delivered by tanker lorry to a cement silo through a pipe, not in bags. It saves the labour of the people carrying the bags but demands an investment in the bunker.

bulk density [s.m.] The weight per unit volume of any material including voids and water contained in it. Normal, well-compacted concrete has a bulk density, air-dry, of 2.3 tonne/m^3. Natural aggregates (not lightweight nor dense) weigh from 1450 to 1750 kg/m^3. See **dry density**, **relative density**.

bulkhead (1) A partition, particularly one in a storage bunker or tunnelling *shield*, or erected in *formwork* to make a stunt end.
(2) A *sheet-pile* wall, usually anchored but occasionally free. It may be a *dredged* or a *fill bulkhead*.

bulking [s.m.] (1) The increase in the volume of excavated material above the volume of the excavation from which it came, often more than 100%.
(2) The increase in volume of dry sand when its moisture content increases. This may amount to 40% when 5% of water is added. This increase disappears entirely when the water content is raised to 20%.

bulk modulus See **elastic constants**.

bulk spreader, powder s. [s.m.] A machine for carrying cement or other material in *soil stabilization*. It also spreads it on the prepared soil.

bulldog grip [mech.] A U-bolt threaded at both ends. The ends pass through a specially shaped washer. The grip is used as a rope clamp since it grips a length of steel rope doubled back on itself. The hold is effective provided that at least three grips are

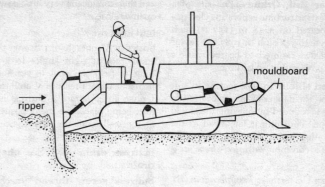

Bulldozer pulling ripper.

used and that the units all face in the same direction. See **rope fastenings**.

bulldozer (see diagram) A *tractor* usually on *crawler tracks*, with a wide blade, the *mouldboard*, mounted in front of it, at right angles to the tracks. The mouldboard is adjustable in height, angle and tilt. It normally moves material by pushing. See **dozer**.

bullhead rail [rly] An old type of British rail rounded at top and bottom, carried in a *chair*.

bull wheel A large wheel at the base of a *derrick* mast, or other heavy plant, which rotates the mast during *slewing*.

bump [r.m.] A noise caused by breakage within the mass of rock, not at its surface as in a *rock burst*. Bumps occur in coal mines or other mines in soft rock and are rarely dangerous.

bump cutter A set of diamond saws fitted on to a machine that is sensitive to upward irregularities in the surface of a concrete or asphalt *wearing course* and cuts them off.

bunker A storage container for coal, ore or stone, etc.

buoyancy, flotation [hyd.] The reduction in weight of a body caused by its immersion in a fluid, equal to the weight of the fluid it displaces. If the body floats, its weight is still equal to the weight of fluid displaced. This is the *principle of Archimedes*. See **flotation, uplift**.

buoyant foundation, b. raft A reinforced-concrete *raft foundation*, usually with walls round the edges, so designed that the total of its own weight and all loads which it carries is approximately equal to the weight of the soil or water displaced. It is used in river estuaries where no foundation is available except nearly fluid silt or mud. See **Vierendeel girder, flotation**.

burden [min.] The burden of the *toe* of a blasting hole is its distance from the nearest *free face* measured at right angles to the hole.

Bureau Veritas A French *classification society*.

Burger Hill, Orkney The site of a 3 MW wind turbine generating electricity, located on a 48 m (157 ft) high concrete tower built in 1987 by Taylor Woodrow, claimed to be the world's most powerful wind turbine.

buried shield In *pipe jacking* in soft ground the danger of inrush of mud or water is reduced by burying the shield or pipe – pushing it far ahead of the excavation inside the pipe. The technique is common with *steel-sleeve jacking*.

burlap US term for *hessian canvas* (*B*).

burn To cut metal with a gas flame.

burnt shale Carbonaceous shale (occasionally used as road-making material) which has been heated either by spontaneous combustion in the tip or by destructive distillation as oil shale.

bury barge After a submerged pipeline has been laid by a *lay barge*, the bury barge, following it, digs a trench and inserts the pipe into it, controlling the plough or other equipment on the sea bed.

bush hammer A compressed-air or electrically driven hand-held hammer weighing about 3 kg. It has a serrated (pyramidal) face which removes the skin from a concrete surface, pleasantly exposing the aggregate.

bush hammering Dressing concrete with a *bush hammer* to remove the outer 1–6 mm skin.

butane A paraffin hydrocarbon gas, C_4H_{10}, usually obtained by refining petroleum, and used in *bottled gas* (*B*).

butterfly valve [hyd.] A circular disc inside a pipe, hinged at two pivots on a diameter. It is often used for controlling the flow in large pipes in hydroelectric schemes between *forebay* and power station. It is perfectly balanced

and therefore needs very little power to open or close it.

butt joint See (*B*).

buttress A concrete or masonry thickening pier at right angles to a wall, built to help the wall to resist earth thrust or water pressure or arch thrust. Unlike a *counterfort*, a buttress is visible, being placed on the opposite side of the wall from the thrust. See also **flying buttress**.

buttress drain [rly] See **chevron drain**.

buttress screw thread [mech.] A screw thread designed to carry a heavy axial load in one direction only. The front face of the thread, carrying the thrust, is perpendicular to the axis of the screw; the back face is at 45° to it. See **square thread**.

butt strap A plate which covers a *butt joint* (*B*) and is welded, riveted or glued to the two butted members.

butt weld A *weld* between two pieces without overlap, usually a *fusion weld* made by *resistance welding* or *flash welding*.

butt-welded tube [mech.] Steel tube made by bending mild steel plate into a cylindrical shape and welding the meeting line.

butyl stearate $C_{17}H_{35}COOC_4H_9$. A colourless, oily damp-proofer for concrete which is practically without smell (*ACI*).

byatt A horizontal timber which supports decking, walkways, etc., in trench excavations.

bye channel, b. wash, diversion, d. cut A ditch along a contour, dug to lead dirty water around, and not into, a reservoir.

by-pass A pipe, conduit or road for directing flow of traffic around, instead of through, another pipe, conduit or road.

C

cabinet projection [d.o.] A way of showing solid objects on a drawing. The object is drawn in plan or elevation. Faces perpendicular to the plan or to the wall elevated are drawn at an angle of 45°, and to half the proportional length of those in the plan. See **axonometric projection, planometric projection**.

cable (1) [elec.] A group of insulated conductors, sometimes *armoured cable*.
(2) Any steel wire *rope*, in particular the *track cable* in a *cableway*.
(3) In *prestressed concrete*, a *tendon* consisting of a number of wires or strands.
(4) One tenth of a nautical mile, exactly 608 ft, or about 185 m.

cable belt A *belt conveyor* in which the belt tension is carried by two steel wire *ropes*, one each side, with the belt merely resting on them. For this reason it can be long and curved. Friction is claimed to be less than with a conventional belt. Cf. **steel cored conveyor**.

cable drill, churn d., percussion d. [min.] A heavy drilling rig used in drilling 7 to 25 cm dia. vertical holes in prospecting, quarrying (and in oil-well drilling down to 1500 m in the USA). The rig consists of a tower known as a *derrick* for handling the tools, a steel wire rope hung from the top of the derrick which raises and lowers the tools into the hole, and the tools themselves which are moved up and down at the bottom of the hole (during drilling) by the walking beam. Drilling speeds vary from 3 m/h in clay or soapstone to 0.3 m/h in hard limestone. Small units are used in *shell-and-auger boring*, using a three-pole *derrick*.

cable duct A protective earthenware, plastic or sheet metal pipe, or a mere hole cast in concrete, through which electric or *prestressing* cables are pulled. After stressing up, prestressing cables may be grouted into their ducts. Cable ducts should be provided with drain holes that remove water, and if grouted also with upward holes to release air.

cable entry (see diagram) Part of the frame of a *flameproof enclosure* and of some other electrical housings, forming a gas-tight hole for the cable.

cable-laid rope Rope twisted with an *ordinary lay*, not *Lang lay*.

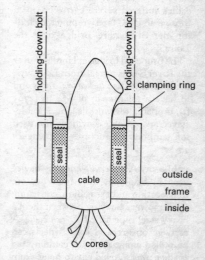

Cable entry. A stuffing box for a pump is very similar except that the clamping ring is not radiused as it must be for a cable, and the seal or stuffing is different.

cable locator A hand-held, battery-operated device which detects buried electric cables.

cable railway An incline up which wagons are pulled by an endless steel wire rope, overhead or beneath the wagons.

cable saddle A metal block carrying a *suspension cable* over the tower of a *suspension bridge* or *aerial ropeway*.

cable-stayed bridge [stru.] Any bridge with straight cables from masts connected directly to the deck girders without *suspenders*. The longest span (457 m (1500 ft)) is over the Hooghly near Calcutta. The *Dartford Bridge* is nearly as long. All these bridges are of striking appearance, especially those with the cables in a single, central plane, such as the Sunshine Skyway Bridge in Florida (1988). But a survey published in early 1988 of 200 built since 1978 showed that most of the cables suffered severe corrosion, even those with PVC tape wrapping, including the St-Nazaire Bridge over the Loire.

The proposed bridge at Honfleur over the River Seine, not far from Le Havre, is expected to use *locked-coil ropes* and to have a main span of 834 m (2765 ft), probably the world's longest cable-stayed span. Cable-staying is also used for temporary structures, e.g. *launch girders*.

cableway A *materials-handling* device used in bridging, dam building, ore transporting and excavation, consisting of two towers carrying between them a heavy steel rope called a *track cable*. A carriage on grooved track wheels can be pulled along the track cable by the *traction rope*. A load may hang from the carriage by the hoisting or fall rope, which lowers or raises the load. Average travelling speeds are about 360 m/min, hoisting speeds up to 120 m/min, and loads 5–10 tons for spans of 100–1000 m. Cableways may have both towers stationary (fixed cableway), one tower stationary (radial cableway), or both towers movable on rails (full-travelling cableway). Cf. **aerial ropeway** and see **Blondin, excavating cableway, slack-line cableway, luffing cableway mast**.

cableway excavator A *slack-line cableway*.

cableway transporter A crane like a *transporter crane* but much more lightly built and having as track for the carrier a steel rope hung between the ends of the girder. The girder is thus not loaded in bending, or only slightly, but mainly in compression, and can be much lighter than a transporter crane girder, since it acts more as a strut than as a girder.

CAD (computer-aided design, computer-aided draughting) A CAD unit can help engineers, architects or other designers to work quickly and efficiently. It usually includes two *VDU*s, a small and a large (the graphics display terminal), as well as the microcomputer generally hidden in a 'black box'. CAD includes CAM (computer-aided manufacturing), CAE (computer-aided engineering), etc.

CADD (computer-aided design and draughting) *Computer graphics* equipment, costing less than a civil engineer's pay for six months' work, can produce a detail drawing in an hour that the engineer could not draw in a week. CADD consequently is profitable for repetitive drawings.

cadastral mapping [sur.] Mapping land for the purpose of recording its ownership.

cage of reinforcement Steel *re-*

inforcement tightly wired or welded to its links on the ground, then lifted by crane to be placed in *formwork*, or for a *bored pile*, in a hole in the ground. Prefabricated cages are available from specialist suppliers.

caisson (1) A *foundation* built partly or wholly above ground and sunk below ground, usually by digging out the soil inside it. See **box c., Chicago c., Gow c., open c., pneumatic c., compressed air, drop shaft, monolith**.
(2) A *ship caisson*.

caisson disease A disease which affects workers in *compressed air* who come too quickly out of the *air lock*. It is caused by bubbles of nitrogen coming out of the blood. The only treatment is to take the sufferer to a *medical lock* or the nearest air lock immediately, for recompression and slow decompression. It is also called the bends, diver's palsy, diver's paralysis, air embolism, compressed-air disease or screws. See **decanting, helium**.

caisson pile A *Gow caisson*.

calcareous Containing calcium carbonate ($CaCO_3$), e.g. chalk, limestone, marl.

calcine [min.] To heat ore or mineral for some time at a high temperature to drive off carbon dioxide and water.

calcite $CaCO_3$. Crystalline calcium carbonate found in marble and other limestones.

calcium aluminate The refractory part of *high-alumina cement* consists of various calcium aluminates, some of them being even more refractory than monocalcium aluminate, which is white and melts at $1608°C$.

calcium chloride $CaCl_2$. An *admixture* used sometimes in proportions up to 0.03% by weight of the cement to accelerate its hardening rate, and therefore added to concrete during frost to accelerate its heat release. Since 1977 in the UK its use in contact with steel has been forbidden. But it is used where there is no steel as in *glassfibre-reinforced concrete*. If the mixing water or aggregates contain salt, or other chloride, the amount of allowable $CaCl_2$ is correspondingly reduced. Because French cements are mostly *blended* and harden slowly, French builders have been reluctant to ban it.

calcium magnesium acetate See **antifreeze**.

calcium silicate hydrate Dicalcium silicate (C_2S) and tricalcium silicate (C_3S) are the strong part of *aerated concrete*, ordinary Portland cement concretes and *sand-lime bricks (B)*.

calfdozer A small *bulldozer*.

calibrate To check the graduations of an *instrument* or machine and if necessary to graduate it correctly.

calibre (**caliber** USA) The bore (internal diameter) of a pipe, or the capacity of other plant.

California bearing ratio (CBR) method A method of designing *flexible pavements* on the basis of the CBR test (below).

California bearing ratio (CBR) test [s.m.] A standardized testing procedure begun by California State Highways Department in 1929 for comparing the strengths of *base courses* of roads or airstrips. The soil is first compacted in a mould and then soaked for four days with a load on its surface. The expansion due to moistening is then measured, a good figure being less than 3%, a bad figure $7–20\%$. The soil resistance to a standard plunger of area 19.35 cm^2 which has penetrated 2.5 mm is measured and the ratio of this resistance to the corresponding resistance in crushed rock is then calculated. This ratio is the CBR. Cf. **Proctor plasticity needle**.

calking See **caulking**.

calliper log A continuous record of the uncased diameter of a borehole, useful in *well logging* for many purposes. Loose or swelling soils can cave, creating a hole of twice the bit diameter.

calliper pig See **pig**.

callipers [mech.] A pair of steel legs pivoted together like a draughtsman's dividers, used for measuring the bore of pipes or their outside diameter.

calorific value, heating v. The amount of heat liberated by the complete burning of unit weight of a fuel, expressed in heat units per unit weight. Thus for solid or liquid fuels it is expressed in kJ/kg. In some countries the kg-cal/kg is used. 1 kJ/kg $= 0.43$ Btu/lb; 1 kcal $= 4.187$ kJ.

camber [hyd.] A *gate chamber*. See also (*B*) and **hog**.

camber rod, belly r. [stru.] The tensioning rod below a *trussed beam*.

camel [hyd.] A large hollow steel float tied to a ship to raise it in the water and float it past a shallow place. See **saucer**.

Camelford, Cornwall A district of 20,000 people whose drinking water was seriously polluted by the supplier, the Southwest Water Authority, in July 1988. At that time the only prosecution possible apart from one by individual sufferers, was for the crime of killing the fish, even though six weeks later people were still suffering

from nausea, vomiting, diarrhoea and skin blisters.

The pollution was caused by a lorryload of aluminium sulphate (*alum*) tipped mistakenly into the water supply instead of into a treatment tank. It acidified the water, dissolving pipe metals and poisoning people with copper and lead as well as aluminium. By November there had been four burst water mains near by. The authority denied the existence of the pollution for a month after it began.

Alum is commonly added to *raw water* in small concentrations but is normally not allowed to acidify the water. To precipitate the hydroxide, $(Al(OH)_3)$ as required for *coagulation*, the water must be alkaline. See also **prosecution of a water authority**.

camouflet [min.] A cavity underground, formed by an explosion that makes no crater. It is usually achieved by *chambering*.

camp sheathing, c. shedding, c. sheeting (1) A *retaining wall* holding back the bank of a river or a *barge bed*. It consists of two connected rows of timber piles 1.5–3 m apart, the space between being filled with earth.

(2) A light *sheet pile* wall.

canal [hyd.] A channel dug or built up to carry water for navigation, water power, irrigation or other purposes.

canalization [hyd.] Dividing a river into *reaches* separated by *locks* and *weirs* to help ships, barges or irrigation, or control flooding, as in the Tennessee Valley, USA. The Rhône, Marne, Seine, Garonne and Oise in France are all canalized. The Loire is not, possibly because it is too steep. In England, the Thames, Trent and other rivers are canalized.

One of the earliest canals, 80 km (50 miles) long and 20 m (66 ft) wide, provided fresh water to Nineveh under

Sennacherib, king of Assyria in the seventh century BC. This stone-lined canal with a bridge 300 m long was completed in 15 months. Two centuries later the *Grand Canal* was begun in China, which with more than 100,000 km of canalizations and canals may have the world's largest network, and almost certainly its first *summit canal*.

canal lift [hyd.] A tank drawn on wheels up or down an incline (or vertically) for passing barges through a *lock* with a lift larger than about 15 m.

cancer cluster In several villages west of Hull before 1988, 13 children became ill with cancer and eight of them died. Radioactivity is one of the known causes of cancer. Their parents suspected the chimney, 180 m (600 ft) high, of Capper Pass Ltd, one of the largest smelters in Europe, which is upwind of them. It received a Government licence to emit radioactive polonium in 1984 though this fact did not become public knowledge for 2½ years. It was generally known that the chimney emitted heavy metal dusts. Civil engineers may unwittingly become responsible for environmental disasters of this sort because HM *Pollution Inspectorate* is not allowed to release facts about air pollution. Clusters of other diseases exist.

canopy A subsidiary roof over an entrance or other feature.

cant [rly] or **banking** *Superelevation*.

cantilever [stru.] An overhanging beam fixed at one end and free at the overhanging (cantilevered) end.

cantilever arm [stru.] In a *cantilever bridge*, the part overhanging from the support into the central span, and carrying at its end one end of the *suspended span*.

cantilever bridge [stru.] (see diagram, p. 60) Generally a symmetrical three-span bridge of which each of the outer spans is anchored down at the shore and overhangs into the central span about one third of the span. The suspended span, resting on the *cantilever arms*, occupies the remaining one third of the central span. The *Forth Bridge* (1890) has the unusual number of two main spans of 520 m clear, flanked by two side-spans of 210 m each. The centre pier therefore carries a bridge element which overhangs on both sides. It has, however, a width at the base of 76 m. The *Quebec Bridge* (1917) is a normal cantilever bridge with central span of 550 m, with 157 m side-spans.

cantilever crane A crane in which the *jib* has no strut for *derricking*. The most familiar is a mobile crane with telescopic jib. It has a lower overall height than a *strut-jib crane*.

cantilever formwork *Climbing formwork*.

cantilever foundation [stru.] A foundation for a column or stanchion which for some reason does not have enough space for a truly central base. This frequently occurs on city sites where a large concentrated load from a stanchion comes down at the edge of the site. The base is therefore built well within the site and a concrete or steel beam is built upon the base to carry at its outer end the overhanging stanchion. The inner end of the beam must be counterweighted by a sufficient proportion of the weight of the rest of the building to prevent it lifting. The whole arrangement of base and cantilever beam is called a cantilever foundation.

cantilever wall (1) A reinforced-concrete *retaining wall* stabilized usually by the weight of the retained material on its heel.

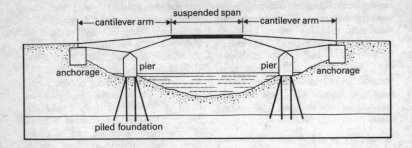

Cantilever bridge.

(2) A *sheet pile* wall stabilized by its length of penetration below ground level on the free side. Cf. **tied retaining wall.**

cantledge See **kentledge.**

cant strip, chamfer s. A triangular or curved metal *insert* fixed inside a corner of *formwork*, which becomes a strong metal corner sticking to the hard concrete and protecting it.

cap (1) A *pilecap.*
(2) A *capping piece.*
(3) A *detonator.*
(4) A mortar made with sulphur or *high-alumina cement* and sand applied to the rough end of a concrete *test core* to smooth it before the core is tested in the compression testing machine.

capacity curve [hyd.] A graph which shows the volume of a reservoir or tank at any given water level.

cape chisel See **crosscut chisel**.

capel [mech.] An eyed steel socket on the end of a steel wire rope, used for *capping* it.

capillarity [s.m.] The rising of fluid in tiny hair-like spaces (capillaries) above the level of the fluid in an open vessel.

In soils, water rises between $\dfrac{1}{eD}$ and $\dfrac{5}{eD}$ cm, where e = voids ratio, D = *effective size* of the soil. The maximum rise (within 24 hours) occurs with soils of D about 0.02 mm but much larger rises may occur with finer soils in several years.

capillary fringe [s.m.] Ground above the *water table* that is continuously wetted with capillary water. Above it is more capillary water, but not continuously. The height of the capillary fringe increases with increasing fineness of the soil and the fringe rises or falls with the water table.

capillary pressure, seepage force [s.m.] In ground which is being drained from outside an excavation (see **wellpoint**), capillary pressures help the excavated earth to stand steeply. However, if the ground is being drained from inside and not from outside the excavation, the capillary pressures will help the earth face to collapse. In *silt* with pores from 0.05 to 0.005 mm in size, the capillary pressure varies from 6 to 60 kPa (0.9–9 psi). In *clays* the pressure is theoretically more (but clay is very much harder to drain) and in sands the pressure is much less than in silts. See **electroosmosis**.

capillary rise [s.m.] In a glass tube of 0.02 mm bore, water will rise about 1.5 m. See **capillarity**.

capillary water [s.m.] Water maintained above the *water table* by capillarity.

capping (1) [min.] US term for *overburden*.

(2) See **cap** (4).

(3) White metal cast around the exposed wires at the end of a steel rope to fix the *capel* to it.

capping piece A horizontal timber placed over the ends of two *walings* butted together. It takes the thrust of a *strut* and transfers it to the walings.

capstan [mech.] A *winch* used in railway sidings for moving wagons or on quaysides for moving ships. It differs from hoists or other haulage engines in having a drum with a vertical shaft, the engine being usually hidden below ground.

capsule anchor A *resin anchor*.

carbon, carbonado See **black diamond**.

carbon-arc welding *Welding* in an electric arc of which one electrode is a carbon rod, the other the piece being welded. Another method is to use two carbon electrodes and to exclude the workpiece from the electrical circuit.

carbonation Chemical combination between the carbon dioxide of the air and the lime in a building material. Cement has much free lime but carbonation reduces it so much that the *pH value* can fall from the normal, desirable value of 13.5 (highly alkaline) to about 8.5 (nearly neutral), making steel in it likely to rust. Carbonation

also causes more than half the drying *shrinkage*. See also (*B*).

carbonation-retardant concrete coating Any protective coating or penetrant applied to concrete to protect it from *carbonation* and the entry of chlorides and water. Penetrant liquids, as would be expected, are clear, very fluid, and may be pore liners or pore blockers. They enter 2–3 mm into the concrete. Sealing coats are much more viscous and may be coloured. Renderings (mortars) may be applied by mortar or machine (CIRIA report).

carbon dioxide (CO₂) flux welding *Metal-arc welding* in which flux from a *covered electrode* is deposited under a shield of carbon dioxide gas.

carbon dioxide (CO₂) welding *Metal-arc welding* using a bare wire electrode. The arc and molten metal are shielded with carbon dioxide gas. See **MIG welding**.

carbon dioxide recorder An instrument which records on a chart the CO₂ content of a *flue gas* and therefore keeps a record of the efficiency of a *boiler* at any instant.

carbon dioxide welding One form of *MIG welding*.

carbon monoxide CO. A highly poisonous gas which is particularly dangerous because it has neither taste nor smell, formed by the burning of fuels in insufficient air, for example, when car engines are idling. There is little agreement about acceptable concentrations. For the second Mersey Tunnel the cost of ventilation to achieve 130 ppm of CO was estimated at £4.20/h, and to achieve 83 ppm of CO, £14.50/h.

carbon monoxide sensor, CO transducer Road tunnels longer than about 400 m need to have forced ventilation unless they have two separate one-way tunnels (to some extent these ventilate themselves). Sensors are needed for this, the worst air pollutant, so that the fans can increase the air flow when the CO level is too high. Typically a *transducer* detecting 15 mg/m³ of CO will switch on another fan or increase the speed of the working fan. At 30 mg/m³ another increase of air flow takes place. At 60 mg/m³ the transducer switches on the alarm hooter and the tunnel is closed to traffic. The World Health Organization guideline is 10 mg/m³ but this is exceeded on calm days in London streets.

No ventilation system can respond quickly. At least two minutes are needed for the effect of any action (switching on a fan, etc.) to become noticeable.

carbon silicide See **silicon carbide**.

carbon steel [mech.] A *steel* whose properties are determined by the amount of carbon present. It contains no chromium, nickel or molybdenum, which are typical elements in *alloy steels*. Its maximum manganese content is about 1½%, that of silicon and copper about ½% each. It generally implies a *high-carbon steel*, *mild steels* being *low-carbon*.

Carborundum A trade name for silicon carbide, a *refractory* and abrasive which is harder than *quartz*. It can be used at temperatures up to 2500°C.

carburizing [mech.] The introduction of carbon into the surface of steel by holding it at a suitable temperature, above the *critical point*, in contact with a source of carbon and nitrogen (often hoof and horn). Carburizing is usually followed either by direct quenching from the carburizing operation or by other suitable heat treatment to produce a hard case of *cementite* and a ductile tough *core*. See **case-harden-**

ing, **cementation**, **gas carburizing**, **nitriding**.

Caricrete Concrete containing a small amount of *polypropylene* in *fibrillated* form. Shell Chemicals patented it in 1966. The maximum that could be put in then was only 1% but even this noticeably improves the impact strength. Caricrete precast concrete 2-ton blocks, nearly cubical outwardly but hollow with holes on all six faces right through, have replaced rock *armour* in the harbour wall at St Helier, Jersey. As energy dissipators they withstand waves 4 m (13 ft) high.

carpet See **bituminous carpet**, **wearing course**.

carriageway (pavement USA) The part of a road which carries vehicles.

carriers Containers or buckets which travel on the *track rope* or a *cableway* or *aerial ropeway*, hung from grooved wheels.

carrier wave In *EDM* the frequency of the radio wave used for measuring distances. For short ranges it is usually visible light or infrared, but for long distances VHF (very high frequency) radio is usual.

Cartesian coordinates, rectangular coordinates *Coordinates* measured perpendicularly from fixed axes of reference which are at right angles to each other. The distances east or west are also called eastings, westings, abscissae or departures. The distances north or south are called northings, southings, ordinates or latitudes.

cartographer [d.o.] One who prepares charts or maps from data supplied by a hydrographical or *land surveyor*. He or she may be a draughtsman who has never worked outside a drawing office, or a qualified surveyor.

cartridge paper [d.o.] A hard opaque

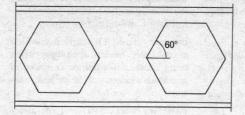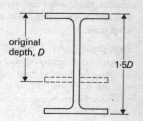

Castellated beam.

white *drawing paper* used for pencil or ink drawing.

cascade A series of steps built into a channel or sewer, separating lengths in which the flow is normal. A cascade prevents *scour* by reducing the energy of the flow. Concreted in below the steps is a pipe which takes the smallest flows at a speed high enough to leave no solids behind.

cascading failure Failure of one supply or service may result in overloading of similar supplies or services and their failure also. This has happened to electrical power supplies in large cities, also to communications networks, and regularly happens to roads in the form of traffic blocks. See **infrastructure**.

case [mech.] The surface of steel which has been hardened by *case-hardening*, leaving a relatively ductile *core* within the case.

cased composite beam A steel beam encased in concrete and acting in conjunction with a concrete slab.

cased pile A concrete pile cast into a steel casing in the ground.

case-hardening [mech.] Surface hardening of steel by *carburizing*, *cyaniding* or *nitriding* followed sometimes by further heatings and *quenching*.

casing (1) *Formwork* for concrete.

(2) [min.] Steel pipe lining to oil or water wells or other *boreholes*.

(3) [mech.] A steel plate enclosure to a *fan* which widens out to a final *volute*.

castellated beam [stru.] (see diagram) Trade name for a steel beam formed by cutting a *rolled-steel joist* along the web in a zig-zag shape. After cutting in two, the two halves are rearranged with the crests of the cuts meeting, and are then welded together at these crests. The resulting beam is 50% deeper and has a moment of resistance about 100% larger

casting yard A *blockyard*.

cast-in-place A more easily understood but less usual term than *cast-in-situ*.

cast-in-situ Concrete or plaster poured in place. The term is particularly applied to *piles* and lintels since these are often *precast*. See **bored pile**.

cast iron [mech.] Alloys of iron and carbon containing more than 1.7% and usually 2.4–4% carbon. Cast-iron articles are made by casting from remelted *pig* iron with cast iron and steel scrap. It has a low melting-point, flows well, and is easier to pour into intricate shapes than steel or wrought iron (which never melts). See **malleable cast iron**.

cast steel [mech.], **crucible s.** Steel which has not been *forged* or *rolled* since casting. All steel is cast during steel-making but most of it is subsequently so worked as to change its shape considerably. Steel castings are expensive because of the difficulty of getting the molten metal to flow properly at a reasonably low temperature. See **ingot**.

cast stone See (*B*).

cast-welded rail joint A welded joint between two butted rails, usually cast by the *thermit process*.

catch basin [sewage] US term for a *catch pit*.

catch drain A *grip* (1).

catch feeder [hyd.] A ditch for *irrigation*.

catchment area, drainage area, drainage basin, gathering ground The area drained by a watercourse or providing water for a reservoir. See **watershed**.

catch pit A pit below the *invert level* of a surface-water ditch or sewer, formed at a place where it can conveniently be pumped or dug out. It catches grit and prevents it blocking less accessible parts of the stream.

catch points [rly] On an upgrade, a cut made through the rails, so that a wagon which becomes unhitched and runs back is harmlessly derailed. See **points**.

catchwater (1) A *grip* (1).
(2) [hyd.] A channel cut along the edge of high ground to divert the streams running off it from the low-lying ground that might be flooded by them. In the UK, the Great Ouse Flood Protection Scheme has a catch-water leading the waters of the Lark, the Wissey and the Little Ouse away from the Cambridgeshire fens.

catenary [stru.] The curve into which a uniformly loaded rope falls when hung between two points. It is seen in *suspension bridges*, *cableways* and *ropeways*.

catenary correction [sur.] *Sag correction*.

catenary suspension An overhead suspension of an electric power conductor by vertical links of different lengths which hang it from a steel wire rope stretched tightly above it. In this way a power conductor can be kept at constant height above the locomotive which it feeds.

caterpillar floor An *articulated floor*.

caterpillar gate [hyd.] A massive steel gate for controlling the flow through a spillway. It is carried on *crawler tracks* with hardened steel rollers bearing on steeply sloping rails at each side of the opening. In England this is often called a caterpillar *penstock*. The name caterpillar gate, used in the USA, is less confusing.

caterpillars [mech.] A popular name for *crawler track* for tractors.

cat-head sheave A *sheave* high up on a pile frame.

cathode [elec.], **kathode** The plate (*electrode*) in *electrolysis* at which metals or hydrogen are released or at which chemical reduction occurs. Cf. **anode**.

cathodic protection Electrical protection of underground or underwater structures such as pipelines from corrosion. The structure is made the *cathode* in a direct current circuit which has a higher voltage than (and an opposite direction from) the estimated corrosion voltage. The *anode* is either *sacrificial* (dissolving away and periodically replaced) or 'inert' with the necessary *impressed current* put in from a power supply. Another method working on a similar principle used for many years is *galvanizing*. In the presence of an *electrolyte* a current flows from the zinc to the steel, dissolving the zinc and protecting bare parts of the steel. This is *sacrificial protection*. To reduce the running costs of cathodic protection it is best where possible to provide other anti-corrosive protection, such as pipe-wrapping, pipe-lining, coating, etc.

Steel *reinforcement* can be cathodically protected if it is electrically continuous. This should also help to protect it against lightning, provided that it is well earthed (grounded).

But cathodic protection cannot help bare steel that is out of the water. The 'splash zone', sometimes out and sometimes in the water, suffers the worst corrosion and needs the best protection. In steel *offshore platforms*, pipe-wrapping may help.

cation [elec.] An electrically charged atom which migrates naturally to the cathode in an *electrolyte*. Most metals and hydrogen produce cations. Cations can exist outside of electrolysis, for instance in clays. See **base exchange, anion, ion**.

caulking (1) The blocking of a seam or joint to make it airtight, watertight or steam-tight by driving in tow, lead, oakum or *dry pack* to carry load.

(2) [mech.] Making boiler plates tight by deforming the exposed edges of plates with a *caulking tool* struck by a hammer, thus driving each exposed edge into contact with its neighbour.

caulking tool [mech.] A blunt *cold chisel* often with an offset shape for *caulking* boiler plates and driving in caulking materials.

causeway (1) A road carried over marsh or water by an earth bank or wall.

(2) (Scots) A road surfaced with *setts*.

caving [r.m.] A common mining method, involving the removal of mineral without leaving support in the void from which it was taken. The rock breaks and increases in volume because of air spaces between the rock pieces. The broken rock in the *goaf* can thus provide some support for the rock above, and delays the appearance of subsidence at the surface.

cavitation [mech.] In pumping at excessive speeds certain parts of the pump may move faster than the water. This occurs with *centrifugal pumps* or turbine *runners* near the *draft tube*. The result is corrosion of metal parts due to the liberation of oxygen from the water.

Similar troubles have been experienced in concrete pipes passing through a dam. High flow volumes with high speeds scour the concrete. One solution would be to provide air inlets to the scoured lengths of pipe, so as to eliminate the negative pressure causing the erosion. Another is to line the pipe with steel, but either of these is difficult after the dam has been built.

cavity construction In a *composite floor*, *ducts* for cables or pipes, formed either by trays below the *profiled steel sheets* (*B*) or by *permanent shuttering* within the concrete or the *topping*. See also (*B*), **cavity**.

cavity tanking *Tanking* (*B*) (waterproofing) of a basement achieved by air gaps rather than by waterproof materials; a very ancient method, the only known one until about 1900. It is absolutely reliable but wastes space, sometimes the whole cellar.

Inside the retaining wall holding back the earth outside the building, is an air gap draining to a sump from which water must be pumped if it is below the level of the drains. If the floor also leaks it must be drained similarly.

The wet floor must be covered with paving flags below which nibs about 25 mm (1 in.) deep are cast, allowing ample drainage space below them. On the flags a *damp-proof course* (*B*) such as a polythene sheet is laid. This is covered with a *topping* of at least 100 mm (4 in.)

CBR See **California bearing ratio**.

celerity [hyd.] The overall speed of a wave.

cellular cofferdam A *double-wall cofferdam* used in very large projects in water. The double wall consists of a succession of cells in contact, each cell being, for example, an 18 m dia. steel *sheet-pile* ring filled with sand. The width of the cell is usually about equal to its unsupported height. The junctions between the 18 m dia. rings are made, in one type of cellular cofferdam, by arcs of steel sheet piling. Another type is the 'diaphragm cell' in which all cells are similar and are joined along a straight line, with circular arcs only on the outer walls. The straight line, or diaphragm, acts as a tie or strut between the outer walls.

cellular concrete US term for *aerated concrete*.

cellulose nitrate *Nitrocellulose*.

cement (1) The bond or matrix between the particles in a rock, particularly that binding the sand grains in a sandstone, quartzite or conglomerate.

(2) A powder that, mixed with water, binds a stone–sand mixture into a strong concrete within a few days. Most cements, except *high-alumina cement*, contain at least some *Portland cement*. Nearly all set well under water. See also **Perspex** (*B*), **hydration**, **hydraulic cement**, **water/cement ratio**.

cementation (1) Injecting cement *grout* under pressure into fissured rocks to strengthen them and make them watertight. It is a form of *artificial cementing* used in shaft sinking and tunnelling, and is also called the *grouting method of shaft sinking*. See **oil-well cement**.

(2) Impregnating wrought-iron bars with carbon by packing them with charcoal and heating them for several days. This was the old method of steelmaking, but the term is also often applied to *carburizing* and sometimes to *sherardizing* (*B*), or similar processes in which steel or iron is packed and heated with zinc or other metal to acquire a protective coating.

cement compounds *Blast-furnace slag* and *Portland cement* contain the same chemical compounds though in different proportions. Cement technologists use the following shorthand for them: C stands for lime, CaO; S for SiO_2, silica; F for Fe_2O_3, iron oxide; and A for Al_2O_3, aluminium oxide. In hydrated cement, H stands for water. Tricalcium silicate ($3CaO.SiO_2$) is written C_3S; dicalcium silicate ($2CaO.SiO_2$) is written C_2S; tricalcium aluminate ($3CaO.Al_2O_3$) is written C_3A; tetracalcium aluminoferrite ($4CaO.Al_2O_3.Fe_2O_3$) is written C_4AF.

cemented carbides, sintered carbides [mech.] Materials used for the tips of very-high-speed tools. Tungsten and molybdenum carbides are the main constituents with some tantalum, cobalt and titanium. See **hard facing**, **sintering**, **Stellite**.

cement grout See **grout**.

cement gun A compressed-air tool for spraying *gunite* or *shotcrete*.

cementing See **artificial cementing**, **oilwell cement**.

cementite [mech.] The very hard but

brittle constituent of white cast iron (Fe_3C), an iron carbide also present in the *case* made during *carburizing* or *cementation*.

cementitious Description of the chemical reaction of cements and *pozzolans* with water and lime to form hard paste – hydrates of calcium silicates.

cement joggle A method of preventing relative movement between concrete blocks in *blockwork* structures by leaving an indentation for the height of each block opposite a corresponding notch in the next block. When the blocks are set, this cavity (in two blocks) is filled with concrete or mortar, poured in. See **joggle**.

cement-latex See **cement-rubber latex** (*B*).

cement mortar *Mortar* of sand and cement, also now often with lime ($Ca(OH)_2$).

cement-mortar lining See **pipe lining**.

cement paint See (*B*).

cement paste The mixture of water, cement, *pozzolan*, if any, and air which surrounds the aggregates in concrete and mortar. It is responsible for many important properties: strength, imperviousness, chemical resistance, etc. But the *shrinkage* as well as the *creep* of cement paste are ten times as great as those of a concrete (the aggregate restrains the shrinkage). Pozzolans improve *workability* because they increase the volume of the cement paste.

cement replacements *Pozzolans*, natural or artificial, including *ground granulated blast-furnace slag*.

cement uniformity Recent publications of the *ASTM* have shown that most US cement makers can without difficulty hold their cement strengths

to within 2.1 MPa (300 psi) of a stated 28 day strength.

cement washer A *pile-cage spacer*.

Cem-FIL Alkali-resistant glassfibre used in *glassfibre-reinforced* or other concrete.

CEN The European Committee for Standardization. It has 18 members: the 12 countries of the European Community and the six EFTA countries. Its standards are prefixed EN, in the same way as British standards are prefixed BS. CEN also publishes *Eurocodes*.

centering Specialized *formwork* for building an arch, shell, etc. which is lowered as a unit when *striking* so as to avoid damage to the newly hardened concrete (*ACI*).

centesimal measure [sur.] Division of a circle into 400 grads or grades, each with 100 minutes, each minute having 100 seconds. A full circle less one second is written $399^g \, 99' \, 99''$, and a full circle *sexagesimal* less one second is written $359° \, 59' \, 59''$.

centi- A prefix meaning 'one hundredth part of'.

centimetre (cm) One hundredth part of the *metric* unit of length the metre. 2.54 cm are equivalent to 1 in.

central reservation, c. reserve, median strip A strip of land between the two *carriageways* of a *motorway*, sometimes with shrubs or a crash barrier along its centre line.

centre (**on center** USA) See **centres**.

centre cut [min.] See **wedge cut**.

centre of gravity, c. of mass, mass c. [stru.] That point in a body at which it will balance if supported. It is the point at which the weight acts, and its location is important for all engineers.

Structural engineers are interested in the centre of gravity of forces or weights because they need to place any column with its *centroid* at their centre of gravity.

centre of pressure [hyd.] The point on an area subjected to fluid pressure, over which the whole force due to the pressures on the area may be taken to act.

centre punch [mech.] A small hard-steel bar with a blunt central point. The point is placed over the centre mark of a hole to be drilled in metal and the other end of the bar is struck with a hammer. The small dent thus made ensures that the bit starts drilling in the correct place. A centre punch looks like a *nail punch* (*B*), except for its point.

centres, centre to centre (on center USA) [d.o.] A description of a dimension; for example, 2 m centres (or 2 m crs) means 2 m between the centres of the pieces in question.

centre to centre *Centres*.

centrifugal blower [mech.] A small low-pressure, high-volume fan with a rotating impeller.

centrifugal brake [mech.] A safety mechanism on hoist drums which throws the brake shoes outwards on to the fixed brake drum when the load begins to run away.

centrifugal compressor [mech.] An air compressor which is not *reciprocating* and usually is made of several *centrifugal blowers* in series.

centrifugal force [mech.] A body carried round in a circle must (by Newton's Laws of Motion) tend at every instant to continue in a straight line, that is at a tangent to the circle. This tendency, its centrifugal force, is equal to its mass × its acceleration, that is the acceleration diverting it from

straight-line motion. It is this force which requires railway and road curves to have *superelevation*. The centrifugal force of a train is considered to act at 1.8 m above rail level.

centrifugally cast glassfibre-reinforced plastic pipes These pipes have been regularly used in Sweden since 1968. In 1985 they dominated the market for diameters from 0.5 to 1.2 m (20–48 in.). For diameters around 400 mm (16 in.) they had more than half the market, while at about 300 mm (12 in.) dia. the main competitor was PVC. Apart from bends, made of GRP, most fittings are of stainless steel.

centrifugal pump A pump with a high-speed rotating impeller. Water enters near the centre of the impeller and is thrown outwards by the blades. These pumps take up less space than *reciprocating pumps*, are often suitable for direct electric drive, and are therefore often preferred, particularly in large plants pumping clean water. See **specific speed**.

centrifuge [mech.] A rapidly turning machine, rotating at several thousand rpm, used in *soil mechanics* (see **centrifuge moisture equivalent**), and for dewatering sewage sludge.

centrifuge moisture equivalent (CME) [s.m.] The percentage of water retained by a soil which has been first saturated with water and then subjected to a force equal to 1000 times the force of gravity for 1 h (ASCE). It is a way of comparing road soils which is generally less used than the *consistency limits*.

centroid (gravity axis USA) [stru.] The centre of area of a section, that point about which the *static moment* of all the elements of area is equal to zero. For a homogeneous beam the *neutral axis* passes through the centroid. See **eccentricity**.

certificate A document written out by the *contractor* and signed by the *consultant*, authorizing payment to the contractor for work done, materials supplied, etc. 'Interim' certificates are usually settled monthly, or if 'final' ones, at the end of the contract.

certification authority See **classification society**.

CESMM *Civil Engineering Standard Method of Measurement*, published by the I.C.E.

cess [rly] The flat area at formation level, adjoining the ballast of a rail track.

cesspit, cesspool A tank, usually underground, of brick or concrete, for collecting sewage where no sewage treatment is available. It is pumped out periodically by the *water authority* for a fee. A cesspool may be overflowing, pervious (leaching) or impervious, but these points are settled by the authority. See also (*B*).

cesspool A *cesspit*.

CFA *Continuous-flight auger*.

chainage (1) [sur.] A measured length.
  (2) Drawings of roads show, every 10 or 20 m, the details at those cross-sections. On the ground these chainages are marked by pegs set by the *site engineer*, before any excavation begins, and are used for placing *profiles*, *batter rails*, lines of *kerbs*, etc.

chain block [mech.] A *differential pulley block*.

chain-bucket dredger [hyd.] A *bucket-ladder dredger*.

chain of locks A succession of lock *chambers* in which the *head gate* of each is the *tail gate* of the one above.

chain pump [mech.] A way of raising water by discs passing up a pipe on a chain. For short lifts it is less inefficient than it appears to be on paper.

chain saw A power saw used for cutting timber in the forest.

chain sling A *sling* of chain made of *wrought iron* or $1\frac{1}{2}\%$ *manganese steel*.

chair (1) [rly] or **rail chair** A cast-iron support 20×37 cm in plan screwed to a *sleeper* in British practice. It held a *bullhead rail* wedged into it with a steel spring or hardwood *key*. See **rail fastening**.
  (2) A bar bent in such a way that it holds up the top steel of a reinforced-concrete slab by resting on the bottom steel.

chair bolt, fang [rly] A bolt which passes up through a *sleeper* and chair from below and holds the *chair* down by a nut screwed on to it.

chalk line A length of bricklayer's line well rubbed with chalk, held tight and plucked against a wall, floor or other surface to mark a straight line on it. It is also used by plasterers, miners, mural painters and others.

chamber, lock bay [hyd.] In a canal *lock* the space enclosed between the upper and lower gates.

chambered-level tube [sur.] A *level tube* with an air chamber at one end from which air can be added to the bubble by tilting. Temperature modifications to the bubble length can thus be corrected.

chambering, springing, squibbing [min.] The firing of successively larger charges of explosive with little *stemming* until the bottom of the hole is sufficiently enlarged (chambered) to take the final charge (which is properly stemmed). The method is much used for heavy blasts in quarrying. See **camouflet, jet drilling, torpedo**.

chamfer Instead of casting the corners of a concrete column or beam square they are often cut off at 45° so as to be less easily scarred. See also (*B*).

chamfer strip A *cant strip*.

change face, reverse f. [sur.] To *transit* a *theodolite*, that is to rotate the telescope through 180° vertically and 180° horizontally, so that the vertical circle is at the opposite side from before, when viewing the same object. See **face left, face right**.

change point, turning point (1) [sur.] In levelling, a point on which two readings of the staff are taken, a *foresight* and a *backsight*.

(2) [mech.] A *critical point*.

channel, c. iron, c. section A rolled-steel section of ⊏-shape.

channeller (channeler USA) A powerful quarrying machine with a row of chisels which cuts a slot in stone at any angle, and without explosive.

Channel Tunnel (see diagram) A project begun and stopped on both sides of the Channel in 1882 and 1975, but begun again in 1987 with the achievement of the break-through of the *pilot tunnel* at the end of 1990 at the planned speed of 1 km/month from both ends. There are three tunnels 49 km long from Fréthun near Calais to Cheriton near Folkestone. The smallest tunnel, of 4.8 m dia., the pilot tunnel for service, access and drainage, was driven first. On each side of it but slightly higher are the running tunnels of 7.6 m (25 ft) dia. All were driven by *tunnel boring machines*.

In France the serious geological difficulties of 1974 caused French engineers in 1987 to sink a vast 55 m (180 ft) dia. shaft 70 m deep at Sangatte to provide access for the driving of six tunnels at 40 m depth, three of them towards

Paris and three northwards. Chalk excavated from the tunnels was dumped below the tracks and pumped as 20% slurry to a settling pit 1.8 km away. The *pipeline transport* used fresh water to prevent pollution of the aquifer below the settling pit.

Transmanche Link, the main contractor, worked for the *client*, Eurotunnel, representing the governments and the shareholders who invested £6000 million.

Financing of the tunnel is based on the tolls received over the 55 years of the *franchise* from its opening in mid-1993. The ten French and British companies of the main contractor were also *promoters*, introducing the contradiction that they were both owner and contractor. To improve the independence of the client, *Eurotunnel*, the two governments stipulated an Anglo-French maître d'oeuvre to be staffed by independent consulting engineers. Engineering managers from Bechtel Corporation were employed by the maître d'oeuvre to manage the project.

Channel Tunnel Group The British half of *Eurotunnel*. Cf. **France Manche**.

characteristic strength The strength of a material on which its *permissible stress* is based; it is equal to 95% of the average of the destructive tests for concrete. For steels it is based on the *yield point*, *proof stress*, etc., whichever is applicable. A *load factor* reduces it to a safe stress.

characteristic stress The stress at the assumed *yield point* or *limit of proportionality* for the material.

charging hopper The part of a *concrete mixer* resting on the ground. The cement, sand and gravel are placed in it. The hopper is then raised and shoots them into the drum.

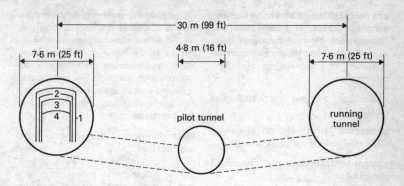

Train sizes
1. Channel shuttle (double deck)
2. SNCF
3. British Rail
4. London tube

Channel Tunnel (typical cross-section).

Charpy test [mech.] An *impact test* in which a notched test-piece supported at both ends is broken by a blow from a striker on the face opposite to and immediately behind the notch.

chartered engineer, professional e. Someone who has been accepted as a full (corporate) member of an *engineering institution*. The exams are now run by the *Engineering Council*. A chartered engineer is qualified usually at least by an engineering degree of a university and by training with a *consultant* or *contractor* approved for the purpose. He or she may previously have been a *student member* and then a *graduate member*.

chaser A *brickwork chaser* (*B*).

check (1) Generally, a verification of a survey, a calculation, etc.

(2) [hyd.] A structure which controls the water level in an irrigation canal or ditch.

(3) [hyd.] An area of land between ridges which confine the irrigation water.

checker [d.o.] An engineer *section leader* who is usually fully qualified and checks structural drawings.

check rail, guard r., safety r., side r. [rly] On curves, a third rail fixed close outside the inner rail to reduce the wear on the outer rail caused by *centrifugal force* and to keep the inner wheel on its rail.

check valve, clack, non-return v., reflux v. [mech.] A valve which allows flow in a pipe one way only. A check valve is always placed on the delivery side of a mine pump or other *force pump*. It protects the pump from the considerable weight of water in the shaft pipes which could otherwise flow back, reverse the pump, or damage it. The usual type is a *flap valve*.

chemical gauging, chemi-hydrometry [hyd.] Measuring the quantity of flowing water by determining the dilution of a chemical solution introduced upstream at a known rate and concentration.

chemical oxygen demand See **COD**.

chemical precipitation [sewage] Chemical sedimentation of sewage.

chemi-hydrometry See **chemical gauging**.

chemise A wall which *revets* an earth bank.

chequer-board construction, alternate-bay c., hit-and-miss c. Concrete on the ground in a factory floor, road, airfield, etc. is cast in alternate rectangles (*bays* (*B*)), leaving half the rectangles as gaps to be filled later. By the end of the first series of rectangles, much of their shrinkage will have taken place, thus reducing the eventual cracking.

chequer plate, chequered p. Steel or cast-iron plate, perforated or patterned to make a non-slip floor in power stations, factories, fire escapes, etc.

cherry picker An *access platform*.

chert Non-crystalline silica which is found in limestones. *Flint* is a form of chert. See **alkali–silica** reaction.

Chesapeake Bay bridge-tunnel Two high road bridges, 20 km (12½ miles) of low precast concrete bridge, four *artificial islands* and two tunnels over a mile long each, completed in 1964. The tunnels provide a clear seaway for shipping, and the high bridges help also. The combined road bridges and tunnels are 28 km (17.5 miles) long and may have inspired the daring *Euroroute* project.

The tunnels, *immersed tubes* laid in trench, are covered with 3 m (10 ft) of fill. The four islands stand in 9–15 m (30–50 ft) of water, 9 m above low water, and measure 450 × 70 m (1500 × 230 ft) at the top. To reduce the bulk and cost of the expensive islands, the tubes continue above ground a short distance.

chevron drain, herring-bone d., counterfort [rly] Diagonal, stone-filled trenches in railway cuttings laid out in herringbone pattern to drain into 'buttress drains' which are laid out along the line of steepest slope.

Chicago caisson, C. well A small *cofferdam* used in medium stiff clays, of about 1.2 m dia. lined with planks added in 1.5 m lengths and sunk to hard ground for pier *foundations*. The vertical plank supports are held in place by steel rings wedged against the sides. See **cylinder**, **Gow caisson**.

Chicago well A *Chicago caisson*.

chief draughtsman [stru.] The chief of a structural drawing office (unless a chief designer is over him or her). He or she is generally a *designer* with high qualifications and much experience.

chilled cast iron [mech.] Iron cast in a metal mould so as to harden the surface of the casting to about 2 cm depth. Railway wheels are sometimes of chilled iron.

Chinaman chute A structure like a *gantry* with a ramp up to it, on to which earth or other material is pushed by a *bulldozer* or dragged by a *scraper* for loading into a lorry below.

chipping [mech.] Removing weld slag or surface defects from steel or iron by *cold chisel* or *chipping chisel*.

chipping chisel [mech.] A *cold chisel* or *chipping hammer*.

chipping hammer, c. chisel A welder's compressed-air tool for cleaning steel after welding. It weighs from 3 to 6 kg. A similar tool is used by the concrete *finisher* for removing fins and other unwanted projections from the face of the finished concrete.

chippings For general construction, crushed stone from 3 to 25 mm; for rail track, chippings are +20 mm in size.

chlorination A *disinfection* process in *water treatment*. It is the most popular one in most English-speaking countries. Usually 0.1 to 0.2 mg/litre of free residual chlorine are left in the water as it flows from the waterworks. Chlorine, though a poisonous gas, is a strong oxidizing agent and therefore an excellent disinfectant, but excessive chlorine on biological residues in water can produce cancer in humans. Some authorities (New York) have therefore ceased to buy gaseous chlorine and use calcium hypochlorite ($CaOCl_2$) or other compounds of chlorine instead (hypochlorination). Hazardous cylinders of chlorine gas are not needed at such waterworks. In the USA chlorine is regularly used in *sewage treatment* to reduce smell and for other reasons, but not in Britain. See **dechlorination**.

chlorofluorocarbons (CFCs), Freons Substances used in foamed plastics, aerosol cans, etc. with a damaging *greenhouse effect*. In construction, the main uses in 1989 were in extruded polystyrene, rigid polyurethane and polyisocyanurate, for which the demand was growing at 3% annually. Harmless substitutes for them are rock fibre, glassfibre, expanded polystyrene.

chord, boom, flange [stru.] The top or bottom, generally horizontal part of a metal, timber or concrete *girder* or *truss*.

C-horizon [s.m.] The parent material, without *humus*, below the topsoil of the *A-horizon* and *B-horizon*, from which these are derived by leaching and deposition respectively. See **subsoil**.

chuck [mech.] A rotating part on a lathe for holding the work, or on a drill for holding the drilling bit.

churn drill [min.] A *cable drill*.

chute A steep channel or *flume*, tube or *dropchute*.

Ciment Fondu *High-alumina cement*.

Cipolletti weir [hyd.] A *measuring weir* with a trapezoidal opening widest at the top, having the sides sloping at 1 horizontal to 4 vertical. It is convenient to use, since a vertical stick can be graduated in such a way as to read discharge directly.

circuit breaker [elec.] A device which automatically breaks a circuit when the current exceeds a certain value. It is a *cutout* for currents which are larger than can safely be interrupted by a fuse. It can also be opened like a switch, but can usually not be reclosed unless the circuit is working correctly without overload.

circular-arc method, cylindrical-surface m., slip-circle m. [s.m] A simple method of determining the stability of an earth slope in *clay* soil. Failure is assumed to occur by *shear* along a circular arc of the length of the earth mass. The resistance to failure is the area of the cylindrical surface of failure times the *shear strength* of the clay. See also **rotational slide**.

circular level [sur.] A *level tube* in which the upper surface of the glass has a spherical curve.

circular mil [elec.] The area of a circle with a dia. of 1 mil (0.001 in.). One circular mil therefore has an area of 0.7854×10^{-6} in^2. Mainly USA.

circulating water The water which circulates in a coal washery or ore concentration plant, etc. To reduce pollution, it has been found possible to completely avoid its release into nearby rivers, streams or sewers.

circumpolar star [sur.] A star which does not set at the latitude in question. For field surveyors it usually means stars close to the pole – those which have a *declination* of 80° or more.

civil engineer One who designs or builds *civil engineering* work, usually in the UK implying someone who holds a qualification of the *ICE*, but *road engineers, municipal engineers, structural engineers*, and some others are also regarded as civil engineers.

civil engineering The whole of non-military engineering at the time of the founding of the ICE in 1818, but now excluding chemical, electrical, electronic, marine, mechanical engineering, etc. It includes airfields, *bridges, canals, docks, foundations, harbours, offshore construction, railways, river basin management, roads, sewage treatment, sewerage, soil mechanics, structural design, traffic engineering, tunnels, water supply*, etc.

The field is so wide that no engineer can easily specialize in more than two of these subjects. Civil engineers prepare drawings after surveying a site, deal with *tendering* and other work connected with construction *contracts*, and supervise construction. In the USA the Massachusetts Institute of Technology recognizes five areas of specialization: structures, materials, hydrodynamics, soils, systems. Water-resources, transport, environmental-control, and information systems are included in 'systems'.

civil engineering assistant One who does the work of a *civil engineer* on the site or in an office but is not in sole charge of his or her work. See below.

civil engineering draughtsman One who prepares civil engineering drawings. Like a *civil engineering assistant*, he or she may or may not be qualified.

civil engineering project Any large construction project begins with agreement between three parties: the promoter (employer, *client* or owner), who conceives it, the *consulting engineer*, who thinks how it can be built, and the *contractor*, who builds it. Once the *contract* is signed, the promoter becomes the employer of the contractor. The engineer is not a party to the contract and has no legal obligations under it, but is bound to the client by the *conditions of engagement*. The engineer closely supervises the contract's progress, helped by a *resident engineer*, who is on site all the time. If necessary, the consulting engineer arbitrates between the contractor and the client.

civil engineering technician See **technician engineer**.

clack See **check valve**.

cladding The non-loadbearing surfacing of a building which keeps out the weather.

clad steel [mech.] *Carbon* or low-*alloy steel* with a layer of some other metal or alloy firmly bonded to one or more surfaces (BS 2094). Cf. **Alclad**.

claim A demand for payment by a *contractor* for essential work for which he or she has not been able to find a sum allocated under the *contract*. The *consultant* always considers and comments on claims before they are sent on to the *client* for payment.

clamp handle [sur.] A hand grip for

Soil description		British Standards Institution after BS 1377	American Society for Testing and Materials
		Size in millimetres	
	coarse sand	2.0–0.6	2.0–0.25
	medium sand	0.6–0.2	—
	fine sand	0.2–0.06	0.25–0.05
frictional soils		0.06 mm is the smallest practicable size for sieving.	
	coarse silt	0.06–0.02	
	medium silt	0.02–0.006	0.05–0.005
	fine silt	0.006–0.002	
cohesive soils	clay	under 0.002	under 0.005

See also **airfield soil classification**, **grading curve**, **soil mechanics**.

tensioning a steel tape when less than a full tape length is being measured.

clamping screw [sur.] On *theodolites*, a screw for clamping a vernier so that the *tangent screw* can be used.

clamshell grab A *grab* shaped like a clamshell.

clapotis [hyd.] The lapping of waves on a wall that rises above the water level. It may double the wave height, raising the sea level and the pressure on the wall.

clap sill [hyd.] A *lock sill*.

claquage grouting See **artificial cementing**, **consolidation grouting**.

clarification Removal of tiny *suspended solids* from *raw water*, usually in small concentrations, often only 100 mg/litre, by settlement in a tank called a clarifier. In the UK the term used for the settling of *sewage* is 'sedimentation' but in the USA, a 'clarifier' is a sewage *sedimentation tank*. See **waterworks**.

class 1 welding Welding of the highest strength and reliability used for important structures including aircraft,

offshore structures, pipelines, etc. where a leak or even a minor failure could be catastrophic. British standards, the *ASME* and other authorities specify tests for welders to pass before they are allowed to do class 1 work.

class 2 welding Welding where the highest quality is not imperative, as for a steel fence.

classification of soils [s.m.] Soil particles are described (after *mechanical analysis* of a soil sample) as sand, silt or clay on the basis of the sizes in the table above. The BSI classification is often used in Europe by civil engineers. *Clays* are defined by their *consistency limits*.

classification society (bureau de contrôle France) Originally, more than a century ago, classification societies surveyed ships only, classing them at various levels of seaworthiness, the highest level being 100A1 at Lloyd's. They now also survey and classify any structure or its design, offshore or on land, and are employed mainly by shipowners, governments and insurance companies. The three main societies are Lloyd's Register of Shipping,

Bureau Veritas (French) and Det Norske Veritas (DNV) (Norwegian).

Certification (*B*) authorities are different, but provide quality levels where no standard exists. They include QSRMC, the quality scheme for ready-mixed concrete; CARES, the UK certification authority for reinforcing steels; WICS, the water industry certification scheme; and BSI quality assurance.

classifier [s.m.] A separator for dividing sand or other pulp into two sizes, the overflow (or slime or undersize) and the underflow (or sand or *oversize*). Water is generally the medium but air is used with fine powders. See **cyclone**.

classify [s.m.] To divide a mixture of particles or lumps of various sizes into products of definite size limits.

clay [s.m.] Very fine-grained soil of *colloid* size, consisting mainly of hydrated silicate of aluminium. It is a plastic *cohesive soil* which shrinks on drying, expands on wetting, and when compressed gives up water. Under the electron microscope clay crystals have been seen to have a platy shape in which for Wyoming *bentonite* the ratio of length to thickness is about 250:1 (like mica). For other clays it is about 10:1. Clays are described for engineering purposes by their *consistency limits*. A 'very soft clay' is one with an unconfined compressive strength of less than 35 kPa (5 psi); soft clay has 35–70 kPa; medium or firm has 70–140 kPa); stiff clay has 140–280 kPa. Clays are further described as organic, intact, etc. See **classification of soils**, **sensitivity ratio**.

clayboard 'Honeycomb' sheets of cardboard from 50 to 150 mm (2–6 in.) thick. It collapses when wetted, losing all but 10–15 mm of its thickness. It can therefore be placed under concrete cast over clay that is likely to swell.

clay cutter (1) In *suction cutter dredgers*, a hydraulically or shaft-driven rotating bit which may be a metre or more in diameter, fixed to the suction pipe and raised or lowered with it.

(2) A steel pipe used for sinking rapid holes with a *cable drill* in clay, particularly the shallow holes needed for *bored piles*. It is about 1.2 m long, of diameter equal to that of the hole required, and drops into the clay under its own weight with sufficient force to penetrate it. After each drop the pipe is hoisted out of the hole and the clay within it is pushed out through a slot in the side.

clay puddle, puddle c., pug Plastic clay used for waterproofing. It is used for lining ponds or ditches, in coffering or in *cut-off walls* to dams. In Britain, clay puddle is a spadeable clay halfway between the *liquid* and the *plastic limits*. In the USA it is a much wetter clay near the liquid limit and it cannot be spaded.

clay sampler [s.m.] A *soil sampler*.

clay spade A *grafting tool*.

cleaning Removing clay, etc., from sand or gravel.

cleanout A hole in *formwork* for removing wood shavings or other refuse before concreting begins (*ACI*).

clearance [mech.] The space between a moving object and a stationary object, particularly that between a rail wagon and a wall or tunnel.

clearance hole See **clearing hole**.

clearing, clearing and grubbing Removal of tree stumps and shrubs before excavation of a site. Graders and other earth-moving plant cannot work on soil containing roots.

clearing hole, clearance h. [mech.] A hole drilled slightly larger than the bolt which passes through it, generally 1.5 mm larger for black bolts or for *high-strength bolts*.

clear span [stru.] The horizontal distance, or clear unobstructed opening, between two supports of a beam. It is always less than the *effective span*.

clear-water reservoir [hyd.] A *service reservoir*.

cleat See **angle cleat**, also (*B*).

cleavage fracture, brittle f., crystalline f. [mech.] Breakage at low strength along cleavage planes with bright facets and little visible *plastic deformation*. Cf. **cup-and-cone fracture**; see **impact test**.

clevis [mech.] A U-shaped iron bar, drilled at the ends of the U. It is used as a *shackle* for connecting steel wire ropes to a load.

client The person or organization who employs an engineer or architect, to whom he or she is responsible and who pays the fees. The client is usually the *building owner* in a building contract. In a *civil engineering project* the client may be the promoter, but from the *contractor*'s viewpoint is often the *consultant*. See **Eurotunnel**.

Clifton Bridge (1864) A *suspension bridge* carried by wrought-iron links over the Avon gorge, built to the design of *I. K. Brunel*, but after his death, using links from his Hungerford Bridge over the Thames in London. In 1977 *weighbeams* were installed in both approaches to bar any vehicle with an axle load above 2.5 tons. Every axle is weighed. A difficulty with lightning, affecting the weighbeam *transducers*, has been overcome by improving the electrical continuity of the earthing (grounding).

climbing formwork, cantilever f. Wall forms which are self-supporting,

being held in place by hook bolts cast into the concrete or by through bolts which are later removed from it. Usually the bolts carry long vertical members (soldiers) which can support consecutively several lifts of *formwork* in cantilever and are therefore moved up less frequently than the forms. In some types there are two wall forms, an upper and a lower. The lower one is leapfrogged over the upper one, and the wall is thus never completely stripped until its full height is reached. See **slipforms**.

clinker (**cinder** USA) Dense ash which has partly melted. Both *blast-furnace slag* and *Portland cement* are clinkers that become cements after fine grinding.

clinograph (1) [min.] A *borehole-surveying* instrument which records the angle of slope of the *borehole* at any point. Several types exist, many being electrically operated, some with an internal camera and gyroscopic orientation.

(2) [d.o.] An adjustable *set square*, not graduated to show angles.

clinometer [sur.] A hand-held instrument used for sighting down or up inclined planes to measure the angle of dip.

clip In *ropeways* a V-shaped steel bar bolted on to the traction rope.

close boarding See **close timbering**.

closed-face shield A tunnelling *shield* which, like the *slurry shield* or the *earth-pressure-balance shield*, has a steel partition (*bulkhead*) protecting the tunnel from inrushes of mud or water.

closed traverse [sur.] A *traverse* forming a closed loop and so finishing at its starting-point, or one which is made between points whose coordinates are known and whose opening and closing *bearings* are known. In either case the accuracy of the traverse can be checked by comparing the calculated closing coordinates and bearings with their known values.

closer A sheet pile cut or made to close a *cofferdam* when a standard pile will not fill the gap. See **creep**, also (*B*).

close timbering, c. boarding Planks placed touching each other against the ground, used in *running ground*.

closing error, e. of closure, misclosure [sur.] In a *closed traverse*, the discrepancy between the starting-point and the finishing-point, calculated from the measured angles and distances. Errors are proportioned among the lengths and angles by *adjustment*.

clough [hyd.] A *sluice* gate in a *culvert*.

clump weight See **guyed tower**.

cluster See **cancer cluster**, **pile sleeve**.

clutch, interlock (1) In *steel sheet piling*, the hook shape at the edge of each pile which grips a corresponding hook on the next pile.
(2) When rolled-steel joists are used as piles, a special section which grips the joists each side of it for their full length.

CME See **centrifuge moisture equivalent.**

coagulant A chemical added to water, *effluent*, etc. for *coagulation*.

coagulant aid A substance such as lime or activated silica added to water in very small quantity with a *coagulant*. It intensifies the settling action and may make the *floc* denser.

coagulation A water-treatment or sewage-treatment process in which a chemical, added to the water or sewage, precipitates something, usually a metal hydroxide *floc* that catches the tiny particles which cause turbidity in water, enabling them to be removed relatively easily by settlement. Coagulation precedes *clarification* and is often used for treating drinking waters, rarely for sewage. *Alum* is the commonest coagulant for drinking water in the UK Cf. **flocculation**; see **waterworks.**

coarse aggregate, stone (1) For concrete, *aggregate* which stays on a sieve of 5 mm square opening. See **fine aggregate.**
(2) For bituminous material, coarse aggregate stays on a 3 mm sieve.

coat See **sealing coat**, **tack coat**, **covered electrode**, and (*B*).

coated chippings/grit Chippings or grit which have been coated thinly with bituminous material for scattering over a wearing course.

coated macadam See **tarmacadam**.

cobbles Rounded stones used for paving. Cf. **sett.**

COD (chemical oxygen demand), dichromate value A test of an *effluent* which can be made in two hours instead of the five days of the *biochemical oxygen demand* (BOD) test. Both biodegradable and chemically oxidizable pollutants are indicated; the BOD value is included in the COD value.

code of practice (CP) A publication describing good practice in the work in question. Unlike American *building codes*, a CP generally does not have the force of law.

coefficient of compressibility [s.m.] The change in *voids ratio* per unit increase of pressure. Cf. **modulus of volume change.**

coefficient of consolidation [s.m.] In the *consolidation* of soils a value expressed in cm²/min, if the permeability is in cm/min. It is equal to:

$$\frac{\text{coefficient of permeability} \times (1 + \text{initial voids ratio}}{\text{coefficient of compressibility} \times \text{density of water.}}$$

coefficient of contraction [hyd.] The ratio of the smallest cross-sectional area of a jet discharged under pressure from an orifice, to the area of the orifice.

coefficient of discharge [hyd.] The ratio of the observed to the theoretical discharge of a liquid through an orifice, weir or pipe. See **effective area of an orifice.**

coefficient of expansion, c. of thermal expansion The expansion of a material per unit length for each degree rise in temperature. For steel, concrete and brickwork the value is roughly 0.00001 per degree C, though for some brickwork it can be as low as half this. The ordinary value involves a change in length of 1 cm in a 30 m long member, when the temperature changes by 33°C. If this change were prevented by complete restraint, it would cause a stress of 7 N/mm² in unreinforced concrete with a modulus of elasticity of 20 kN/mm², and of course a higher stress in concretes with higher *E* values. Such high stresses explain why *movement joints* are built into concrete.

coefficient of friction [mech.] The ratio between the force causing a body to slide along a plane and the force normal to the plane.

coefficient of imperviousness US term for *impermeability factor*.

coefficient of internal friction [s.m.] The tangent of the angle ϕ, the *angle of internal friction*.

coefficient of permeability [s.m.] The imaginary average velocity of flow through the total (voids and solids) area of soil under a *hydraulic gradient* of 1. See **permeability**.

coefficient of traction [mech.] See **tractive resistance**.

coefficient of uniformity, modulus of u. [s.m.] The ratio between the grain diameter which is larger than 60% by weight of the particles in a soil sample, to that diameter, the *effective size*, which is larger than 10% by weight of the particles. It is more briefly expressed as $\frac{D_{60}}{D_{10}}$. Uniform soils have uniformity coefficients of less than 3. Non-uniform soils have a relatively flat *grading curve*, uniform soils a steep one.

coefficient of variation In statistics an estimate of the variability of, for instance, the *crushing strength* of a brick from a certain kiln. It is the ratio of the *standard deviation* of a series of values to its mean. A simpler figure is used in the British Standard for sand-lime bricks (BS 187). It is the ratio of the strength of the seven weakest to the average strength of the sample of 12. This ratio is never allowed to be lower than 0.8, and for the best bricks a minimum figure of 0.9 is required.

coefficient of velocity [hyd.] Of an opening through which fluid is flowing; the ratio of the measured discharge velocity to the theoretical discharge velocity.

coefficient of volume decrease [s.m.] The *modulus of volume change*.

coffer A canal lock *chamber*.

cofferdam A temporary *dam*, either *sheet piling* driven into the ground or a dam built above the ground to exclude water and thus give access to an area which is ordinarily submerged or water-logged. Cofferdams are used down to 10 m below water level in deep foundation work. For greater depths a *caisson* or *cellular cofferdam* is needed. See **half-tide cofferdam**.

cogging [mech.] The start of the *hot rolling* of steel from *ingot* to *billet*, the purpose being to reduce the cross-sectional area of the ingot as fast as possible to a bar which can be rolled in a finishing mill to a *rolled-steel section* or forged to the final product.

cohesionless soil, non-cohesive soil A *frictional soil*.

cohesion of soil [s.m.] The stickiness of *clay* or *silt*, absent from sands, characteristic of clays. It is the *shear strength* of clay, generally about half its strength in the *unconfined compression test*. See **Coulomb's equation**.

cohesiveness of concrete, stickiness This is controlled by the volume of *cement paste* relative to that of stone. *Fly-ash* is much less dense than cement; its *relative density* is 2.1 compared with 3.1 for cement. Being proportioned by weight it increases the bulk of the cement paste in whatever proportion it is added, and so improves cohesion, *workability* and pumpability. Cohesive concrete does not segregate. See **thickening agent**.

cohesive soil [s.m.] A sticky soil like *clay* or clayey silt. Some authorities define it as a soil with a *shear strength* equal to half its strength in the *unconfined compression test*. See **cohesionless soil, Coulomb's equation**.

Colcrete *Grouted aggregate concrete*.

cold bend test [mech.] The *bend test*.

cold chisel [mech.] A fitter's chisel used for cold-cutting mild steel (or similar soft metals) when struck with a hammer.

cold drawing, wire drawing [mech.] Making steel wire by drawing it through successively smaller round holes in steel blocks called *dies*. This hardens the steel, raises its *ultimate tensile strength*, and reduces its diameter. By this means steel wire for *prestressing* and for mine winding ropes and haulage ropes is made. The strongest metal is that which has passed through the largest number of dies and is therefore of the smallest diameter. Thin wire at a diameter of 2.03 mm has an *ultimate strength* of 2300 MPa (330,000 psi), but thick wire at 7.01 mm only 1500 MPa (220,000 psi). See **extrusion, patenting, standard wire gauge** (*B*).

cold expansion [rly] *Cold working* of *fishplate* holes with an expanding mandrel. This leaves a *residual stress* of compression around the holes to prevent *fatigue* cracks.

cold joint (see diagram) A weak and leaky joint between two lifts of concrete, caused by delay between them.

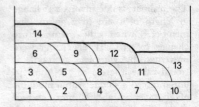

Concrete casting sequence to avoid cold joints.

To avoid it, no upper surface should be allowed to wait for an hour (less in hot weather) before the next lift is placed over it.

cold planer A machine which prepares the resurfacing of roads, by removing a surface strip sometimes as thick as 15 cm (6 in.), at a rate up to 100 m²/h, more quickly and quietly than with jackhammers.

cold rolling, c. forming [stru.] Cold bending of steel sheet from 1.5 to 5 mm thick to make very light structural sections (channels and angles). They are widely used for building. See **Lally column, Stran-steel**.

cold sett A smith's chisel held with a pair of tongs by one person while another, the *striker*, hits it with a *sledge hammer*.

cold shortness [mech.] Brittleness at room temperatures. In iron or steel this is due to too much phosphorus.

cold-weather concreting The object of cold-weather concreting is to prevent the mixing water freezing in the concrete. So long as the water remains liquid it will continue to hydrate the cement and to release *heat of hydration*. But freezing expands it by 9%. When the thaw comes, any hard concrete that contained ice is found to be porous and weak. Precautions must be taken in any weather colder than + 5°C.

cold working [mech.] The shaping of metals at room temperature by *cold drawing*, *cold rolling* or pressing or stamping. It results in work *hardening* for such metals as iron, copper and aluminium. This involves generally an increase in strength but if carried too far may make the metal brittle. Since they have no yield point, cold-worked steel bars are specified by their 0.2% proof stress, as in BS 4461. See **hot working, twisted deformed bars**.

Colgrout The cement–sand mix pumped through *trémie* pipes in *grouted aggregate concrete*.

collapse design [stru.] *Plastic design* of steel structures based on the research initiated by Prof. Sir John Baker at Cambridge, England, from 1935 onwards.

collar [min.] The mouth of a drill hole or shaft.

collecting system [sewage] Every drain or sewer in a sewerage system between the house and the outfall or sewage disposal works.

collimation error [sur.] Error in surveying instruments caused by the line of sight not being horizontal or being otherwise out of line.

collimation line [sur.] The line of sight of a surveying instrument. It passes through the intersection of the *cross hairs* in the *reticule*.

collimation mark, fiducial m. [air sur.] A mark on the register glass of an air survey camera, usually at each corner of the glass. Images of the marks appear on each air photo. The diagonals between these marks meet at the principal point (plumb or nadir point), P, on the photo. For a *vertical photo* the point P on the ground was vertically below the camera at the instant of exposure. P in the camera is on its vertical axis or centre.

collimation method [sur.] In levelling, also known as the 'height of instrument' method as opposed to the '*rise and fall*' method. Throughout the fieldwork the instrument height is always known by taking the first sight on a point of known level. At any time, therefore, the level of a point can be quickly worked out by subtracting its staff reading from the level of the instrument (instrument height). The main difference from 'rise and fall' is that the calculations of level are usually made during the survey. The method is convenient for obtaining the levels of many points from one set-up.

collision load BS 5400 demands the following allowances in the event of a car crashing into a bridge column; at guard-rail level 0.75 m above the road, 150 kN (15 tons) perpendicular to the road or 50 kN (5 tons) parallel to the road. A further blow at the most unfavourable point between 1 and 3 m above the road is added. It is looked on as a force of 100 kN (10 tons) parallel or perpendicular to the road.

colloidal grout See **Colgrout**.

colloids [s.m.] Particles smaller than 0.002 mm (the European definition of the largest size of *clay* particles) and larger than 0.000001 mm, ten times the diameter of an atom. Particles smaller than 0.0002 mm do not settle in water and those between 0.002 and 0.0002 mm settle only very slowly. *Mechanical analysis* of clays is thus not fruitful and X-ray studies are often more useful than microscopic examination, since with the microscope an object of 0.002 mm size can only just be seen. Colloids make up most of living matter.

collusion, to collude (bid-rigging USA) An improper practice which, if proven, allows a *contract* to be set aside. A group of *contractors* may collude to choose one of their number for a particular contract and make their prices high so that that contractor has the lowest tender. In March 1988, in the UK Restrictive Practices Court, 48 firms agreed to cease this activity. Any future price fixing or cartel arrangements by them would be contempt of court and severely punished.

column [stru.] A post carrying compressive force. See **long column, stanchion**.

column analogy [stru.] An *analogy* due to *Hardy Cross* between the equations for slope and deflection of a bent beam and those for load and moment in a short eccentrically loaded column. It cannot be so widely used as *moment distribution* but the particular cases of fixed-base *portals* and arches can be rapidly analysed by it.

column clamps Steel quick-release clamps wedged round column *forms*.

column head In reinforced-concrete *mushroom construction* an enlargement (thickening) of the column where it meets the slab.

column strip In the design of *plate floors*, *mushroom slabs*, etc. a part of the *reinforcement* layout which includes a quarter of the width of the slab centrally over the column, extending one eighth of the width each side. Cf. **middle strip**.

combined stresses [stru.] Bending or twisting stresses combined with direct tension or compression.

combined system [sewage] A system of drainage by which *soil* and surface water are carried in the same *drains* and *sewers*. In the combined or *partially separate system*, rainwater may be connected through a trap to a foul drain. But in the *separate system*, sewage must not flow into a purely rainwater drain.

comminutor [sewage] In the *preliminary treatment* of sewage, this is a device with a screen that catches the solids and shreds them small enough to pass through. It eliminates the need for handling the solids.

commissioning Starting up a power station or other large plant, a task which may be very slow. Recommissioning a shut-down cement kiln can take nine months and cost £1 million. Filling a pipeline may be called charging, gassing up or bringing on stream. Before an oil pipeline is filled it must be purged of oxygen and water. This may be done by injecting nitrogen or other inert gas, separated by *pigs* from the air in front and the oil behind, as in the *Alaska pipeline*.

committal rate The percentage of the

working week for which a concrete pump is available for work. *Down time* is set aside daily for cleaning, lubrication and inspection. The committal rate is 100% minus the down time per cent.

communications program A program which, with a *modem* connected to a *microcomputer*, allows it to send or receive data from a *mainframe* or other computer.

compacting factor test A test of the workability of freshly mixed concrete made by weighing the concrete which will fill a container of standard size and shape when allowed to fall into it under standard conditions of test (BS 1881). More precise and sensitive than the slump test, its fairly elaborate apparatus is used more in the laboratory than on the site.

compaction [s.m.] Reduction of the *air content* of fresh concrete or granular soil with increase of *dry density* by *vibration* or rolling, or for deep compaction by driving *sand piles, vibroflotation* or *dynamic consolidation*. AASHTO recommends for embankments less than 15 m high that a dry density of at least 1450 kg/m³ should be obtained. For embankments over this height at least 1900 kg/m³ is required. Apart from the method of running earthmoving plant over the area to be compacted, there are many methods and six main types of compacting plant: (i) *pneumatic-tyred rollers*, in which the rear wheels cover the gaps left by the front wheels; (ii) *tamping rollers*; (iii) *sheepsfoot rollers*; (iv) *vibrating rollers*; (v) *frog rammers* (trench compactors); (vi) *vibrating plates*. The last two, for confined spaces, are operated by one person. See also **optimum moisture content**, and cf. **consolidation**.

compaction grouting *Consolidation grouting*.

compact material Material which can be dug with a pick. See **loose ground**. It is usually a granular soil with a *relative compaction* of 90% or more. This means that its *dry density* is more than 90% of that obtained in the *compaction* test in use.

compactor-finisher Part of a *fixed-form paving train* which includes rotary paddles for striking off the concrete, a vibrating compacting beam (an elaborate screeder) and an oscillating finishing beam. It can achieve any *camber* (*B*) or cross-fall.

comparator [mech.] An instrument for accurately measuring short lengths. A reading telescope is arranged to travel along a scale and to observe in turn the points whose distance apart is to be measured. It is used in *photogrammetry* for measuring the two rectangular coordinates of a point on a photograph.

comparator base [sur.] A carefully measured horizontal distance, usually one tape-length long, used as a means of checking and comparing the tapes in the USA. 'Standardization length' is the British term.

compass, magnetic c. [sur.] An instrument carrying a steel magnetized needle pivoted so as to be free to turn in a horizontal plane. The needle automatically orients itself in the magnetic north–south direction, and thus gives a reference line from which any *bearing* can be measured. The bearings can be accurate to within 1° of arc.

compasses [d.o.] A pair of compasses is an instrument for drawing circles.

compass traverse [sur.] A *traverse* in which the magnetic bearings of all the lines are recorded.

compensated foundation [stru.] A common way of calculating the necessary depth, D, of excavation for a building *foundation* on clay or other compressible soil. Most cities are built on such soils. The likely average total weight of the loaded building with its foundation is first calculated. The weight of the soil dug out must equal this.

With an excavation depth of D m, a building area of A m^2 and a soil density of (say) 1.8 tonnes/m^3, the weight of soil to be dug out is

$$D \times A \times 1.8 = 1.8DA \text{ tonnes}$$

and this must equal the building weight. D is thus easily found. The procedure generally reduces the sinking of the building to a minimum, especially the dangerous *differential settlement*. The taller the building, therefore, the deeper will its *basements* (B) and *sub-basements* (B) be.

compensating diaphragm [sur.] A fitting to a telescope in *stadia work* which alters the interval between *stadia hairs* when a sloping sight is made. In this way the horizontal distance of the staff from the surveyor can be directly calculated from the staff *intercept*.

compensating error, accidental e. [sur.] One of the three kinds of *error* in measurement, the others being *gross* and *systematic errors*. Compensating errors are small and equally likely to be $+$ or $-$ in sign. See **probable error**.

compensation water The water which must be allowed to pass a *dam* so as to satisfy those people who used the water before the dam was built.

competent strata [min.] Rocks that are stable and relatively uniform, without many faults, breaks and other weaknesses. They can be held by *roof bolts*, unlike incompetent strata.

composite action [stru.] Combining two or more materials in such a way that the resulting structure is stronger

than would be obtained by merely adding their strengths. With a *composite floor* it can be achieved by dimples rolled on the steel, by welding *shear* studs to it or by using *profiled steel sheets* (B) rolled with re-entrant angles to grip the concrete.

composite box beam [stru.] A steel box beam acting compositely with a concrete slab. In an open box, the steel box is closed by the concrete slab. In a closed box, the concrete slab is cast on top of the steel of the box.

composite cement See **blended cement, fillerized cement**.

composite column A hollow steel section filled with concrete or a steel section encased in concrete. Either may use *composite action*.

composite construction (see diagram) Different materials used together, such as steel beams in reinforced-concrete floors; precast with in situ concrete; carbon-fibre-reinforced polymers; glassfibre-reinforced concrete (GRC) or polymers (GRP);

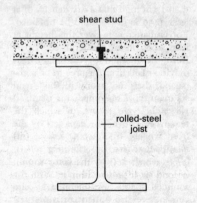

shear stud

rolled-steel joist

Composite construction: rolled-steel joist acting compositely with concrete slab. (D. Nethercot, *Limit States Design of Structural Steelwork*, Van Nostrand Reinhold, 1987.)

metal-web timber joists; *preflexed beams*; *pultrusions*; etc. Composite column design is described in ACI publication 381–83. BSI and the American Institute of Steel Construction also publish texts on concrete–steel composites.

composite floor, c. slab *Profiled steel sections* (*B*) used as *permanent shuttering*. The absolute minimum thickness of steel is 0.75 mm and of the concrete is 90 mm (3½ in.) overall. Where there is no *composite action*, the steel is said to be 'non-participating' (BS 5400, BS 5950).

composite materials *GRC*, *GRP*, etc. See **composite construction**.

composite plate A concrete slab having *composite action* with a steel plate in contact with it underneath.

composites [stru.] New materials such as cement or gypsum or concrete reinforced with glassfibre, *fibre-reinforced concrete* or *plastics*, etc. A new composite for making yacht hulls weighs 30% less than aluminium, costs more, but has higher impact resistance and has been used by the Royal National Lifeboat Institution. It is made of *epoxy resin* reinforced with fibres of kevlar, glass and carbon. 'Advanced composites', like this one, usually have fibre reinforcement surrounded by a protective resin matrix.

compound air lift An *air-lift pump* modified for use where the available depth of immersion is not enough. Two air-lift pipes are used. The first pipe is made only as high as it can pump, and discharges into a second pipe, which acts as the *sump* for the second air-lift pipe within it. With this large submergence the second pipe pumps twice the height of the first.

compound curve [sur.] A *curve* consisting of two or more arcs of different radii curving in the same direction and having a common tangent or transition curve at their point of junction (BS 892).

compound dredger A *bucket-ladder dredger* which is also provided with a clay-cutter like a *suction-cutter dredger*.

compound engine [mech.] A steam or compressed-air engine in which the working fluid expands from the small high-pressure cylinder to the larger low-pressure cylinder. A compound engine uses steam more efficiently than a simple engine but is more complicated. Both types were used for winding, pumping, air compression and other duties about mines. See **compounding**.

compound girder, plated beam A *rolled-steel joist* with plates fixed to the flanges by welding or riveting.

compounding [mech.] The expansion of steam or compressed air in two or more cylinders in series. If the working fluid is compressed air it is often passed through a reheater, but this is not usual with steam. See **multiple-expansion engine**.

compound pipe [hyd.] A pipe consisting of several lengths of different diameter in series.

compound pump [mech.] A pump driven by a steam *compound engine*.

compressed air (1) [mech.] A source of power for drills and motors; like steam it must be raised to a high pressure. This is done in *reciprocating* or turbo-compressors in several stages with *intercoolers* and *aftercoolers*. Compressed-air power, although important in *civil engineering*, is indispensable in mining, where for certain very deep mines it is the only practicable source of power, since it both improves the ventilation and cools the air at the working face. It is also completely safe in gassy mines. Compressed air must be cleaned after compression. The nuclear accident at Three Mile Island began with water in the compressed air. *Hydraulic power* can now do many jobs that could formerly be done only by compressed air.

(2) A compressed-air atmosphere in a tunnel or shaft excludes water from it even under 30 m (100 ft) of water, but this depth requires 3 atm of air pressure to exclude the water. Someone working in this pressure needs 2.5 hours of decompression and a short working shift.

Fine silts or soft clays are best for compressed-air work. In gravels, air losses apart from being expensive may damage the surface. To prevent such troubles, any *shield* should have ample soil cover – at least its own diameter. To avoid these as well as the medical hazards of compressed air (*caisson disease*, *bone necrosis*), *slurry shields* and *earth-pressure-balance shields* have been developed. But at gauge pressures of less than 1 atm, medical troubles with compressed air have been rare. See also **decanting**, **diving bell**, **medical lock**.

(3) In *diamond drilling*, *air-flush drilling* is used as a substitute for water in caving ground which would soften and cave even more with water. The rock cuttings are brought up by the air, and the cutting-bit cooled by it. Although consumption of diamonds may be slightly increased, the costs per metre drilled are not necessarily higher.

(4) See **air-lift pump**, **bubble curtain**.

compressed-air disease *Caisson disease*.

compressed-air energy storage (CAES), underground compressed-air storage (see diagram) Profitable 'lopping the peak' of power demand, used by electrical generating authorities and akin to *pumped storage*. Underground cavities usable for CAES include *aquifers*, *solution cavities* or mined caverns. At Huntorf, (formerly West) Germany, a 290 MW installation has been working since 1978, storing air at 73 atm pressure in solution cavities. At slack generating periods at night and weekends, air is compressed and sent underground. During peak loads the air is released into a gas turbine burning any convenient fuel and driving a generator. Steam and therefore fuel are saved and consequently air pollution is reduced.

compression [mech.] A force which tends to shorten a member; a push; the opposite of a tension.

compression boom, chord A *compression flange*.

compression failure [stru.] See **long column**, **short column**.

compression flange, c. boom, chord [stru.] That part of a beam or girder which is compressed. It is the upper part at the mid-span and the lower part at the support of a continuous beam.

compression testing (1) [s.m.] For clays the compression test is an important laboratory measure of strength. See **unconfined compression test**, **triaxial compression test**.

(2) The crushing of bricks, stone, concrete, etc. to determine their *ultimate compressive strengths*. See **cube test**, **cylinder test**.

compressive strength [stru.] The resistance expressed in force per unit area of a structural material at failure in a compression test, expressed in the USA in psi, and in Britain in megapascals, newton/mm^2, meganewton/m^2, etc.: 1 MPa = 1 N/mm^2 = 1 MN/m^2 = 145 psi.

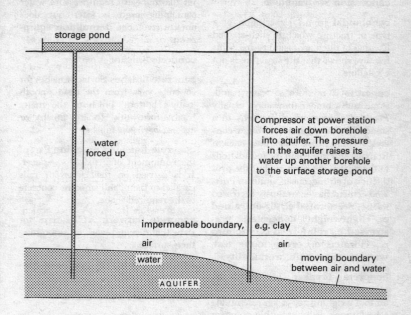

Compressed-air energy storage. The use of an aquifer eliminates the need to excavate a cavern but its depth controls the air pressure. For each 10 m (33 ft) of depth, the pressure increases by 1 atm.

compressor, air c. [mech.] A machine for compressing air to about 7 atm for rock drilling and many other uses on site. Compressors are reciprocating, centrifugal or *free-piston compressors*.

computer-controlled tunnel drilling A method of drilling claimed to save 20% of excavation because it eliminates wasteful *overbreak*, and thus allows 20% faster advance. One operator with computer control can manage three heavy rock drills.

computer graphics Using computers to make drawings, diagrams or even animation ('Mickey Mouse pictures'). For civil engineers it began with *plotting instruments* and developed to *CAD*, *CADD*, etc. using *digitizers*, *graphics packages*, *light pens*, *wire frames* and sometimes *interactive graphics*.

concentrated load, point l. [stru.] A load which is not spread over a large area, the contrary of a *distributed load*. A knife-edge load is a particular concentrated load.

concentric tendon A tendon whose layout coincides with the *centroid* of the concrete section (*ACI*).

concession See **franchise**.

conchoidal [min.] Description of a type of fracture which is shell-shaped like that of pitch, glass and resins. This fracture shows that the structure is not crystalline.

concrete A mixture of water, sand, stone and a binder (nowadays usually *Portland cement*) which hardens to a stone-like mass. Lime and other concretes were used by the ancient Romans and in Britain for foundations in the nineteenth century, but the production of strong, cheap, uniform Portland cement has enormously increased its use. See **aerated c., air-entrained c., lightweight concretes, prestressed c., reinforced c., vacuum c., vibrated c., creep, cube test, water/cement ratio, workability**.

concrete breaker (1) **road b., ripper, jackhammer** A compressed-air tool weighing 20–50 kg, resembling a heavy rock drill but with a point for breaking roads and concrete, not with a rotating rock-drill steel. Much heavier units can be mounted on an *excavator* boom – *hydraulic hammers*.

(2) **concrete pulverizer** A unit that takes the place of the bucket on a hydraulic *backhoe*. With its powerful teeth it breaks beams, slabs, etc.

concrete cancer *Alkali–aggregate reaction*.

concrete cart See **buggy**.

concrete coatings See **carbonation retardant**.

concrete cutting Concrete can be cut by *concrete breakers* or *hydraulic hammers*, both of which break rather than cut. *Diamond saws* are fast and effective but only to 1.2 m depth. *Diamond drills*, however, have no depth limit, however much steel there is. *Thermic boring* is effective but a fire hazard.

Jet blasting with high-pressure water containing sand is safer but does not cut steel. See **demolition equipment**.

concrete finisher A *finisher*.

concrete finishes Surface finishes for concrete vary from the dead-smooth 'baby's bottom' finish to the many 'visible-formwork' (board finish) or *exposed-aggregate* finishes.

concrete-finishing machine Part of the equipment of a *fixed-form paver*. It is carried on *road forms* or rails parallel to them, and smooths concrete to the required shape.

concrete flatwork *ACI*'s term for slabs that demand work on smoothing their surface.

concrete grades See **grades of concrete and steel**.

concrete insert See **insert**.

concrete mixer A machine, usually with a rotating drum, in which aggregates, cement and water are mixed for 2–3 min to make concrete. Mixers are described by stating only the volume of wet concrete produced, either in litres, for those smaller than 1 m^3, or in m^3, for those larger than 1 m^3. See **batch m., continuous m., forced-action m., non-tilting m., tilting m., truck m.**

concrete paver, asphalt p. (1) A *slipform* or a *fixed-form paving train*.
(2) A concrete paving brick (*ACI*).

concrete pile (1) **driven pile** A reinforced-concrete precast *pile* driven into the ground by a *pile driver* or crane.
(2) A reinforced-concrete *pile* cast in a hole bored in the ground (*in situ* or *cast-in-situ* pile). This type is common in central London. See also **driven cast-in-place pile**.

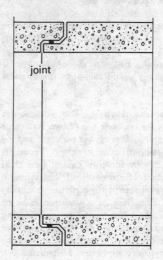

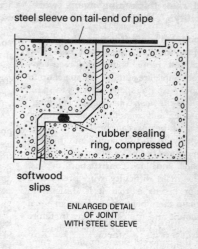

steel sleeve on tail-end of pipe

joint

rubber sealing
ring, compressed

softwood
slips

ENLARGED DETAIL
OF JOINT
WITH STEEL SLEEVE

Concrete pipe for pipe jacking.

concrete pipe (see diagram) Pipe made of concrete, varying in bore from 100 mm to several metres, porous for use as *subsoil drains* or capable of withstanding considerable water pressure. Diameters under 0.4 m usually have no reinforcement. See **tube**.

concrete placer A device for pushing concrete along a pipe by compressed air. It can be a much smaller piece of plant than the *concrete pump*, and it is therefore convenient for tunnelling. See also **placing plant**.

concrete properties Concrete properties are measured under the many parts of BS 1881 which cover *slump*, compaction, *V.-B. consistometer* test for *workability*, *flow-table test*, air content of air-entrained concrete, *cube test*, *beam test* and other strength tests including tensile tests, saturated and dry densities, dynamic or static *modulus of elasticity*, length changes on wetting and drying, water absorption, also the analysis of fresh or hardened concrete for cement content and type, aggregate content and grading, and original water content.

concrete pump Typical pumping rates are 50 m³/h to 500 m away or to 50 m height. Very large pumps can move 150 m³/h. They may for a large job be permanently installed, but are usually truck-mounted *mobile concrete pumps* with long placing booms. Pumpability can be achieved without *admixtures* but with careful grading of the aggregates, possibly with additions of *fly-ash* replacing part of the cement. See also **commit-**

concrete roofs

tal rate, **thickening agent**, **workability**.

concrete roofs Concrete can be used for building many sorts of roof. The simplest roofs, solid reinforced-concrete slabs, are suitable for spans up to about 6 m. For spans from 4.5 to 10 m, *hollow-tile floor* construction is lighter, cheaper and sometimes also thinner. Precast beams can be laid dry very rapidly to form a roof, and precast, *prestressed* beams can be used for large spans and low construction depths. The main difficulty with prestressed floor beams is that it is difficult to obtain them bent upwards by an equal amount and the ceiling may therefore be uneven. For large spans of about 30 m the concrete dome or *shell* is completely smooth underneath. The flat *Diagrid floor* is popular now though expensive, and spans 15 m without difficulty. Prestressed or *reinforced-concrete* girders often cover large spans.

concrete spreader, asphalts. A road-building machine which may be part of a *slipform* or *fixed-form paving train*. It uniformly spreads concrete or asphalt poured into it or dumped in front of it. Apart from the roller, the mixer and the trucks, it is the main unit needed for *rolled concrete*.

concrete technology Designing, testing and analysing concretes and their components, the aggregates, *cements*, *pozzolans*, *admixtures*, *release agents*, etc. A knowledge of *concrete mixers*, *formwork* and the production equipment for the components is essential, as well as of concrete defects and the chemical reactions between cements, water and aggregates. Most concrete technologists are civil engineers or chemists, not necessarily fully qualified. Leading concrete technologists are members of the *Institution of Concrete Technology*.

concrete tensile strength The *tensile strength* of concrete is usually between 20% and 5% of its compressive strength. For strong concretes it is nearer 5% and for weak ones nearer 20%.

concrete testing The commonest tests for concrete are of two types: for the *workability* of wet concrete and for the strength of set concrete. Testing is used both for mix design (see **trial mix**) and for site control. The strength of a concrete in any structure is usually less than that of a standard cube or cylinder taken from the mix during casting and cured for 28 days, but see **Seattle**. *Curing* conditions for the cube or cylinder are better. *Destructive tests* can rarely be used on a structure and when possible they are expensive and slow. *Semi-destructive tests* may help, for example the *internal fracture test* of the *BRE*.

The main *non-destructive testing* (*NDT*) methods for investigating concrete strength are the *rebound hammer* and pulse velocity determination by *ultrasonic testing*. For other properties of concrete, many NDT methods exist. Magnetic devices (cover meters) can find *reinforcement* or show if it has not enough cover. *Gamma radiography* discovers steel bars as well as voids or their opposite – dense, usually strong concrete. Simple electrical probes can measure the content of chloride or moisture, or whether steel is corroding, or the thickness of a concrete slab on the ground. Acoustic methods find cracks.

The cement content of hard concrete can be found by the *RAM test*. Chemical testing also can find the chloride and aggregate contents and the presence of some *admixtures*, See **core test**.

concrete-vibrating machine A machine which travels like the *concrete-finishing machine* or *spreader* and vibrates up to 30 m³ of concrete per hour. See **vibrated concrete**, **vibrator**, **slip-form paver**.

concreting boom A light metal *truss* supported at one end on a frame near a *concrete mixer* and at the other end on another frame on wheels. On its underside is a rail along which a concreting bucket can travel, carried by a pair of overhead wheels. A 200 litre capacity bucket can be carried on a 12 m long boom weighing 180 kg. A *placing boom* is much more elaborate.

condensate Water which condenses from air, flue gas, etc. cooled below the *dewpoint*.

condensed silica fume (CSF) See **microsilica**.

condition monitoring Checking the condition of a machine by such techniques as vibration analysis and *oil debris analysis*.

conditions of contract, general c. of c. Conditions of contract are published by many bodies, including the *Joint Contracts Tribunal*, *FIDIC*, the *ICE* and the Department of the Environment. They describe in detail how *contractors* should view the *contract* and do the work. In addition to the conditions themselves, books exist to explain them, such as the ICE's *Guide to the Fifth* of 120 pages, A4 size. The first edition of the 'ICE conditions' was published in 1945. Most contractors are familiar with the ICE conditions and they are drawn up and periodically revised by agreement between *consulting engineers* and *contractors* (ACE, FCEC).

conditions of engagement The agreement between a *consulting engineer* and his or her *client*.

conductor pipe The outermost casing section of an offshore oil or gas well. It extends from a short distance below the seabed, up to the deck of the *offshore platform*. Its diameter is from 700 to 910 mm (2.3–3 ft).

conductor rail [rly] A rail carried on electrical insulators and parallel to running rails. It provides the power for an electric train through the locomotive's metal shoes pressing against it.

conduit [hyd.] Any *open channel*, pipe, etc. for flowing fluid. See also (*B*).

cone of depression The cone shape of the *water table* around a well being pumped. A similar cone or crater is formed around a structure built on clay or other compressible soil as the soil slowly squeezes under the load of the structure. See **subsidence**.

cone penetration test, deep p.t. [s.m.] The testing of soils by pressing a standard cone into the ground under a known load and measuring the penetration. These methods are used in Scandinavia and the Netherlands. In the Dutch deep-sounding test, which is used in the Netherlands to deeper than 30 m, an inner mandrel is driven separately from the outer casing. This enables the toe resistance to be measured separately from the skin friction of the casing. This method is used for forecasting the resistance to driving of *bearing piles*, and can supply rapidly and cheaply the information for a preliminary site exploration. The cone diameter is 36 mm, so the force required to push it in is not very great. The method is not suitable for clays, for which the *vane test* is preferred, nor for stony soils. Cf. **dynamic penetration test**.

confined aquifer See **artesian well**, **confined water**.

confined compression test See **triaxial compression test**.

confined concrete Concrete containing close transverse *links* to restrain it in directions perpendicular to the applied stresses (*ACI*). See **anchorage zone**.

confined water [hyd.] *Groundwater* that is overlain by impermeable ground. It can be under a pressure higher than atmospheric, like a flowing *artesian well*. When a pipe is driven into it, the water can then rise above the bottom of the impermeable bed that confines it. Certain types of confined groundwater are not annually renewable. See also **unconfined water**.

congestion See **infrastructure**.

conglomerate, pudding stone [min.] A cemented rock containing rounded stones.

conservative pollutants Persistent *pollutants*, such as dissolved salt, nitrates or detergents, that do not decay.

consistence Of concrete, its ease of flow or *workability*, measured by the old *slump test* or the newer *compacting factor test*.

consistency index [s.m.] A figure for comparing the stiffness of *clays* in their natural state. It is calculated as follows:

$$\frac{(liquid\ limit) - (\text{water content of sample}) \times 100\%}{(\text{liquid limit}) - (plastic\ limit)}$$

It may rise above 100% but such values indicate a stiff clay.

consistency limits, Atterberg l. [s.m.] The *liquid limit*, *plastic limit*, *shrinkage limit*, and sometimes also the sticky limit of a clay. These are all water contents of a clay, each in a certain condition defined in Britain by BS 1377. They are the standard way of describing clays and correspond to *mechanical analysis* for sands.

consistometer See **V. -B. consistometer**.

consolidated quick test [s.m.] A test of the shear strength of a *cohesive soil* made in the laboratory after full consolidation under load. The *triaxial* or *shear test* is carried out quickly without drainage or further consolidation. See also **drained shear t., quick t., unconfined compression t.**

consolidation [s.m.] The gradual, slow compression of a *cohesive soil* due to weight acting on it, which occurs as water or water and air are driven out of the voids in the soil. Consolidation occurs only with *clays* or other soils of low *permeability*. It is not the same as *compaction*, which is a mechanical, immediate process and occurs only in soils with at least some sand. His theory of consolidation of clays under increased pressure was first published about 1925 by Terzaghi. With this paper Terzaghi founded the new science of 'Erdbaumechanik' which became *soil mechanics* in English. See also **consolidation settlement, dynamic consolidation, effective/pore-water pressure**.

consolidation grouting, compaction g., claquage g. Unlike *permeation grouting*, this grouting method uses high grouting pressures and changes the soil structure by forcing tongues of grout into the soil. The appreciable surface movement may be troublesome, but the soil may be strengthened.

consolidation press, consolidometer, oedometer [s.m.] A laboratory apparatus for obtaining the data necessary for plotting the curve of pressure to *voids ratio* of a clay sample. In this way the *coefficient of consolidation* of the clay can be determined and sometimes its variation in *permeability* with increasing *consolidation*.

consolidation settlement [s.m.] The settlement of loaded clay which takes place over a period of years, but can sometimes be accelerated by *vertical sand drains*. The Leaning Tower of Pisa is an example of unequal consolidation settlement. See above.

consolidometer A *consolidation press*.

consortium A *joint venture*, often of *consultants*, *contractors* and financiers.

constant-velocity grit channel [sewage] An advanced type of *detritus tank*.

constructional engineer A fabricator or contractor working on steel frames. Cf. **structural engineer**.

constructional fitter and erector See **steel erector**.

construction joint A surface in *reinforced concrete* along which concreting stops one day to be resumed later. *Starter bars* are left projecting so that *reinforcement* can be lapped when the next concrete is cast. *Laitance* on the cast face may have to be removed by *scabbling*, *bush hammering* or wire brushing. Cf. **dry joint**; see **stunt end**.

construction load In addition to any fluid concrete, stacked materials, etc., BS 5850 requires an allowance of 1.5 kN/m^2 (31 lb/ft^2).

construction spanner A *podger*.

construction way [rly] Temporary track for building the *permanent way*.

consultant A registered architect, *chartered engineer* or other specialist who acts for a *client*, and whose functions often go much further than consultation. The consultant, and his or her staff, provide the complete design and supervision of the construction until completion.

consulting engineer A *chartered engineer* who is approached by an architect or client or another engineer for the purpose of designing a dam, railway, sewage-treatment plant, building, etc. The engineer advises the client on the choice of project. Once the project is agreed on, the engineer ceases to advise and begins to draw out a scheme, expands it in detail after the client's approval, and supervises to completion.

consumable electrode In *arc welding*, a *filler rod*.

consumptive use [hyd.] The quantity of water lost by transpiration and evaporation from fields.

contact aerator A tank in which sewage is aerated by compressed-air injection. See **activated sludge**.

contact bed [sewage] A forerunner to the *trickling filter*. It worked intermittently.

contact ceiling A ceiling formed by the underside of a floor slab (ACI).

contactor [elec.] An electrically operated switch, used for controlling the motors of coalcutters or other powerful machines. The main switch is opened or closed by the operation of the *pilot circuit*.

contact pressure under foundations [s.m.] Although foundation slabs are usually calculated on the assumption that their load is uniformly spread over their area of contact, this is not so in reality. In footings on sand, the actual contact pressure decreases from centre to rim, while on clay the contact pressure at the centre is less than at the rim. This means that on clay the real bending moments are greater than those calculated, while on sand they are lower.

contact print [d.o.] A print on light-sensitive paper, made by placing a drawing in opaque ink or pencil on transparent paper in contact with the light-sensitive paper and exposing it to light for a period. See **blueprint**, **dyeline**.

containment material *Pollutants* from *landfill*, such as poisonous *leachate* (water) or *landfill gas*, can often be contained by sheets of *polythene* film welded to each other. Clay can be gastight; even though it demands much greater thicknesses and more labour than plastic film, it has been used in *nuclear decommissioning*.

contaminant *Pollutant*.

contiguous bored piles *Bored piles* in contact with each other. They will support the side of an excavation but may be less waterproof than *diaphragm walls* and *secant walls*.

continuity, fixity [stru.] The joining of floors to beams, of beams to other beams and columns so effectively that they bend together under load and so strengthen each other. This is easily done in concrete or welded metal, less so in other materials. See **continuous beam**; also (*B*).

continuous beam [stru.] A beam of several spans in the same straight line joined together so effectively that a known load on one span will produce an effect on the others which can be calculated. A continuous beam generally has at least three supports. This sort of *continuity* is economical and safe where the supports are unlikely to settle. *See* **end span**.

continuous filter [sewage] A *trickling filter*.

continuous-flight auger (CFA) A powerful lorry-mounted *auger* for

making *bored piles* of from 300 to 750 mm (12–30 in.) dia. in water-bearing sands, gravels, stiff clays and soft chalk. The auger, welded to a central tube, is first screwed into the ground without removing soil. Concrete is then pumped into the central tube to come out at the bottom while the auger is slowly removed with the excavated soil. The reinforcing cage is inserted as soon as the auger is extracted; a vibrator attached to the cage helps to force it down. The hole is never left open to collapse but the high speed claimed is rarely achieved. It is probably the most silent method of building a pile. The depth of pile is limited by the length of the auger, but 18 m (59 ft) long augers exist.

continuous grading See **gap-graded aggregate**.

continuously welded rail [rly] A main line railway in which the rails are joined by welding rather than by *fishplates*. Lengths of 91.5 m (300 ft) may be welded in the shop and taken to the site on special trains, where the rails are joined by *thermit* welding in position and then tensioned (*destressing*). Welded track increases rail life, reduces track maintenance, especially rail breakages, and improves running speeds and passengers' comfort, but is expensive and needs excellent organization. By 1987 65% of British Rail track had been continuously welded and another 400 km (250 miles) were being welded each year.

continuous mixer A very large *concrete mixer* from which concrete flows in a continuous stream.

continuous rating [elec.] See **rating** (3).

continuous ropeway An *aerial ropeway* in which the loaded *carriers* travel one way and the empty carriers return on the ropes the other side of the ropeway towers.

contour See **contour line**.

contour check [hyd.] Compartments of a field made by borders following the contours, a form of terracing. See **contour ploughing**.

contour gradient [sur.] A line set out on the ground at a certain constant slope. (A contour line is set out at zero slope.)

contour interval, vertical interval [sur.] Contour lines are drawn at a vertical distance apart, which is called the vertical interval. For British Ordnance Survey maps to a scale of 1:50,000, formerly 1 in. to 1 mile, it is 15 m, formerly 50 ft. Contours are drawn at Ordnance Datum and every 15 m or 50 ft above.

contour line, contour [sur.] A line on a *map* drawn between all points at the same level, as for instance the high-water line.

contour ploughing, terracing Soil conservation on slopes by ploughing horizontally or with furrows at very gentle slopes along which water can scour only very slightly.

contract A legally enforceable agreement between two people that one of them, the *contractor*, shall do some stated work for payment. The employer (*client*) aims for a contract based on a *bill of quantities*, or, if that is not possible because of shortage of time, for one based on a *schedule of rates*. The worst alternative for the client is a *lump-sum contract*. See also **cost-reimbursement contract, measurement contract**.

contract documents The contract documents form the legal *contract*. In a *measurement contract* they include the drawings made by the *consultants*, the *specification*, the *bill of quantities* or *schedule of rates*, the *conditions of contract* and a legal deed making these binding on the *contractor* and the *client*. Other contract documents include the instructions to tenderers, the form of *tender* and its appendix, and the *daywork* schedule.

contracted weir [hyd.] A *measuring weir* which is shorter than the width of the channel, and is therefore said to have side or *end contractions*. Cf. **suppressed weir**.

contract hire See **contractors' plant**.

contraction in area, necking [mech.] In the *tensile testing* of metals, the reduction from the original area of the bar to the cross-section at the point of fracture. For ductile metal like rivet steel the contraction may be more than 50%, and is accompanied by *elongation*.

contraction joint, shrinkage j. In concrete work, a break in a structure made to allow for the drying and temperature shrinkages (of concrete or masonry) and thus to prevent cracks forming at undesirable places. Since all materials containing cement shrink appreciably on drying, contraction joints are needed in every long structure. See **dummy joint, movement joint**.

contractor A person who signs a *contract* to do certain work for payment, usually within a specified time. In France, possibly also in Germany, the contractor has a much higher social status than in English-speaking countries because of the effect of Napoleon's *Code Civil* in 1804, which placed the full liability for the stability of a structure on the builder, not on any *consultant*. See also **responsibility for construction**.

contractor's agent See **agent**.

contractors' plant Equipment used for building or earthmoving, from the humble scaffold board or wheelbarrow to *earthmoving* equipment of several thousand horsepower. Plant is now increasingly driven by *hydraulic* power rather than by compressed air. There is thus often no need for a compressor; a power pack is available on the nearest *excavator*, *loader* or Land-Rover.

Leasing or hiring may be more profitable than buying plant, especially if it is needed for only a short time. But these are not the only options; the seller may agree to 'extended terms' (hire purchase) but will not then be willing to reduce the price. Your bank may allow you an overdraft, especially if the bank manager can see that the equipment will increase your profits. Rather than an expensive overdraft you may be able to negotiate a special bank loan with the manager. But specialist companies which finance leasing are more interested than banks in lending money to buy equipment. Leasing is hire over a period fixed under a contract, but several types of lease exist. A 'finance lease' provides nothing more than the equipment. Under an 'operating lease' or 'contract hire', the payment includes some maintenance and repairs and is fully deductible from profits for tax purposes.

Contractors' plant includes *access platforms*, *backhoe-loaders*, *breakers*, air *compressors*, circular saws for wood, metal or concrete, *concrete mixers*, *concrete finishing machines* and *placing plant*, *vibrators*, *dredgers*, *dumpers*, *dump trucks*, *excavators*, *forklift trucks*, free-standing-tower hoists, goods hoists, *generator sets*, grinders for concrete floors or for metal, *hauling plant*, heaters, *hydraulic hammers*, paint or metal spray equipment, pumps, pipe benders, pipe threaders, roads, railways, sanding machines, *shields*, *tunnelling machines*, woodworking plant, etc.

contract sum, c. total The total amount payable to a *contractor*, usually based on the measurement of the completed work. Cf. **tender total**.

contraflexure [stru.] Contrary flexure, a change of direction of bending. A point of contraflexure is called in the USA a point of inflexion. It is a point at which the *bending moment* is zero, changing from hogging on one side of the point to sagging on the other side.

control [hyd.] A part of a channel where bed and bank conditions make the water level a good indication of the flow, for example a weir or a waterfall or a hard bed to the stream. See **Venturi flume**.

controlled rolling Control of temperatures and deformations during the *hot rolling* of steel, to make it well and cheaply.

controlled tipping See **landfill**.

control network [sur.] A network of accurately surveyed points obtained by *traversing*, *triangulation*, *trilateration*, *satellite-Doppler* or similar methods. From these accurate points, surveys of lower accuracy, such as the details of the countryside, can be plotted by *plane-tabling*, *stadia work* or other faster but less accurate surveys.

control point [sur.] A point on the ground of accurately known position (and usually altitude) which is a starting-point or check on a *plane table* survey, *traverse* or *photogrammetry*. See **vertical**, **horizontal control**.

control valve [mech.] A *discharge valve*.

conversion factor [d.o.] See p. xiv.

converter, signal processor An electronic circuit for changing *analogue* to digital data or the reverse. In *transducers* it modifies an electrical signal received from a *sensor* to make it usable by a *microcomputer*, display, etc. It often implies amplification.

conveyor [mech.] Equipment for moving sand, stone, ore, coal, etc. continuously over relatively short distances. See **belt/helical/pneumatic conveyor, elevator**.

conveyor loop, loop take-up [min.] In an advancing tunnel, the belt conveyor which removes the excavated rock must be extended every shift. This is conveniently done by inserting a belt loop in a gantry carrying the belt high enough above ground to accommodate the loop. It may be at either end of the belt. In 1988 the Neuchâtel tunnel, the first tunnel to use a *curved conveyor*, had a gantry carrying 140 m of belt, enough for a week's advance of the tunnel. A new length of 140 m was added every weekend, which took eight hours including vulcanizing the new belt to the old. An automatic hydraulic loop take-up can maintain the belt tension.

Conwy tunnel An *immersed tube* 700 m long on the North Wales coast road, designed by Travers Morgan & Partners. Site work began early in 1987 by Costain-Tarmac. It is due for completion in mid-1991 and is intended to last 125 years. The rectangular tube, 24 m (79 ft) wide, will meet conventional tunnels at each end, making a total tunnel length of 1090 m (1200 yd) with *dual carriageways*. The six precast tubes each weighing 35,000 tons are set in a *suction-dredged* trench from which 6 million m³ of mud must be taken. The bottom and sides of the tube are lined with 6 mm thick steel plate as permanent shuttering, with *shear* connectors welded on. The steel is cathodically protected. The carriageways are 10.45 m (34 ft) wide by 5.1 m (16½ ft) high.

coordinates [d.o.] Coordinates are distances measured in a certain way from fixed straight lines called axes of reference which intersect at the *origin*. The purpose of coordinates is to locate a point. The system invented by Descartes, using *Cartesian* coordinates, is the most commonly used and convenient system. See **graph**.

copper-bearing steel *Weathering steel.*

copper brazing *Brazing* of steel or other metals of high melting-point with copper as the *filler* metal.

copper welding *Fusion welding* of copper (not *bronze welding*).

cordage rope *Fibre rope.*

Cordtex, Cordeau [min.] *Detonating fuse.*

corduroy road A road built of 7 cm dia. saplings, equal in length to the width of the road, wired tightly together at the ends. This is an excellent temporary road, very quickly laid, which will float in liquid mud, but it is expensive in timber. It can, however, be picked up and reused elsewhere.

core (1) A *cut-off wall* of clay or concrete or other material.
  (2) A piece, often of expanded polystyrene, inserted into a mould for concrete before pouring the concrete, so as to form a hollow in it for a bolt, *prestressing* cable, etc. Cores for prestressing cables are laid in the cable path, wired to the reinforcement. They may be tubes which are inflated before the concrete is poured and a few hours later deflated and withdrawn. Another type, also of rubber, contains a steel cable withdrawn after pouring the concrete. In both types the rubber core when pulled reduces in diameter and detaches itself from the concrete.
  (3) **kern** [stru.] The middle part of a wall or column, the area through which

core barrel

the resultant compression on the whole section must pass if there is to be no tension anywhere in the section. For a wall, the core is the *middle third*.

(4) [stru.] In a spirally reinforced *column*, the concrete within the centre line of the spiral reinforcement.

(5) The cylinder of rock or soil or concrete cut out by a *diamond drill* or *soil sampler*, etc. See **test core**.

(6) [mech.] The soft but tough steel within a *carburized* steel case.

(7) [elec.] A conductor within a cable including its insulation.

core barrel [min.] Next to the cutting bit of a *core drill* is a length of pipe known as the core barrel which contains the core. In very soft rock it may be double so that the inner lining does not rotate or damage the core. See **Sprague and Henwood core barrel**, **wire-line core barrel**.

core box [min.] A large wooden box, divided up into narrow strips by partitions. It contains the *cores* from a borehole in the order in which they were extracted from the ground.

core catcher [s.m.] A steel spring used for keeping samples of sand from dropping out of a *soil sampler*.

core cutter, core lifter (1) [min.] An attachment at the foot of the *core barrel* which grips the core and breaks it at the root when the core is withdrawn for examination.

(2) [s.m.] A *soil sampler*.

(3) A rotary drill for cutting a cylinder from a road for test purposes.

cored hole A hole cast by leaving a *core* in a concrete (or metal or plastics) piece. When the core is removed shortly after the concrete is poured, it leaves a neat hole. The contrary of a drilled hole.

core drill [min.] A power-driven tool for extracting a continuous core of rock

for inspection, usually a *diamond drill* or shot drill. Water is pumped down the hollow drilling rods and returns outside them with the rock cuttings to the surface.

cored slab A *voided slab*.

core lifter A *core cutter*.

corers, coring tools [min.] In *rotary drilling* for oil wells, cores are not usually made, since speed of drilling is the aim and the debris of drilling is brought up by the drilling fluid. Corers are the special tools used for extracting cores when an oil horizon or other important bed is being approached.

core test A *crushing test* on a concrete cylinder cut from a structure by a *diamond drill*. See **cap** (4), **test core**.

core wall (1) A *cut-off wall*.

(2) **shear wall** Interconnected reinforced-concrete walls running the full height of a block of flats or offices, and centrally located in the block, usually surrounding the plumbing services, particularly the hot and cold water and the drainage, also the stairs and the lift shafts. The core walls stiffen a tall building against earthquake or wind forces and can be designed to be built very rapidly by *slipforms*.

A slipformed rectangular core of a concrete-framed tower block can be built at the rate of two storeys per day and provides most useful support against wind as well as locations for pipes, cables and lift at an early stage of a contract.

In the USA, in 1987, a *lift-slab* block collapsed during construction, killing many people, probably because the core walls stiffening it were in arrears. The five-storey-high slender steel columns were topped with floor slabs for upper floors when they collapsed like a house of cards.

coring tools *Corers*.

corporate member of an engineering institution A *chartered engineer*.

correction [sur.] See **adjustment**, **tape corrections**.

corrosion The gradual removal or weakening of metal from its surface by chemical attack. It can be of two types but the high-temperature types that occur in fires will not be discussed here. The low-temperature type, very much more widespread, requires the presence of water and oxygen and is helped by sulphur dioxide and carbon dioxide, and probably by other materials in small quantities in the water or air. It is always electrolytic, an oxidation, rusting for iron or steel. In Britain the loss of thickness of steel sheet piles submerged in fresh water on both faces is about 0.1 mm yearly. In salt water, the loss is about 50% more. But below groundwater level, submerged in the calm conditions of undisturbed soil, steel will suffer so little corrosion that no protection is needed. Above groundwater level or in disturbed soil (thus in the presence of both air and water) some protection is needed, whether metal, paint, concrete coating or *cathodic protection*. Copper, lead, zinc and aluminium when not in contact with other metals form a thin film of oxide, the *patina* (*B*) which protects them from further oxidation, but if they – especially zinc or aluminium – are in contact with iron through an electrolyte (which may be merely an invisible film of dirty water on the surface of the metal) they will dissolve in sacrificial protection of the iron. They are therefore used in *metal coating* (*B*) to protect iron and steel. See **corrosion rate**, **pH scale**.

corrosion fatigue [mech.] A weakening of steel by small *fatigue* cracks which are entered and corroded by water during reversals of stress. Even

relatively pure water like tap water lowers the *endurance limit* considerably.

corrosion-inhibiting admixture An *admixture* claimed to reduce the corrosion of metal in concrete, e.g. sodium benzoate, chromate or nitrite.

corrosion rate The thickness of metal lost in a year. Mild or galvanized steel should not be used in acid water with a *pH* below 6.5 (7 is neutral). At pH 6, steel loses at least 0.5 mm thickness yearly, but see **corrosion**.

corrugated steel pipe Pipe made from steel sheet; not described in any BS; it is widely used for building *culverts*, usually galvanized. These pipes exist in diameters from 150 mm (6 in.) to 2.6 m ($8\frac{1}{2}$ ft). Some are helically corrugated with locked helical seams. Others have annular corrugations.

corundum [min.] Al_2O_3, alumina, a very hard mineral used as an abrasive, since its *hardness* is less only than that of *diamond*.

costing The frequent (usually monthly) calculation of the expense needed on a contract to produce a m^2 of roofing or m^3 of excavation or other *item*, to help the *contractor* in *tendering* and in keeping site costs down. Items in the *bill of quantities* often do not correspond exactly with those costed.

cost-plus-fixed-fee contract, prime-cost c. A *cost-reimbursement contract* in which the contractor's *overheads* and profit are covered by a fixed fee.

cost-plus-percentage contract A *cost-reimbursement contract* in which the contractor is paid a percentage to cover *overheads* and profit.

cost-reimbursement contract Any *contract* based on the costs of labour and materials plus an allowance for *overheads* and profit, including *cost-plus-fixed-fee, cost-plus percentage* or *value-cost contracts*. Such contracts are used only for very small jobs or very urgent work which cannot await completion of the drawings. *Clients* and *consultants* try to avoid them where possible because they provide an incentive for the *contractor* to enlarge the scope of the work and delay completion. *Measurement contracts* are preferable.

Coulomb's equation [s.m.] A simplified statement of the *shear strength* of soils, given by the equation:

$$S = C + P \tan \phi$$

in which S = shear strength, C is the *cohesion*, P is the pressure at right angles to the plane of shear, and ϕ is the angle of shearing resistance of the soil. Coulomb thought this out in 1776 and it is therefore about 100 years older than any other equation used in *soil mechanics* today.

Council of Engineering Institutions (CEI) A federation of the engineering institutions formed in 1965 but superseded by the *Engineering Council* in 1983.

counter-arched revetment A brickwork *revetment* to a *cutting* with arches between *counterforts* like a *multiple-arch dam*.

counter bore [mech.] To enlarge a hole by drilling.

counter bracing, cross bracing [stru.] Two *diagonal braces* provided in each *panel* of a *truss* to stabilize it and carry wind loads.

counter drain A drain running along the foot of a canal or dam bank to remove leakage and strengthen the bank.

counterfort (1) A thickening to a *retaining wall* on the side of the retained

material, therefore not visible. Cf. **buttress**.
(2) A *chevron drain*.

counterpoise bridge A *bascule bridge*.

coupling, coupler A tube threaded inside for its full length, which joins two correspondingly threaded reinforcing bars. For *deformed bars*, wedged splices exist. See also (*B*).

coupon A *test piece* for *destructive testing* or for insertion into a water main for checking the corrosiveness of a water, but in any case cut from the metal or other material being investigated.

course [sur.] The direction and length of a survey line. See also **base course**, **wearing course** and (*B*).

coursed blockwork, c. masonry In *breakwater* construction where precast concrete blocks of 10–50 tons are common, *blockwork* laid like masonry in horizontal, bonded courses. Cf. **sliced blockwork**. See **Titan crane**.

coursed masonry *Coursed blockwork*.

cover The thickness of concrete between any bar and the nearest face of the concrete member. The cover is always at least equal to the bar diameter and usually more, especially in exposed conditions. For all outside concrete, 40 mm is the minimum cover unless the concrete is coated with some protection such as *asphalt* or *carbonation retardant*, or the steel is *epoxy-coated*. On a sea front the minimum is 50 mm and under the sea 75 mm. Contradicting these values are those for *ferrocement* – 2 mm maximum so as to keep the wire mesh well spread out across the section. See also **effective depth**.

covered electrode, coated e., stick e. A metal *electrode* used in *arc welding*, which is coated with *flux* or other material to protect or improve the weld metal or stabilize the arc.

cover meter, electromagnetic c. m. A *non-destructive testing* intrument that can locate steel in concrete up to a distance of about 70 mm from the surface.

cover plate, c. strap A *fishplate*.

crab The moving hoist of an *overhead travelling crane*, running on rails at the top of the gantry. It may be called a crab because it travels at right angles to the crane rails. Cf. **telfer**. See **Titan crane**.

crack inducer A *dummy joint*.

cracking in concrete Cracking is always expected in *reinforced concrete*, since it has such a high *shrinkage* on hardening. Additional cracks will occur on the stretched side of a beam but reinforcement should be inserted sufficient in quantity and closeness to make the cracks invisible to the naked eye and very close together. If a contraction or expansion joint is inserted, this will also reduce cracking near it. Rusting *reinforcement* also causes cracks in concrete. See **autogenous healing**, **prestressed concrete**.

cradle (1) On a slipway the low framework running on rails which is sunk below a ship at high tide to take its weight and pull it up the slipway. It is also used for launching a ship after repair. See **sue load**.

(2) A temporary framework to carry something during construction.

craftsman See **tradesman** (*B*).

crane [mech.] Generally means power-operated *lifting tackle* with a *jib* which can move loads a considerable distance horizontally as well as lift and lower them. See **crab**, **derrick**, **excavator**, **telfer**, **creeper c.**, **floating c.**,

Goliath c., jib c., level-luffing c., mobile c., portal c., revolver c., Titan c., tower c.

crane gantry See **gantry**.

crane post [mech.] The upright mast of a *jib crane*, which at its top end holds the upper end of the jib by a tie rod or ropes and at its lower end is pivoted on the ground with the lower end of the jib.

crane safe working load (see diagrams, pp. 102–3) Cranes have overturned in spite of their efficient *safe-load indicators*. The safe load varies with the radius of the jib, i.e. whether the jib is low (greatest radius) or high (least radius). The safe load at the lowest position of the jib may be only a quarter of that at its highest position. With a longer jib or *fly jib* the difference can be even more. Electronic safe-load indicators adapt themselves to jibs of any length.

crane tower, king t. Of the three towers supporting a raised *derrick* (6), that which carries the mast and the crane machinery. The other two are anchor towers.

crater of depression See **cone of depression**.

crawler track, caterpillars [mech.] An endless chain of plates used instead of wheels by *tractors* which travel over soft ground. The ground pressure is only about 50 kPa (7 psi) instead of three times this amount with tyres, so crawlers can travel and work much better on slippery clay or soft ground than wheeled vehicles. Special machines with tracks 1 m (3 ft) wide have ground pressures much less even than 50 kPa.

creep (1) [stru.] Gradually increasing permanent deformation of a material under stress, well known in metals as

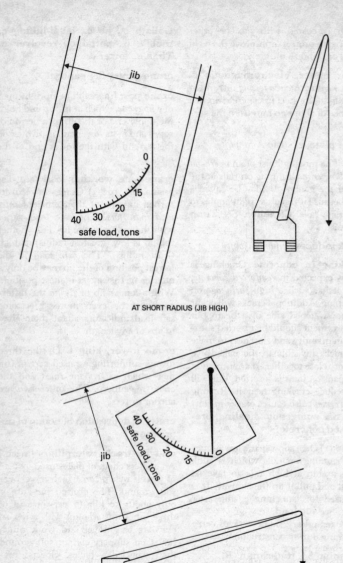

AT SHORT RADIUS (JIB HIGH)

AT LONG RADIUS (JIB LOW)

Safe-load indicator for crane jib. See also diagram opposite.

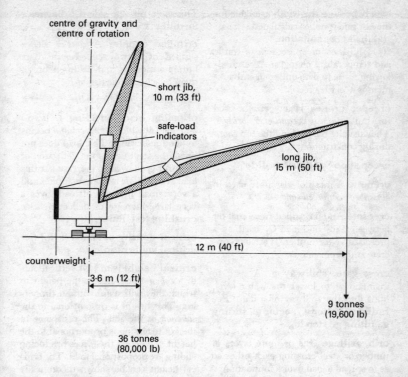

centre of gravity and centre of rotation

short jib, 10 m (33 ft)

safe-load indicators

long jib, 15 m (50 ft)

counterweight

12 m (40 ft)

3·6 m (12 ft)

9 tonnes (19,600 lb)

36 tonnes (80,000 lb)

Reduction of safe working load using long jib at maximum radius (US 40 ton crane; 40 US tons = 80,000 lb = 36,300 kg). See also diagram opposite.

high-temperature creep. In civil engineering, however, creep can be useful to the building owner because it tends to load the whole structure more uniformly. Hard-drawn steel at high stresses extends, whereas the concrete compresses, under creep. The numerical values of creep are important in *prestressed concrete*. Up to 40% of its compressive strength, concrete has a creep usually proportional to the stress on it. As for *shrinkage*, half the creep takes place in the first month and three quarters in the first six months.

(2) The slow movement of a *free retaining wall* which holds up a bank of

shrinkable clay. If the clay shrinks every year and the cracks between the clay and the wall are filled with debris during the drought, the wall may be forced forward every year in the wet season by the swelling of the clay. This may eventually overturn the wall.

(3) When *sheet piles* are driven singly they have a tendency to accumulate a forward leaning movement. This may be counteracted by driving the piles in panels of 12 or so at a time, driving each pile only slightly ahead of its neighbours at any one time. In closing a cofferdam a special *closer* is often needed even if there has been no creep,

merely because the width available for the last pile is not as calculated.

(4) [hyd.] See **saltation**.

(5) [rly] Rails creep because of traffic and temperature changes. An arch allowing cattle to pass under an embankment is a 'cattle creep'.

creeper cranes Heavy *cranes* used for building steel *cantilever bridges*. They usually travel along the top *chord* during construction.

creep slide See **detritus slide**.

crenette A mesh of glassfibre used in *glassfibre-reinforced concrete*.

creosote An oil distilled from coal tar between 240° and 270°C, used as a preservative for timber. It attacks concrete and is poisonous.

crest gate, spillway g. A gate to maintain or to lower the water level, built into the spillway of a dam. See **radial g., roller g., sector g., sliding g., tilting g., stop log**.

crib, grillage One or more layers of timber or steel crossing each other so as to spread a load over a foundation.

crib dam, c. wall A *retaining wall* built of stone-filled *gabions* or precast concrete units or timbers stacked on top of each other, and built to a batter of 1 in 6 to 1 in 8. See **gravity retaining wall**.

cribwork Large timber cells which are sunk full of concrete to make a bridge foundation.

Crimp and Bruges' formula [sewage] A formula connecting rate of flow (*v*), *hydraulic mean depth* (*m*), and the slope of sewer (*i*).

$$v = 56m^{0.67}\sqrt{i}.$$

The units are m/s and metres. It is now believed that the value of *v* obtained from this formula is 20% lower than experiments on new pipes show.

This is on the safe side. Cf. **Barnes's formula**.

crimped coupler A *coupler* sleeve slipped over two *reinforcement* bars butted inside it, and crimped on to them by a hydraulic press.

crimper An *indenting roller*.

crippling load, buckling l. [stru.] The load at which a *long column* begins to bend noticeably. This load does not break the strut unless the deflection of the strut is allowed to increase. At this load *Hooke's law* no longer applies and a very small increase of load gives a very large increase of deflection. See **crushing test, Euler**.

critical density of sands [s.m.] See **critical voids ratio**.

critical height [s.m.] Of vertical cuts in a *cohesive soil*, this is the height to which they will stand without timbering. This height is proportional to the *cohesion* of the soil. The resistance of the cut to sliding is proportional to the height *h*, whereas the forces producing sliding are proportional to h^2. The critical height can be shown theoretically to be equal to twice the strength in *unconfined compressive test* of the soil, divided by its density. Sands have no critical height, since they will stand safely to any height at their angle of repose.

critical hydraulic gradient [s.m.] The *hydraulic gradient* at which a sand becomes a *quicksand*. This can happen to any sand if the upward velocity of water flowing up through it is high enough to make it 'float'. See also **boil, piping**.

critical path scheduling, c. p. method (CPM), project network analysis A valuable development of the progress chart on urgent, complicated contracts that allows site managers to assess rapidly those tasks on which resources should be concentrated. The time that each operation demands is first written down. The successive times are then added together to make the expected time of completion. Operations which are not forced to precede or follow any particular operation, but can take place at any time, are not critical, so can be disregarded in the first draft. If the total time is too long, the operations are scanned to see which could be most effective in shortening the total time.

For instance, if the fabrication of the steel building frame takes longer than the preparation of the site for it by demolition, excavation and casting of foundations, it is the fabrication which is critical. A decision is therefore made to investigate the time needed to erect a reinforced concrete frame, partly precast. It is found to be much quicker than steel and is adopted, so demolition, excavation and foundations may then become critical. These usually all depend closely on the site, which may be impossible to change at this stage. So any further shortening of the time has to be done at later stages of construction. The scheduling is done on computer. See **management by exception**, **master programme**.

critical point, arrest point, critical temperature [mech.] A temperature at which, on heating or cooling a plain *carbon steel*, a change in its molecular structure is shown by a noticeable delay (arrest) in the heating or cooling. A steel warming up will darken and on cooling down will brighten in colour. Magnetized bars do not lose their magnetic properties until the critical point is passed, hence magnetic indicators can show blacksmiths the critcal temperature. Drill steels must be forged at a temperature slightly above the critical, and *tempered* and *quenched* at a slightly lower temperature just above

the critical point with rising temperature. The critical point on heating (decalescent point) is about 750°C, the critical point on cooling (recalescent point) is about 700°C. Both temperatures vary with the carbon content.

critical sewers About 20% of UK sewers are likely to fail. These are the critical ones. Failure implies collapse of the sewer, usually its blockage and sometimes collapse of the road above, with flooding. It is therefore best to prevent sewers becoming critical, and this costs less than any catastrophic failure.

The most expensive 10% of collapses account for 80% of total failure costs. Sewer conditions likely to cause catastrophic failure are: running full or nearly full; large enough for man entry; in bad ground often with a high water table; of brick or stone and near buildings or underground structures. See **pipe renewal**, **surcharged sewer**.

critical velocity [hyd.] (1) In an *open channel*, that velocity at which the *Froude number* is 1. In a pressure pipe, Reynolds's critical velocity is at the change point from laminar to turbulent or vice versa, where friction ceases to be proportional to the first power of the velocity and becomes proportional to a higher power, practically the square. Reynolds's number for this is about 2000 with water.

(2) Kennedy's critical velocity in open channels is that which will neither deposit nor pick up silt.

(3) Belanger's critical velocity is that condition in open channels for which the velocity head equals one half the mean depth (ASCE).

critical voids ratio of sands [s.m.] During shear tests it has been noticed that dense sands expand and loosely packed sands contract. The intermediate state, at which, during the shear test, a sand will neither expand nor contract, is described as its critical *voids ratio*. The importance of this value is that there need be no fear of *flow slides* or sudden *liquefaction* of a sand which has a voids ratio less than the critical. (The *shaking test* proves this expansion and contraction of sands and silts.)

cropper A tool for cutting steel bars (bar cropper).

cross bracing See **counter bracing**.

crosscut chisel (cape chisel USA) [mech.] A *cold chisel* of rectangular section with a cutting edge at a steep angle for hard cutting.

crosscut file, double-cut f. [mech.] A file with two intersecting rows of cuts forming a repeating pattern of cutting points. See **cut** (4).

crossfrogs See **frog**.

cross hair, spider line [sur.] A straight vertical or horizontal line which fixes the line of sight in a surveying telescope. It is set in the *reticule*.

cross hatching [d.o.] *Hatching*.

crosshead (1) [mech.] A steel block which slides between steel guides in a steam engine and helps to convert the to-and-fro motion of the piston into rotary motion.
(2) [min.] A steel frame which slides between the guides in a sinking shaft. It is carried up and down by, but is not attached to, the bail of the *kibble*. It thus prevents the kibble from swaying during hoisting and is demanded by law in some states of the USA.

crossing [rly] At a rail intersection, a steel casting containing U-shaped grooves for the wheel flanges, usually of manganese steel to reduce wear.

crossover [rly] A track which joins two parallel tracks and is therefore S-shaped in plan. See **scissors crossover**.

cross poling In trench excavation, short *poling boards* placed horizontally to cover a gap between *runners*. They are inserted behind the runners next to the earth where the runners cannot be driven.

cross-section (1) [d.o.] The shape of a body cut transversely to its length, sometimes called a transverse section. A cross-section of a pipe is therefore an O. Hence a scale drawing of a body cut across.
(2) A vertical *section* of the ground suitable for calculating earthwork quantities.

cross-sectional area (CSA) [mech.] The area of a cross-section, often referring to the area of steel in mm^2 in a bar, joist, stanchion, etc.

cross staff [sur.] A brass box with slits in opposite faces forming sight lines perpendicular or at $45°$ to each other. See **optical square**.

crowbar, bar, crow A round or hexagonal steel bar 1 to 2 m long usually with a point at one end and a chisel shape or claw at the other end. When shorter than 1 m it becomes a *pinch bar* (*B*).

crowd force In a *backacter* or *backhoe* the greatest force available at the bucket teeth. Cf. **break-out force**.

crown, vertex The highest part of the underside of an arch shape, e.g. of a drain or tunnel. Cf. **invert** and **soffit** (*B*).

crushed aggregate Although angular and therefore less workable than rounded aggregate, crushed aggregate can give higher concrete strengths because of its higher surface roughness and better bond.

crusher (1) [min.] See **breaker**.

(2) *Excavators* or even relatively small *loaders* can now be provided with crushing buckets or special crushers with a crushing force up to 60 tons, useful where a *hydraulic hammer* would be too noisy.

crushing strength The load at which a material fails in compression, divided by its cross-sectional area, properly called its crushing *stress*. This figure is used for comparing the strengths of concretes, bricks, stones, mortars and similar walling materials.

crushing test Any test in which a material is made to fail as a *short column*, such as the *cube test* or *dump test*. In fact with strong concretes 'crushing tests' usually do not end with the crushing of the cube, since they merely reach the concrete target strength (about $1.3 \times$ design stress) to avoid the mess of the crushed cube.

crush plate, wrecking strip A replaceable piece of wood that protects the main *formwork* during *striking*, or an easily replaceable panel that helps in the striking of the main formwork (*ACI*).

cryogenics The study of cold, including **permafrost**.

crystalline fracture [mech.] All metal fractures are crystalline but this term is reserved for *cleavage fracture*.

CSF (condensed silica fume) See **microsilica**.

C/S ratio The ratio of the amount of calcium oxide (CaO) to the amount of silica (SiO_2), usually referring to *binder* cured in an *autoclave*. See **cement compounds**.

cubature A term used by builders of sea walls. It is the volume of water that enters through a gap in the wall on each tide, twice daily.

cube mould A milled steel mould for making accurately sized concrete cubes for the *cube* test.

cube strength The strength of a concrete cube when crushed. See below.

cube test (1) A test of the strength of a *Portland cement* made by testing to destruction, in standard conditions, a cube of mortar made with this cement and standard sand.

(2) In *concrete testing*, cubes are cast in a 150 mm (6 in.) cubical mould and subjected to a *crushing test* at 7 or 28 days. See **cylinder test**, **wet cube strength**.

culmination, transit [sur.] The moment that a star or the sun crosses the *meridian*, used by surveyors for determining (i) the geographical meridian and (ii) the latitude and longitude. These are calculated by suitable combinations of observations of angle and of time at or near transit using astronomical tables. Each star crosses the meridian twice in 24 hours, at its upper and lower culminations. Both are visible for *circumpolar* stars; for other stars only the upper culmination is seen.

culvert (1) A covered channel up to about 4 m width or a large pipe for carrying a watercourse below ground level, usually under a road or railway. It is usually laid by *cut and cover*.

(2) A tunnel through which water is pumped from, or flows into, a *dry dock*.

cumec [hyd.] 1 m^3/s, analogous with *cusec*.

cumulative errors [sur.] *Systematic errors*.

cup-and-cone fracture [mech.] A typical *plastic fracture* of a *ductile* material in tension. One side of the break is cup-shaped and the other cone-shaped, fitting into it. This is the normal *plastic fracture* of mild steel, accompanied by *necking*.

cuphead [mech.] The shape of the head of a rivet or bolt which is rounded like the inside of a shallow cup or deep saucer.

cupola [mech.] A vertical cylindrical furnace in which pig iron is melted to make iron castings.

curb See **kerb**.

curing, maturing Keeping concrete or mortar damp for the first week or month of its life so that the cement is always provided with enough water to harden, either by immersing in water, spraying on a *curing compound* or covering with polythene sheet. This greatly reduces *shrinkage*, improves the final strength of concrete, particularly at the surface, and should reduce surface cracking or *dusting*. The curing time is shorter for *rapid hardening* than for ordinary cement, and is longer for low-heat cement. Similarly it is shorter when accelerators are used and longer with retarders. See **accelerated curing**.

curing compound Synthetic resins, tarry liquids, etc. sprayed over freshly placed concrete to prevent it drying. When the compound has dried it becomes a *curing membrane*.

curing membrane Polythene sheet, building paper or other sheet material laid over fresh concrete to prevent it drying out. See above.

curing period The amount of time for any given weather during which concrete members must be kept damp after casting. In Britain, usually a week is required.

curling of a screed Lifting of a *topping* at its edges or corners. Usually caused by *shrinkage*, it may be visible on a polished floor surface as undulations, and is most likely on an unbonded topping, e.g. a *floating screed* (*B*). It can be overcome at the design stage, by using a thick topping of at least 100 mm (4 in.) containing small aggregate to make a concrete instead of a mortar. See **screed tester**.

current meter, rotary m. [hyd.] An instrument with a vane like a windmill which rotates when a fluid passes through it. The distance travelled by the flowing water is recorded on a revolution counter geared to the vane. The observer works out the velocity from the counter reading and the time of submergence. This instrument is used for measuring the current in wide rivers. The velocity is measured at a number of points and depths systematically spaced across it so that the mean velocity can be obtained. The mean velocity is the velocity which multiplied by the cross-sectional area of the water gives the quantity flowing. An anemometer for measuring air flow works on the same principle. See also **float**, **flow meter**, **Pitot tube**, **rating**, **Venturi meter**.

curtain See **grout curtain**.

curtain wall See (*B*).

curve (1) A bend in a road or railway. Railway curves are set out usually to circular arcs; they may be *horizontal curves* or *vertical curves*, the usual sense being a horizontal curve. See **compound c.**, **reverse c.**, **transition c.**

(2) [d.o.] A transparent plastic or pearwood shape used by a *draughter* for drawing curves.

curved conveyor Until 1963 all *belt conveyors* had been built straight in plan, though often with appreciable vertical curvature. In that year a French company, Rei, built its first long conveyor (800 m (2600 ft)) to a horizontal curve of radius 700 m. In 1980 a Rei Curvoduc steel-cored belt conveyor was erected in New Caledonia to carry nickel ore 11 km (7 miles) to a discharge point 557 m below the loading-point. Because of the drop in level, the power required at the drive motor was only 50 hp in spite of four curves of radius 1080 m (3500 ft) and eight changes of gradient. See also **conveyor loop**.

curve ranging [sur.] Setting out points on a curve.

cusec [hyd.] ($1ft^3/s$,), a unit of flow of water. 35.3 cusecs. = 1 cumec.

cushion head A very descriptive US term for a *pile helmet*.

cut (1) See **cutting**.

(2) **lock cut** [hyd.] A short canal beside a river which enables boats to by-pass a *weir* and go through a lock.

(3) [min.] Any pattern for drilling a round of shot holes in tunnelling or shaft sinking, such as the *bottom cut*, *pyramid cut*, *wedge cut*. See **cut holes**.

(4) [mech.] The shape of the cutting teeth of a file, whether made by one row of parallel incisions or by two rows of incisions which intersect at about 45°. The first is single cut or float cut, the second a double cut or cross cut. Cut also refers to the coarseness or fineness of the teeth. Short files are more finely cut than long ones, as the following table shows. The table gives the average number of teeth per linear centimetre (except for saw files, which are much finer).

Description of file	10 cm long files	50 cm long files
Bastard	16	6
Second cut	17	7
Smooth	24	13
Dead smooth	35	22

(5) *Flame cutting*.

cut-and-cover Trenching to excavate a tunnel and then relaying the earth over it. Originally in UK cities this was the only feasible method because

tunnelling under a house was allowed only if it was one's own. The first *tube railway* of 3.2 m ($10\frac{1}{2}$ ft) finished dia., from the City of London to Stockwell (opened 1890), was tunnelled under streets throughout its length, which was much cheaper and less troublesome for surface vehicles than cut-and-cover.

cut and fill Road, railway or canal construction which is partly embanked and partly below ground in cut.

cut holes [min.] US term for a set of holes fired first in tunnelling or shaft sinking so as to break out a wedge of rock and form a *free face* for the outer holes. See **cut, lifter/relief/rib holes, sumpers**.

cut of a file See **cut** [mech.].

cut-off (1) A construction below ground level intended to reduce water seepage; for example, a *cut-off-wall* or a *grout curtain* under a cut-off wall.

(2) The drainage of rainfall into the soil, as opposed to *run-off*, more often called *infiltration*.

cut-off depth The depth below excavation level to which *sheet piling* or a *cut-off trench* reaches.

cut-off trench In dams, a trench excavated well below the foundation to an impervious layer and then filled with clay or concrete to make a watertight barrier. See below.

cut-off wall, core w. A watertight wall of *clay puddle* or concrete which is built up from the *cut-off trench*. See **grout curtain**.

cutout [elec.] A circuit-breaking device such as an electric fuse or *circuit breaker*.

cutter-boom machine [min.] A *boomheader*.

cutter-dredger A *suction cutter-dredger*.

cutting, cut An excavation for carrying a canal, railway, road or pipeline below ground level in the open.

cutting curb, drum c., shoe [min.] A steel ring on which a *drop shaft* is built. The sharp metal edge on its vertical blade cuts the ground.

cutting list, summary of reinforcement A list of steel bars showing diameters and lengths only, from which the reinforcement is ordered. This list is prepared by the contractor from the *bending schedules* issued by the reinforced-concrete designer together with the detail drawings.

cutting-out piece A short timber in trench timbering which can be sawn out to ease the striking of the timbering.

cutwater The streamlined head of a bridge pier. Cf. **starling**.

cyanide hardening, cyaniding [mech.] The introduction of carbon and nitrogen into the surface of low and medium *carbon steel* by holding it at about 800°C in contact with molten cyanides. Cyaniding is followed either by direct *quenching* from the cyaniding operation or by other suitable heat treatment to produce a hard case. It is a form of *case hardening*. Cf. **nitriding**.

cyberphobia Fear of computers.

cyclone [min.] A cone-shaped air cleaner. It removes the dust from the air by centrifugal separation. It has been developed for use as an air classifier for particles small enough to be airborne, and has been successfully used, with water or heavy medium, for washing various minerals.

cyclopean concrete *Mass concrete* in a dam or other large structure, containing *plums* of 50 kg or more (*ACI*).

cylinder (1) In the USA, particularly New York, steel tubes of 0.25–1.5 m dia., 3 mm or more thick, driven through bad ground to bedrock, excavated inside, filled with concrete and used as a pile foundation for skyscrapers and in underpinning. See **pile core**.
  (2) A *monolith* of circular cross-section. Small cylinders resemble the *Chicago caisson*, and even smaller ones resemble *bored piles*. See **screw pile**.

cylinder caisson A *drop shaft*.

cylinder prestressed concrete pipe A *prestressed concrete cylinder pipe*.

cylinder test A concrete cylinder of 152 mm dia. and 305 mm long is used in the USA for testing site concrete. When tested in similar conditions to the 150 mm cube used in Britain, the cylinder, for geometrical reasons, gives about only 0.75 of the strength of the same concrete crushed as a cube. The BS 1881 (part 117) split-cylinder test is a useful measure of the tensile strength of the concrete. See **cube test**.

cylindrical slide [s.m.] A *rotational slide*.

cylindrical-surface method See **circular-arc method**.

cylindrical vault See **barrel vault**.

D

dam (1) A wall to hold back water. For some different types see **arch dam**, **earthen dam**, **gravity-arch dam**, **multiple-arch dam**; also **design life**.
(2) Horizontal, cast steel fingers in meshing pairs built into the wearing surface of a bridge expansion joint to allow traffic to pass over the gap.

damping [stru.] A force which tends to reduce vibration as friction reduces ordinary motion. See **resonance**.

damp-proof membrane A vertical, horizontal or sloping waterproof skin, that may be of asphalt 20 mm thick, copper sheet, polythene film, etc., or many other materials such as blue brick or slate. See **damp-proof course** (*B*), **vapour barrier** (*B*).

Darby float A two-handled wooden or light-alloy float 1–1.5 m long (3–5 ft), about 13 cm (5 in.) wide, used for levelling ceilings, and in the USA for floors. There it may be up to 2.5 m long (*ACI*).

Darcy's law [s.m.] For the velocity of percolation of water in saturated soil; states that

Velocity = *coefficient of permeability × hydraulic gradient.*

Dartford Bridge A *cable-stayed bridge* completed in 1991 over the Thames at Dartford, spanning 450 m at a height of 55 m (180 ft) above the water. This first large British cable-stayed bridge is also the first recent British bridge built privately, as a *franchise*.
Trafalgar House, its promoter, will own it until 2010, levying tolls from motorists and eventually giving it to the Ministry of Transport. The two existing tunnels below carry four lanes of northbound traffic. The bridge carries four southbound lanes. Essex and Kent County councils in 1988 were demanding 3 m (10 ft) high walls each side as windshields to enable vehicles to cross at 80 kph (50 mph) in all but the severest wind. Since this would probably triple the wind load and sideways bending on the bridge, resulting in increased cost and delay, the need for it was disputed. Construction began early in 1988. The four cable towers are of steel above the deck, concrete below it. The £200 million needed to buy the tunnels, build the bridge, and repay the Kent and Sussex County Councils' debt (£55 million) is guaranteed by a financial group headed by the Bank of America.

dash-bond coat (*ACI*) (**spatterdash** UK) A thick *slurry* of *Portland cement*, sand and water, flicked on to a surface with paddle or brush either to provide a good bond for cement *rendering* or as a finish. If these hardened rough splashes of mortar do not provide enough bond, another bonding mix may have to be tried, e.g. *polyvinyl acetate* mixed with mortar.

data logger [sur.] An electronic calculator for use in the field or for other scientific work. It accepts observations from *theodolites* and from *EDM* units, by either manual or direct electronic input. It may be programmed to calculate coordinates, heights or other data at the discretion of the surveyor.

datum [sur.] Any level taken as a

reference point for levelling. The datum for any building site in Britain is usually the nearest *Ordnance bench mark*.

day joint A *stunt end*.

daywork Payment to a *contractor* based on his or her costs of materials and wages plus a percentage for *overheads* and profit. The contractor submits invoices for materials with daysheets of pay rates and hours of labour to the *resident engineer* or *inspector of works* for them to be signed in approval. The method is *cost reimbursement* on a small scale.

dead end (*ACI*) For a *tendon* stressed from one end only, the end opposite the stressed end.

dead load, d. weight [stru.] The weight of a structure and any permanent loads fixed on it. Cf. **live load**.

dead man, anchorage, anchor block/wall A buried plate, wall or block, some distance from a *sheet pile* or other *retaining wall*, which serves to anchor back the wall through a tie between the two. The dead man is held in place by its own weight and by *passive pressure* from the soil, resulting in a passive anchorage. Nowadays the tie is often tensioned for added security. See **anchor, anchorage distance, cantilever bridge, suspension bridge, land tie, stay pile**.

dead-mild steel, dead-soft s. [mech.] Steel with 0.07–0.15% C, used for bending, drawing, pressing and flangeing. See **mild steel, wrought iron**.

dead-smooth file [mech.] The finest grade of file *cut*.

debris dam [hyd.] A barrier built across a stream channel to store sand, gravel and so on. See **drift barrier**.

deca- A prefix of the *decimal system* meaning ten times.

decalescent point [mech.] *Critical point*.

decanting In *pneumatic caisson* sinking where accommodation in airlocks, particularly *manlocks*, is very limited, the process of locking people through from high pressures (pressures much above 1.3 bar) in 10–15 min. Immediately the workers leave the lock they go into a special, more spacious manlock, where they are recompressed to the full working pressure and decompressed at the correct rate (from 1 to 5 min for each 70 millibars above atmospheric pressure). Because decanting is open to abuse which may lead to *bone necrosis* and *caisson disease*, some engineers believe it should be forbidden.

decenter To remove *centering* (*ACI*).

dechlorination Removal of most of the excess chlorine (leaving the *residual chlorine*) after *chlorination* of drinking water, mainly to remove the chlorine taste, usually by adding sulphur dioxide, thus:

$$HOCl + SO_2 + H_2O = H_2SO_4 + HCl.$$

deci- A prefix in the *decimal system* meaning one tenth.

decimal system Any system of counting or measuring in tens or tenths, hundreds or hundredths, and so on. The *metric system* of measurement is decimal, the feet and inches system *duodecimal* (*B*).

deck (1) **decking** A flat roof or a quay, jetty or bridge floor, generally a floor with no roof over.
(2) *Formwork* for a level surface.

deck bridge A bridge in which the top *chord* carries the deck and traffic. Cf. **through bridge**.

declination [sur.] (1) The angular distance (or elevation) of a star from the celestial equator.

(2) The variation in degrees of the magnetic compass needle from true north at any point. The declination is variable from place to place and at each place from year to year. See **annual variation**.

decommissioning Putting out of action. A nuclear power station may be stopped but decommissioning implies also the sometimes expensive removal of hazards. Normally to remove danger to shipping, an *offshore platform* must be cut off at least 55 m (180 ft) below water, costing perhaps £60 million. Floating it away could cost £100 million. After the fire on Piper Alpha platform in mid-1988 it was proposed to cut the legs with explosives 75 m below water, to make the area safe for shipping. The *environmental impact* is important. See **nuclear decommissioning, nuclear waste**.

dedicated street, adopted s. US term for a street administered by a *local authority* (B).

deep blasting, explosive compaction [s.m.] A *ground improvement* method suitable for free-running soils, in which successive small explosive charges are detonated underground at or below the depth of the foundation which is to be loaded.

deep compaction [s.m.] See **compaction, sand piles, Vibroflot**.

deep foundation [stru.] A foundation, usually on some type of *pile* or *caisson*, generally more than 3 m below ground. Deep foundations are often needed in conjunction with *ground engineering* work.

deep manhole An *inspection chamber* built with an access shaft above it. The access shaft is considerably smaller in plan than the manhole. See **shallow manhole, side-entrance manhole**.

deep penetration electrodes *Covered electrodes* for *arc welding* which make a penetrating arc to melt the root of the weld deeply.

deep-penetration test [s.m.] *Cone penetration test.*

deep shaft system (see diagram, p. 114) An intensive *activated-sludge process* in which the *aeration tank* is a shaft of from 0.4 to 5.8 m (16 in. to 20 ft) dia. and 30 to about 60 m (100–200 ft) deep. It is divided into a downcomer, or downflow, section and an upflow, or riser, section. The downcomer may be merely a pipe. To start the fluid circulation, air is first fed into the riser by a pipe going down nearly half-way. When circulation is well established, air is fed into the downcomer, dissolving easily in the *mixed liquor* as the pressure increases to several atmospheres, greatly increasing the solubility of gases. As the *mixed liquor* moves up the riser, nitrogen and carbon dioxide bubble away, maintaining an *air-lift pump* effect. The method is highly efficient in oxidizing carbonaceous *pollutants*. At Tilbury, Essex, a deep-shaft system claimed to be the world's largest was started in the summer of 1987, 60 m deep, 5.8 m in diameter. There were then 35 throughout the world, three quarters of them in Japan. Shaft sinking is expensive but the land area needed is relatively tiny; the oxidation is efficient so the system may be economical.

deep well A well passing through shallow impermeable strata (which may yield water) but drawing its water only from beneath them.

deep well pump A centrifugal pump which is driven by a long shaft from a surface electric motor, the pump being

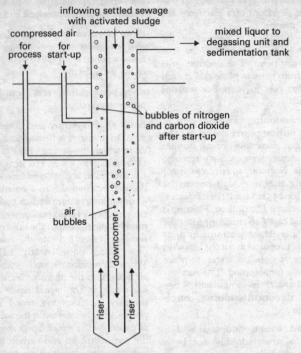

inflowing settled sewage
with activated sludge

compressed air

for   for
process   start-up

mixed liquor to
degassing unit and
sedimentation tank

bubbles of nitrogen
and carbon dioxide
after start-up

air
bubbles

downcomer

riser   riser

Deep-shaft process of activated-sludge treatment.

at the foot of the borehole. The motor is easily accessible for maintenance, unlike the *submersible pump*.

defects liability period, maintenance p. A period after the handover of a completed structure during which the *contractor* is obliged to repair any shortcomings in his or her work.

deflected tendon, draped t. A *tendon* which curves and may therefore be outside the *centroid* of the concrete section (*ACI*).

deflection (or **deflexion**) (1) [stru.] The *elastic* movement of loaded parts of a structure. The word often refers to the sinking of the mid-span of a beam which in British housing generally is

not allowed to exceed 1/325 of the span. See **deformation**.

(2) [sur.] The angle between a line and the extension of the preceding line of a *traverse* is a deflection angle. Deflection angles are also used in setting out circular curves. The curve is marked out by pegs at equal spacings with the same deflection angle (intersection angle) from each other.

deflection curve See **elastic curve**.

deflectometer An instrument for measuring the deflections of structures, usually of beams under load. One type has an *invar* wire tightly strung between the loaded structure and a *dial gauge*.

deflocculant See **dispersant**.

deformation [stru.] A more general term than *deflection*, which includes the plastic, non-recoverable movement of a structure.

deformed bars Concrete reinforcement consisting of steel bars with projections or indentations, which increase the *mechanical bond* between the steel and the concrete.

deformeter [stru.] An instrument used in the *model analysis* of a structure to help in drawing out its influence line. A model of the structure is made and carefully cut at the required point. The deformeter applies shears or rotations at the cut and the resulting deflected shape of the structure is the *influence line* at the cut for shear or rotation, whichever is applied.

degreased wire See **as-drawn wire**.

degree of a curve [sur.] US method of describing circular curves by the number of degrees subtended at the centre of a circle by a chord 100 ft long. As the radius increases, the number of degrees decreases. In Britain the radius of the curve is usually stated instead of the degree.

degree of compaction, d. of density [s.m.] The tightness of packing of a soil sample, estimated by the formula

$$\frac{\text{(voids ratio in loosest state) minus (voids ratio of sample)}}{\text{(voids ratio in loosest state) minus (voids ratio in densest state)}}$$

degree of saturation [s.m.] The percentage of the volume of water-filled voids to the total volume of voids between the soil grains. It gives a measure of the air in the voids, since the air content is 100% minus the degree of saturation.

degritting [sewage] See **detritus tank**.

Dehottay process [s.m.] A refinement of the ground *freezing* process for shaft sinking or foundations. Instead of circulating brine in the pipes installed in the ground, liquid carbon dioxide is pumped in. An advantage of carbon dioxide is that when it passes out through a leak in the pipe into the freezing ground it has no effect on its freezability. But if brine escapes, the freezing point of the ground is so much lowered that it may be impossible to freeze it again.

de-icer (deicer ACI **)** US term for salt or calcium chloride spread on the road to melt snow or ice.

delamination Separation of a layer from its neighbour in any construction, especially any surface layer of concrete, parting sometimes along a layer of rusting steel. Delamination also may be caused by repeated freeze-thawing. One way of testing for unseen delamination is by tapping with a steel tool. Unlaminated, solid concrete does not 'drum'. But 'chain dragging' has been found to be a more reliable way of sounding bridges or ground slabs. A 2 m length of chain with links measuring 50 mm (2 in.) long and weighing 2.2 kg/m is dragged over the slab. Delaminated concrete makes an unmistakable sound. Cf. **lamination**.

delay-action detonator [min.] A *detonator* which explodes at a suitable fraction of a second after the passing of the firing current from the exploder. In shaft sinking or tunnel driving this is very convenient, as the *cut holes* can be arranged to fire first without delay and delay-action detonators can be used for the *relief holes*. The complete round of shots can thus be fired once without the shot firer being obliged to return to the face after the cut holes have been fired, to connect up to the relief holes in the unpleasant, often poisonous fumes.

delivery [mech.] The volume of air or water delivered per minute or per second by a compressor or a pump.

de-mineralized water (usually the same as **de-ionized water**) Feedwater for high-pressure steam boilers – e.g. at a power station – has to be raised to such a purity that its total dissolved and suspended solids do not exceed 1 ppm. In the past this was done by distillation but now *ion exchange* is used.

demolition equipment Circular saws with diamonds set in the blade can cut concrete to a depth of about 50 cm (20 in.); wire saws, consisting of a 9 m (30 ft) loop of stainless-steel wire set with diamonds, can cut much more deeply. Both types cut steel bars as well as concrete. Alligator shears or croppers can cut steel beams or pipes. *Hydraulic hammers* can break concrete where noise and dust do not matter. A pile breaker is a heavy unit lifted by crane on to the top of a concrete pile to be cut off. Its chisels break into the concrete. All this equipment is normally powered by the *hydraulic system* of an excavator or other convenient hydraulic unit on the site. See **concrete cutting**.

de-moulding (demolding *ACI*) Removing moulds or forms from cast concrete.

denitrification [sewage] See **nitrification–denitrification**.

dense aggregate *Aggregate* for a *biological shield* usually with a *relative density* above 3.5, including iron ore, barytes or scrap iron.

dense concrete Concrete is regarded as dense (though not necessarily

strong) if it weighs more than 1900 kg/m^3. If its density is over 2250 kg/m^3 it is almost always strong. Generally strength is proportional to compaction.

densification [s.m.] Many methods of *ground improvement*.

density The weight per unit volume of a substance (at a certain temperature stated for solids and liquids only when great accuracy is required). The density of water is 62.4 lb/ft^3 or 1 kg/litre or 1 $tonne/m^3$. The density of dry air at 29.92 in. of mercury (760 mm) and $0°C$ is 1.23 kg/m^3. The density of air falls as it becomes wetter.

density current, gravity c., turbidity c., internal flow, layered f., subsurface f., stratified f. Currents of fluid of different density occur when colder or sediment-laden (dense) water enters a reservoir and creeps along the bottom. When saltwater meets freshwater in an estuary, the freshwater generally stays separate and floats above, though eventually the two may mix together. The same occurs when other gases mix with air. In sewage *sedimentation* tanks a difference of only $0.2°C$ can thus disturb the settlement process.

density/moisture relationship [s.m.] See **dry-density r., moisture-content r.**

dental [hyd.] A tooth-like projection on an *apron* or other surface to deflect or break the force of flowing water, a form of baffle (ASCE).

dentated sill [hyd.] A notched sill to break the force of a stream and reduce *scour*.

deoxidized steel *Killed* or *balanced steel*, not *rimmed steel*.

departure [sur.] The distance of a point east or west of the north–south reference line. A point is completely located by its departure and its *latitude*, which are its *coordinates*.

depressant, bathotonic reagent, surface-tension d. A flotation reagent which so lowers the *surface tension* that finely powdered worthless material can sink through floating bubbles covered with valuable mineral. Soap is a well-known depressant.

derrick A lifting device which may be hand-operated by one man or worked by several powerful motors. The main types are described below, starting with the simplest and smallest. Strictly, a derrick is a stationary crane, but some Scotch derricks move on two or more rails.

(1) The **pole, gin-pole, guyed-mast** or **standing d.** is a pole held in a nearly vertical position by four or more guy ropes. A hoisting rope passes over a pulley at the top of the pole and raises the load by a winch fixed at the foot of the pole. Though generally hand-operated for small sizes the pole may be from 5 to 30 m long.

(2) **shear legs, shears d.** has two poles lashed together at the top, from which the hoisting tackle is hung. Two guy ropes only are needed but the shears are rather less manoeuvrable (and therefore more stable) than the pole, which can be moved towards any of its guy ropes.

(3) The **guy d.** (see diagram, p. 118) has a pole or mast from which is suspended the upper end of a boom or lifting jib, pivoted at the foot of the mast. The boom is shorter than the mast and this enables it to turn a full circle, passing under all the guy ropes when it is in a vertical position. The mast is held up by guy ropes in the same way as type (1). Both the guy derrick and the simple pole are much used for erecting steel-framed buildings, because of their manoeuvrability,

the small space they take up, and the relatively great heights which they can lift to.

(4) The **three-legged d.** is a development of the shear legs, and having three legs needs no guy ropes. It is used for vertical lifting only, such as well sinking, *diamond drilling*, and sinking *bored piles*.

(5) The **oil-well d.** is a square latticed tower of timber or steel for vertical lifting only from a pulley (the crown block) hung at the top. It stores the drill pipe drawn from the hole and needs to be as tall as possible to accommodate the biggest lengths of drill pipe. In this way the amount of screwing and unscrewing of pipe can be minimized.

(6) The **Scotch d.** or **d. crane (stiff-leg d.** USA) has no guys and is usually a stationary crane, diesel or electrically driven. It works on the principle of the guy derrick but the permanent steel structure holds the mast in a vertical position, tied back to two horizontal legs which meet at the foot of the mast. These legs (called sleepers) must be held down by heavy weights (*kentledge*) at their ends, proportional to the load lifted. The jib, the top end of which hangs from the top of the mast, is longer than the mast and can therefore turn through only about 240°, since the horizontal legs occupy an angle of about 90° to each other in plan. Cf. **swing-jib crane**.

derricking, luffing Altering the radius of the hoisting rope, that is lowering or raising the jib of a crane or derrick. See **level-luffing crane**.

derricking jib crane See **luffing jib crane**.

derrick tower gantry Strong wooden or steel staging consisting of three towers, one *crane tower*, for the mast and jib of the derrick, and two

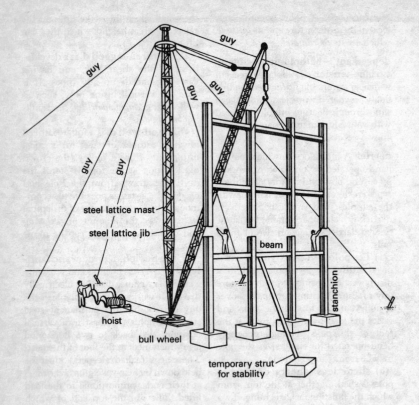

Guy derrick lowering into place one bent of a steel frame. In practice, before adding the upward extension, at least one beam or frame at right angles to the frame depicted would be connected to prevent overturning. Wherever a tall *mobile crane* can be hired, this method is obsolete.

anchor towers for the legs and their *kentledge*. The towers are tied together at their upper ends by the derrick legs.

desalination, desalting Removal of dissolved salts from water, usually for making drinking-water from seawater or brackish water. Drinking-water should have less than 500 ppm of dissolved salts, and seawater has 3.5% (35,000 ppm). There are two types of process: first employed were the distillation processes, which are still being used, and now the *membrane processes* are being developed. Distillation processes at sea are economic because they use waste steam. Any sort of desalting is a last resort and is not used if drinking-water can be found.

design [stru.] Although design is not drawing, all designs are expressed by *drawings* to which the builder works. See **structural design**.

design-and-build contract, design and construct, all-in c., turnkey c., package deal An agreement between a *client* and a trusted *contractor* for the contractor to both design a job and build it. The best work is done by contractors who have all *consultants* on their staff: architects, *civil, structural*, mechanical and electrical engineers, and *quantity surveyors*. It may be a *lump-sum contract*. One real advantage for the client, especially one whose professional staff are overloaded, is the need to speak to only one responsible person, not to four or five consultants. So the turnkey contractor must be both trustworthy and competent.

designated mixes A series of BS concrete mixes intended for most purposes from trench filling to prestressed concrete. For the specifying consultant they have the advantage that they simplify specification writing and eliminate the need for site testing since they are backed by BRMCA quality assurance. These standard mixes also make work easier for ready-mix suppliers.

designed mix A concrete mix designed by a specialist mix designer, usually a concrete technologist or a civil or structural engineer. It is difficult to design concrete mixes exactly, given the normal variations of aggregates. The design process therefore includes the making and testing of *trial mixes*, followed by *destructive* and other tests. The mix is then adjusted, especially with *lightweight aggregates*.

designer [stru.] An engineer who works in a drawing office and ensures that a job is safely calculated but will not personally draw the details if, as usual, this is left to a *detailer*.

designer-detailer [stru.] An engineer who both calculates the size of structural members and draws out their details.

designer-draughter A person who makes his or her own designs and drawings. One successful *consultant* has no technicians or pure draughters. Every designer is a designer-draughter who sees his or her work built, attends necessary site meetings, etc.

design life, designed l. The life intended for a structure may vary from five years or less for a site hut or prefab house to the 120 years for a bridge. The *Conwy tunnel* is designed for 125 years. All large dams except low-head irrigation dams (*barrages*) fill with silt and eventually become useless. The Tarbela Dam in Pakistan loses 1.5% of its capacity every year by siltation.

design load [stru.] The weight, or other force for which a structure is designed, that is the worst possible combination of loads. The term is also used in a similar sense by mechanical engineers for air-conditioning plant, and by other engineers. *ACI* considers 'design load' obsolete and prefers 'factored load'.

design spectrum [stru.] All the stress ranges caused by the worst loadings summed in the most adverse way for assessing the chances of failure.

destressing [r.m.] (1) If a *prestressed concrete* beam or other member has to be moved, altered, demolished, etc., it may be advisable first to remove some or all of the prestress. This must be done under the control of a civil or structural engineer.

(2) [rly] Newly installed *continuously welded rails* in cold weather need to be pulled to extend them to their length in warmer weather. A hydraulic stretching tool called a rail stressor pulls the

rails the specified amount. See **stress-free temperature**.

(3) To reduce rockbursts in South African gold mines, a proportion of the blasting holes in the face are drilled to 3 m instead of the usual 1.5 m. The inmost 1.5 m of these destressing holes is loaded and fired first. All the 1.5 m long holes are then fired together with the outermost 1.5 m of the destressing holes. This has significantly reduced the accident rate, which has also been helped by the timing of the shotfiring. Shots are fired from the surface when no one is underground. Rockbursts often occur soon after shotfiring.

destructive tests Tests which break a test specimen or structure but do provide a guide to its strength, such as the *cube test*, *cylinder test* and *impact test*. Cf. **non-destructive**, **semi-destructive tests**.

detachable bit [min.] A threaded piece of metal screwed to the end of a rock-drill steel, as a cutting point. The steel thus remains permanently at the working place and only the bit needs to be sent to the smith's shop for retipping or resharpening. Its cutting edge is coated with *hard facing*.

detail, d. drawing, working drawing A drawing which has enough detail on it for the *contractor* to build the work correctly and for the site dimensions to be true to the drawing.

detailer A draughter of any sort (structural, architectural, mechanical, etc.) who works out and draws *details* of construction, sometimes under the supervision of a *designer*, sometimes on his or her own responsibility.

detail paper [d.o.] Lightweight, nearly transparent, cheap *drawing paper*, used for making a first rough drawing but not normally for printing purposes.

detector pad A *vehicle detector pad*.

detention tank, tank sewer Flow *attenuation* provided in a *sewer* by a large-diameter pipe, either by enlargement of the sewer itself (in-line) or by another pipe connected to the sewer at both ends (off-line). Detention tanks reduce river pollution by reducing *stormwater* overflows.

detention time, holding t. [sewage] The length of treatment of a *sludge* or *effluent*.

determinate [stru.] See **statically determinate**.

Det Norske Veritas The Norwegian *classification society*.

detonating fuse, Cordeau, Cordtex, Primacord [min.] Fuse used in quarrying in *well holes* and similar types of blasting where several distant charges must be fired simultaneously and surely. Its firing speed is 6000 m/s. It has a textile wrapping round a detonating core such as *TNT* or *PETN*. No *primer* or *detonator* need be used, since the fuse detonates every cartridge which it touches, but may be too powerful for use near housing.

detonation The rapid explosion of very *high explosives* such as are used in *detonating fuse* or *detonators*.

detonator, cap [min.] A small sealed copper or aluminium tube, at one end containing an explosive mixture, usually fired electrically by a *blasting machine*, but in small-scale work by a *safety fuse* inserted into the detonator, crimped on to it, and lit with a match or taper. If it is electrically fired, the detonator is sealed at the factory with two copper wires leading out of it which are connected to the exploder through the neighbouring detonators in series. Whether electrically fired or otherwise, every detonator is inserted into a *primer* which fires all the other cartridges in its hole. See also **delay-action detonator, detonating fuse**.

detritus chamber, pit See **detritus tank**.

detritus slide, creep s. [s.m.] Slow movement of surface soil downhill, a generally harmless *landslip*.

detritus tank, d. chamber, grit c. A tank in which grit is removed by settlement from sewage.

develop-and-construct contract A *contract* based on a *client*'s rough scheme which the contractor develops with *detail drawings* and eventually builds.

deviation (1) [stat.] The difference between one value of a set and the *average* of the set. It is a figure used in estimating the reliability or variability of a test, such as a *crushing strength* of brick or of a *cube* of concrete. See **standard deviation**.

(2) **drift** [min.] Any departure of a *borehole* from the straight, usually measured by the magnetic bearing and the dip value. All boreholes depart from the straight, particularly long ones drilled by rotary methods (diamond or shot drill). See **borehole surveying**.

devil (1) An iron firegrate (often wheeled) used for heating asphalting tools or for softening a small area of bituminous road.

(2) **lifter** A stretcher carried by two men, for loading stone, in US quarrying.

dewatering (1) [min.] The pumping out of a drowned shaft or caisson with a *submersible pump* or an *air-lift pump* or by any other means.

(2) *Groundwater lowering*.

(3) [sewage] Reduction of the water content of *sludge*, always a difficult task. Although a reduction of the water content of a sludge from 98% to 96% water seems a minute improvement, it is in fact a real commercial success. It means there are now 4%, instead of 2%, dry solids and the solids/water ratio has been raised from 1 in 49 to 1 in 24. There is thus less than half as much water as before, and the bulk of the sludge for expensive disposal has been more than halved.

dewpoint The temperature at which air of a given absolute *humidity* begins to give up its water vapour as drops of dew. It is the temperature at which the *relative humidity* rises to 100% and can be found from tables. It is important in buildings to keep the air temperature above the dewpoint, since *condensate* is formed at all points where the air is cooled below it.

diagonal brace [stru.] A sloping member which carries compression or tension forces or both at different times and is generally used to stabilize a frame against wind or other horizontal forces. See **counter bracing**.

diagonal eyepiece [sur.] The eyepiece of a *prismatic telescope*.

diagonal spacer A short tube or iron casting with slotted ends to fit the diagonally opposite *main steel* bars in the *cage* of a concrete column or pile to keep them in place during lifting and before concreting.

diagonal tension In reinforced or prestressed concrete the *principal* tensile stress due to horizontal tension and vertical shear.

Diagrid floor [stru.] A network of diagonally intersecting ribs spanning a rectangular space. The ribs may be of metal or concrete and are often *prestressed*. As a roof it can admit a large amount of light, is not heavy, and occupies a small depth. The span/depth ratio in concrete may be about 30.

dial gauge [mech.] An instrument which shows, by its needle indication on a graduated dial, displacements of its plunger equal to one hundredth of a millimetre or less. The plunger is accurately geared to the needle. It is used in conjunction with *deflectometers* and *proving rings*.

diamond (1) [min.] One of the carbon minerals, a gem formed in volcanic necks. The hardest mineral, apart from its value as a gemstone it is extremely useful as a cutting agent. (See **black diamond**.) The diamond fields of India, now exhausted, yielded every diamond known before 1725, when diamonds were discovered in Brazil. In 1867 the first diamonds were discovered in the Orange river, South Africa, followed a few years later by the discovery of the world's richest diamond fields in the same country. Since the USSR and South Africa are the main producers, both of gold and of diamonds, it is not surprising that they secretly agree to keep prices up.

(2) **diamond crossing** [rly] The rails at the intersection of two rail tracks.

diamond drilling [min.] Rotative boring in rock with a hollow cylindrical bit set with diamonds, usually in a tungsten-carbide alloy crown, so as to obtain a *core* for geological or mineralogical examination. In spite of the cost of diamonds worn out or lost in the hole, diamond drilling is the only method for drilling small holes near to the horizontal or upwards. It is also the most popular drilling method for small downward holes. Diamond drilling speeds may be up to 3 m/h; core diameters are up to 500 mm.

diamond pyramid hardness test See **Vickers hardness test**.

diamond saw A diamond saw is a thin steel disc set with diamonds on the tungsten-carbide matrix of the rim so as to cut stone, concrete, steel or other hard material. Diamond saws cut like a *circular saw* (*B*), except that they must be cooled by a water spray. *Pretensioned* precast floor slabs or beams are cut to length by diamond saws. They will saw 8 cm thick granite at 8 cm/min, and the sawn granite surface is smooth enough to be polished immediately without rough grinding. The rim of the wheel moves at about 15 m/s so that a 50 cm dia. saw rotates at about 1800 rpm, while a 75 cm dia. saw rotates more slowly, at about 1150 rpm, to provide the same rim speed. A reciprocating saw can cut 1.2 m deep.

diamond-wire saw See **demolition equipment**.

diaphragm (1) [stru.] In general a stiff plate or partition such as a *bulkhead*. The temporary wall built across each end of each unit of an *immersed tube* to enable it to be floated into position and sunk next to its neighbour is also a diaphragm. When the next unit is jacked up to it, the water in the space between the diaphragms is pumped out and air is admitted. The force of the water then pushes the two units more closely together and the diaphragms can be broken out without flooding the tube. In a hydropneumatic *accumulator*, the diaphragm is flexible.

(2) [sur.] A brass fitting in a surveying telescope which carries the *reticule*.

diaphragm pump A reciprocating pump with neither ram nor piston but a flexible rubber, canvas or leather partition moved to and fro by a rod. It is used by contractors more than any other pump, since it can handle gritty water containing 15% solids, and even small stones, with little wear.

diaphragm shield See **blind shield**.

diaphragm wall (1) **ICOS wall, slurry wall** A concrete retaining wall underground, which may be as much as 24 m deep, built in a mechanically excavated trench that has been filled with bentonite-loaded or ordinary mud to support it during excavation. Reinforcement is dropped into the mud and the concrete is lowered into the bottom of the trench by *trémie*. The method is relatively silent and vibration-less compared with driving sheet piles. In 1973 a French method of converting bentonite slurry to concrete by adding cement (Sepicos) was said to be economical, because there was no need to buy concrete nor to pay for the removal of waste mud. About 150 kg of cement were added per m³ of mud, with some lignosulphite retarder. The 28 day strength was very low, varying from 1 to 3 N/mm^2, but the material was impermeable and crack-free and some-times this is all that is needed.

To prevent dilution of the slurry it must always be at least 1 m (3 ft) above the *standing-water level* in the surrounding ground. The grab may be supplemented by an *air-lift pump* when cleaning solids from the bottom of the trench before concreting. The slurry density should be less than 1.1. At more than 1.1, the slurry should be desanded to prevent it mixing with the concrete. The concrete should be sloppy with a slump of between 15 and 20 cm (6–8 in.), but this should be achieved by a *superplasticizer* to ensure a low W/C ratio. The wall is usually cast in lengths (barrettes) of 4–6 m (13–20 ft). Precast, post-tensioned, prestressed wall units have also been used. See **masthead gear**, **slurry trench**.

(2) See **brick diaphragm wall**.

diatomite See (*B*).

die A very hard metal block with a

hole through which *ductile* metal is *cold drawn* to convert it to wire or pipe. See **wire gauge** (*B*). Another sort of die is used for forming a *screw thread* (*B*) on a bar by forcing it into and twisting it through the die.

die-formed strand [stru.] A *strand* for *prestressed concrete* which began to be commercially available in the UK about 1964. When strand is passed through a die, its bulk is reduced some 27–37%, enabling more strands and thus more force to be packed into a given area of concrete.

Diesel engine [mech.] An *internal-combustion engine* which burns a relatively cheap oil of about the consistency of light lubricating oil. The oil fuel is pumped into the cylinder (solid injection or airless injection) by a pump, or in the original types of engine by a blast of compressed air. The fuel is ignited solely by the high compression in the cylinder, without electrical spark.

Diesel hammer See **pile hammer**.

die-shrinking See **swaging**.

differential-pressure flowmeter Any meter like the *orifice plate, Pitot tube, Venturi meter*, Dall tube, etc.

differential pulley block, chain block [mech.] A builder's *lifting tackle* consisting of an endless chain threaded over two wheels of slightly different diameters turning together on the same shaft. The lifting power increases as the diameters become closer. The chain cannot run back and someone can safely lift 500 kg or more alone.

differential settlement, relative s. [stru.] Uneven sinking of different parts of a building. The allowable difference between one column and the next is less for a building with marble facings and plastered interior walls than

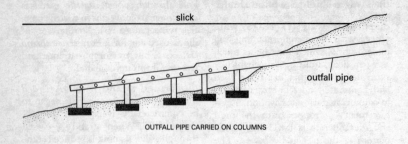

OUTFALL PIPE CARRIED ON COLUMNS

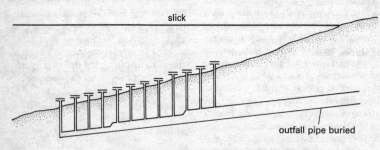

OUTFALL PIPE IN TRENCH WITH UPWARD DIFFUSER PIPES

Two types of diffuser at sea outfalls.

for a steel-framed single-storey factory with hardboard facings to its inside walls. The worst differential settlement is unlikely to exceed half the total settlement in a normal *settlement crater*. If the foundations are designed to limit the total settlement to 5 cm, the 2.5 cm maximum expected differential settlement is unlikely to damage even the most sensitive building seriously. Uniform settlements of several metres occur in some cities (Mexico City, Chicago), but they do not damage the buildings and may even not sever its connections with the outside world (pipes, drains, cables). See **compensated foundation**.

diffuser (1) [hyd.] In centrifugal pumps, compressors, fans or turbines,

the gradually increasing cross-section at the inlet or outlet, which reduces the speed of the air or water and thus increases its pressure. See **draft tube**.

(2) (see diagram) [sewage] A porous plate or similar device through which air or occasionally oxygen is blown to aerate sewage, or at an *outfall*, holes in the pipe to spread the sewage.

diffusion (1) [min.] The movement of the molecules of gases in all directions which causes them to intermingle, with no ventilation current, in a way which is often contrary to gravity. See **Graham's law of diffusion.**

(2) Wood preservation by laying a stiff paste or a concentrated solution on to green timber, and allowing it to penetrate gradually.

digestion [sewage] A way of treating sewage *sludge* in the absence of air in closed, heated tanks. With effective treatment it gives off a gas that is mainly methane (CH_4), with one third carbon dioxide, but the process takes three weeks even with heating. Because of the expense and care needed, the process is not now favoured in the cold climate of the UK. Even after digestion, the bulk of the sludge still has to be disposed of.

In the vast new underground *sewage treatment* plant for Marseille, completed in 1988, and covering 10 hectares (25 acres), most of it beneath the city athletic ground, heated digestion of sludge lasts 20 days followed by a further heating under pressure at 190°C to kill viruses and bacteria. No chemicals are used. Some 3000 m³/day of sludge become 124 tons of dry sludge for fertilizing forests round the city. About 1000 m³/day of *effluent* from sludge *dewatering* are used industrially or for irrigating the forest.

digital Concerned with digits, numbers; for the computing sense see **analogue**.

digitizer An electronic device used in *CAD*, *CADD*, *computer graphics* and elsewhere in computing for converting an image into *Cartesian coordinates*.

dike See **dyke**.

dilatancy [s.m.] A property of *silt* which distinguishes it from *clays*. If a pat of wet silt is shaken in the hand it exudes water and becomes shiny. If the pat is then pressed, the water re-enters the silt, leaving a matt surface on the silt because of a dilation or increase in volume. See **shaking test**.

dimensional analysis [hyd.] In the *model analysis* of rivers, ports, ships, breakwaters, aircraft and other problems involving the flow of fluids, *dy-* *namic, kinematic* or *geometric similarity* (components of dimensional similarity) help to simplify the study. These relationships enable the experimenter to see quickly the effect of variations in flow rate and size of model compared with similar variations in the full-size structure (prototype). See **Reynolds number**.

DIN Deutsches Institut für Normung e. V. meaning German standards body.

dip [min.] The angle of maximum slope of the beds of rock measured from the horizontal at any point. The dip is shown at *outcrops* on geological maps by an arrow pointing downwards with a figure giving the dip in degrees. The dip is at right angles to the *strike*.

dip compass See **dip needle**.

dip meter A *well logging* instrument with three or four pads pressing against the borehole sides. It finds the *azimuth* of the *strike* of the strata as well as their *dip*.

dip needle, d. compass, dipping needle [sur.] A magnetic needle with a horizontal pivot which allows the needle to swing only in a vertical plane. The needle is set in the magnetic meridian and the inclination to the horizontal of the needle is read off. This angle of dip is the dip of the earth's magnetic field at that point.

dipper The far end of the jib of a *backhoe*.

dipper dredger See **grab dredger**.

direct-acting pump [mech.] A compressed-air or steam-driven *reciprocating* pump having the power cylinder and water cylinder at opposite ends of the same piston rod. There are therefore no important rotating parts.

directional drilling, guided d. (see diagram, p. 21) The drilling of a hole with an intentional *deviation* (2). Single holes can now be drilled with several others branching off them all round so as to increase their gas or oil output. In offshore drilling, any hole has to be used to the maximum because of the high cost of the drilling- or production-platform built over it. The technique has been greatly helped by improved *turbodrills* and *borehole surveys*. The turbodrill, with stationary drill pipe behind it, can be turned to any desired angle by a short length of bent pipe between it and the main drill pipe. *Sidetracking* is more difficult with rotary drilling, and the accuracy possible with turbodrill sidetracking was previously unheard of. For *microtunnelling*, many accurate, reliable directional drilling methods have come into use, able to drill 600 m under a river and to arrive well within 1 m of target, so as to place e.g. a 76 mm (3 in.) *polythene* service pipe. Some methods depend on a surface radio receiver held above the tunnel and receiving signals from an underground transmitter at the face. Another method, but suitable only for depths up to 2.5 m (8 ft), involves two loops of wire, carrying signals of a few milliamps, laid in the form of four parallel wires about 1 m apart on the ground above the microtunnel. Good directional drilling depends on accurate *borehole surveying*. See **horizontal drilling**.

Director-General of Water Services
A civil servant in the Department of the Environment, the chief of OFWAT, the 'consumer watchdog' of the privatized water industry, supervising the ten newly created *water service plcs*, the 29 old water companies, their relations with customers, the investment needed to improve water quality, demands for payment for water, the *National Rivers Authority*, etc. See **k factor**.

direct reading tacheometer [sur.]

A *tacheometer* from which the plan length (from the staff *intercept*) and the difference in level between staff and instrument can be read directly without measuring the vertical angle.

direct stress [mech.] A *stress* which is wholly compression or tension and involves no bending or shear.

dirty money See (*B*).

discharge The volume of fluid per unit time flowing along a pipe or channel, or the output rate of plant such as a pump.

discharge coefficient [hyd.] See **coefficient of discharge**.

discharge curve [hyd.] A curve relating the water level (*stage*) of a stream to its discharge.

discharge head [mech.] The height between the intake of a pump and the point at which it discharges freely into the air. Strictly speaking this is the static discharge head, and the total discharge head is this height together with another height representing the friction in the pipe and pump, plus the *velocity head*.

discharge valve, control v. [hyd.] A valve for reducing or increasing the flow in a pipe, as opposed to a *stop valve*.

discount A reduction from a published price. Of the many sorts of discount, two are important to builders and civil engineers: trade discounts and cash discounts. A cash discount is a reduction because the buyer pays quickly. A trade discount is based on the buyer being in the building trade, therefore likely to be a frequent customer. In *daywork* payments the contractor has the benefit of the common $2\frac{1}{2}\%$ cash discount, but the *client* should benefit from trade discounts above 5%.

disease See **caisson disease, cancer clusters**.

disinfection Destruction of micro-organisms that harm humans. Conventional *waterworks* treatment without disinfection eliminates 90% of the bacteria and viruses in water. The remaining 10% are removed usually by chlorination in the UK, in France by ozonizing, elsewhere also by other methods, e.g. ultra-violet radiation, gamma radiation and heat. Disinfection is always the last process in water treatment.

disk See **floppy disk, rotating biological contactor**.

dispersant, dispersing agent, deflocculant Chemicals for dispersing (deflocculating) finely ground materials. Sodium oxalate is used in the *wet analysis* to prevent *flocculation* and the settling of particles more quickly than they should. *ACI* recommends them for thinning mortar, concrete or *grout* or as an aid to grinding.

dispersion A suspension of very fine particles (often *colloids*) in a liquid medium. Most paints and some varnishes are dispersions. See **emulsion**.

dispersive clays Clays which disperse (deflocculate or 'melt') in the presence of clean water and can therefore be easily scoured or allow *piping*. The *index properties* give no indication about these treacherous soils.

displacement (1) [mech.] The volume displaced (swept) by a piston or ram moving from top to bottom of its stroke. For a pump this is equal to the theoretical amount of water delivered per stroke. For an engine (compressed air, steam or internal combustion) the product of displacement and the mean pressure in the cylinder is equal to the work done per stroke. From this the engine power is calculated.

(2) [hyd.] The volume of water displaced by a floating vessel. By the *principle of Archimedes* it weighs the same as the vessel and its contents.

displacement pile Any solid or hollow driven *pile* that is closed at its lower end by a shoe or plug and displaces ground when it is driven can therefore be called a displacement pile. This displacement may weaken the ground or affect nearby structures unfavourably. It may even lift the ground or piles nearby, involving the need to re-drive those that were driven first. *Bored piles* do not displace ground but replace it, and may therefore be preferable in the neighbourhood of sensitive structures. Small-displacement piles include open-ended tubes, box-sections or H-sections, and *screw piles*.

displacement pump Any ram-operated or piston-operated pump; but the term is generally kept for *diaphragm pumps*, or for *air-lift pumps* in which compressed air displaces the water. They can pump very dirty water or mud or corrosive liquids without excessive wear on their metal parts.

displacer A *plum* in concrete.

disposal well, injection w. A well (usually drilled, not dug) into which polluting liquid is poured and left. Checks should first be made that *groundwater* will not be polluted or earth movements caused (in an arid area).

dissipator (1) [hyd.] Any device for reducing the speed and abrasive power of a stream, for example a *stilling pool* or a jet disperser for converting a jet of water into a spray which will not scour the river bed.

(2) [sewage] Any *sewerage* layout that reduces peak flows and thus the pollution of *stormwater* overflowing into a river. A *detention tank* is one device. In electronics, heat dissipators (heat sinks) exist to keep transistors cool.

distance piece, separator A part used for maintaining the position and spacing of bars or rails and check rails, or built-in members or formwork during concreting, or permanently. See **bolt sleeve**.

distancing unit, EDM u. [sur.] An *electronic distance-measuring instrument*.

Distomat [sur.] An early *EDM* unit, made by Wild of Switzerland, that is suitable for engineering work and is designed for fast, accurate measurement of short to medium ranges. Many other makers produce similar instruments with a maximum range of 3 km and an accuracy of 5 mm ± 3 ppm. In a *total-station instrument* or an *EDM* mounted on a *theodolite* with a *data logger*, the slope length, plan length, difference in height and coordinates of a point may be displayed.

distributed load [stru.] A *design load* uniformly distributed along a beam. See **live load**.

distribution curve [stat.] A *frequency curve*.

distribution reservoir [hyd.] A *service reservoir*.

distribution steel In a reinforced-concrete slab, the subsidiary reinforcement placed at right angles to the main steel to hold it in place during concreting and to spread concentrated loads over a large area of slab. See **one-way slab**.

distribution tile US term for clay *field drains* which distribute the overflow from a septic tank to the soil.

ditch Either a drainage or an irrigation channel, or an *oxidation ditch*.

ditcher, trencher A *trench excavator*.

ditching by explosives Holes are punched or bored to about 15 cm from the bottom of the ditch along its centre line, and about one cartridge of explosive is charged per 0.6 m of hole. A line of holes along the ditch is fired simultaneously. The rubbish is thus scattered effectively and a rough ditch is made.

diver Most divers wear *SCUBA* or *standard diving gear*. In SCUBA a diver is very mobile but also more vulnerable to the cold. An underwater current of 1 knot (0.5 m/s) has the same effect on the diver as an 80 kph wind on the surface would, and this is the strongest current that a diver can work in. Consequently most dives are made at slack water (high or low) not during running tides. Pneumatic tools used under water need compressed air at a pressure that is larger than would be needed on the surface by the amount of submergence. Thus at 30 m depth of water (3 atm) the compressors at the surface must provide air at 3 atm higher pressure. Since breathing alone is hard work under water, it is inadvisable to expect divers to work hard, but theoretically they can do underwater cutting, welding, guniting, carpentry, steel plating and shipwright's work. Divers normally do not work below 60 m depth but 90% of dives are less deep than 10 m. A diver who comes up from 60 m must spend a large part of the shift decompressing in the air lock at the surface. See **caisson disease**, **decanting**.

diversion cut A channel dug for a stream or river to by-pass civil engineering work.

diversion dam [hyd.] A barrier across a stream built to turn all or some of the water into a *diversion*.

diversion requirement See **gross duty of water**.

diver's paralysis *Caisson disease.*

divide US term for a *watershed.*

dividers [d.o.] An instrument used for transferring equal lengths from one point to another on a drawing. Like the compass it has two legs hinged together at one end, but unlike the compass it has both legs pointed.

divider strips Plastic or nonferrous metal strips embedded 10–40 mm in the top of *terrazzo* (*B*) to mark out the panels and localize cracks.

diving Divers wearing *standard diving dress* are often needed for harbour maintenance and construction, or for the laying of pipelines or cables. A diving crew consists of the diver and the linesman, plus two air-pump operators, or one if the air is pumped by power-driven compressor to the diver. Other workers are needed for taking tools down to or for bringing material up from the diver. The absolute minimum crew thus numbers three or four, so that diving work is necessarily expensive. It is therefore not surprising that *SCUBA* diving has become popular, since it involves one person only, at least to shallow depths.

diving bell An open-bottomed metal tank lowered by crane to the seabed, riverbed, etc. *Smeaton*'s diving bell at Ramsgate in 1784 measured only 1.38 m (4½ ft) high and long and 0.9 m (3 ft) wide. A stone cut to shape, hung on chains in the bell, was lowered with the men inside who placed it when the bell reached bottom. Men in the boat above pumped air down to the bell through a hose.

In 1807 *Rennie's* diving bell, also at Ramsgate, was larger: 1.8 m (6 ft) high and 1.8 × 1.37 m (6 × 4½ ft) in plan. It had eight cast glass bullseyes in the roof to give the men some light; the air supply hose was of 63 mm (2½ in.) dia.

Modern diving bells have an air lock through which rock can be hoisted, and are much larger. Clean, cool air must be pumped in at the rate of 0.3 m³ per person per minute. A standby blower and drive engine are also needed, also electric light at 50 V maximum, a telephone for the divers to tell the crane driver when to raise or lower the diving bell or the muck skip, and, if *compressed-air* tools are used, a high-pressure compressor. For a bell working at 10 m (33 ft) depth, the compressor must provide air at a pressure 1 atm higher than is needed above water. A lifeline to the boat allows signalling if the telephone is out of order. The bell must be raised slowly to conform to the decompression rates allowed by law. Divers working in a bell must be supervised by a doctor as for other work in compressed air. See also **helium, Hydreliox, limpet**.

dock A basin for shipping which is cut off from the tides by dock gates (except for tidal docks). See **dry d., floating d., self-docking d., wet d.**

docking blocks Blocks which support the underside of the hull of a ship in *dry dock*. The centre row are called keel blocks; a row is also provided at each side. They are made of oak or other hardwood, preferably with a softwood cap which does not damage the hull.

dog A square-section spike driven into wood to hold heavy timbering or a flat-bottomed rail. See also (*B*).

dolly A block of hardwood or other material placed over the *pile helmet* to receive the shock of the *pile hammer* and thus reduce the damage to the head of the pile from driving. See **grommet, follower**.

dolo A type of precast concrete *armour* block.

dolomite [min.] $CaMgCO_3$. A source of basic *refractory* as well as of limes for *jointless flooring* (*B*).

dolphin A mooring in the open sea or a guide to help ships to come into a narrow harbour entrance. It is usually built of *raking piles* of steel or timber driven into the sea bed. See **bell dolphin**.

dome [min.] A locally spherical shape of the strata, which, if it contains opentextured sandstones beneath impervious shales or salt, may well be an excellent natural oil reservoir, as are the salt domes of Texas.

Doppler shift The tone of a railway engine whistle approaching a listener is higher in pitch than when it recedes from the listener. The difference in frequency in the two tones, known as the Doppler shift, could be used to calculate the speed of the railway engine and is used for measuring the speed of a river, in the *ultrasonic flow meter*. See **satellite-Doppler system**.

Dornoch Firth road bridge See **incremental launching**.

dosing chamber A *dosing tank*.

dosing siphon [sewage] An automatic siphon that discharges the contents of a *dosing tank*, usually to a *trickling filter*.

dosing tank, d. chamber A tank into which raw or partly treated sewage flows until the desired quantity has accumulated, after which it is discharged automatically for treatment. It ensures that the flow rate is never too small for the process to work properly.

double-acting [mech.] A description of reciprocating pumps, compressedair or steam engines, which means that both sides of the piston are working under pressure, so that every stroke of an engine is a power stroke and every stroke of a pump delivers fluid. A double-acting *pile hammer* forces its hammer down by steam or compressed air. Cf. **single-acting**.

double-cut file See **crosscut file**.

double-drum hoist [min.] A mining haulage engine consisting of two drums which can be separately driven and connected or disconnected by a clutch.

double-headed nail (duplex-headed nail USA) A round wire nail on which two heads are formed, so that although the nail can be driven home to fix concrete *formwork*, a second head 12 mm higher than the first enables the nail to be quickly withdrawn with a claw hammer.

double joint To halve the number of outdoor welds on a pipeline, pipes are often welded into pairs ('double joints' in the USA) at a convenient point between the pipe maker and the site.

double-layer grid, two-way g. [stru.] A space frame, often of steel tubes joined in square pyramids with their apices and bases interconnected by other tubes forming two grids, one above the other. Thus two horizontal layers of steel tubes are held apart by occasional diagonals to form pyramids. The first roof of this type was the 138 m span roof of the jumbo jet hangar at London Airport, erected in 1960 by jacking from the column heads with *lift-slab* techniques. All grids of this type have two layers, one above and one below; therefore the term two-way grid is easier to understand. Doublelayer grids are occasionally curved. See **grid** (3).

double lock [hyd.] Two parallel canal-lock chambers with a sluice between them that halves the water loss.

double-rope tramway US term for an *aerial ropeway* with two *track cables* and one endless *traction rope*.

double-seal manhole cover A *manhole cover* with two tongues projecting down into its frame instead of one, to improve its gas-tightness. This extra expense is usual for manhole covers inside a building. See **manhole cover** (diagram).

double sling A *chain sling* or rope sling, also described as a *two-leg sling*, which is less confusing.

double tee See **pretensioned floor beams** (diagram).

double-wall cofferdam A *cofferdam* consisting of two rows of sheet piling held at the right spacing by tie bars. The space between the rows is filled with gravel or other permeable material. This width should be not less than 0.8 to 1.0 times the retained height of water or soil. Such structures as the *cellular* or *Ohio cofferdams* are used in water when a single skin of sheet pile could not be strong enough to hold the depth of water required. Drainage holes need to be provided near the foot of the inner wall.

double-webbed beam A *box girder*.

doubly reinforced beam A *reinforced-concrete* beam with steel in the top and the bottom.

doughnut A large steel or plastic or concrete washer (spacer) threaded on to *reinforcement* to maintain the concrete cover over it.

dowel A downward projection (usually one of many) from the base of a *gravity platform*, deeper than the *skirt*, aimed in part at preventing the buckling of the skirt as the platform sinks. See also (*B*).

dowel-bar placer Equipment in a *slipform* or *fixed-form paving train* which sets *dowel bars* (*B*) in the slab joints.

dowelled joint In a concrete road slab or other slab on the ground, a joint in which short bars are cast into one side of the joint as *dowels* (*B*). The bars cross the joint and on the far side they are enclosed in a metal or plastic sleeve or they may be coated with a debonding compound to allow movement of the concrete. Many types of dowel exist.

downdrag *Negative skin friction.*

downstand A beam projecting down below a slab which it forms part of.

down-the-hole drill Drilling a deep vertical hole with a *rock drill* becomes slow when the length and weight of the drill rods are enough to absorb most of the energy of the blow. Down-the-hole rock drills have short drill rods because the machine enters the hole it is drilling, consequently saving power and drilling fast. The term is usually confined to air-driven drills, so *turbodrills* are not included.

For drilling holes of from 85 to 127 mm ($3\frac{1}{2}$–5 in.) dia., a 15 ton machine is needed. Larger machines can drill holes of up to 200 mm (8 in.) dia.

down time The time for which any unit of equipment is not usable because of essential maintenance or other technical needs. See **committal rate**.

dowsing Searching for water and ore deposits by the feeling of a branch or a pendulum held in the hand. Most engineers prefer the evidence of boreholes.

dozer An *angledozer, bulldozer, calf-dozer*, etc.

Dracone A plastics 'sausage' that can be folded up when empty and carried on a lorry. It can be towed in the sea carrying hundreds of tons of fresh water, oil, etc., or may be used as a storage tank or as a *floating boom*. See **Fabridam**.

drafting machine [d.o.] See **draughting**.

draft tube [mech.] The metal casing by which the water leaves a *turbine*. It corresponds to the *diffuser* of a centrifugal pump.

drag (1) A simple towed implement with tines, blades or chains for scraping or levelling a surface of loose material. See also (*B*).
(2) [hyd.] The drag on a particle in flowing water or air is the force of skin friction plus the 'form drag' from the pressure difference in front of and behind the particle. As in aerodynamics, it is accompanied by an upward lifting force opposed by the particle's submerged weight. In a channel, the drag of all the particles together is its *roughness*.

draghead, trailing suction-cutter dredger, trailer dredger A large and extremely powerful *suction dredger*, so called because it steams ahead while dredging and so trails the suction pipe behind it. They are thought to have been first developed by the US Army Corps of Engineers before 1900, but have since been much used in Europe. A typical 'trailer' has a hold (hopper) capacity of 18,000 tons, that it can fill in one hour with its two pumps of 22,000 m³ hourly capacity from 22 m depth. The total engine power is 21,500 hp, overall length 143 m, and it can dredge to 35 m depth. The suction pipes are both 40 m long and of 1.2 m dia. Overflow pipes carry away surplus water. Dumping is through 26 bottom valves, 13 on each side, each 3.6 m in diameter, distributed along the 54.6 m length of the hopper, controlled from the bridge. The main disadvantage of these dredgers is that they must interrupt dredging when they move away to dump their load. This can be avoided by delivery into a hopper barge along-side, or a *floating pipeline*. A 6000 hp hopper dredge made in the USA in 1969 cost about $15 million, with swell compensators, air conditioning, helicopter landing pad, etc. Some 49 m long, it belonged to the US Army Corps of Engineers.

dragline, d. excavator (see diagram) An *excavator* which works by pulling a bucket on ropes towards it. The bucket is hung on the end of a long jib and a skilful operator can throw the bucket considerably beyond the end of the jib. It is generally used for digging below the level of its tracks. See **walking dragline**.

dragline scraper Rope-controlled equipment for withdrawing piled material like coal, coke, sand, or broken stone from a stockyard and pulling it up on to a loading platform from which it drops into lorries or wagons. At the platform end the two drums of a *double-drum hoist* pull the ropes controlling the *scraper loader* bucket. The main rope pulls it full of material towards the loading platform. The return rope passes over a return sheave at the far side of the stockyard and pulls the empty scraper bucket back to the pile of material. The name is misleading, since in conception the dragline scraper is closer to the mining *scraper loader* than to the *dragline*. It is a stationary piece of plant, though sometimes both ends are built to travel a short distance on rails laid perpendicular to the direction of the ropes.

drag shovel A *backacter*.

drain A channel, pipe, or duct for conveying surface or subsoil water or sewage. Drains may be *house drains*, *surface-water*, *field* or *highway drains*; see also **absorbing well/box/ French/rubble drain** and below.

drainage The removal of water by

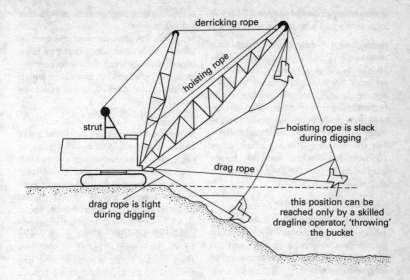

Dragline during digging. (F. C. Harris, *Construction Plant*, Granada, 1981.)

flow or pumping from the ground or from its surface or from buildings. It may include *sewerage*.

drainage area, d. basin A *catchment area*.

drainage tunnel [min.] A tunnel driven mainly for drainage. Its cost is justified by the reduction in the total head which the mine pumps must work against and their resulting lower first cost and working charges.

drained shear test, slow t. [s.m.] A *shear test* or triaxial compression test, applied to a specimen of cohesive soil after completed *consolidation* under normal load, carried out in drained conditions slowly enough to allow further consolidation due to shear during the test. Drained tests give higher strengths than *quick* or *consolidated quick tests*.

drain pipes Pipes, usually below ground, that remove *sewage*, other waste water or rainwater. They may be of clay ware, plastics, cast iron, asbestos cement or concrete. Joints may be *spigot and socket* (*B*), made tight with mortar pushed hard in against a gasket of old untwisted rope or hemp, or a rubber O-ring joint, etc. In any case the spigot faces uphill, to make the jointing easier.

drain rods Rods made of cane or tightly coiled steel springs, with threaded brass end fittings which screw together and can be pushed to and fro in a drain to remove a stoppage (*rodding*). See **inspection chamber**.

drain tile US term for *field drains*.

drain well An *absorbing well*.

draped tendon A *deflected tendon*.

draughter, draughtsperson [d.o.] A man or woman who makes drawings.

draughting machine (drafting machine USA) Equipment clamped on the edge of a *drawing board* that replaces the older tee-square and set square. It is lighter and quicker to use but much more expensive. The earliest type may have been the brass parallel ruler with its heavy cylindrical rollers, but the modern ones are much more sophisticated, with straight edges graduated in mm, built-in protractor, etc.

draught tube A *draft tube*.

drawbar [mech.] The steel bar by which a locomotive pulls its train, or by which a wagon receives its pull. It passes through the wagon from end to end, transmitting the pull through each wagon to the one behind it.

drawbar pull [rly] See **tractive force**.

drawbridge [stru.] A *movable bridge* lifted at one end by chains or ropes either to let vessels pass under or to stop traffic passing over it. Modern *bascule bridges* may lift from the middle, with hydraulic jacks.

draw cut [min.] See **bottom cut**.

draw-door weir A *weir* with gates which can be raised vertically.

drawdown Lowering of a water level. See **sudden drawdown**.

draw-file [mech.] To grasp a file with both hands and push it across a surface with its length at right angles to the direction of movement. This produces a smoother surface than the usual method of filing, but removes less metal.

drawgear [min.] Chains, shackles, etc. used in hoisting and haulage.

drawing (1) [d.o.] The work of a *draughter*, putting on to paper pic-

torial working instructions, perspective views, etc.

(2) [stru.] A dimensioned and usually true-to-scale elevation, plan or section, sometimes showing *details* to be constructed by a builder. Drawings which are not to be worked to are generally called sketches.

(3) See **cold drawing**.

drawing board A flat softwood sheet made of several boards fixed edge to edge, measuring about 5 cm more each way than a standard size of *drawing paper*. A common size used by engineering drawing offices is for A1 paper, 594 × 814 mm, making a board of about 650 × 920 mm. Modern boards are usually surfaced with plastic sheet 1 mm or so thick that is better to write on than wood, being smoother and less easily deformed. Most drawing boards are fixed to a stand, which adds to their cost but makes them much easier to use. Some means of drawing accurate parallel lines is always provided for engineering drawings. The most primitive device is the *tee-square*, which is appropriate for small boards and until about 1950 was the only equipment available. *Parallel motion equipment* is used in many offices. It is easier to use, perhaps more accurate than a tee-square, and very much less expensive than a *draughting machine*.

drawing paper The paper and film, or textile, used for making drawings in drawing offices is of many sorts but is either translucent, and used for making negatives, or is opaque, and not usable for this purpose. Most papers or films can be drawn on and written on with pen or pencil. Paper is commercially described generally by its weight in g/m^2, the best quality usually being the heaviest. For comparison, very good notepaper weighs about 85 g/m^2, solicitor's notepaper about 100 g/m^2. The three main types of material are: (i)

Translucent tracing paper (weighing from 63 to 112 g/m²). The negative formed by drawing on the translucent material is put into a dyeline process from which *contact prints* are produced. This gives a white background that enables the print to be further worked on. More durable but more expensive materials for the same purpose are acetate film or other plastics. Linen has been used for any years. Acetate film is made in thicknesses from 0.075 mm to 0.18 mm. (ii) *Detail paper*, sometimes called layout paper, can be used for tracing or for drawing preliminary sketches that take little time and not much erasing. It is a lightweight paper, of 45–50 g/m². (iii) Cartridge paper is an opaque, white, very strong paper suitable for making drawings or maps that need much work and erasing. It therefore is very heavy and may weigh from 120 to 200 g/m² When prints are needed from a drawing that has been made on cartridge, a sheet of tracing paper or film is laid all over it and the entire drawing is traced through, by hand or photographically. The standard-sized 'A' sheets listed below are designed to help in the filing of drawings, so that the drawings can be laid flat in the drawer without difficulty. They are internationally standardized. Their sides are in the ratio of $\sqrt{2}$. Photographic enlargement and reduction are simplified because of this ratio and so are folding and insertion into envelopes. The same ratio, $\sqrt{2}$, is used between each size and the next in the series. Larger sizes exist, thus 2A is twice Ao, but Ao, the basic size, covers an area of 1 m².

drawn strand [stru.] Die-formed strand (Dyform) or other compacted strand.

dredge, dredger [min.] A vessel fitted with *bucket ladder*, *grab* or *suction dredging* machinery for underwater excavation or for mining alluvial deposits, generally of tin, the noble metals or gems. Most harbour dredges are seagoing vessels, but mining dredges are more like rafts and may pass their whole life in one small muddy pond. The mining dredge moves the soil from one side of the pond to the other through its concentrating machinery and thus the pond and the dredge move together, often many miles in the course of years. When the deposit is worked out, the dredge is dismantled and moved by road or rail to another small muddy pond. See also **stationary dredger, hydraulic dredger, mechanical dredger**.

dredged aggregate, marine a. Sand or gravel dredged from the sea under government licence in areas from the Thames to the Tyne, near the Isle of Wight, the Bristol Channel and Liverpool Bay. Beach sands also are regarded as marine aggregate. They may contain salt or shell or coal and need washing to remove them. In 1986 some 16 million tons were produced, or about 15% of UK sand and gravel output, but 3 million tons of this were exported.

dredged bulkhead An anchored *sheet-pile* wall (bulkhead), from which the soil near the toe is dug or dredged.

dredging Excavation, below water level, with a *dredge*.

dredging well The opening in a dredger through which the ladder or suction cutter passes for excavation.

Sizes of A-sheets, mm

Ao	841 × 1189
A1	594 × 841
A2	420 × 594
A3	297 × 420
A4	210 × 297
A5	148 × 210
A6	105 × 148
A7	74 × 105

drift (1) [hyd.] The speed of movement of a body of water.

(2) [air sur.] The angle between the fore-and-aft line of an aircraft and its actual course. The difference between these two lines is caused by the wind.

(3) **drift bolt** [mech.] A tapered steel bar driven into rivet holes before riveting, to bring them into line, or used for expanding the end of copper tube before joining.

(4) An entry into a mine from the surface, usually dipping at about 1 in 5, enabling people to walk in or out, usually provided with a conveyor to bring out minerals, often also with rail track. It is safer and more flexible, though five times as long (at 1 in 5), as a vertical shaft.

(5) Superficial, geologically recent, loose deposits of gravel, boulder clay, sand, etc., which are often present in river valleys (Thames gravel). They are not shown on *solid* geological maps.

drift barrier An open structure of ropes or chains across a stream, to catch driftwood for removal at low water. See **debris dam**.

drift bolt See **drift** (3).

drift test [mech.] A test of metal plates or tubes in which a *drift* (3) tapered at 1 in 10 is forced into a hole in the plate until it cracks or until a certain specified increase in diameter is obtained.

drill carriage, jumbo A wheeled or tracked self-propelled mining or quarrying vehicle carrying one or more rock drills, now often hydraulically powered. Modern vehicles of this type can also do the *scaling* of the face needed after *blasting*, with the driver under cover, safe in the cabin controlling a *hydraulic hammer*. They may also drill holes for and even place *roof bolts*. One of the most powerful machines can auger two parallel holes 1 m (39 in.) apart, each of 0.4 m (16 in.)

dia. See **computer-controlled tunnel drilling**.

drilled pier A *bored pile*.

drill extractor [min.] A *fishing tool* for pulling a drill out of a borehole.

drill feed [mech.] (1) The mechanism for pushing a drilling tool into a hole.

(2) The speed at which a drilling tool is pushed into a hole, and therefore the speed at which it drills rock.

drilling fluid, d. mud, mud flush [min.] The mud which is pumped into the drill pipe in *rotary drilling* and rises up outside the pipe, filling the hole, ensuring that gas or oil does not escape and plastering the walls at the same time. It was developed together with rotary drilling about 1890, but can also be used, though with less refinement, in cable drilling. Mud flush was designed to reduce the disastrous *blow outs* of the early holes. Thanks to mud flush, blow outs are now rare. A typical water mud contains *bentonite* to make it *thixotropic*, barytes to increase its density, and water. Specific gravities may be as high as 2.1 but are generally below 1.7. Oil-based muds with less than 5% water are used where the water in the mud might penetrate and damage the oil pool. Salt-saturated muds are used for drilling through the Texan salt domes to prevent the mud dissolving the salt and causing the hole to deviate. An ideal mud is thixotropic but not very viscous, able to cool the drilling bit, unaffected by high temperature (which at 5000 m depth may rise to 160°C), and has good wall-building (plastering) properties to prevent loss of mud from the hole.

The sophistication of drilling fluids has reached such a level that it is normal for an exploration platform to have a mud engineer, often a chemist, on the rig who can decide e.g. whether in the event of serious losses of mud it might be advisable to use a foam mud to reduce its weight and to improve the penetration rate. See **drill rods**, **heaving shale**, **oil-well cement**, **turbodrill**.

drilling mud *Drilling fluid.*

drill rods [min.] In *rotary drilling*, the hollow rods which form a continuous string in the hole, carry the cutting bit screwed on to the bottom end, and act as pipe for the *drilling fluid* to be injected on to the fresh rock under the bit.

drill steel [min.] In *rock drilling*, the round or hexagonal steel bar (not less than 18 mm across the flats) which enters the hole in the rock. It has a drill bit forged or screwed on to its end. See **detachable bit**.

drip See (*B*).

drive (1) [mech.] The means by which mechanical power is transmitted to an implement or piece of mechanical plant such as a conveyor, rock crusher, vibrator, etc. See **power pack**.

(2) In the USA a *dedicated street* or road, but in Britain an access to a private house.

driven cast-in-place pile, compressed concrete p. A *pile* made by driving a steel casing into the ground and ramming it full of concrete. The casing may be withdrawn and reused. The Franki pile is one type.

driven pile A *pile* jacked or hammered or vibrated into the ground. See also **bored p., jacked p., jetting**.

drive pipe [s.m.] A pipe driven into the ground, often for *soil survey* purposes, within which it may sometimes be possible to obtain an *undisturbed sample* of the soil at the bottom of the hole. In sands or gravels it may be quickly sunk by *wash boring*. Like *casing*, it is often reusable.

driveway US term for a private access, that is a British *drive*.

drive wheels A 4 by 4 vehicle usually has four wheels, all of them driven, but it may have two rear axles – four rear wheels – only two of which are driven. An 8 by 4 vehicle has eight wheels of which four are driven. See also **rubber tyres**.

driving See **pile hammer**.

driving band, pile hoop, p. ring A steel band fitted round the head of a timber pile to prevent *brooming*.

driving cap, d. helmet A steel cap placed over the head of a steel pile to reduce the damage during driving. See **pile helmet**.

drop (1) [hyd.] A steep part of a channel or pipe.

(2) The part of a reinforced-concrete *mushroom slab* which is immediately round the column head and is deeper than the rest of the slab.

drop-bottom bucket, bottom-opening skip A container for placing concrete.

dropchute, articulated d. A succession of tapered metal or plastic cylinders, each with its lower end fitting into the top of the one below. Used in the UK as refuse chutes, in the USA they also place concrete.

drop hammer, beetle head, monkey, ram, tup A metal block which is raised by a hoist and allowed to drop freely on to a *pile head* to drive it into the ground. For steel or timber piles a drop hammer should, for easy driving, either equal the weight of the pile or be rather lighter. For concrete piles a drop hammer may weigh up to three quarters the weight of the pile, if the *dolly* effectively protects them, and should weigh at least one third of the weight of the pile and dolly. Steam hammers should weigh roughly the same as corresponding drop hammers. See **pile hammer**.

drop on [min.] A portable rail crossing which can be laid on top of two parallel tracks which are set at a distance apart equal to the gauge of the track. By this means the inner two rails can be used as a track and wagons can be transferred from one track to the other.

drop panel The deepened panel of a *mushroom* floor around a *column head*, reducing the concrete stress in this part of the slab.

drop penetration test [s.m.] A *dynamic penetration test*.

drop shaft, cylinder caisson, open c. A *caisson* for shaft sinking (or excavating bridge piers, etc.) in soft waterlogged ground, consisting of building up at ground level, on a *cutting curb*, a massive concrete, brick or iron ring (the drop shaft or caisson) which can serve as or contain the permanent shaft lining. Soil is removed by *grab* within the drop shaft, which gradually sinks under its own weight without dewatering.

drowned weir, submerged w. [hyd.] A weir whose *tailwater* level is above the weir crest. Modern *measuring weirs* can be used for measuring in this condition, which is desirable in several ways. It lowers the cost of the structure and reduces the obstacle that can prevent fish moving upstream.

drum [mech.] A large cylinder (or part cone) on to which steel rope is wound, used for hoists and haulages in mines and elsewhere.

drum curb [min.] A *cutting curb*.

drum gate [hyd.] A spillway gate shaped like a sector of a circle. It is opened or closed by admitting or releasing water through appropriate valves.

drum screen [sewage] A screen

shaped like a cylinder or cut-off cone, turning on its centre line.

dry-bulb thermometer An ordinary thermometer. It indicates the dry bulb temperature. Cf. **hygrometer**.

dry density [s.m.] The weight of dry material in unit volume of a soil sample after drying at $105°C$. See **compaction** and below.

dry-density/moisture-content relationship [s.m.] The relationship between *dry density* and *moisture content* of a soil for a given amount of *compaction*. The relationship is usually drawn on a graph from which the *optimum moisture content* can be deduced.

dry dock, graving d. A *dock* into which a ship floats. The dock gates are closed behind it, the water is pumped out, and the ship rests on the *docking blocks* ready for its hull to be repaired or cleaned. See **flotation structure**.

dry galvanizing [mech.] A *galvanizing* process in which the steel is first fluxed in hot ammonium chloride solution and then dried in hot air before passing through the bath of molten zinc. See **wet galvanizing**.

drying shrinkage See **shrinkage**.

dry joint [stru.] A plane of contact between two structures or parts of a structure, to allow relative movement caused by shrinkage, expansion or settlement. There is no connection between the adjacent parts which may, in fact, be separated by *building paper* (*B*). See **expansion joint**.

dry lean concrete *Lean concrete* with a very low *water/cement ratio*, used for *blinding concrete, base courses, roller-compacted concrete*, etc. See **forced-action mixer**.

dry mix (1) Concrete or mortar sold dry in bags. Only water has to be added but the *mix* is likely to segregate in the bag, so must be remixed before use.

(2) **dry spraying, dry shotcreting** A *shotcreting* method in which the aggregate and cement are premixed, fed dry by an air blast to the nozzle, and only there mixed with water and *accelerator* if need be.

dry pack Concrete or mortar which is just damp (described as earth-dry), used as filling to join up two load-bearing members by ramming it in with a hammer and cold chisel or piece of wood. It is used between the head of a *cast-in-situ* pile and the building above it which it *underpins*. It is also used for joining precast members in *prestressed* structures. Dry pack gives much less shrinkage than fluid *grout* and can be loaded immediately.

dry weather flow (dwf) The rate of flow in a sewer in prolonged dry weather. It includes *infiltration*, but apart from infiltration the dwf should equal the water consumption of the population of the area drained. Foul sewers should be designed to carry all quantities between one third and four times the dwf, occasionally six times the dwf but not more than this in a truly *separate system*. A partially separate sewer may have to carry fifty times its dwf after a rainstorm.

dry well (1) An American term for *soakaway* (in Scotland a rummel).

(2) That part of a pump house which contains the machinery, as opposed to the *wet well* or sump, which contains water.

dual carriageway (d. highway USA) Two separate roads, one for each direction of traffic.

duct A passage or protective tube along which pipes, cables or people can pass through ground or a structure. See **cable duct** and (*B*).

duct bank A group of *ducts* together.

duct former Any device of steel, timber, plastic sheet or *inflatable* tube, for making a rectangular or rounded passage through concrete for a cable or other purpose. The duct former is fixed to the *formwork* before the concrete is placed, and removed after it has set. Inflatable duct formers must be deflated and removed within 4–12 h of concreting. They must be solidly anchored to prevent them floating in the concrete. Ducts may be used also to reduce the weight of a concrete unit.

ductile [mech.] That which can be drawn into wire. Many metals are ductile, the commonest being mild steel, wrought iron, copper, lead, light alloys.

ductile iron pipe Pipe that is almost as strong and ductile as steel but cheaper. It is more ductile than *malleable cast iron*. Its minimum *elongation* is 10% and its minimum *tensile strength* is 420 MPa (60,000 psi). The 0.2% proof stress is 300 MPa (43,000 psi). The commonest inside lining is cement mortar – in France often with *high-alumina cement*. Outside protection may be by zinc sprayed on in the works or in corrosive ground by a polythene sleeve.

ductility [mech.] The ability of a metal to undergo cold *plastic deformation* without breaking, particularly by pulling in *cold drawing*.

ductility plateau After the *yield point*, the *stress/strain curve* has a short level length, the ductility plateau, whose length is a measure of the *ductility* of the material. This is important in any structure because it allows an overloaded member to adjust safely to some overloads.

Ductube An inflated tube used for forming *cable ducts* in concrete. It is tied to the reinforcement, left until the concrete is a few hours old, then deflated and withdrawn.

Duff Abrams' law See **Abrams' law**.

dumb barge, d. dredger, scow A towed barge or *dredger*.

dummy gauge An *electrical resistance strain gauge* which is set in a direction perpendicular to the direction of strain. Being perpendicular it carries no strain but its resistance may change because of a change in temperature. If so, the value of this change may be subtracted from or added to the change in resistance of the active gauge to compensate for the temperature change.

dummy joint, crack inducer A *contraction joint* in a concrete road slab, consisting of a groove formed either through the top half of the slab with a fibrous *joint filler* or through the bottom half with a piece of wood or proprietary plastic crack inducer. The plane of weakness thus formed ensures that the break is vertical and does relatively little damage.

dumper A rubber-tyred vehicle with four wheels. The driver is at the back, with the load in front, and dumps it forwards or sideways. Dumpers exist in very small sizes, down to 500 kg load. *Dump trucks* are larger.

dumpling In a large excavation such as a railway cutting or dry dock, a dumpling is a mass of ground with excavation on two or more sides, left untouched until the end of the dig. It is used as an abutment for timbering the sides of the dig and this saves the expense of long, heavy timbers.

dump test, upending t. [mech.] A test for the detection of surface defects which shows the suitability of billets, bars and so on for hot or cold forging. A bar 2 diameters long is shortened by cold-squeezing to 1 diameter and should then show no cracks.

dump truck (hauler USA) (see diagram) Like a *dumper*, dump trucks are usually four-wheeled and move earth or other bulk loads, but the driver sits in front and the machine may be very large and not used on public roads. A large dump truck carrying 250 tons of earth has a 3000 hp electrical generator supplying the electrical power to each of the separate motors on the six driving wheels.

dumpy level [sur.] The simplest levelling instrument, now more or less superseded by *automatic* or *tilting levels*.

dune [hyd.] A sandhill formed by wind (or in water by current), usually with a long straight crest, often with *ripples* on its surface. Underwater dunes are much smaller than wind-formed desert dunes, which can be 100 m high. See **antidune, saltation**.

duplex engine [mech.] A steam-driven pump or compressor engine arranged with two steam pistons, each of which drives one pump (or air) piston on the same piston rod. The machinery is simple as there is no rotating motion. See **free-piston compressor**.

duplex-headed nail See **double-headed nail**.

duplex track [rly] A track built on a quayside to carry heavy crane loads. Two *flat-bottom rails* are laid in contact with each other at the flanges. The crane wheels are also double, with the flanges placed centrally and in contact. Each wheel can thus travel on its own rail and the wheel flanges pass down the gap in the middle.

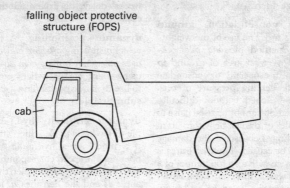

falling object protective structure (FOPS)

cab

Dump truck – rear-dump, side-dump or bottom dump, rigid or articulated.

duralumin A common *light alloy* of aluminium, with originally (1909) 4% copper and 0.5% each of magnesium, manganese and silicon. It age-hardens at room temperature after *quenching* to reach a 0.2% *proof stress* of 240 MPa. Its working stress in tension is about 110 MPa. This compares with the *yield point* at 230 MPa of mild steel. Copper-free 'duralumins' have good resistance to corrosion and proof stress of at least 130 MPa.

duration curve, rating c. [hyd.] A curve which shows the amount of flow through a river for power genera-tion. It gives for a certain period, the abscissa, the percentage of the time that the flow equalled or exceeded the ordinate. The area under the curve represents the total quantity (the *run-off*) which flowed down the river in the period under consideration. The Q 95 flow is a low rating, the flow which occurred for 95% of the time. The Q 60 or median flow is the middle rating. The average rating is the arithmetic mean flow.

dusting A description of a concrete surface that is disintegrating to dust. Dusting may result from too much *release agent*, lack of *curing*, excess water in the mix, *laitance* floating to the surface, or because of dirty sand. Dusting may be reduced by painting or spraying special chemical hardeners on to the concrete.

dustpan dredger A *draghead dredger*.

Dutch deep sounding [s.m.] See **cone penetration test**.

dyeline [d.o.] A *contact print* which shows brown-to-black lines on a white-to-pink background. These prints can be made on opaque paper or cloth, or tracing paper or tracing cloth. They are generally not so easy to read as a *blue-print* when printed from faint pencil negatives, but the white ground has made them more popular, since notes and corrections can be written on them.

dyke, dike (1) [min.] A tabular-shaped *igneous intrusion*.
(2) [hyd.] A mound of earth along a river bank and at some distance from the river to retain floodwater; but see **polder**.

(3) A large ditch.

dynamic consolidation, ground bashing [s.m.] The use of the weight-dropping method of *sand piles*, extended all over an area of ground to strengthen it, especially to consolidate *fill*. A safe bearing pressure of 0.16 N/mm^2 is reached without difficulty but the method cannot be used close to pipes, cables or drains, or within 20 m of a building, because of the damage that could be caused. The whole area is covered twice with a 10 ton tamper of 4 m^2 at two weeks' interval. Total settlement of the surface is 50–60 cm.

dynamic factor An *impact factor*.

dynamic loading [stru.] Loads which include *impact*.

dynamic penetration test, drop p. t. [s.m.] Tests such as the *Raymond standard test* as opposed to *static penetration tests*.

dynamic pile formulae Pile-driving formulae are an unreliable basis for calculating the loadbearing capacity of a pile. Piles in or on clay can go on settling for years, while others may be driven to refusal and then after two weeks' delay, be driven another 5 m easily.

dynamic positioning A drilling vessel or floating *offshore platform* that uses no anchor but stays in position by using water jets, propellers, etc. is said to use dynamic positioning. See **semi-submersible**.

dynamic similarity [hyd.] A principle of *model analysis*, which states that if a model of a hydraulic structure operates at a speed corresponding truly with the full-size project then the resistances R and the densities d, lengths l and velocities v are in the following relationship:

$$\frac{R_1}{R_2} = \frac{d_1 \times l_1{}^2 \times y_1{}^2}{d_2 \times l_2{}^2 \times y_2{}^2}$$

Dynamic similarity implies that the forces on model and prototype are in proportion with the scale of the model. See **dimensional analysis**.

dynamic strength [stru.] Resistance to suddenly applied loads.

dynamite [min.] Old term for any *high explosive* consisting of *diatomite* (*B*), as an absorbent containing *nitroglycerin*.

E

E [stru.] The *modulus of elasticity* of a material, the value which tells what its *stiffness* is.

early civil engineers The first civil engineers came from four trades: millwrights like Brindley, Murdoch, Rendel and Rennie; land surveyors like Rendel and many railway engineers; masons like Telford; or instrument makers like Smeaton. James Watt, at first an instrument maker, had to survey for and set out Scottish canals for a living. Rendel was lucky: his father was a surveyor and he was apprenticed to his uncle, a *millwright*.

earth auger [s.m.] See **auger**.

earth borer A *truck-mounted drilling rig*.

earthen dam A *dam* built of earth, sand, gravel or rock (*rock-fill dam*), having a core of clay, concrete or other impervious material or an impervious skin on the water face, made of concrete, steel plate, etc. Where the soil available is sand or silt and there is a considerable amount of water, a *hydraulic fill dam* is usually economical. If hydraulic fill is not feasible the earth is usually transported and tipped by heavy rubber-tyred vehicles which help to compact the dam by travelling over those parts which need *compaction*. Special *rollers* which weigh up to 200 tons (with water ballast) may also be used for compaction. Earth is a suitable material for a dam whose foundation is likely to move. See **pore-water pressure**.

earth flow See **flow slide**, **detritus slide**.

earth-leakage protection [elec.] *Protective equipment* which switches off the power as soon as its *relays* detect current passing through the *earthing lead* (*B*). A very small leakage is enough to operate the relay and break the circuit. Since a fraction of a second is needed to break the circuit, the man on the machine may get an electric shock but it is unlikely to be dangerous since it cannot last for long.

earth-moving plant, muck-shifting p. *Bowl scrapers, dozers, dumpers, excavators, graders, loaders.*

earth pressure [s.m.] Earth pressure is a push from retained earth that varies between two extremes: the minimum, or *active earth pressure*, which is the force from earth tending to overturn a *free retaining wall*; and the maximum, the *passive earth pressure*, which is the resistance of an earth surface to deformation by outside forces. The word 'pressure' in this sense is wrong since it implies a force, but it is customary. There are three phases of earth thrust: (i) its gradual application to the new wall; (ii) a slight movement of the wall reducing the thrust; (iii) the final *active earth pressure*.

earth pressure at rest [s.m.] The thrust from earth on to a *fixed retaining wall*. It is intermediate between the *active* and the *passive earth pressures*.

earth-pressure-balance shield (EPBS) A *shield* that eliminates the need for a *compressed-air* atmosphere, and is used with a *tunnelling machine*. The volume excavated is mechanically controlled. A screw conveyor delivers

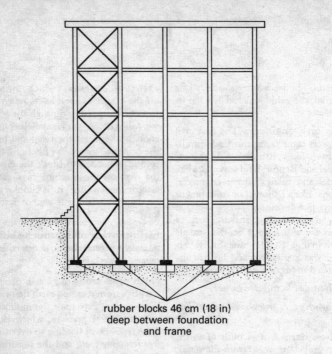

rubber blocks 46 cm (18 in)
deep between foundation
and frame

Earthquake-resisting structure of the courthouse of San Bernardino, California (see also **lateral-force design**).

it through a pipe passing through the *bulkhead* on to the belt conveyor or wagon. The 'sand plug' thus formed in the pipe by the conveyor can withstand the pressure of a collapsing face. At the delivery point on to the conveyor, an additional control may be provided in the form of a rotary discharger. One EPBS type controls the pressure of the cutting wheel on the face rather than the volume of excavation. Removal of muck may be by belt, pipeline, etc. EPBSs have become a Japanese speciality.

earthquake (see diagram) Sinking of part of the earth's surface along a fault plane, i.e. a crack in the earth's crust.

Buildings designed to stand up in an earthquake are usually nowadays calculated according to *lateral-force design*. Earthquake intensities are recorded on *seismographs*.

They occur at cracks in the earth's crust on the edges of its 15 or so vast 'tectonic plates'. Any abrupt fault movement is an earthquake. See **life-safety design**, **Rossi-Forel**.

earthwork (1) Any digging or artificial raising of the ground.

(2) The volume of dig or banks which is paid for.

easement curve, easement A *transition curve*. See also (*B*).

easers [min.] British term for *relief holes*.

easting [sur.] An eastward *departure* from the north–south axis of a survey. A westing is a negative easting.

ebb channel [hyd.] A channel in an estuary made by the river at low water, distinguished by its S-shape in a long estuary. Sand bars have a seaward direction in an ebb channel. Cf. **flood channel**.

EC Prefix to a *Eurocode*. Cf. **EN**, **Euronorm**, **ENV**.

eccentric [d.o.] Away from the centre, whether purposely or not.

eccentricity [stru.] The distance between the point of application of a direct load to a column or tie and the *centroid* of the member. It is sometimes called the *arm*.

eccentric load [stru.] A load on a column applied at a point away from the column centre and therefore putting a *bending moment* on the column equal in amount to the load multiplied by the *arm*.

echo sounder [sur.] An instrument used by ships and in harbour surveys for determining the depth of water. It measures the time taken for a sound to be echoed back from the seabed. It is much quicker than *sounding* with a lead and line.

ecology The relationships between living organisms and their surroundings, including vegetation, fishes and animals, their foods and populations. What is good for humans may be bad for the rest of the *ecosystem*. In hot countries large reservoirs for human drinking-water are better than small ones because, per unit volume of water, they have shorter shore lines and fewer shallows for snails and mosquitoes to breed. But these large, deep reservoirs

are less good for the many birds and fish which eat snails and mosquitoes. Snails harm humans because they carry disease from one host to another. See **Aral Sea, vector**.

economic ratio [stru.] A design of a reinforced-concrete beam is said to have the economic ratio of steel if the concrete and steel are both stressed to the maximum. Such a design is generally not the cheapest. If cheapness is the aim, the steel should be fully stressed; the concrete should usually be understressed.

economizer [mech.] In any but the smallest boiler plant, a bank of tubes through which the boiler *feedwater* passes. It is placed across the path of the flue gases as they leave the boiler and enter the chimney. By warming the feedwater it saves considerable fuel.

ecosystem An ecological system, a system in which there is a steady cycle of interchange between living organisms and their surroundings. In a tropical rain forest, for example, the dead leaves and branches created by *photosynthesis* drop to the ground where they rot and provide foods needed by tree roots. The forest flourishes even though the soil may be poor.

ecosystem development strategy (EDS) Any plan worked out to help life. An EDS for a lake could be designed to reduce the *pollutants* flowing into it in sewage *effluent*. An EDS might include aeration and stocking with fish. Elimination of one substance is easier than eliminating several. The phosphorus content of sewage effluent, for example, can be reduced by more than 90%. The land surrounding the lake and the streams flowing into it should be forested where possible and kept free of fertilizers. Soil erosion should be controlled by soil conservation methods such as *contour ploughing*. Single-crop farming should be eliminated if possible.

eddy flow [hyd.] *Turbulent flow*.

eddy loss [hyd.] The energy lost by eddies as opposed to that lost by friction. Both are converted into heat.

Eddy's theorem [stru.] In an arch the bending moment at any point is equal to the product of the horizontal thrust and the vertical distance between the arch centre line and the *line of thrust*.

edge preparation The edge of metal before *welding* usually has to be cut or ground to suit the welding process. Edge preparation may be a large part of the cost of making the weld.

edger, curb tool US terms for an *arrising tool*.

EDM [sur.] An *electronic distance-measuring* instrument.

eductor A *hydraulic ejector*.

effective area of an orifice [hyd.] The actual area of an orifice multiplied by its *coefficient of discharge*.

effective depth [stru.] In the design of reinforced-concrete beams and slabs, the depth from the outer face of the compression flange of the concrete to the centre line of the stretched steel. The *cover* plus half the bar diameter plus the effective depth are together equal to the overall depth.

effective height of a column [stru.] In the calculation of *slenderness ratio*, a value varying from 0.70 × the actual column length, for a column fully restrained in position and direction at both ends, to 2 × the actual column length for a column fully restrained at one end and free at the other end.

effective intergranular pressure [s.m.] *Effective pressure*.

effective length of a strut [stru.] See **effective height of a column**.

effective pressure, intergranular p., effective stress [s.m.] The pressure in a soil between the points of contact of the soil grains. In a soil system in equilibrium it is equal to the *total pressure* minus the *neutral pressure* of the water in the pore space. It increases during the *consolidation* of the soil to a maximum at complete consolidation. In conditions of *hydrostatic excess pressure* the effective pressure is equal to the total pressure minus (neutral pressure plus hydrostatic excess pressure).

effective size [s.m.] Hazen's definition of an effective size of a soil is the following, used in *Hazen's law*. It is the grain size which is larger than 10% by weight of the soil particles, as seen on the *grading curve* of the soil, and is described for short as the D_{10} size. See also **coefficient of uniformity**, **graded filter**.

effective span [stru.] The distance between the centres of the supports of a beam. This length, which is larger than the *clear span*, is used in calculating the *bending moment* of a beam.

effective stress (1) [s.m.] *Effective pressure*.

(2) [stru.] In *prestressed concrete*, the prestress remaining after subtracting all *losses of prestress*.

effective thickness of a wall [stru.] In the calculation of *slenderness ratio*, (i) for plain brick or masonry walls it is the actual thickness; (ii) for *cavity walls* (B) it is two thirds of the thickness of the two leaves added together.

efficiency, mechanical e. [mech.] A very widely misused term which means the work output of a *machine* as a percentage of its theoretically perfect performance. For an engine or motor it is the power output divided by the power input (energy input per unit of time). Efficiency is always less than 100%. See **mechanical advantage**.

efflorescence See (*B*).

effluent Liquid or gas that flows away. So far as *sewage* is concerned, it is some 99.9% water, but a desirable standard of purity was set by the *Royal Commission*.

effluent stream [hyd.] A stream that receives *groundwater*, unlike an *influent stream*.

egg-crate structure [stru.] A grid layout with closely spaced cells, e.g. a concrete *raft foundation* with intersecting beams.

egg-shaped sewer A sewer shaped like an egg with its small end down, a shape chosen for its satisfactory flow whether empty or full.

E I A (environmental impact assessment), EIS (e. i. statement) See **environmental impact**, also EEC document 85/337/EEC.

Eider jacket An economical *jacket* for the marginal Eider oilfield, 190 km north-east of Shetland. See **pile sleeve** (diagram).

ejector (1) [sewage] A pump for raising sewage by injecting compressed air into a pipe containing the sewage, on the principle of the *air-lift pump*.
(2) See **hydraulic ejector**.

elastic [stru.] A material is said to be elastic if it expands or contracts by foreseeable amounts when it is pulled or compressed, and regains its shape when these forces are released. See **elasticity**, **Hooke's law**, **modulus of elasticity**, **resilience**.

elastic constants [stru.] The *modulus of elasticity*, the *shear modulus* and the bulk modulus (the change in stress per unit change in volume). A fourth constant, *Poisson's ratio*, is sometimes included.

elastic curve, deflection curve [stru.] The curve showing the deflected shape of the neutral surface of a bent beam, an essential part of the theory of bending. Its shape was first appreciated by Jakob Bernoulli in 1694 and later by Euler as a circle of radius equal to $\dfrac{EI}{M}$.

elastic design [stru.] The design of a structure to *permissible stresses* which are about half to two-thirds of the elastic limit. This was the usual method of design, but for redundant frames it is being replaced now by design based on *limit states*.

elasticity [stru.] The property of materials that returns them to their original shape and size after loading and deflection – the opposite of *plastic deformation*. Most structures behave elastically throughout their lives, but in the extreme conditions of *limit states*, parts of a structure are assumed to be loaded beyond the *elastic limit*.

elastic limit [stru.] The stress beyond which further load causes *permanent set*. In most materials the elastic limit is also the *limit of proportionality*.

elastic moduli [stru.] The *elastic constants*.

elastic rail spike [rly] A rail fastening.

elastic strain [stru.] A *strain* (or deformation per unit of length) produced by a force acting on a body, which disappears when the force is removed. See **Hooke's law**.

elastomer Any natural or synthetic rubber.

elastomeric laminate Layers of rubber stiffened by steel plates.

elbow A sharp corner in a pipe, roadway, etc. See also (*B*).

electrical interference See **electronic noise**.

electrical-resistance strain gauge [stru.] A flat coil of fine wire (0.025 mm dia.) wound round an insulating plate between two sheets of insulating paper, of total size about 1 × 2 cm. The gauge is glued to the surface of metal or cast into concrete and the ends of the wire connected to a sensitive electrical-resistance measuring instrument. As the length of the part increases so also does the resistance increase. An advantage of these gauges is that they can be oriented in any direction to indicate the *strain* in that direction.

electric-arc welding [mech.] See **arc welding**.

electric drills Drills driven by an electric motor having a rotating bit and no percussive action are used for drilling coal, wood and steel, and have also drilled rock. Percussive, rotary-percussive and hydraulic electric drills are well established. See **rotary drilling**.

electric eye See **photo-electric cell**.

electric motor A *machine* which is forced to rotate by electric power, brought usually by a cable. Electric motors, first used about 1890, are the most compact source of power available and are now used for almost every conceivable purpose in industry and construction. See **hydraulic power**.

electric shock Injury due to contact with high-voltage electricity. The loss of consciousness which often follows severe shock may prevent the shocked person from disengaging himself or herself. The power should be cut off or short-circuited, whichever is the quicker, the victim removed from the electrical conductors, and artificial respiration applied. If the power supply cannot be interrupted quickly the victim should be dragged away from the contact by a trouser leg or other loose garment so that the rescuer is not shocked also.

electric traction Haulage of vehicles by electric motor. The motor may be supplied with electric power by *catenary suspension*, storage battery or diesel-driven generator on the vehicle.

electro-chemical hardening, stabilization of soil [s.m.] Certain clays, especially those that contain monovalent cations (potassium, sodium, lithium), can be strengthened by passing a direct current through them with an anode consisting of a polyvalent metal such as iron, aluminium or calcium. Dissolved additives at the anode may also help the base exchange reaction in the clay; calcium chloride is a common additive. The polyvalent cations form clays that adsorb water less easily.

electrode [elec.] Generally a conductor leading electric current into an electrolytic cell, furnace or welding implement. In welding, the meaning is specialized as follows: (i) in *metal-arc welding*, the *filler rod*, bare or *covered*; (ii) in *carbon-arc welding*, a carbon rod; (iii) in *resistance welding*, a bar, wheel or clamp which presses together the metal parts to be welded.

electrodialysis (ED) An electrical *desalination* process relying on ion-selective membranes impervious to water, using e.g. 100 A at 400 V if there are 200 cells. The membranes allow ions through in the form of electric current and partially desalt the water. ED can desalt water down to 200 ppm economically but does not remove anything except ionized solids and is fouled by some other impurities. ED works most economically desalting a feed water at up to 7000 ppm dissolved solids but preferably from 2500 ppm down to 500 ppm and can compete with RO (reverse osmosis). Packaged systems can purify from 100 to 4000 m^3/day, working at any pH between 1 and 13 and up to 43°C. It wastes less water than RO and like RO is clean, demanding no fuel, but it does need skilled people.

ED, or electrodialysis,
Is no longer restricted to palaces.
Its cost can be low
Compared with RO
On a suitable water analysis.

electrogas welding A fully automatic type of *MIG welding* in which a gas-shielded or flux-cored *consumable electrode* deposits metal into a molten pool which is surrounded by water-cooled copper shoes moving upwards as the joint is made. It makes vertical joints in thick steel from 12 to 75 mm thick. The usual shielding gas for steel is carbon dioxide or argon.

electrolysis [elec.] The conduction of electric current through *electrolytes*. Direct current causes metal or hydrogen to be liberated at the *cathode* and acid radicals or oxygen at the *anode*.

electrolyte [elec.] A liquid which conducts electricity, or a salt which makes water highly conductive when dissolved in it. Most soluble salts, acids or alkalis and many fused salts are electrolytes. See *aluminium*.

electrolytic copper, e. lead, e. zinc [min.] Copper, lead or zinc obtained by electrolytic refining, 99.9% pure.

electrolytic corrosion [mech.] *Corrosion*.

electromagnet [elec.] A soft iron bar with thick copper wire wound round it, through which a heavy magnetizing (direct) current passes. An electromagnet is often used for separating tramp iron from ore or coal before it is crushed. See **lifting magnet, solenoid**.

electromagnetic length measurement [sur.] *Electronic distance measurement.*

electronic distance measurement (EDM), electromagnetic, electro-optical, geodetic-distance, length measurement or **metering** [sur.] In 1989 all major makers of surveying instruments had a wide variety of EDM instruments with different *carrier waves*, ranges and accuracies. In the past, before EDM, distance measurement was avoided, but fast, accurate EDM has made distance measurements faster than angular ones and therefore preferable. For short ranges and high accuracies, EDM instruments like the *Mekometer*, Geomensor and *Distomat 2000* can measure accurately to within less than 1 mm ± 2 ppm over distances from a few metres to 5 km. At longer ranges, instruments like the *Tellurometer*, Electrotape and *Geodimeter* are accurate up to 100 km within 3 cm ± 3 ppm. Except over very short ranges, the atmospheric conditions at the time of measurement most limit the accuracy. Because of EDM, *traversing* and *trilateration* are now preferred to *triangulation* for survey networks.

electronic noise, electrical interference, crosstalk, disturbance, etc. Unwanted fluctuations in the mains power supply may trouble the workings of radio, television, computers or electronic control equipment. They may come from the nearby radio transmitter of the police or taxi company or from neon lighting, lightning discharges, etc., which can be picked up by the power mains acting involuntarily as antennas (aerials).

Radio signals are high-frequency electromagnetic waves and can be kept out by connecting a low-pass filter into the circuit; this allows only low-frequency waves to pass. Any filter must be as close as possible to the circuit to be protected.

Electrostatic discharges also may cause trouble. Someone walking in rubber shoes over a carpet who then touches a steel filing cabinet may suffer a usually harmless shock of several thousand volts. Such a discharge can affect electronic equipment. An electrically operated lift has caused computer trouble because, drawing power from the same mains, it caused voltage sags when the lift started and surges when it stopped. (*BRE Digest*, 335, 1988.)

electronic tacheometer [sur.] A combination of a *theodolite* and an *EDM* which enables distances and angles to be measured together. The term is usual where the EDM and theodolite can be used separately. See **total-station instrument**.

electro-optical length measurement [sur.] *Electronic distance measurement.*

electro-osmosis [s.m.] A *groundwater lowering* process, used in silts to speed up natural drainage and to produce a flow of water away from an excavation. (When a direct current is passed through wet soil, water flows to the cathode.) Small quantities of water are pumped away at the cathodes, which are about 10 m apart, with anodes intermediately. The process is expensive but greatly increases the strength of silts owing to the change in direction of *capillary* forces in them and is useful since other pumping methods cannot be applied to silts. The cost in electrical energy varies from $\frac{1}{2}$ to 10 kWh/m^3 of silt excavated. The field of use at present is in those silts which are too fine for the vacuum method of *groundwater lowering*. Voltages of 40–180 V direct current have been used, with currents from 15 to 25 A per well (BSCP 2004). The principle has also been used: (i) to reduce the voids in concrete and thus increase its strength, somewhat similarly to the *vacuum concrete process*; and (ii) to dry out walls that have a defective *damp-proof course* (*B*). See **osmosis, thermo-osmosis**.

electroplating [elec.] The deposition of a thin film of one metal on another by *electrolysis*, for instance in electrotinning or electrogalvanizing. The noble metals, nickel, chromium, copper, cadmium, and others, are also put on by electrolysis.

electroslag welding An *automatic welding* method suitable for welding steel plates usually from 20 to 460 mm thick, with high rates of metal deposition and few weld defects, provided that welding is not interrupted. A *consumable electrode* is used in a conducting bath of molten slag. The molten metal and slag are held by water-cooled shoes moving upwards. The process begins with arcing, but once the slag and metal are fluid, the heat comes from the electrical resistance of the weld pool.

electrostatic precipitator, dry scrubber A common dust-catcher in power stations burning refuse or *pulverized coal*. Though 99% of the dust is often caught, any increase from 99% to 99.9% may mean doubling the size and cost. This equipment on coal-fired furnaces provides *fly-ash* for civil engineering.

elephant's trunk A *hydraulic ejector*.

elevated railway [rly] A rail track carried on a bridge supported above road level by columns passing through the road.

elevating grader A *grader* equipped with a disc or plough collector and a belt elevator at right angles to its direction of travel. It digs earth or loose materials and discharges it to a height. It is very suitable for excavating road or railway cuttings or wide trenches. Outputs of as much as 300 m³/h have been obtained with these machines in the USA, but they have had little success in the wet climate of Britain, where soils are probably too sticky. See **multi-bucket excavator**.

elevation (1) [d.o.] A view of (a part of) a machine or a structure drawn without perspective as if projected on to a vertical plane. See **sectional elevation**.

(2) [sur.] The altitude of a point, its height above the sea.

elevation head, position h., potential energy Energy due to position. It is the product of the density of a fluid and its height above the point of reference. For water it is usually stated in metres or feet of height without mention of density.

elevator [min.] A contractor's or mining *conveyor* which raises material up a steep slope usually at about 60°, the material being contained in a series of buckets. It is usually no longer than 15 m and can move from 5 to 120 tons of material per hour.

elevator dredger [hyd.] A *bucket-ladder dredger*.

ellipse of stress [stru.] An ellipse which, drawn proportional to the *principal stresses* in a plane at a point, shows the resultant stress at any angle through the point in magnitude and direction. See **Mohr's circle of stress**.

elliptical trammel [d.o.] A *trammel*.

elongation (1) [stru.] *Elastic* or plastic extension of a structural member, particularly the plastic elongation of a piece under tensile test. For mild steel the elongation is often 20% of the *gauge length*. The elongation is a good measure of the *ductility* of a metal, so BS 5400, for bridges, demands an elongation not less than 15% on a 20 cm (8 in.) *gauge length*. See **contraction in area**.

(2) [sur.] The extreme eastern or western position of a star, sometimes observed (in *circumpolars*) for determining the true meridian or *azimuth*.

elongation index A number that gives an idea of the general shape of the stones in an aggregate. It shows how long they are in relation to their nominal size. The index is found by weighing the pieces that are longer than their nominal aggregate size and dividing their weight by the total weight of the sample. The sample must be *representative* and is sieved with a full series of screens from 5 to 63 mm, aiming at 100–200 pieces on each screen. The procedure is laborious but useful. See also **angularity number**.

elutriation [s.m.] Size classification of airborne or liquid-borne particles by their settlement speeds. See **elutriator, Stokes's law**.

elutriator [s.m.] A *classifier* which works on the principle that large grains sink faster through a fluid than small grains of the same material. Industrial elutriators for mineral washing or classification usually work by a continuous upward current. Laboratory elutriation is often done by allowing the material to settle in a tall beaker and decanting according to a standard method or measuring the density with a hydrometer. Used in *mechanical analysis* of soils. See **Stokes's law, vel**.

embankment, bank A ridge of earth or rock thrown up to carry a road, railway, canal, etc., or to contain water (*levée*). Cf. **cutting**.

embankment wall A *retaining wall* at the foot of a bank to prevent it from sliding.

embedment length *ACI* term for *grip length*.

emery [min.] A mixture of *corundum* and magnetite or hematite used as an abrasive on emery paper, emery cloth or emery wheels.

empirical formula A formula based on one or many series of observations but with no theoretical backing.

empirical method A method based on experience rather than science.

emulsifier A chemical agent added to a mixture of two or more fluids to help form an *emulsion*.

emulsion A relatively stable suspension of one or more liquids minutely dispersed through another liquid in which it (or they) are not soluble, for example milk, which is a *dispersion* of fat in water. See **breakdown**, also (*B*).

emulsion injection A process of *artificial cementing* in which a *bituminous emulsion* is injected into soils which have a particle size equal to that of coarse sand (2–0.6 mm).

EN Prefix to the number of a European standard (**Euronorm**), just as the 'BS' precedes a British standard's number.

encastré, encastered [stru.] Said of a beam which is built in at the ends, i.e. *end-fixed*.

encroachment, intrusion [hyd.] Seawater that travels inland into a freshwater *aquifer* and spoils it. Excessive pumping has caused encroachment in California and other dry areas bordering the sea, and even in wet countries like England.

end-bearing pile, point-bearing p. A *bearing pile* which carries its full load down to hard ground at its point. In fact, nearly all end-bearing piles are thought now to be partly supported by friction. Cf. **friction pile**; see also **under-reaming**.

end block [stru.] The concrete at the end of a *tendon*, containing the *anchorage* and reinforced to resist bursting.

end contraction [hyd.] Contraction of the water area flowing over a *measuring weir*. See **contracted weir**, **effective area of an orifice**.

end-fixed [stru.] Said of the end of a beam which is so held that it can develop a *fixing moment*. The term is also applied to columns in the calculation of their *effective height*. An end-fixed column is effectively shorter than a *pinned* column.

endless belt A *conveyor belt*.

endoscope An *optical probe* which costs less than the *fibrescope*. It also gives a better picture of the inside of a pipe, but is rigid so cannot be inserted without dismantling a hydrant to gain access to the pipe.

end span [stru.] A *span* which is a *continuous beam* or slab only at its interior support and is for this reason often shorter or more heavily reinforced than the *interior spans*. See **exterior panel**.

end thrust [mech.] A push from the end of a member, in particular the thrust of a *centrifugal pump* towards the suction end, which must be resisted by a special bearing called a thrust bearing.

endurance limit [mech.] In *fatigue testing*, the maximum stress for any material below which fractures do not occur, however many reversals of stress take place. For steels the endurance limit can be determined at 6–10 million cycles of stress. It is roughly three-quarters of the *yield point* in *mild steel*.

energy [mech.] A capacity for doing *work*, usually expressed in work units (N-m or kg-m), sometimes in heat units (kJ or cal). Energy may belong to the speed of a body (*kinetic energy*) or to its height (*elevation head*), or, for a fluid, to the pressure in it. See **Bernoulli's theorem**.

engine [mech.] A *machine* driven by electrical, hydraulic, compressed-air, steam, internal-combustion, animal or other power to do *work* such as traction, hoisting, pumping, sawing, ventilation.

engineer One who contrives, designs or constructs electrical or mechanical plant, public works or mining work, or a *tradesman* (*B*), such as a *mechanic* or a fitter and in the USA also the driver of any engine. See **civil engineer**, **heating and ventilation** (*B*), **mechanical** (*B*), **site engineer**, **structural engineer**.

Engineering Council A federation of some 30 *engineering institutions* which took over the functions of the Council of Engineering Institutions in 1983. Its Part 1 exam is at the level of the end of the first year of an engineering degree. Its Part 2 exam is at university degree level.

engineering geology Geology applied to civil engineering. Closely concerned with such matters as the foundations of dams, their stability, *permeability*, etc., it includes *hydrology*, *soil mechanics*, *rock mechanics*, *geomorphology*, *geophysics*, mineralogy, petrology and related subjects.

engineering institution A group of engineers who unite to approve methods of admitting members, standards of behaviour and quality of work. There are about 30 UK engineering institutions; civil engineering includes structural, municipal and highway engineers.

engineering technician A man or woman registered with the *Engineering Council* but at a lower technical level than an *incorporated engineer* or *chartered engineer*. Generally he or she works under a chartered engineer and aims to become one eventually, but without the necessary university degree there is no natural progression upwards.

engineer's level [sur.] A telescope with a level tube attached. It is often a *dumpy level*.

Engineers' Registration Board See **Board of Engineers' Registration**.

engineer's transit, surveyor's t. [sur.] US description of a *theodolite* having a vertical graduated arc and a telescope bubble, as opposed to a plain *transit*, which measures only horizontal angles since it has no telescope bubble or vertical circle.

enlarged-base pile A *bored pile* base can be enlarged by hammering a plug of concrete at the bottom of the hole or by *under-reaming* the base.

enrockment *Rip-rap*.

entrained air Tiny bubbles usually from 0.01 to 0.1 mm in size, purposely brought into the *cement paste* by an *air-entraining agent* (*ACI*). Cf. **entrapped air**.

entrance head [hyd.] The head required to cause flow into a conduit or other structure; it includes both *entrance loss* and *velocity head*.

entrance lock [hyd.] A *lock* which

provides access to a *dock* in which the water is at a different level from the water outside.

entrance loss [hyd.] The head lost in eddies and friction at the inlet to a conduit.

entrapped air Irregular voids in concrete, 1 mm or larger, less useful than *entrained air* (*ACI*). They can be reduced by the use of *trémies* or by pumping from below as at *Seattle*.

ENV The prefix for a draft European standard. Cf. **EN**.

environment Engineers have been responsible for great improvements to the human environment if we think only of the London underground railways. In the world's first, opened in 1863 with steam locomotives, and running from Paddington to Farringdon St, there were three classes of passenger – first, second and third – but the differences concerned only the interior furnishings. Smoking was prohibited but anyone who wanted to read had to bring his own candle and stick it on the nearest windowsill. The two gas jets in each compartment gave enough light only to make the doors visible. Windows were shut tightly so as to prevent smoke coming in. It was said that no gentleman would take a lady into the underground. These conditions lasted for more than 40 years, until 1905, when the Inner Circle was at last electrified.

It is true that various electric tube railways had been driven before this but the first one, the City and South London, which reached as far as Stockwell and was opened by the Prince of Wales in 1890, was grossly underpowered. There was electric light in the carriages but the stations had to be gaslit.

Attempts to reduce the air pollution had included the burning of coke in-stead of coal but this failed. Coal burned better. Another failure was to fill the boiler with steam before the journey and then to remove the fire, but this also did not work. Smoke outlets into the street were provided and the periodic puffs of black smoke as a train chuffed by underground would frighten the horses in the street.

environmental disasters See **Aral Sea, Camelford, cancer clusters, North Sea**.

environmental engineering A term that has two senses in the USA: (1) sewage treatment, sewer networks, and prevention of pollution by gases, dust or noise, or other 'insult to the environment';
  (2) all the engineering work concerned with central heating and air conditioning – the improvement of the enclosed environment. In Britain the US senses are being accepted.

environmental health officer (EHO) Originally called sanitary inspectors, later public health inspectors, these local officials were given wide responsibilities for *pollutants* by the Health and Safety at Work Act, 1974.

environmental impact Effect on the environment, whether the smoke of a factory or diesel engine or the loss of a breeding site for rare birds or animals. In the USA the National Environmental Policy Act, 1970 (see **Alaska pipeline**), and later laws require that for each major federal development a statement of its effect (impact) on the environment must be submitted for approval (environmental impact statement (EIS)) including at least two possibilities: (i) what will happen if nothing is done (the null alternative); and (ii) any alternatives to the project proposed. In the UK an EIS is not needed unless the Department of the Environment insists on one, but see **EIA**.

Ephemeris [sur.] See **American Ephemeris and Nautical Almanac.**

epoxide, epoxy, ethoxylene resin A synthetic, usually two-part material that can set and harden under water or be used for bonding *roof bolts* or for repairing concrete in heavily trafficked areas, etc. One disadvantage is that its fire resistance is low.

epoxy-coated steel Steel coated with epoxy resin is compulsory on bridges in the USA. Bare steel corrodes badly. A 560 m long, 406 mm dia. epoxy-coated steel pipe was installed under the Medina River, Cowes, for a sewer in 1987. If it has to bond with the concrete, epoxy-coated prestressing strand must include grit in the resin, but epoxy-coated *deformed bars* bond well. Three bridges built in Florida from 1979 to 1982 suffered severe corrosion of epoxy-coated main column bars, but one with twice the cover (10–15 cm, 4–6 in.) did not suffer. Bitumen has also been used to protect steel.

epoxy concrete *Epoxide resin* mixed with fine and sometimes also coarse aggregate can quickly repair roads or bridges, though it is more expensive than other concretes.

epoxy glue Steel plates may be glued to each other or to concrete with *epoxide resin*. See **plate bonding**.

epoxy mortar Mortar made with *epoxide resin* and sand resists corrosion. It was used for acid-resisting pointing to the blue brick linings of the precast concrete sewers built in Cairo in 1988.

equal angle An *angle section* with legs of equal length. Cf. **unequal angle.**

equal-falling particles [s.m.] Particles of equal *terminal velocity*. They may be found in the *underflow* of a *classifier*.

equalization of boundaries [sur.] A method of calculating areas which have irregular boundaries. The irregular lines are replaced by straight lines which cut off on the one side an amount equal to what they put on on the other side. The area can then be calculated by adding the areas of the triangles so formed. See **give-and-take lines**.

equalizing bed A bed of ballast or concrete on which pipes are laid in the bottom of a trench.

equilibrium [stru.] The state of a body which does not move. A body is in stable equilibrium when any slight movement increases its potential energy so that when released it tends to fall back to its original position. A body in unstable equilibrium, when moved slightly, tends to move farther away from its original position.

equilibrium moisture content [s.m.] The moisture content of a soil in a given environment, at which no *moisture movement* occurs. See also (*B*).

equilibrium of floating bodies [hyd.] See **principle of Archimedes**.

equipotential lines [s.m.] Contours of equal water pressure in the soil mass round a water-retaining structure such as an earth dam or river bank. They can be determined by building a scale model of the bank with the water levels to scale. The soil mass is connected to glass *stand pipes* which show, by the water level in them, the pressure at the points to which they lead. From these elevations the equipotential lines can be plotted. See **flow lines, flow net**.

erecting shop A large open workshop or yard where steel frames are joined up for *trial erection* after *fabrication* to make sure that they fit, before being sent in separate pieces to the site.

erosion *Scour.*

erratic [stat.] Said of values which seem to vary excessively from the *average*.

error [sur.] A difference from an *average* value. See **closing e., collimation e., compensating e., gross e., probable e., systematic e.**

escape [hyd.] A wasteway for discharging the entire flow of a stream.

Escherichia coli, E. coli, colon bacillus (formerly **Bacillus coli, Bacterium coli**) A bacterium of exclusively faecal origin, it forms 90% of the coliform bacteria in the human intestine. Its presence is a sign of faecal pollution and its absence is sometimes taken to indicate the absence of bacteriological pollution in water.

estimating draughtsman/draughtswoman [d.o.] An experienced draughtsman or draughtswoman with or without diplomas who estimates quantities and the costs of work from drawings made by himself/herself or others.

estimator A *quantity surveyor*, working usually for a *contractor*, who can price a job from his or her own or someone else's drawings or other documents. He or she need not be qualified.

ethane [min.] C_2H_6. A gas found in *natural gas* from oilwells, normally absent from coal mines.

Euler crippling stress [stru.] The *crippling load* (P) of a strut divided by its cross-sectional area (A). It is calculated from the formula

$$\frac{P}{A} = \pi^2 E \frac{k^2}{l^2}$$

where E is the *modulus of elasticity*, k is the *radius of gyration*, and l is the *effective length* of the *strut*. See **slenderness ratio**.

Eurocodes (EC) Eurocodes are codes of practice intended to create a single construction market in the countries of the European Community, with their population of 330 million people, by 1992. When related matters such as European standards for building materials have been unified, the codes should increase the competitiveness of European companies compared with those outside. These common codes of practice for Europe in 1988 were: 1. General Principles; 2. Concrete; 3. Steel; 4. Composite steel–concrete structures; 5. Timber; 6. Masonry; 7. Foundations; 8. Earthquake design; 9. Loadings not elsewhere listed. The first drafts of EC1, 2, 3 and 4 had appeared by mid-1988. CEN, the European Committee for Standardization, publishes them.

Eurocodes are not yet mandatory. *Euronorms* (standards), on the other hand, may have to be obeyed. Eurocodes are documents of the European Commission and can have no statutory effect in the UK until the commission makes its formal proposal to the Council of Ministers.

EC1, 2, 3, and 4 are likely to be published in the main European languages in 1989 and will for two years resemble a *BSI* 'draft for development'. In 1991 they will probably be updated and become legal documents. Some time later national codes will be withdrawn and EC will supersede them. See **limit states**.

Euronorm (EN) A European Community standard. By 1992 most British building standards (BS) will have been revised to correspond to Euronorms. In 1988 there were about 400 EN numbers, most of them corresponding to a BS or part of one, probably also to AFNOR and DIN numbers. Cf. **EC, ENV**.

European engineer (Eur. Ing.) A

chartered engineer registered with the *Engineering Council*, who has in addition eight years of engineering experience and training acceptable to FEANI, the federation of national engineering institutions of 20 countries in Europe.

Euroroute Before the *Channel Tunnel* was accepted by the governments as the surest link, an imaginative combined tunnel and bridge scheme was proposed for vehicles to drive to France, using two or more *artificial islands* in the channel to provide access between tunnels and bridges. The attractions of this costly scheme probably led *Eurotunnel* to agree to consider a road tunnel before the year 2000. See **Chesapeake Bay**.

Eurotunnel The *promoter* of the *Channel Tunnel*, a Franco-British partnership between the *Channel Tunnel Group* and *France Manche*. Responsible to the governments and shareholders, it employs the *contractor, Transmanche Link*. By 1989 it had ceased to be a promoter and was wholly project manager. In 1993 it should again change its role and become the tunnel operator. The *client* is also, in part, the contractor. Ten of the 15 main founder shareholders are the contractors (*Transmanche Link*) building the tunnel; the other five founders are banks. See **Euroroute**.

evaporation pan [hyd.] A container, often of about 1.3 m dia. and 25 cm deep, from which the loss of water by evaporation can be determined by measuring its depth at various times. Pans serve mainly to determine evaporation losses from lakes or reservoirs. Reservoir losses are less than pan losses so that the evaporation from the pan should be multiplied by 0.7 or 0.75 to determine the reservoir loss. Pans may

be sunk in the earth, be 30 cm above it, or may float in the reservoir or lake.

evaporation retardant Cetyl alcohol or other waxy substance spread over fresh concrete after casting to delay evaporation (*ACI*).

evapo-transpiration [hyd.] Combined loss of water from soils by evaporation and plant transpiration.

excavating cableway A *cableway* fitted with a clamshell or *orange peel* bucket for digging.

excavation Digging, breaking and removing soil or rock.

excavator A self-propelled, crawler-mounted, rarely wheeled digging machine which can slew 360° without moving its tracks. The *backacter, dragline, grab* and *skimmer* are different sorts of front-end equipment which can be attached to the base machine instead of the usual digging arrangement, a *face shovel*. But some less common, specialized excavators do not slew, including the *dredge, elevating grader* and *trencher*. It can be used also as a crane. Cf. **loader**.

exception reporting See **management by exception**.

exciter (1) Lime, alkali, sulphates, etc., which, added to a crushed blast furnace slag, cause it to set when mixed with water. *Portland cement* acts as an exciter in the *Trief process*.

(2) [elec.] A direct current generator which energizes the field magnets of an alternator or similar electric machine.

exfiltration Leakage outwards. The opposite of *infiltration*. See **sewage**.

expanded clay See **lightweight aggregate** and (*B*).

expanded metal (XPM) Steel or other metal mesh formed by slotting a metal sheet and widening the slots to a diamond shape. It is used for *metal* *lathing* (*B*), as a base for plaster and for concrete reinforcement.

expanding cement (expansive c., sulphoaluminate c. ACI) *ACI* mentions three types of expanding cement, each with different proportions of tricalcium aluminate, calcium sulphate and tetracalcium trialuminate sulphate. The expanding component is generally an anhydrous calcium aluminate or sulphoaluminate with a source of sulphate and of free lime.

By their expansion these cements can either merely compensate for and thus eliminate *shrinkage* or can stretch the steel, effectively *post-tensioning* the concrete. The difficulty in practice with bonded steel must be to ensure that expansion does not take place before full bond is available.

On a small scale expanding cement is regularly used in *underpinning* buildings. In this application, if as usual the concrete is unreinforced, bond is unimportant.

expanding pile A type of *driven pile* devised by J. B. Burland, Professor of Soil Mechanics at Imperial College, who believes that a mere 10% expansion yields a six-fold increase in carrying capacity; driving destroys the soil's structure, but it is rebuilt by expansion, using wedges in the pile foot.

expanding soil, expansive s. See **swelling soil**.

expansion bend A loop in a pipe which enables the expansion or contraction due to temperature change to be taken up without danger to the pipe. See **expansion joint**.

expansion bolt An *anchor bolt* into masonry consisting of a split malleable-iron cone inserted thin end first into a drilled or *cored* hole. A bolt between the split halves projects outside the hole. When the head of the bolt is turned, a nut between the halves of the cone is drawn out and tightens the sides of the cone against the walls of the hole. Other expansion bolts work by a *toggle* mechanism or other means. See **Rawlbolt**.

expansion joint (1) In concrete work, a gap in the steel and the concrete to accommodate, at different times, both expansion and contraction. Confusion can arise because of this. A *contraction joint* provides mainly for shrinkage. Both are *movement joints*.

(2) An *expansion bend* or similar device.

expansion rollers Rollers provided at one support of a bridge or truss to allow for thermal movement. The other end is usually fixed.

expansive cement *Expanding cement*.

expansive soils [stru.] See **swelling soil**.

expansive use of steam, e. working [mech.] The expanding of steam in the cylinder or cylinders so that the highest practicable amount of energy is extracted from it. This is usual for stationary steam (also for compressed-air) engines, but to ensure proper expansion, engines must be kept in good order.

expending beach A beach designed to use up the energy of the waves. See **tetrapod**.

expert system A computer *program* compiled by close consultation between an experienced programmer (the knowledge engineer) and a human consultant, in e.g. the *structural analysis* of building frames.

The best applications of expert systems are to problems within relatively narrow, closely defined domains. They can be run on *microcomputers* provided that enough memory is available, and can save the engineer much time and some boring work, freeing him or her for those problems for which no expert system exists. Because they can state and sometimes solve that part of a problem which is routine, they are useful for familiarizing an engineer with an unfamiliar subject. They can be interrogated and usually answer helpfully, so they might be used in developing countries to clear up the background of a problem before the more expensive human expert is called in. All expert systems are limited and therefore suspect. Their conclusions must always be questioned.

If future co-operation between industry and universities or polytechnics could generate expert systems that could be linked together instead of, as now, working separately, this would be a marvellous development.

Expert systems are being developed or already exist for: the analysis of contractors' claims; the choice of the right number and type of cranes for a site; storm sewer management and sewer rehabilitation; geotechnical designs (spread footings); tutoring of civil engineering; and many other subjects.

exploder See **blasting machine**.

exploration See **site exploration**.

explosive compaction [s.m.] *Deep blasting*.

explosive dust Many dusts apart from coal dusts are explosive, especially the finer ones such as flour. Grain silos therefore must have explosion-relief panels at the top; the use of flame, including welding, is forbidden.

explosive limits See **lower explosive limit**.

explosives [min.] Substances which are used for demolition or for *breaking ground* by *blasting*. See **detonator**,

dynamite, gelatine explosives, high explosive.

exposed-aggregate finish A concrete surface from which the surface *cement paste* has been removed by *bush hammering*, or by light brushing if the concrete has been cast with a *retarder* in the *release agent*. A gentle water spray after stripping the *formwork* is also effective.

extended prices, e. bill, e. rates Prices multiplied by quantities to make a *priced bill*.

exterior panel A panel of a slab of which at least one edge has a discontinuous support. Sometimes called an *end span*.

external vibrator A vibrator for concrete placing, fixed to the *formwork*, as opposed to an *internal vibrator*. This method of vibration has the disadvantage that the shuttering must be unusually strong and it is therefore little used.

extrapolate [d.o.] To continue a curve beyond those points for which data have been obtained for it. Hence to make inferences which are little more than guesses. Cf. **interpolation**.

extruded sections The commonest *light-alloy* structural sections, formed by *extrusion*.

extrusion [mech.] Forming rods, tubes or *sections* of intricate shape by pushing hot or cold metal or plastics through a *die* shaped to the required section. The *lead sheaths* of electric cables as well as structural sections of *light alloy* or plastics are made in this way. Cf. **cold drawing**; see **ductility**, **pultrusion**, **wirecut brick** (*B*).

eye bolt [mech.] A *bolt* with a steel loop forged at one end instead of a head, and a thread formed at the other end. It is used for lifting any heavy unit of plant into which it is screwed.

F

fabric See **expanded metal, geotextile, wire-mesh reinforcement**.

fabrication [mech.] Preparing the steel members of a building framework or part in the workshop by such operations as shearing, drilling, plate bending or straightening, sawing, flame cutting, planing of castings, squaring the ends of stanchions with ending machines, *notching*, grinding, *automatic welding, forging*, etc.

Fabridam An inflatable plastic 'sausage', anchored in a river bed, which can be filled with water and/or air and thus rise to its full diameter. In floods it has the advantage of being completely collapsible and not blocking flow when deflated. See **Dracone, floating boom**.

face left [sur.] The position of a theodolite when its vertical circle is to the left of the telescope, seen from the eyepiece. See **face right**.

face piece A *face waling*.

face right [sur.] The position of a theodolite when the vertical circle is to the right of the telescope when viewed from the eyepiece. See **change face, face left**.

face shovel, crowd s., forward s. An attachment fitted to an *excavator* by which it digs away from itself into a bank (the face) with a toothed bucket fixed to a rigid arm. Being crawler-mounted it is more stable and can dig more powerfully than a *loader*.

face waling, f. piece A *waling* across the end of a trench. It is held by the ends of the main walings and supports the end of the trench together with the end strut.

facing A protective covering to sea walls, dykes or cuttings, or a decorative wall surface of brick or stone. See also (*B*).

facing points [rly] See **points**.

facing wall A concrete, precast or *in situ* lining used instead of timber sheeting against the earth face of an excavation. It is held by the main timbering (which is later removed) and used as a base for asphalt tanking against which the main retaining wall is eventually built. Timber sheeting cannot be used because it would not be a good base for asphalt and would rot.

factored load The actual load on a structure increased by the appropriate *factor of safety*.

factor of safety, load factor The failure load divided by the *design load*.

Faculty of Building An association of people in all building disciplines; entry is by invitation only.

failure A condition at which a structure reaches a *limit state*. It may be due to leakage, deflection, cracking, etc., but it usually does not involve rupture (fracture) because most structures are considered to be unsafe, therefore unusable, before they collapse.

fairlead (1) A metal fixture to a quay or ship deck which guides a rope and helps a ship to berth smoothly.
  (2) A swivel pulley on the drag rope of a dragline. This pulley is held on the cab.

fall (1) The *gradient* of rivers, roads or railways, described as a fall of so many metres per kilometre or as a percentage.

(2) **fall rope** The free hoisting rope used with *lifting tackle*.

fall block In *lifting tackle*, a pulley *block* which rises and falls with the load. The load hangs from a hook or eye under the fall block.

falling apron A *revetment* consisting of a *mattress* which is launched by the *scour* beneath it.

falling velocity, fall velocity [hyd.] See **Stokes's law**.

fall rope See **fall** (2).

false leaders A steel mast set on the ground, held upright by guy ropes and used for guiding a pile during driving and for holding the weight of a *pile hammer* over it. Cf. **hanging leaders**, **leaders**.

false set (**premature stiffening** and **hesitation s.** are deprecated) Very early stiffening of mortar or concrete which is not harmed by adding water or remixing (*ACI*).

falsework Support for concrete *formwork* or for an arch during construction.

fan [mech.] A ventilator for delivering large volumes of air at a low pressure. For a colliery main fan, the pressure difference is usually a suction of about 20 cm of *water gauge*.

farm duty [hyd.] The quantity of water delivered to a farm for irrigation (USA). See **net duty**.

fast track Site organization aimed at work with minimum delays. The detail design drawings available at any moment are purposely limited to those most urgently needed, reducing delays from over-conscientious design. Information on structure, finishes and mechanical services is shared early between all *consultants*.

Precasting of concrete stairs, beams and columns also helps. Concrete walls are slower to build than brickwork or blockwork. Even better, wet trades using mortar, concrete or plaster can be avoided by using *dry lining* (*B*) or suspended metal tile ceilings.

Time is always lost between one trade leaving an area and the next trade coming into it. Instead of waiting for the roof to be on before allowing the finishing trades to start, it is better to 'dry up' the building floor by floor as each floor is cast. Temporary waterproofing is provided all round, with upstands at stair wells and other holes through slabs, to allow rain through only at chosen points.

fatigue failure [stru.] Failure caused by a repeated stress that does not cause failure when applied once. See **corrosion fatigue**, **endurance limit**, **fatigue test**.

fatigue test [mech.] The testing of metal testpieces under repeated reversals or fluctuations of stress to determine the *endurance limit*. See **stress-number curve**.

fault [min.] A break in the bedding of rocks. It displaces any deposit vertically by the 'throw', and horizontally by the 'heave' or lateral shift. *Earthquakes* result from the movements at faults.

faying surface [stru.] A contact surface at a joint between steel members.

feasibility study The first duty of a *consultant* appointed to a *civil engineering project* is to find whether the project can be built and if so how. This decision, in the form of the consultant's drawings and text, is the promoter's feasibility study. It must be understandable to the lay person yet technical enough to appeal to an engineer. If the promoter accepts the feasibility study, he or she may submit it to a *quantity surveyor* for preliminary costing. The next step is the preliminary design followed by *tendering*.

feather See **levelling compound** (*B*), **plug and feathers**.

feed (1) [mech.] The rate of advance of a cutting tool, drill bit, drilling rod, etc.

(2) [mech.] The water supply which has to be pumped into a boiler at the boiler pressure by the *feedpump*.

feeder (1) [elec.] A cable of high current *carrying capacity* (*B*), which connects power stations to substations.

(2) [mech.] A device for delivering coal, ore or other loose material at a controllable rate.

(3) [hyd.] A channel which supplies a reservoir or canal with water.

feedpump [mech.] A pump which injects *feedwater* into a *boiler*.

feedwater [mech.] Water which in the best practice is *demineralized*, heated nearly to boiler temperature, and de-aerated before being pumped into a steam *boiler* by the *feedpump*.

Fellenius's circular-arc method [s.m.] See **rotational slide**.

Fellowship of Engineering (F. Eng.) A body which came into existence in 1976 with 126 eminent engineers, Fellows of the Royal Society, to encourage excellence in engineering. It furthers the interests and skills of professional engineers at the highest levels, and elects up to 60 distinguished engineers every year. Total membership is not allowed to exceed 1000.

fence [mech.] (1) A guard round machinery to protect people from being drawn into it.

(2) The guide on a circular saw to keep a constant width of cut.

fender A rope mat or ball, an old rubber tyre, etc. which protects a vessel from impact with a pier (and the pier from the vessel).

fender pile An upright, generally free-standing, wooden pile driven into the ground just clear of a berth. It absorbs some of the impact of vessels and thus protects the berth.

fender post, guard p. A *bollard*.

ferrocement Cement mortar, rarely thicker than 25 mm (1 in.), containing up to 12 layers of wire mesh, called *ferciment* in 1848 by Monsieur Lambot in France, who built more than one boat with it. Very roughly the mix by weight is: steel 1, water 1, cement 2, dry sand 4. By volume the steel is some 4–8% of the concrete, 4–8 times as much as in *reinforced concrete*. Made without *formwork*, using light vibration or *shotcreting* the material can be much stronger than Canadian spruce. It was rediscovered in 1946 by the Italian engineer Pier Luigi Nervi, who first built his firm's warehouse with it, 11 × 22 m (35 × 70 ft) in plan, with walls and roof 30 mm (1.2 in.) thick. In the same year he built himself a 165 ton motor-sailing boat with a 36 mm (1½ in.) ferrocement hull, weighing 5% less than a wooden hull. Nervi made himself a name for fast construction with *Turin Exhibition Hall*, which was followed by several equally spectacular structures.

In Red China, by 1978, some 30 boatyards had built more than 2000 ferrocement sampans 12–15 m long, about 1.5 m in the beam and 1 m deep, with a capacity of 10 tons of goods, moved by poling or rowing. Frame ribs for fixing the reinforcement are precast 25 mm (1 in.) thick and are about 0.9 m (3 ft) apart. About 50 man-shifts are needed to make one boat. Rusting has been overcome, it is claimed, by coating the ferrocement with glassfibre-reinforced plastic.

In the USA many ferrocement canoe races have been organized, some with hulls only 1 mm thick. Fibre-reinforced *shotcrete* has many similarities with ferrocement.

ferroconcrete Hennebique's name for *reinforced concrete* in the 1900s and earlier, introduced by his London partner, Mouchel.

ferro-prussiate paper [d.o.] Paper treated to make *blueprints*. It is almost obsolete and is superseded by *dyeline* prints.

ferrosilicon dust A substitute for *microsilica*, claimed to be equally useful.

ferrule A connection on a water main through which water flows to a consumer.

fetch The free distance to any point over which the wind can travel in raising waves, i.e. the distance from the nearest coast in the direction of the wind. See **Stevenson's formula**.

FGAN See **ANFO**.

fibre-reinforced concrete Concrete reinforced with randomly oriented short steel, glass or plastic fibres, often 50 mm (2 in.) long and of 0.5 mm dia., which improve the bending and impact strength and ductility compared with other concrete. *Shotcreting* can introduce a higher percentage of fibre than can be put in by ordinary mixing. See **fibrillated polypropylene film**, **glassfibre-reinforced concrete**, **steel-fibre-reinforced shotcrete**.

fibre-reinforced plastics (FRP) Polyester, vinyl ester or epoxy resins reinforced with glass or carbon or aramid fibre or high-tensile steel wire have begun to be used for building

bridges. In the early 1980s several footbridges in Europe and North America were built of it including a 10 m (33 ft) span bridge at Ginzi, Bulgaria in 1982. In the Chinese capital Beijing, a 20 m (65 ft) span bridge was also built that year. FRP will probably always be more expensive than steel but its lightness and corrosion resistance are certain to provide good reasons for its economical use, possibly with *pultruded* sections. See **Polystal**.

fibre rope, cordage *Manmade fibre* ropes, less liable to rot, are gradually superseding the old vegetable fibres, some of which are hempen. Cordage has in the past been described by its circumference, unlike steel ropes which are described by their greatest diameter.

fibrescope, borescope (in medicine called an **endoscope**) A flexible probe for inspecting the inside of a cavity wall or even a water pipe under pressure (4 bar max.). Glassfibre bundles in the steerable probe transmit light down the pipe and back to the observer. A suitable size is an 8 mm dia. probe, 3 m (10 ft) long, because this can be inserted about 1.5 m (5 ft) into the pipe each way. If the water pressure is too high the supply can be closed down for a few minutes without excessively annoying customers.

fibrillated polypropylene film Plastic film in which slots have been cut to make fibres connected at each end. It has excellent bond with cement mortar and is used in *Netcem*, *Caricrete*, etc. As many as 60 sheets can be fitted into a 6 mm ($\frac{1}{4}$ in.) thickness of Netcem, making 8% by volume. This is much more than the usual 1% maximum of fibre reinforcement. But random fibres may be added in lengths from 5 to 50 mm to ordinary concrete at rates varying from 0.1% (0.9 kg/m^3) to 2%, improving the toughness, imperviousness and tensile strength.

FIDIC conditions of contract *International Conditions of Contract*.

Fidler's gear Lifting tackle for laying large blocks at any angle, in *blockwork* below water level.

fiducial line, f. point [sur.] A reference line or point.

field book [sur.] A surveyor's book for recording field measurements.

field drain, agricultural d., land d. Unsocketed, earthenware, porous concrete, perforated plastics or *pitch fibre pipes* (*B*), about 8 cm bore, laid end to end unjointed so as to drain the ground. The trench is usually backfilled with coarse sand or gravel rather than clay next to the drain so as to prevent blockage. See **Trammel drain**.

field moisture equivalent [s.m.] The minimum *moisture content* at which a water drop placed on the smoothed surface of a soil will not be absorbed immediately by the soil but spreads over the surface, giving it a shiny look.

field tile US term for *field drain*.

filament reinforcement, fibre r. See **fibre-reinforced concrete, fibre-reinforced plastic, fibrillated polypropylene film, glassfibre-reinforced concrete, reinforced earth**.

files [mech.] Hand tools more used for cutting metal than wood. They may be of several cross-sections, flat, round, half-round, square and triangular (also called three square). See **cut** (4).

fill *Earthwork* in *embankment* or *back filling* (*B*).

fill bulkhead An anchored *bulkhead*

which is backfilled to make the foundation for a *quay*.

filled bitumen *Bitumen* containing *filler*.

filler Fine mineral powder added to road tars and bitumens to make them stiffer. See also (*B*).

filler beam floor, f. joist floor, f. concrete slab *Rolled-steel sections* or *built-up beams* in a concrete floor slab, acting compositely with it. This development from the *jack arch* is useful where the location and size of holes are not known before the floor drawing is made. It may include 152 × 89 mm or smaller *rolled steel joists* (filler joists) spaced at intervals of 0.45–0.75 m. The intervals may be filled with plain or reinforced concrete or *hollow tile* covered with a concrete topping. Generally the filler joists are carried on larger steel beams. See **jack arch**.

fillerized cement, filler c., composite c. *Ordinary Portland cement* mixed with ground, selected, pure limestone, usually with less than 3% coarser than 80 microns. French experience since 1979 with about 20% filler has shown it to be a low-cost, energy-saving cement of medium strength that produces workable, durable concrete. See **blended cement**.

filler rod A metal rod, often a *covered electrode*, which provides the metal to make a weld.

filler tile, slip t. Flat rectangles of burnt clay placed in a *hollow-tile floor* to close the openings at the ends of *hollow blocks* (*B*).

fillet weld [mech.] A weld of roughly triangular cross-section between two pieces at right angles.

filling Material used for raising the ground to a higher level. See also (*B*).

filter (1) [hyd.] An arrangement for straining harmful matter, including bacteria, from water to make it drinkable or usable for other purposes. Filters can also clean *flue gases* to remove polluting dusts. See **filtration**.
(2) [s.m.] See **graded filter**.

filter bed [sewage] A *trickling filter*.

filter blocks Hollow vitrified clay blocks which may be salt glazed and are designed to carry a *trickling filter* (USA).

filter drain The approved term for a *French drain*.

filter material (1) [s.m.] Sand and pebbles like those in a *graded filter*, usually for cleaning water.
(2) [sewage] **filter medium** The strong metallurgical coke, clinker, plastics or broken stone in a *trickling filter*.

filter medium *Filter material*.

filter well [s.m.] A bored well of about 30 cm dia., used as an alternative to *wellpoints*, in which each borehole has its own pump installed at the top. It is more expensive and slower to install, but there is less disturbance to the ground than with wellpoints because there are fewer, though more efficient, wells. The suction lift of each pump is also appreciably better. If the pumps are submersible they can be installed at the foot of each well, and there is then no limit to their depth.

filtrate Liquid (usually water) that has passed through a filter.

filtration Essential to *water treatment*, filters remove from *raw water* the solids that are too fine to be removed by *sedimentation*. Some of the filters used are: *sand filters*, multi-media filters (like sand filters but with anthracite on the top, sand in the middle and garnet or other dense mineral below), *microstrainers* and a wide variety of others, even *ultra-filtration*.

final finisher A unit in a *slipform* or *fixed-form paving train*.

final grade US expression for *formation level*.

final setting time The time, determined by ASTM or BS standard tests, needed for a *cement paste* to stiffen a certain amount more than its *initial set*.

fin drain A *geogrid* sheet sandwiched between two layers of *geotextile*, sometimes as thin as 27 mm (1 in.). It is a substitute for a *French drain* which has the advantage that it needs only a narrow slot of excavation. Fin drains are commonly supplied in rolls 600 mm (2 ft) wide. To dig the narrow trench, either a special narrow bucket or a trenching machine is needed. A porous plastic pipe may be set at the base.

fine-adjustment screw [sur.] A *tangent screw*.

fine aggregate (1) Sand or grit for concrete which passes a sieve of mesh 5 mm square. This is coarser than 'sand' in the *classification of soils*. See also **fine sand**.
(2) Sand or grit for bituminous road-making which passes a 3 mm square mesh.

fine cold asphalt A *wearing course* of bitumen and fine aggregate which is spread and compacted while cold or warm.

fineness modulus A number which indicates the fineness of a sand, pigment, cement, etc. It is calculated by determining the percentage residues on each of a series of standard sieves from 37.5 mm opening downwards, each opening being half the preceding one.

The percentages are then summed and divided by 100, the quotient being the fineness modulus. The series is as follows: 37.5 mm; 19 mm; 9.5 mm; 4.75 mm; 2.36 mm; 1.18 mm; 0.6 mm; 0.3 mm; 0.15 mm. The test sieves are of perforated plate down to 1.18 mm; below that size they are of woven wire mesh. The largest BS test sieve is 125 mm (previous sizes were in inches). The *grading curve* is often used in Britain and the USA instead of the fineness modulus, and for fine powders, like cement or fly-ash, the *specific surface* is another common description.

fines [s.m.] The smaller particles in a *mechanical analysis*.

fine sand The *classification of soils* uses a different dimension, but for the concretor fine sand is what passes through a 0.6 mm sieve, roughly equivalent to *soft sand*.

fine-wire drag, f.-w. sweep [sur.] In *hydrography*, to ensure there are no upstanding reefs or wrecks on the bed of a waterway, two boats about 100 m apart have to sweep the bed before or after it has been sounded for depth. A light wire, the drag, is held between them, scraping the bed, and wound from a friction-braked drum at one end. When the rope pulls out, the horizontal and vertical angles from the boats along the wire are noted. The obstruction is thus located and, later, mapped.

finisher, concrete f. A person who repairs minor defects in the concrete after removal of the *formwork*. If the concrete is to remain fair-faced, the finisher will also *sack rub* or otherwise dress it to the client's requirements.

Fink truss, Belgian/French t. [stru.] A common, steel roof truss suitable for spans up to 15 m.

fin wall A wall stiffened with *buttresses*.

fire setting [min.] A method of *breaking ground* practised in ancient Egypt and medieval Europe up to the time of gunpowder. A hot fire is lit next to the rock. As soon as the rock is judged to be hot enough, it is rapidly cooled by pouring water over it. See **jet drilling**.

fire-tube boiler [mech.] A steam boiler in which the smoke passes through tubes surrounded by water, e.g. locomotive boilers, or the Lancashire, Cornish or other relatively small boilers. See **water-tube boiler**.

fire welding *Forge welding*.

firing [mech.] The charging of fuel into a furnace generally of a steam boiler.

firm clay, f. silt [s.m.] A *clay* (or silt) which can be dug with a spade and moulded by firmly squeezing in the hand.

first moment [stru.] See **static moment**.

first order, primary triangulation, trilateration [sur.] A *triangulation* or *trilateration* with sides 20 to 60 km or more in length. See **second order**.

fished joint A joint made with *fishplates*.

fishing (1) Bolting up *fishplates* to rails or other members.

(2) [min.] In oilwell or exploratory drilling, letting down recovery tools (fishing tools) into the hole to extract broken tackle.

fishing tools [min.] Tools for cutting off rope or for recovering or drilling round, into or through a tool lost in a borehole.

fish ladder, f. pass, fishway A channel along which fish can travel up or down past a weir or dam.

fishplate The end of a rail is joined to the next *rail* in the track by a pair of specially shaped steel plates called fishplates, one each side, which are bolted through the rails. Fishplates of a simpler, rectangular shape are used for joining a stanchion extension to the stanchion below. For joining timbers they may be called *splice* plates, splice pieces, cover straps or flitch plates, and may be of steel or wood.

fish screen [hyd.] A barrier to prevent fish entering a channel.

fishtail bit [min.] A bit used in the oilwell *rotary drill* for drilling through soft ground.

fishtail bolt An *anchor bolt* with its tail split and cast into concrete or masonry.

fishway A *fish ladder*.

fissured clay [s.m.] A *clay* which like the London clay has a network of joints that open up in dry weather. See **intact/stiff-fissured clay**.

fitchering [min.] US description of the jamming of drill steels in a hole.

fitted bolt See **turned bolt**.

fitter [mech.] A *skilled man* who can assemble engines in an engineering shop.

fitting [mech.] Skilled work in an engineering shop which involves assembly by a fitter.

fit-up A description of *formwork* which is framed so as to be struck without destroying it. It is therefore suitable for repeated erection.

fix [air sur.] A determination of an aircraft's position when taking photo-graphs of the ground; a form of *ground control*.

fixed beam [stru.] A beam with a *fixed end*.

fixed end [stru.] A fixing to an end of a beam or column which can develop, without movement, at least as much *bending moment* as the *moment of resistance* of the beam or column. See **fixing moment**, and cf. **hinge**.

fixed-end moment See **fixing moment**.

fixed-form paving train, f.-f. paver A succession of roadbuilding machines which can concrete a road, airstrip, etc., carried on flat-bottomed rails, *road forms* or concrete strips cast carefully beforehand to correct line and level. Its many units include: (i) side feeder to receive and pass on the mixed concrete; (ii) *concrete spreader*; (iii) road forms at the edges; (iv) *compactor-finisher*; (v) placer for putting dowel bars across the joints; (vi) joint-grooving machine; (vii) joint-groove finisher; (viii) final finisher; (ix) texturing equipment; (x) sprayer to spread *curing* compound; (xi) protective sheeting (tents). Cf. **slipform paving train**.

fixed retaining walls [stru.] Basement and similar walls which are rigidly supported at top and bottom are subjected to a pressure much higher than that due to *active earth pressure* on *free retaining walls*. The coefficient for the equivalent fluid pressure of sands increases to about 0.5 at 2 m depth instead of the 0.27 value usual for free retaining walls holding up sand. In clayey materials the pressure is higher if the clay can get wet and swell.

fixing moment, fixed-end m. [stru.] The *bending moment* at the support of a beam required to fix it in such a way that it cannot rotate, so that it has a *fixed end*.

fixity See **continuity**.

flagstone See **paving flag**.

flake ice Crushed ice used in hot weather, instead of mixing water, to cool a concrete mix.

flame cutting Cutting steel, iron or other metal up to 1 m (40 in.) thick with an oxyacetylene, oxypropane or proprietary gas mixture. *Laser cutting* can be accurate and faster. Cutting with explosive is less accurate but fast, though destructive.

flameproof enclosure (FLP) [min.] Electrical switchgear, transformers, motors, cable couplers, etc. must be certified as flameproof before use in fiery mines, and with slightly different, possibly even stricter requirements, in the oil industry. FLP enclosure for mining can be provided by a steel box with bolted *flanges* at least 25 mm (1 in.) wide, with a maximum gap in the flanges of 0.5 mm (0.02 in.). *Methane* ignited inside such an FLP enclosure is cooled by the flanges so much that no flame or explosion can spread outside. This is the only enclosure permitted for powerful motors in hazardous atmospheres. Cf. **intrinsic safety**.

flange (1) A ring-shaped disc forged or screwed to a pipe end to enable it to be bolted to the flange on the next pipe. The flanges have matching holes drilled in them to receive the bolts. Other flanges in the form of narrow metal projections, exist on other equipment.
(2) **chord** The wide top and bottom strips (compression and tension flange) of a *rolled steel joist* or other beam.
(3) [rly] The projecting rim of a railway wheel which holds it on to the rail.
(4) [rly] The flat part of a flat-bottomed rail is for railwaymen the 'foot', not the 'flange'.

flangeway [rly] The space between a *check rail* and the gauge face of a running rail in contact with the wheel flange.

flanks The outer quarters of a carriageway, also called shoulders or haunches.

flap trap [sewage] An *antiflood* valve. See **flap valve**.

flap valve [mech.] A *check valve* on an *outfall* pipe, with a hinged disc which opens when the water flows out to the sea or the river but is closed by gravity or by the backward flow when the tide rises.

flared column An enlargement of the top of a concrete *column* immediately below the slab.

flashboard A *stop log*.

flash-butt welding See **flash welding**.

flash distillation *Desalination* by *distillation* under vacuum. It produces several times more drinking-water than the steam consumed.

flashing See (*B*).

flash set (quick s., grab s. *ACI*) Inconveniently fast setting of cement.

flash welding, resistance f. w. (BS 5400, part 8:1978 calls it **flash-butt welding**) British Rail's flash welding machines can weld a *continuously welded rail* in 2 min. Rails are received in 36 m (118 ft) lengths and are welded into strings up to 91.5 m (300 ft) long. The rail ends are brought together; high currents pass as they touch, causing these points to melt. One rail is advanced towards the other and a series of small arcs heat the rail ends up to melting-point. The rails are then forced together to push out the impurities and consolidate the joint.

flat [mech.] A thin rectangular iron or steel bar.

flat-bottomed rail [rly] A rail resembling a T upside down, with a thickened foot to the T, on which the wheels travel. This type is universally used on the continent of Europe and is now standard in Britain, having displaced the *bullhead rail* in new track. See **rail fastenings**.

flat jack A nearly flat, hollow, steel cushion made of two discs in contact, welded round the edge. It can be inflated by injecting oil or cement grout under a pressure which can be maintained, increased or reduced as required. Flat jacks were used at *Plougastel* by Freyssinet and are still used particularly at the abutments and crowns of arches to relieve the formwork of load at the moment of striking the *formwork*, or in *prestressing*.

flat slab A *plate floor* or *mushroom slab*.

flat-slab deck dam, f.-s. buttress d. [hyd]. A reinforced-concrete slab with a flat upstream face sloping at about 45°, carried on parallel buttresses. It resembles the *multiple-arch dam* except that its wetted surface is smooth.

flattened strand rope [min.] Steel-wire circular ropes built up from strands laid *Lang's lay* round a hemp core. Each strand is made of wires laid round an oval or triangular soft-iron core wire which gives the strands their flattened outer surface. This shape increases the amount of wearing surface and probably, therefore, lengthens the life of the rope.

flaw detection Several *non-destructive testing* (NDT) methods exist for finding flaws, usually in metals rather than in concrete, though there is a demand for flaw detection in *bored piles*. NDT flaw detection includes the use of penetrant liquids, magnetic particles, radioactive or *ultrasonic* testing, or eddy

currents. See many BS and ASTM standards.

flexible armoured revetment (FAR) Many revetments which combine weight with flexibility. One FAR type is made of precast *armour* blocks linked by plastic cord or cable and underlain by plastic filter film, either *geotextile* or punched *nonwoven*. Sometimes the blocks are interlocked; geosynthetics may be impregnated with bitumen.

flexible membrane [s.m.] Sheet, usually of synthetic rubber, that can be laid over the ground under a reservoir, tank or swimming pool to reduce leakages. It is usually less than 1 mm thick and protected against punctures by synthetic foam or felt below and may be protected above by other means. Sometimes the membrane is pervious to hold back silt but not water, when used on a beach, and then it may be of woven polystyrene. Other types of membrane are used for *curing* or waterproofing concrete – bitumen emulsions, rubber emulsions, *epoxy* resins, etc. They can be applied inside steel or concrete tanks to waterproof them.

The first butyl-rubber-lined ponds were made in the late 1940s. PVC linings followed in the 1960s, and polythene in the 1970s. In the 1980s HDPE (high-density polythene) became available, in rolls up to 8 m (26 ft) wide.

flexible pavement Road or airstrip construction with a waterproof wearing surface of bituminous material which is assumed to have no tensile strength. The load is transferred to the foundation soil by the *base course*, designed from experience of the *California bearing ratio* method. Flexible pavements are considerably cheaper than rigid (concrete) pavements. See **pavement, prestressed concrete pavement**.

flexible pipe Pipes that bend before they break, i.e. those made of plastics, steel or *ductile iron*. As with rigid pipe, the joints may be rigid or flexible.

flexible wall A reinforced-concrete wall in which the stem is designed as a cantilever or as a beam or both. It requires much less concrete than a *gravity retaining wall* but it can be damaged by wave impact.

fleximer See (*B*).

flexural rigidity The *stiffness* (2) of a beam, column, etc.

flexure Bending.

flints Lumps of *chert* bedded in the chalk.

float (1) [hyd.] An object which shows the direction and speed of flow of the water carrying it. See **rod/subsurface float**, **travelling screen**.

(2) A body floating in a water tank. It opens the water supply to the tank when the water level falls and closes it when the level rises.

(3) Any tool like the plasterer's *float* (*B*) for smoothing mortar, concrete or hot asphalt. See **power float**.

float-cut file [mech.] A single-cut file. See **cut** (4).

floater US term for the plasterer's *float* (*B*).

floating boom, f. breakwater Any device for making a *floating harbour*, or for calming waves, controlling an oil slick, etc. Some are made of expanded polystyrene surfaced with *glassfibre-reinforced concrete*, or of *prestressed concrete* with expanded polystyrene *void formers*. The latter, by Harris & Sutherland, consulting engineers, were 48 m long and 10 m wide (157 × 33 ft). Several types are threaded together by a plastic rope threaded through a plas-

tic pipe along their centres. See **bubble barrier**, **Dracone**.

floating crane, semi-submersible crane vessel (SSCV) (see diagram) Since every dock is fitted with one or more powerful cranes, a floating crane, especially if *semi-submersible*, resembles a *floating dock*. In 1989 the largest SSCV was an Italian structure launched by Monfalcone in Trieste, in 1987, mainly for repairing or building *offshore platforms*. The Micoperi 7000 is 175 m long and 87 m wide, built as a catamaran with two large cranes, each of 7000 tons lift, which together can lift 14,000 tons. The onboard power supply is 57,000 kW. It can cruise at 9.5 knots (17.5 kph) and needs 16 anchors of 40 tons to moor it. It has living-quarters for 800 staff, a swimming-pool, gymnasium, cinema, library, etc. In 1988 it worked at the mouth of the Amazon for Petrobras, the Brazilian state oil company, and was then expected to sail to the Gulf of Mexico and South Africa, eventually with hopes of building the bridge from Sicily to Italy over the Strait of Messina, the largest Italian civil engineering project.

floating dock, f. dry dock A steel hollow box which can be sunk or floated at will. When a vessel at sea needs underwater repairs to its hull, the floating dock sinks below the vessel, rises underneath it, and thus lifts the vessel for repair in the dry. Lift is achieved by pumping water out of its tanks, and letting air into them. See **floating crane**.

floating foundation A *buoyant foundation*.

floating harbour [hyd.] A *breakwater* of *pontoons* connected end to end.

floating pipeline [hyd.] A pipeline

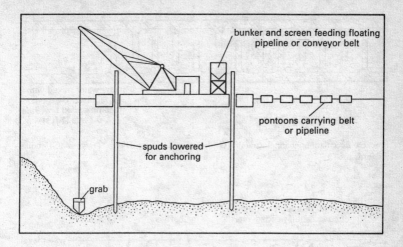

Floating crane digging gravel. When pile-driving, the barge needs several spuds fore and aft.

carried on pontoons used for removing the material dredged by a *suction dredger* which pumps fluid sand or silt into the pipeline. The dredged material is deposited on land so as to raise its level, to make a *hydraulic-fill dam*, etc. Pipelines 3 km long have been successfully used like this to make *artificial islands*.

floating screed See (B).

float switch [elec.] A pump-house switch for starting or stopping the pump motor when the water level rises or falls. It is operated by a *float*. A float switch may be used for automatically opening or closing a spillway when the water level rises or falls.

floc A woolly looking accumulation of solids in a liquid. It settles with difficulty and is the opposite of granular solids like sand, which sink and are therefore easily removed from water in treatment processes.

In the *mixed liquor* of the *activated-*

sludge process, the flocs are millions of tiny gelatinous *activated-sludge* particles produced by the *biological treatment*.

flocculating agent An *admixture* used in a *lean concrete* which must not *bleed*. No European or North American standard exists but a bleed-evaluation test may help. Flocculating agents include polyelectrolytes, wax emulsions, or suspensions of clay or gums in water. They reduce the *slump* and increase the cohesion (stickiness) of the concrete but reduce its *workability*.

flocculation (see diagrams) Gently stirring sewage *effluent* or *raw water*, in which *floc* has formed, to encourage the flocs to enlarge and settle. Rough stirring destroys them. Flocculation is a physical process preceded by *coagulation*, which is chemical.

flood channel [hyd.] The channel in an estuary formed by the flood tide. It is generally wide at the seaward end,

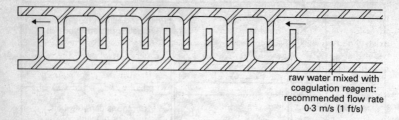

Sinuous flocculation channel (plan). (G. Smethurst, *Basic Water Treatment*, Thomas Telford, 1988.)

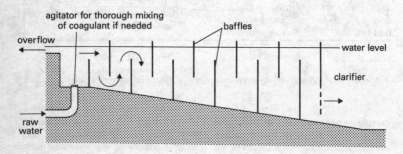

Another way of flocculation (diagrammatic vertical section). (G. Smethurst, op. cit.)

narrowing to a point at the upstream end, where there is a *bar*. Generally in a narrow estuary the flood channels branch upstream off the meandering *ebb channel* so that it looks like a poplar tree when mapped.

flood routing [hyd.] Reduction of a flood or its peak by arranging for dams or overflow channels, storage basins, and channel widening or deepening.

floor arch Term used in the USA for a *jack arch*.

floor centres Steel *formwork* for a floor slab above ground.

floor saw A portable power-driven saw which cuts a groove in hard concrete. Cf. (*B*).

floor scabbler, f. scabber A powerful wheeled machine which can cut off bumps or roughen the surface of a concrete slab either to receive a *topping* or for cattle to walk on.

floor slab A reinforced-concrete floor, particularly the thin part between reinforced-concrete (or steel) beams.

floppy disk, magnetic d. Plastic disk with a magnetic surface coating, housed in a square cardboard jacket about 20 or 13 cm (8 or 5 in.) square, and used for storing either data or computer programs. In 1988 the commonest were not these but minidisks (microfloppies) in a stiff plastic case measuring 10 × 8 cm (4 × 3 in.), but three or more other minidisk sizes exist. The two sides of a minidisk can contain at least 120 pages of text. Any type of magnetic disk should be out of the disk drive when the power is switched on or off, and they should not be brought near any strong magnetic field.

flotation [stru.] (1) The floating of a structure is a condition that has to be checked by calculation whenever a deep basement is built below water level. After the basement walls and floor have been built, and before the building load is put on to them, lifting of the structure by the water pressure under it must be prevented by one of two means: either the water level outside the walls must be kept down by pumping, or the structure may have to be loaded with additional weight. In a *buoyant foundation*, in an emergency the extra weight may have to be water.

Submerged oil pipelines can be weighted with concrete held on by chicken wire or steel mesh wrapped around them.

(2) [min.] Many *mineral separation* treatments.

flotation structure A steel raft built of tubes that can be of 9 m dia., used for transporting an oilwell production *jacket* from the dry dock where it is built to the sea site where it will be used. It must be built before the jacket because the jacket is built on it in the dry dock and slipped off it at the sea site, after which the flotation structure is towed back to the dry dock and the dry dock is pumped out to enable another jacket to be built. The flotation stucture may contain 10,000 tons of steel and its tubes are divided by steel bulkheads. The sections of tube so separated from each other can be emp-

tied or flooded at will, enabling the raft to be sunk or floated as required, or rotated under complete control (upended) through 90°, allowing the jacket to slide under control on to the seabed.

floury soil [s.m.] A fine-grained soil which looks like *clay* when wet but is seen to be a powder when dry. It is therefore a silt or rock flour, not a clay.

flow curve [s.m.] A straight-line graph of the points obtained in the *liquid limit* test showing number of blows on the horizontal, log scale and water contents on the vertical, arithmetic scale. The point where the curve intersects the 25-blows vertical line is the *liquid limit*.

flow enhancer, polymer injection Equipment for automatically injecting doses of lubricating chemical into a sewer in which the level tends to rise hazardously. It increases the flow speed, thus reducing the level and limiting any *stormwater* overflow. The polyacrylamides used do not harm *biological treatment* or fish or other water life. They can also be used for reducing the power cost of pumping.

In 1987 a flow enhancer cost £20,000 and the polyacrylamides also were expensive, but for the Wessex Water Authority the cost was justified by comparison with the replacement of a sewer at £100,000.

Polyacrylamides cost £2.2/kg. Even at the extreme dilution of 10 ppm (sometimes 50 ppm are used) they would be expensive in a sewer flowing at 500 litres/min – roughly £15 per day or £5500 annually. See **surcharged sewer**.

flow index [s.m.] The slope of the *flow curve*. Since the abscissae are plotted logarithmically it is equal to the difference between the water content at 10 blows and 1 blow or at 100 and 10 blows.

flowing concrete Concrete that flows like a liquid can be obtained by adding (in percentage of the cement) from 1 to 1.5% of *superplasticizer* to a concrete of 100–120 mm (4–5 in.) *slump*, increasing the slump to 150–225 mm (6 in. to collapse). To prevent *segregation* there should be a high percentage of fines smaller than 0.3 mm. Added to the cement content these fines should amount to 450 kg/m³. Similar strengths to the unmodified mix are obtainable. *Fly-ash* can be used as fines but should not be considered as part of the *minimum cement content* from the viewpoint of durability. See **flow-table test**.

flow lines (1) [s.m.] Lines in a *flow net* which show the direction of flow of water through a soil mass near a dam or cofferdam. They intersect the *equipotential lines* at right angles.

(2) [hyd.] The paths traced by particles in flowing water.

flow meter An instrument for measuring the quantity of fluid such as water, air or gas which flows in a unit of time, e.g. a *Venturi meter*. The amount of fluid paid for is also measured by a *meter* (water, gas or air meter), sometimes called an integrating flow meter. A *current meter* is a flow meter which has not been calibrated to read amounts of fluid or flow, but reads distances.

flow mole See **slurry shield**.

flow net [hyd., s.m.] (see diagram) A two-dimensional picture of *groundwater* flow, consisting of *equipotential lines* intersecting *flow lines* at right angles in *isotropic* soil. If the soil is not isotropic the lines will not be perpendicular; the flow lines are then likely to follow soil laminations. The flow net shows: (i)

the *neutral pressure* at any point, from which the *uplift* under a dam can be calculated; (ii) the *effective pressure*, for calculating dam stability; (iii) the ordinary *seepage* flow; (iv) the increased pressure due to seepage after rain.

flow slide, earth f., mud f. [s.m.] A slide of a liquefied mass of loose sand or silt which spreads out to a very flat slope after the slide. Those slides which have been examined by the methods of *soil mechanics* have had densities below the critical, that is their void ratios were above the *critical voids ratio*. See **landslip**, **liquefaction**.

flow-table test A standard test for assessing the stiffness of fresh concrete by measuring its spread, usually for concrete with a *slump* greater than 100 mm (4 in.). cf. **V.-B. consistometer test**.

flue gas [mech.] The smoke from a boiler fire, mainly CO_2, CO, N_2, O_2 and water vapour. Its composition, particularly the CO_2 content, gives a very good idea of the furnace efficiency, and this can be permanently recorded on a continuous *carbon dioxide recorder*.

flue-gas desulphurization (FGD) Most coals and many oil fuels contain at least 1% sulphur and each ton of sulphur burning to sulphur dioxide, SO_2, becomes 2 tons of SO_2. UK law requires any power station that is new or has been shut down for 12 months or more to have FGD plant installed. The USA has had such a law since the 1970s. One method sprays lime water into the flue gases to react with the SO_2 and produce mud of calcium sulphate (gypsum) waste (wet scrubbing). Dry absorption is possible with activated carbon or with copper oxide to make copper sulphate. Pollution gets worse as standards of living rise, so more and more FGD plants will be needed. The first UK FGD plant was

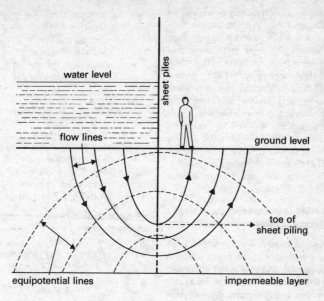

water level

sheet piles

flow lines

ground level

toe of
sheet piling

equipotential lines

impermeable layer

Flow net at sheet-pile cofferdam.

authorized in 1988 at the 4000 MW plant at Drax, Yorkshire, for completion in 1995. It was expected to make *gypsum plasterboard* (*B*) – 500,000 tons in 1993, possibly a million tons in 1996.

An agreement in 1988 between the UK and the European Community's Environment Council requires UK power stations to reduce SO_2 emissions in three stages before 1993, 1998 and 2003, by 20%, 40% and 60% respectively. FGD must be installed in every new plant over 50 MW, but this would not apply to the proposed 1000 MW generating station at Barkingside if it burns natural gas, since this produces no SO_2.

fluidifier *ACI* term for a *water reducer* used in *grout*.

flume, launder [hyd.] A concrete, wooden or metal *open channel* to carry water, slurry or other fluid, or to measure fluid flow. See **Venturi flume**.

flushing manhole A manhole which periodically fills with water that is discharged suddenly into a sewer to prevent it becoming septic. In hot countries, on flat land near the coast, even seawater may be used.

fluxes In *soldering* (*B*), *brazing* (*B*) and *welding*, fusible substances like borax, which cover the joint and prevent oxidation.

fly-ash, pulverized fuel ash (PFA) Extremely fine ash from the burning of pulverized coal. When it has more than 90% silica it may be used as a *pozzolan* in concrete. With about the same fineness as cement, it usually benefits concrete by increasing the volume of *cement paste*, but not all flyashes are suitable. Permissible PFAs for use in concrete are described in British standards. With its relative density of 2.1, PFA is much lighter than cement at 3.1. After pelletizing and sintering it can become a first-class *lightweight aggregate*. Like *ground granulated blast-furnace slag*, it is water reducing and resists sulphates and some *alkali–aggregate reactions*. 'Flyash' is a term disliked by electrical generating authorities because they do not wish attention to be drawn to the fact that their tall chimneys emit pollution – flying ash, apart from *sulphur dioxide*, NO_x, etc.

fly cutter In *microtunnelling*, a *reamer* to enlarge the tunnel.

flying buttress A *buttress* which supports a wall only at the point where it carries a roof, and that is not in contact with the wall at lower levels. In Tudor times this improvement allowed vast stained-glass windows to be built with complete safety.

flying shore See (*B*).

fly jib (see diagram) An extension provided at the top of some crane *jibs* to extend their reach. When not in use it is telescoped or folded under, or to the side of the main jib.

fly-off, interception The catching of rain by plants, followed by evaporation – the opposite of *run-off*.

foam concrete A material of density between 400 and 1800 kg/m³ and strength from 1 to 24 MPa (145–3500 psi), used as a trench infill. Made by mixing a preformed foam with cement and sand, it is a suitable porous fill above a buried drain in a road. It is much more expensive than fill, being slightly cheaper than readymix, but as it is self-compacting it can be placed with confidence in narrow trenches only 10 cm (4 in.) wide.

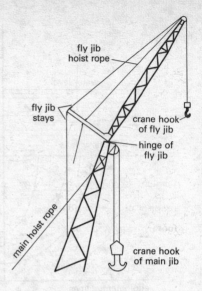

Fly jib in use.

foamed slag, expanded s. A *lightweight aggregate* made by foaming *blast-furnace slag*.

folded-plate roof, polygonal shell r. [stru.] (see diagram) A roof, built of reinforced-concrete slabs at an angle to each other. Such roofs can cover long spans. Other shapes exist.

folding wedges See (*B*).

follower (1) **long dolly, puncheon, sett** A long timber by which the blows of the *pile hammer* are transmitted to the *pile head* when it is below the *leaders* and thus out of reach of the hammer. See **pile driver**, etc.
(2) [sur.] See **leader**.

fondu *High-alumina cement.*

foot block, footpiece, sleeper A timber used as a base for carrying a *side tree* or similar post.

footing A widening of any structure at

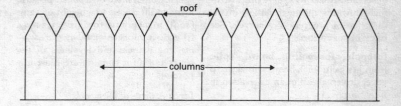

Folded-plate or polygonal shell roofs in cross-section.

the foot to improve its stability, in *breakwaters*, earth or other dams, or simple walls. See **foundation**.

foot iron See **step iron**.

footpiece See **foot block**.

foot-pound A unit of *work* or *energy*, 1 lb lifted 1 ft, so called in order to avoid confusion with the *pound-foot*, the unit of *bending moment*.

foot screws, levelling s., plate s. [sur.] Three screws which connect the *tribrach* of the *theodolite* (and some levels) to the plate which is screwed on to the tripod head. They are used for levelling the instrument at each *set-up*.

foot-ton A unit of *work* or *energy*, 1 ton lifted 1 foot. See **foot-pound**.

foot valve [mech.] A *check valve* at the foot of a length of pipe, generally above a pump suction.

footway (sidewalk USA) That part of the road reserved for pedestrians. See **carriageway**.

FOPS (falling object protective structure) The steel protection over the cab of a *loader, excavator, bulldozer, dumper*, etc. Cf. **ROPS**.

force [stru.] That which tends to accelerate a body or to change its movement; for example, the weight of a body is a

force which tends to move it downwards. Cf. **pressure**.

force account US term for *daywork* or any type of *cost-plus* payment.

forced-action mixer A type of *concrete mixer* which is essential for very stiff mixes, especially *lean concrete*. One type is a *twin-shaft paddle mixer* (*B*).

forced drop shaft [min.] An obsolescent method of shaft sinking by jacking down a cast-iron shaft lining inside which the soil is removed by *grab* or *trepan*. Cf. **Betws ventilation shaft**.

forced vibration [stru.] Vibration of a structure, due generally to engines or machines, occasionally wind. See **damping, free vibration, resonance**.

force pump [mech.] A pump which delivers liquid appreciably above its own level, as opposed to a *spout-delivery pump*.

forebay (1) [hyd.] A reservoir at the end of a pipeline, particularly above a hydroelectric power station.

(2) The area over which water is taken into the ground, eventually to flow or be pumped from a well. In an *artesian well*, the forebay is higher than the outflow.

fore observation, f. sight [sur.] Any

observation made by a surveying instrument towards the next station in the direction of progress of a survey. See **back observation**.

forepole, forepoling board, spile [min.] A 5 cm thick board with a sharp edge, driven ahead of an excavation in *forepoling*.

forepoling, spiling [min.] In tunnel timbering, driving *forepoles* ahead over the caps of the last *four-piece set* erected. Forepoling is used in tunnelling or mining loose ground in any state between hard, unsafe rock to nearly fluid mud. See **jet grouting**.

foreshaft [min.] The first 33 m (100 ft) or so of a shaft's depth in loose surface ground and above rock, built with thick concrete walls. But if strong rock should occur at a depth of 2 m the foreshaft is only 2 m deep.

fore sight [sur.] A *fore observation*, particularly one made during levelling. The ASCE prefers, for levelling, the term 'minus sight'. See **back sight**.

forge [mech.] (1) To shape hot metal (wrought iron or steel generally) by pressing or hammering like a blacksmith. See **forging**.
(2) The fire where the metal to be forged is heated. See **forge welding**.

forge welding, smith w. Joining two pieces of red-hot iron by hammering them together, the oldest *welding* method, in use for several thousand years.

forging Work with power hammers, drop stamps or hydraulic forging machines to shape red-hot steel or other metals. It generally does not include *forge welding*.

fork-lift truck A power-driven truck with a steel fork projecting forwards which can lift, travel with and stack heavy packages at a height. The packages are often lifted on a *pallet*. Because it has to lift 2 tons to a height of 12 m (40 ft) in a very narrow aisle, 1.3 m (4 ft) wide in modern *high-bay warehousing*, the flatness and levelness of the concrete slab it runs on are most important.

form A piece of *formwork*.

formation (1) (**grade** USA) The surface of the ground in its final shape after completion of *earthwork*, but before concreting.
(2) Any recognizable sediment in geology. It is the smallest unit in *stratigraphy*.

formation level (final grade, grade level USA) The surface level (or elevation) of the ground surface after all digging and filling but before concreting.

form lining Materials such as hardboard or plywood which are placed next to the concrete in *formwork* to give it a smooth or textured surface. Being easily bent, they are particularly useful for curved surfaces.

form oil, mould oil *Release agent*.

form scabbing Loss of fragments from a concrete surface which sticks to the *formwork* when this is stripped. It can be avoided by using the right *release agent*.

form stop A *stunt end*.

formwork, casing, shuttering Temporary boarding or sheeting erected to contain concrete during placing and the first few days of hardening. The face texture of the formwork and its stiffness can greatly improve the look of the finished concrete slab. Formwork may be of steel, wood boards, hardboard, plywood, and so on. See **form lining, mould, permanent formwork**.

Forth Bridge A cantilever bridge over the Firth of Forth near Edinburgh, built in 1890 of plates riveted into shapes like boiler shells. It has two central spans of 520 m clear, slightly less than the Quebec bridge. This railway bridge is 1630 m long overall.

Forth Road Bridge A motorway *suspension bridge* completed in 1964 and close to the *Forth Bridge*. The main span is 1006 m and the overall gap 1660 m. The towers are 156 m high.

FORTRAN (FORmula TRANslation) A computer *language* suited to problems in science and engineering.

forward shovel A *face shovel*.

foul sewer (sanitary sewer USA**)** A *sewer* that carries *sewage*, whether in a *separate* or *combined system*.

found (1) To make a *foundation*.
(2) [mech.] To make a metal casting.

foundation An ambiguous word meaning either (1) The soil or rock upon which a building or other structure rests, or
(2) The structure of brick, stone, concrete, steel, wood, or iron which transfers the building load to the ground, sometimes called, together with the rest of the building below ground, the substructure. See **buoyant f., deep f., grillage f., pad f., pile f., draft f., shallow f., caisson, permafrost, pretesting, strip footing**.

foundation bolts *Anchor bolts*.

foundation cylinder See **cylinder**.

foundation failure [stru.] Foundations of buildings can fail in one of several ways, first by *differential settlement*, secondly by shear failure of the soil. See **circular arc method**.

foundation pier A foundation pier for a bridge is a solid concrete block several metres wide and usually at least the width of the bridge in length.

foundry A works with a cupola for melting pig iron (or other metal) and making metal castings.

four-in-one bucket, Drott b., multipurpose b. A digging bucket which is hydraulically controlled in such a way as to be used as a shovel, *bulldozer*, *grab*, *clamshell* or *scraper*. It can make *loaders* competitive with *excavators*.

four-leg sling [mech.] A *chain sling* or rope *sling* with four hooks hung by chains or ropes from one link or *thimble*.

four-piece set [min.] A frame of squared timbers used when *forepoling* in bad ground with a weak floor. It consists of a cap carried on two posts resting on the sill.

four-stage compression [mech.] Air compression in four stages with *intercoolers* between stages, an *aftercooler*, and sometimes also an antecooler. Three- or four-stage compression is necessary for reaching the pressures required for compressed-air locomotives, about 70–80 bars.

four-way reinforcement *Reinforcement* of a *plate floor* with *main steel* parallel to both sides as well as to both diagonals.

four-wheel drive *Loaders* with four-wheel drive have all four wheels the same size but each axle may be separately driven. With only the front axle driven, traction when carrying a full bucket is better. With only the rear axle driven, the loader digs better. Power is supplied through a *torque converter*. *Hydrostatic drive* is available, sometimes also *articulated steering*.

fraction [s.m.] Soils which have been subjected to a *mechanical analysis* are described in terms of their weight percentages of each component, the sand fraction, silt fraction and clay fraction. See **classification of soils**.

fractional-horsepower motor [elec.] An electric motor with a rating of less than 1 hp. It may be a *universal motor*.

fractional sampling Sampling a granular mass by a sampling machine which quarters or decimates a sample without segregation, more quickly than coning and *quartering*.

fractography Examination with a microscope of a broken surface, especially of metal but also of polymers, carbon fibre and other *composites*, to find the cause of the break, or at least the point where it began, and the relationship of the break to the structure of the material.

fracture Breaking.

frame (1) Two or more structural members joined together so as to be stable. A *plane frame* is two-dimensional and only stable in its own plane. A *space frame* is three-dimensional and stable in all directions. Frames may be *redundant* or *perfect*. The typical frame is the portal which is redundant. See **bent**, **steel frame**.

(2) In a timbered trench excavation, the *struts* which separate the boards on the opposite sides of the trench, together with the *walings* which they hold. Thus, in a shaft, all the walings or struts at one level. A frame may include *poling boards* (see **setting**). A ground or top frame is a frame of walings or struts set about 0.5 m below ground as a guide for the first setting of *runners*. A guide frame is built above ground to guide the runners and as a stage for the workers who drive them.

frame weir, framed dam [hyd.] A weir built of timber and steel or cast iron of which some types are propped up by struts against the bed of the river in low water and lowered to the bed or hung above the water when it is in flood. See **needle f. w.**, **rolling-up** curtain f. w., **sliding-panel f. w.**, **suspended-f. w.**

framework [stru.] A loadbearing frame may be of timber, metal, reinforced concrete or other *composite* material.

France-Manche The French half of *Eurotunnel*; cf. **Channel Tunnel Group**.

franchise, concession, build-operate-transfer (BOT) In civil engineering the right to operate a railway, tunnel, bridge or other public service on a commercially profitable basis. The commercial incentive has attracted useful promoters to *civil engineering projects* which otherwise might have been delayed or not built at all. *Dartford Bridge* is one example. Under Hong Kong harbour an 1860 m long, large *immersed tube*, with two rail tracks, two 2-lane roads, and a ventilation and services tunnel, is also a commercial investment. The *Channel Tunnel* is to be commercially operated for 55 years from 15 May 1993, when it opens, until 2042. *Eurotunnel*'s revenue will come from two main sources: (i) fares from people and vehicles crossing the Channel; and (ii) fees paid by the railways for the right to send their trains through the tunnel.

Francis turbine [hyd.] A low-head to medium-head *water turbine* used on many large hydroelectric schemes. The water enters it radially inwards and leaves it, passing downwards. The turbine shaft is usually vertical. See **draft tube**.

Franki pile Trade name for a *driven cast-in-place pile* which has, like others, the advantage of a bulbous toe (clubfoot). Its main disadvantage is the large amount of space needed for the pile frame, which cannot be accommodated on small city sites. *Bored piles* on the other hand can be driven with a 4 m high, three-legged derrick made of steel pipe, which can be carried by two men.

frazil ice, slush i. [hyd.] Granular or spiky ice formed in rapids or other agitated water during long cold spells (USA).

FRC *Fibre-reinforced concrete.*

freeboard [hyd.] The height between normal water level and the crest of a dam or the top of a *flume*, a height which allows for small waves to splash without overflow.

free end [stru.] A *hinge* of a beam.

free face [min.] In rock blasting, an exposed surface. Generally each *round* has one free face. Holes with more than one free face need very much less explosive. See **burden**, **line of least resistance**.

free-falling velocity See **terminal velocity**.

free flow [hyd.] Flow over a weir or dam which is so high as not to be affected by the *tail bay* level.

free haul The maximum distance which excavated material is transported without extra charge (BS 892). This distance is generally fixed by the item in the *bill of quantities* (*B*) under which the particular excavation is paid for. See **overhaul**.

free-piston compressor [mech.] A modern compressor which differs from centrifugal and *reciprocating* compressors in having no important rotating parts. The force of explosion in a Diesel cylinder drives the piston out to compress the air in an air cylinder at the other end of the piston rod. See **duplex engine**.

free retaining wall [s.m.] A *retaining wall* which tilts slightly about its base

or slides slightly, so that the movement of the top is in the neighbourhood of $\frac{1}{2}\%$ of the wall height. By this means the earth force in granular material is reduced to the fully *active earth pressure*. This value is one half to one third of the *earth pressure at rest*, which occurs at a sub-basement or other fixed retaining wall. See **creep**, **fixed retaining wall**.

free vibration [stru.] The vibration which occurs at the *natural frequency* of a structure when it has been displaced, released and allowed to vibrate freely. In structures this can occur under wind load but is relatively unimportant. *Forced vibration*, however, must sometimes be considered.

free water, gravity w. [s.m.] US terms for *held water*.

freeway In the USA a road for fast through-traffic to which abutting owners have no automatic right of access. See **parkway**.

freezing (see diagram) For shaft sinking in fine-grained, waterlogged soil, where ordinary excavation and the *grouting method of shaft sinking* are found impracticable, freezing is one way of making strong dry walls within which the ground can safely be excavated in the dry. A ring of cased holes is sunk to the foot of the waterlogged soil, a second tube inserted in each hole, and cold brine circulated at a temperature of about $-18°C$ until the permanent shaft lining is built. This method has been used for shaft sinking in European collieries since the first shaft was sunk by freezing in South Wales in 1862. It was also used for sinking escalator shafts to the Moscow Metro. From 10 to 24 months are needed for drilling and forming the ice wall round the shaft, and the lowest temperature achievable by these methods is $-35°C$. Liquid nitrogen, which

outline of freeze chamber at
275 m depth, 19·2 m dia.

shaft dia.
7·32 m (24 ft)

16·1 m (53 ft) dia. circle
for 39 freeze holes
at 1·29 m (4½ ft) spacing

Layout of freezing holes at shafts of British Coal's Asfordby mine. When surveyed in the freeze chamber, 275 m below ground, the average deviation of the 39 holes was 225 mm (9 in.), the maximum was 830 mm (33 in.), and the minimum was 20 mm ($\frac{7}{8}$ in.). Three additional holes were drilled purely for observation of temperature, water, etc. See also p. 21 (diagram).

boils at $-196°C$, is more expensive since it is not recirculated. See **Dehottay process**.

French chalk [d.o.] Finely ground talc.

French drain (properly called a **filter drain**) A *field drain* surrounded by a *graded filter* or gravel. See **fin drain**.

French truss A *Fink truss*.

frequency [stat.] How often an event occurs.

frequency curve, distribution c. [stat.] A curve representing ideally the form to which the frequency distribution tends as more and more observa-

tions are obtained. See **Gaussian curve**.

frequency diagram, histogram [stat.] A diagram which shows a *frequency distribution* and is so drawn that the areas under the curve correspond to frequency.

frequency distribution [stat.] The relation which exists between the magnitude of an observed variable and the frequency of its occurrence. This can be expressed in tabular form by grouping the observed data according to the magnitude of the variable and it can be represented either graphically as a *frequency diagram* or *frequency curve*, or mathematically as an equation.

fretting, ravelling The breaking away of aggregate from a road surface. See **scabbing**.

Freyssinet, Eugène, 1879–1962 Founder of *prestressed concrete*, well known for his *flat jacks*, and for six prestressed concrete, *segmental bridges*, built over the River Marne immediately after the 1939–45 war in a time of shortages. Luzancy, the first, was built in 1946 with a 55 m (180 ft) span, and was a two-hinged arch, like the five others of 74 m (243 ft) span at Esbly (1950), Anet, Changis, Trilbardou and Ussy. Each structure was placed in 120 hours after steam curing in a factory on the bank.

friction [mech.] A force which always opposes motion. See **angle of internal friction, coefficient of friction, rolling resistance**.

frictional soil [s.m.] A clean silt, sand or gravel, i.e. a soil whose shearing strength is mainly decided by the friction between particles. In *Coulomb's equation* its shear strength is given by the statement $S = P \tan \Phi$, since it has no *cohesion*. See **classification of soils**.

friction head [hyd.] The energy lost by friction in a pipe, sometimes considered to include eddy losses at bends and elsewhere.

friction loss [stru.] In *prestressing*, a *loss of prestress* caused by friction between a curved *tendon* and its surrounding concrete or *cable duct*.

friction pile [stru.] A *bearing pile* supported wholly by friction with the earth surrounding it. It carries no load at its point, unlike an *end-bearing pile*.

friction welding Rotating a usually circular metal piece in close contact with another piece of metal to heat the area of contact and weld them together, sometimes used for welding shear connectors to steelwork.

fringe water [s.m.] US term for *held water* just above the water table, which may or may not be permanently held.

frog [rly] In a *crossing*, a hard-wearing steel casting with an X-shaped cross in it to accept the wheel flanges. See also (*B*).

frog rammer, trench compactor A manhandled compacting tool, weighing about 500 kg, which lifts itself by the internal combustion of a Diesel or petrol engine. See **power rammer**.

frontage See (*B*).

front-end equipment (1) Fittings to a crawler or wheeled *tractor* to make of it an *excavator*, *crane*, etc.
 (2) Equipment at the input end of a computer.

front-end loader A *loader*.

frost Weather during which dew is deposited as ice. The danger to construction caused by frost is that water expands by about 9% of its volume when it freezes. Therefore concrete or mortar which have not set and contain free water are disintegrated by it. Some wet bricks will chip and lose their arrises in hard frost. If a concrete or mortar has begun to set it will generally continue to set without freezing, because of the heat generated by the action of setting. New concrete or brickwork should be covered by at least a tarpaulin in frosty weather.

frost boil The softness of a soil which has thawed after *frost heave*.

frost heave [s.m.] Swelling of soil due to the expansion of water in *frost*, usually upwards. The ice in freezing expands, forces the soil particles apart, increases the void space, and draws more water up from below if the capillary spaces are small enough and the water near enough. In this way layers of ice parallel to the ground surface can be formed in some silts, but not in clean, coarse sands, gravels or clay. Soils with less than 1% of grains smaller than 0.02 mm never form cumulative ice layers in this way. Silts are the soils most likely to suffer from frost heave or *frost boil*.

Froude number [hyd.] In an *open channel*, a ratio which should be the same for the *model analysis* as in the full-size project. It is the velocity squared divided by depth times the acceleration of gravity. See **Reynolds number**.

frozen ground [min.] See **freezing**, **frost**, **permafrost**.

FRP *Fibre-reinforced plastic*.

fullering [mech.] *Caulking* riveted joints such as boiler plates to make them steam-tight.

fuller's earth [s.m.] A clay composed, like *bentonite*, of montmorillonite, originally used for fulling, i.e. absorbing the fats from wool. It is used in paints as an extender, for its *thixotropy*.

fullface diaphragm See **blind shield**.

fullface tunnelling machine A *tunnelling machine* which, unlike a *boomheader*, breaks out a face of circular cross-section with its cutting wheel. Probably the first fullface machines were those that in 1882 bored circular 7 ft (2.1 m) dia. tunnels 6000 ft (1800 m) long from Dover and from Calais in the first attempt at a *Channel Tunnel*. They were known as Beaumont-English machines after the inventor and developer.

full-tide cofferdam, whole-tide c. A cofferdam, usually in an estuary, built high enough to keep out the water at all tides. See **half-tide cofferdam.**

fully divided scale [d.o.] A *scale* in which the main divisions are fully subdivided for its full length.

fully fixed [stru.] A description of an end of a member in a structural frame which is a *fixed end*. Cf. **partially fixed.**

fume Solid particles so fine that they behave like a gas, in the same way as another fume, cigarette smoke. Being easily breathable, fume is more polluting than dust, which is coarser. Fume can enter the lungs and the blood.

fungicidal admixture A poisonous *admixture* such as copper sulphate, added in very small quantities to a concrete mix to discourage the growth of *algae* or moss.

funicular railway A mountain railway pulled by rope.

fuse [min.] See **detonating fuse**, **safety fuse.**

fusible link [mech.] A plug of low-melting-point metal screwed into the part of a boiler just above the furnace, under the water. If the water level drops below the fusible plug, this will melt, steam will blow down into the fire and put the fire out. A similar plug is used in *sprinklers* (*B*).

fusion welding Joining metals (or plastics) by any method that involves melting them, without pressure and with or without filler metal. The main fusion welding processes are *gas*, *arc* or *resistance welding* but not *braze* or *bronze welding*. *Pressure welding* and *forge welding* are examples of *plastic welding*, not fusion.

G

gabion A rectangular steel mesh basket filled with rocks, used with others for building a free-draining retaining wall.

GAC (granular activated carbon) A useful filter material for removing smells or even *radon* from drinking-water. It can be regenerated by steam at 1000°C but with about 5% loss each time.

gad, moil A short pointed steel bar or wedge for breaking out mineral in sampling, mining or *broaching*. The US bull point is longer, 0.3–0.45 m.

gage US spelling of *gauge*.

gale [stru.] A wind of 75 kph (at 9 m above ground).

gallery [min.] A mine roadway or a tunnel for collecting water in rock or in a concrete dam.

gallium arsenide An expensive substance used for making *integrated circuits*; electronic switching can be about three times faster than with silicon, so it is used in *lasers*, *EDMs* and high-quality computing equipment.

gallon The British Imperial gallon contains 4.55 litres of water; one US gallon contains 3.79 litres of water, but see p. xiv (conversion factors).

galvanic anode A *sacrificial anode*.

galvanize To dip into molten zinc (hot-dip galvanizing) or to coat with zinc electrolytically. Not all zinc coatings are galvanized (see (B) sherardizing or metal coating). Galvanizing may have caused hydrogen embrittlement of high-tensile prestressing bars, with

resulting breakages. Sherardizing is much safer. Galvanized steel is also not advisable in some lightweight concretes which may contain *admixtures* that cause corrosion. See **cathodic protection, dry galvanizing, wet galvanizing**.

galvanized iron [mech.] Not usually iron but steel or steel sheet coated with zinc.

gamma radiography, g.-ray testing, γ-ray testing *Non-destructive testing* by γ-rays, which can show steel bars as fuzzy lines on radiographs, verify whether there are voids in concrete or whether *tendons* are grouted, etc. The method has long been used for checking *weld* quality.

ganat, ghanat See **khanat**.

gang (1) A group of workmen, particularly of labourers or skilled labourers like *navvies* or concretors.
(2) [mech.] A prefix which indicates that several similar machines operate simultaneously and automatically. A *gang-saw* (B) has several saws, a gang-drill several drills, a gang-mortise several mortising machines. See **ganged forms.**

ganged forms Prefabricated panels of *formwork* joined to make a large unit (up to 9 × 15 m (30 × 50 ft)) for convenience in erection, *striking* and reuse. Because of their large size they must be stiffened with *girders*, *strongbacks* (B) or other helps for lifting by crane.

ganger The person in charge of a group of concretors, *navvies*, or other labourers or skilled labourers (not

gang mould

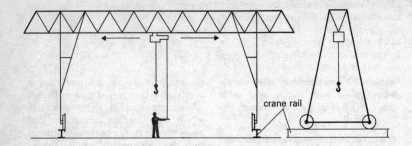

Gantry and crane.

tradesmen). He or she may be a working ganger who stays and works with one gang, or a travelling or *walking ganger* who supervises several gangs.

gang mould A mould for casting simultaneously several similar concrete units. See **gang.**

gantry (see diagram) (1) A temporary staging for carrying heavy loads such as earth or stones, usually built of square timbers or steel joists.

(2) A permanent gantry (crane gantry) may carry the rails for an *overhead travelling crane* or service a stockyard as shown in the diagram.

gantry crane [mech.] A *portal crane.*

gap-graded aggregate *Aggregate* with which an attempt has been made to form a dense concrete by adding a proportion of sand, grit or stone rather than by continuous grading, which implies materials of all sizes in roughly uniform proportions. Excellent dense concrete can be made by both methods, but gap grading is usually easier in practice, since two sizes of aggregate are more easily found than a large number. See **grading** (3).

garland drain, water ring A gutter

cut into the side of a mine shaft, often of cast iron, to catch the shaft water and lead it away to a sump for pumping. In civil engineering excavations garlands have been installed on the face of a cut to prevent water reaching the *formation* and softening it.

gas carburizing [mech.] *Carburizing* steel by heating it in a current of gas containing carbon, for instance carbon monoxide, or hydrocarbon gases.

gas concrete *Aerated concrete.*

gas engine [mech.] An internal-combustion engine which uses as fuel a gas such as *natural gas* or town gas.

gasholder, gasometer A large steel cylinder, closed at the top, usually in a masonry or concrete tank formed deep in the ground, containing water to prevent gas escaping. The steel tank floats high because of the gas pressure inside it, and sinks when the gas pressure drops. Gasholders may be eliminated by high-pressure gas-storage methods: underground in *aquifers* or in the gas mains. But the gas mains are old and leaky, so British Gas must replace 1000 km of mains every year. See **pipe renewal.**

gasket [mech] Generally any prefabricated jointing material such as the asbestos-copper sheet between the cylinder head of an engine and its cylinder block, the joints between *drain pipes* or in the *stuffing boxes* of pumps, or the specially made rubber joints between the concrete units of an *immersed tube*. See also (*B*).

gas-shielded metal-arc (GMA) welding *Metal-arc welding* using either bare wire or cored electrode, in which the arc and molten metal are shielded by inert gas from an outside supply. It is either *MIG* or *MAG welding*.

gas tungsten arc (GTA) welding US term for *TIG welding*.

gas welding [mech.] Welding of metals (or plastics) by the oxy-hydrogen, oxy-coal gas or oxy-acetylene flame, generally the last. See **bronze welding**.

gate [hyd.] A barrier across a water-channel which can be removed so as to regulate the flow of water. See **crest gate**.

gate chamber, camber [hyd.] A recess in a lock wall to take a *ship caisson* or other lock gate when the lock is open.

gate valve [hyd.] A *stop valve* which closes the flow in a pipe by a plate at right angles to the flow. The plate slides in its own plane and beds on a seating round the bore of the pipe when it is closed. When open it offers less resistance to the flow of fluid than any other valve type, since the full bore of the pipe is available for flow.

gathering-arm loader [min.] A loader which pushes its wedge-shaped 'duckbill' into a heap of mineral on the floor and loads it with its rotating arms, pulling the mineral to a central steel-link conveyor for discharge to a belt conveyor or wagon. These loaders may form part of a *boomheader*, loading the rock as it falls. Modern types have a vibrating 'duckbill' which pushes itself in easily.

gauge (gage USA) (1) [mech.] See **Standard Wire Gauge** (*B*).

(2) [rly] The distance, normally 1.435 m (standard gauge), between the inner faces of the rails of railway track.

(3) [hyd.] A water-level measuring device which may be an elaborate recording instrument or merely a stick with metres of depth painted on it.

(4) [mech.] An instrument which indicates fluid pressures by a needle moving over a graduated face. See **Bourdon pressure gauge**.

(5) [mech.] A *dial gauge*.

gauge length [mech.] The length under test in metal test pieces, on which the percentage *elongation* is calculated. It is frequently 200 mm, and must be marked off before testing.

gauge water, gauging w. The water needed to mix a batch of concrete.

gauging pig The simplest gauging *pig* is a sheet steel or aluminium disc made to 90–95% of a pipeline's bore. It is pulled through the pipe and examined afterwards. A deformed or undersized pipe will have buckled the pig. The location of the fault in the pipe can usually be found from the pressure chart made of the pig's journey.

gauging station [hyd.] A point in a stream channel fitted with a *gauge* and means of measuring the flow, from which continuous records of discharge may be kept.

Gaussian curve, normal c. [stat.] A particular type of *frequency curve* with a known equation. It fits a large number of those *frequency distributions* which most often occur.

gel *Colloids*. The cement in concrete is largely gel.

gelatine explosives [min.] Explosives which have a jelly-like texture but contain no gelatine. All contain *nitroglycerin*, some contain ammonium nitrate. They can be used in wet boreholes but must be treated with great care when frozen as they are then liable to detonate if dropped or broken.

general contractor See **main contractor**.

generator [elec.] A machine turned by a *prime mover* such as a steam turbine, steam engine or Diesel engine. It generates electric power either as alternating current (an alternator) or as direct current (a dynamo).

genset, generating set A *motor-generator set*.

geocontainer [s.m.] A bottom-dump barge, lined with strong geotextile (the geocontainer), is filled with sand or other fill. The geocontainer is then stitched up and released from the barge at an appropriate site over the sea bed. The method comes from Holland.

geodesy *Geodetic surveying*.

geodetic construction [stru.] *Stressed-skin construction*.

geodetic surveying, geodesy [sur.] Surveying areas of the earth which are so large that their curvature must be allowed for in calculations. This is specialized work which rarely concerns civil engineers. Cf. **plane surveying**.

Geodimeter, geodetic distance meter [sur.] The AGA company's EDM instruments, which use both visible and infra-red light for measuring distance. Their accuracy is high, typically ± 5 mm ± 1–5 ppm.

geogrid A *nonwoven geosynthetic* with large holes on a rectangular layout.

geological map A map showing either the *outcrops* of all *formations* of igneous rocks, or only the outcrops of the *solid* with the overlying *drift* removed. Maps of Britain exist on the scale of 1 in. to 1 mile (1:63 360), many have been published, and all can be inspected at the Institute of Geological Sciences, London.

geomatrix (plural **geomatrices**) A hollow honeycomb, a semi-rigid grid 100 mm (4 in.) deep with sides 200 mm (8 in.) long, hence about 0.3 m (11 in.) across the diagonals. In an area without vegetation the empty honeycomb is filled with soil or stones to prevent erosion. It has helped to stabilize irrigation ditches fed by the Rio Grande River in New Mexico.

geomembrane An impermeable film of *nonwoven*, sometimes strengthened by a *geotextile*. In the USA a thick impermeable geomembrane, e.g. *polythene*, has to be used to enclose any waste tip whether for domestic or industrial waste. This holds in *leachate* and to some extent *landfill gas*.

geometric similarity A *dimensional analysis*, if it shows that corresponding dimensions in model and prototype are to the same scale, indicates their geometric similarity.

geomorphology The study of the surface of the earth and its relation to the geology of the soils and rocks beneath. Geomorphology explains, among other things, how rockfalls, landslips, flow slides, etc. occur or can be forecast, or whether a river is likely to flood and change its course. It is of great interest to the builder of a road, railway, pipeline, dam or canal, who must make sure that digging will not make the ground unstable.

geophone, seismometer A *geophysical* listening instrument planted in the ground or in water to detect the arrival times of sound waves (shocks) in *seismic* surveys. These instruments were used in Japan from AD 136 onwards to detect earthquake shocks.

geophysical prospecting, g. exploration, g. surveying, applied geophysics Searching usually for mineral deposits by making geophysical surveys, i.e. mapping variations in the earth's elastic properties (*seismic* surveys) or gravitational field or magnetic fields. Temperature surveys reveal salt domes that might contain oil. The *resistivity* method has been used in civil engineering to locate bedrock, the thickness of gravel overlying the London clay, the position of old mine workings, etc. *Well logging* makes use of many geophysical methods, but many other methods do not use boreholes at all, e.g. *seismic prospecting*, apart from a small shot hole.

geophysics The study of the earth. It uses information from geodesy, meteorology, oceanography, geography, electromagnetism, the tides, and so on. See **geophysical prospecting**.

geosynthetics All types of plastic sheets, *nonwoven* or woven plastics (*geotextiles*), or *fin drains*, *geogrids*, *geomatrices*, *geomembranes*, etc., used in drainage, filtration, *reinforced earth* or erosion control, or as a separator between *armour* blocks and underlying fine material. In 1988 the market was expanding at 15% yearly.

geotechnical engineer An engineer who specializes in *rock mechanics*, *soil mechanics*, *foundations*, *groundwater*, etc.

geotechnical processes [s.m.] *Ground engineering*.

geotextile [s. m., hyd.] (see diagram, p. 190) Corrosion-resistant, i.e., *nonbiodegradable*, tough woven plastic sheet. Unlike a *geomembrane* it is permeable. The term is surely wrong for other *geosynthetics* − *nonwoven* film of polypropylene, polythene, etc. Geotextiles are often laid below the *armour* of a *rubble-mound breakwater* to protect the underlying sand from *scour*. They are also used in *reinforced earth* or in areas of mining subsidence under foundations, after any fissures opened by subsidence have been filled with *bentonite* or other *grout*.

Geotextile or nonwoven film reinforced with steel net costs about one quarter as much as *rip-rap* but to become an effective *mattress* exposed to waves needs to be weighed down. Precast concrete blocks strung together with plastic rope cost about the same as rip-rap. Any geotextile or nonwoven is simpler, faster and much cheaper to place than any *graded filter*, but normally needs a top layer of heavy armour. Many of these materials are obtainable in rolls 5 or even 8 m wide. See **fin drain**, **river basin management** (diagram).

geothermal energy, hot rocks Getting heat from the ground depends on the geothermal gradient (downward *temperature gradient*). In Alsace, eastern France, near Soultz-sous-Forêts, the temperature in the granite at a depth of 2000 m is nearly 150°C, corresponding to about 7°C/100 m depth. Near the Nevada hot springs, USA, the temperature rises by 6.2°C for every 100 m depth increase, but by only 0.55°C in the Rand, South Africa. In Britain the gradient is between the extremes. The granites of Cornwall seem to be the most favoured UK source of heat from the ground.

ggbfs *Ground granulated blast-furnace slag*.

giant, monitor [min.] A nozzle for

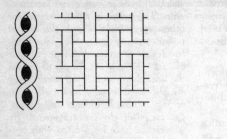

PLAIN WEAVE

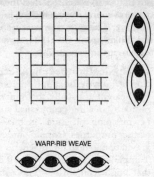

WARP-RIB WEAVE

Geotextile weaves.

projecting water at (and thus breaking up and washing downhill) sand or gravel. Supply pipe diameters vary from 18 to 45 cm, for heads from 30 to 120 m; nozzle diameters from 5 to 25 cm; discharges from 0.04 to 2 cumecs. For all but the very smallest giants, a mechanism called a deflector is needed for turning the jet. One or even two men cannot hold them.

Gibraltar The Strait of Gibraltar is only 16 km wide but 950 m deep, which for a tunnel would imply too steep a gradient. Slightly west of Gibraltar is a 37 km crossing that is only about 380 m deep. Spain and Morocco, in 1988, were considering a project for a power cable between them to cost between $60 and $180 million but no tunnel was in prospect. A suspension bridge over the strait hung from cables made of glassfibre-reinforced plastic has been discussed.

gin pole See **derrick** (1).

girder [stru.] A large beam, originally of wood or iron, now usually of steel or concrete, though light alloys have occasionally been used. Apart from the *bowstring girder* its *chords* are paral-

lel. See **bowstring/box/compound/plate/stiffening/Warren girder**.

girder bridge (sometimes called **beam bridge**) A bridge carried by girders or large beams.

girt (*ACI*) A small beam, usually one between outside columns to carry the wall of an industrial building.

give-and-take lines [sur.] In the calculation of land areas, the straight lines used for the *equalization of boundaries*.

Gladesville Bridge The *Parramatta Bridge*.

gland [mech.] A sleeve or washer which compresses the packing in the *stuffing box* of a pump.

gland bolt [mech.] One of two or more bolts used for tightening or slackening an unthreaded gland.

glassfibre A reinforcing material for plastics and concretes which does not burn and has high *tensile strength*, weight for weight exceeded only by the much more costly carbon fibre.

glassfibre-reinforced concrete, glass-reinforced cement/concrete

(GRC), fibre-reinforced concrete, etc. A rustless material that can replace *asbestos cement* (*B*), cast iron or sheet steel. Pilkington's Cem-FIL 2, dating from 1979, is a glassfibre with 20% zirconia that can be used with ordinary Portland cement, being fairly alkali-resistant. Other alkali-resistant glassfibres exist.

By weight a typical GRC is 5% glassfibre, 14% water, 40.3% *rapid-hardening cement*, 40.3% fine sand and 0.4% cement *admixture*. The first mixes had no sand but their *shrinkage* was 0.3%, extremely high because of the high cement content. Adding sand halved the shrinkage. *Polymer* additions may help to reduce *moisture movement*. *Fly-ash* or *blast-furnace slag* can replace up to 40% of the cement to reduce shrinkage and improve fire resistance, but their slow hardening rate may be a disadvantage. Ordinary Portland cement is usual but *high-alumina* cement is less alkaline and can increase the ten-year strength by 20%. A weight percentage of 5% of fibre is 4% by volume at the usual density of good GRC – 2 tons/m³. All glass is slowly attacked by cement. The result is that GRC does not get stronger like conventional concrete, but in damp surroundings very slowly weakens, though its *limit of proportionality* does increase.

GRC *cladding panels* (*B*) have the fire resistance of conventional concrete at very much lower weight, and their surface is smoother without rust stains. The normal 10 mm thickness of GRC is considered impervious to rain. An *exposed aggregate* finish is made by placing a *face mix* (*B*) in the mould immediately before the 10 mm of GRC.

glassfibre-reinforced plastic ferrocement (GRPF) *Ferrocement* in Red China has been coated with *glassfibre-reinforced polyester*, which is suitable for building irrigation *flumes* because it is smooth, resists cracking and rusts less than ferrocement.

glassfibre-reinforced polyester, g.-r. plastic (GRP) The commonest reinforced plastic. Glassfibres reach very high strengths and are stable up to 600°C. Most GRP is thin, therefore to resist cross-thrust it must be folded or curved. *Folded-plate* structures are very suitable. *Cladding panels* (*B*), church spires, window frames or sills, caravan roofs, boat hulls, sports car bodies and translucent shell roofs spanning 15 m (49 ft) have all been made of GRP. Cladding panels can weigh as little as 5 kg/m² (1 lb/ft²) uninsulated or 50 kg/m² (10 lb/ft²) if insulated. GRP pipe is accepted for water supply and sewerage and can be made in diameters up to 2.5 m (8 ft). Being non-combustible and smooth it is suitable for ventilation ducts in tunnels.

global positioning system (GPS) The latest system for locating points in three dimensions. GPS was delayed by the Challenger space shuttle disaster in 1986 but the full constellation of 21 Navstar satellites, orbiting at altitudes of about 20,000 km every 12 hours, should be available in 1993, providing global coverage 24 hours a day. If several radio receivers are used, simultaneously measuring the ranges to four satellites, the observer's position, using *translocation*, can be fixed with an accuracy of a few cm, from observations made over 30 minutes. Cf. **satellite-Doppler**.

global safety factor [stru.] The combined factor of safety resulting from all the partial safety factors.

glulam Glued laminated wood for making beams of large span. 'Laminated' means that the *grain* (*B*) of the boards is parallel, unlike plywood in which the thin adjoining veneers are laid with the grain crossing. This strengthens plywood but for the large span of glulam beams it would not be feasible. For a thorough description see C. J. Mettem in *Structural Engineer*, Sept. 1988, pp. 287–94.

goaf, gob (**cundy**, Scotland; **self-fill** USA) [r.m.] In mining, a void from which mineral has been removed, usually with *caved* roof.

go-devil See **pig**.

going See (*B*).

Golden Gate Bridge Formerly the largest *suspension bridge*, built in 1937, of main span 1280 m and side spans each 343 m. During an 80 kph gale which lasted for four hours this bridge caused some anxiety by vibrating with a double *amplitude* of 3.4 m, 8 times per minute. See **Tacoma Narrows Bridge**, **Verrazano Narrows Bridge**.

Goliath crane A heavy *portal* frame, usually of about 50 tons capacity, with a crane *crab* which travels along the beam at the top of the portal. The frame of the portal usually has four legs and at least eight wheels running on 2, 4 or more rails. It is used for shop erection of heavy steel and for other heavy lifting jobs such as harbour or nuclear-power-station construction. Cf. **Titan crane**.

goniometer [sur.] An angle measurer, e.g. *compass, sextant* or *theodolite*.

go-out [hyd.] A sluice in a tidal embankment which impounds tidal water. The water can pass out through the sluice at low tide.

Gow caisson, Boston c., c. pile A device for sinking small shafts through soft clay or silt to prevent excessive loss of ground. A short cylinder of steel plate is driven into the clay, which is then excavated within it until the

clay in the bottom begins to *heave*. Another short cylinder of slightly smaller diameter is then driven inside the first, and so on until the shaft reaches firm ground. The shaft is then filled with concrete and the steel tubes are withdrawn as the concrete rises to their level. See **Chicago caisson, loss of ground, underreaming**.

GPS *Global positioning system*.

grab, g. bucket An excavating attachment hung from a crane or *excavator*. It is a split and hinged bucket fitted with curved jaws or teeth (tines) which dig while the bucket is being dropped and move together to pick up while it is being raised. Grabs are classified according to the shape of the bucket as a mud grab, clamshell grab, dumping grab (for rehandling aggregates), whole-tine grab (for heavy digging in hard clay, dense gravel or earth), *orange-peel grab*, and heavy-weight grab (for ore, difficult dredging or general dredging). It is the oldest type of excavator and is superseded in most uses by the dragline, but is still used for specialized work. See **excavating cableway**.

grab dredger Two main types of *dredger*: (*a*) the dipper dredger, which is a mere *face shovel* mounted on a barge and consequently has very limited depth of excavation; (*b*) the grab proper, hung on ropes or chains, which therefore has unlimited depth of excavation. Grab dredgers have small output compared with *suction* or *bucket-ladder dredgers* but they can lift solids or pieces of wreckage and can work close to ships or to a dock wall and in a heavy swell of 2.5 m.

grab sampling [min.] *Random sampling* to obtain a general idea of the characteristics of a deposit or shipment when there is not enough time, labour or space to make a representative sample. The sampler takes care with each sample to have the representative proportions of coarse and fine material.

grade Term used in the USA for (*a*) *gradient*, (*b*) formation, (*c*) ground level.

grade beam US term for a *ground beam*.

graded aggregate Aggregate containing selected proportions of different particle sizes, usually chosen to form a concrete or soil of maximum density, but see **graded sand**.

graded filter [s.m.] Layers of coarse gravel, fine gravel, coarse sand and fine sand arranged over each other so that water flowing through one material does not carry it into the next to clog it. Graded filters are placed at the downstream toe of an earth dam to weight it, adding stability while allowing drainage, or at the bottom of an excavation which is *boiling*, or round French drains. In every case the finer material is placed on the side which the water reaches first. To ensure that one layer of the filter will not clog the next, Terzaghi devised the following rule: the grading curve of each layer must be plotted. The grain sizes corresponding to 15% of the total of each layer, their D_{15} sizes (see **effective size**) are read off. If the D_{85} size of the finer soil is larger than a quarter or at least one fifth of the D_{15} size of the next coarser soil, there is no fear of the fine material clogging the coarser.

graded sand [s.m.] A *sand* which is not merely coarse or fine, as defined in the *classification of soils*, but contains a mixture of at least two of the coarse, medium and fine sizes. To geologists, however, a 'graded' sand is what civil engineers would call an 'ungraded' sand. Geologists think of the 'grading' (size-classifying) action of the river

which has sorted the sand from the pebbles into a single size.

grade level US term for *formation level*.

grader, motor grader (see diagram) A self-propelled machine usually now with three axles and an engine of up to 300 hp with a wide, hydraulically adjustable blade for cutting, moving and spreading soil smoothly to the shape required. In modern machines the driver is able to sit. The grader is more efficient than the *bulldozer* in smoothing rough ground for roadbuilding but it also makes ditches or windrows or backfills trenches.

grade-separated junction A road

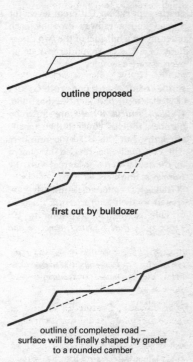

outline proposed

first cut by bulldozer

outline of completed road –
surface will be finally shaped by grader
to a rounded camber

Roadbuilding in sidelong ground.

Steels Tensile strength, MPa		Grades (names) BS 4360: 1986	Euronorm and future BS	ISO
340–500 ⎱ 430–580 ⎰	⎰ carbon ⎱ steels	40 43	235 275	360 430
490–640 ⎱ 550–700 ⎰	⎰ alloy ⎱ steels	50 55	355 450	510 —

intersection such as an underpass, with one road at a different level from the other. If one is a motorway or express-way the junction is called an inter-change. *Grade separation* allows roads to cross each other without traffic inter-ference.

grade separation Different levels for a road and a railway crossing each other or, on a hillside, of the two direc-tions of a road. An underpass or an overpass are examples.

grades of steel and concrete Grades of steels, according to BS 4360: 1986, *Weldable structural steels*, are shown by numbers roughly equal to their tensile strengths, but the BS 4360 numbers will eventually be revised as shown in the table above to agree with the *Euronorm* numbers.

Grades of concretes, similarly, are related to their *characteristic strengths* or cube strengths, according to BS 5400: 1984 *Part 4: Steel, concrete and*

Cube strength (28 days) MPa psi		Concrete name	E concrete, GPa (short-term loading)
20	2900	20	25
25	3600	25	26
30	4250	30	28
40	5800	40	31
50	7250	50	34
60	8700	60	36

Other authorities give *E* values which are as much as 15% different.

composite bridges. Below, the probable modulus of elasticity (*E* value) is added, as a matter of interest, in gigapascals. The long-term *E* value is half the corresponding short-term value.

gradient (grade USA) The fall or rise per unit horizontal (or slope) length of a pipe, road, railway, *flume*, etc. The slope can also be expressed as the number of degrees from the horizontal or as a percentage (USA). Some typical gradients are: locomo-tive haulage about $\frac{1}{2}$% in favour of the load and against the empties; belt conveyors preferably not more than 30%, never more than 45%; gold sluicing 4–8%. See **limiting gradi-ent**, and for drainage, **self-cleansing gradient**.

gradienter [sur.] A *micrometer* fitted to the vertical circle of a theodolite or level, which allows the telescope to be moved through a known small angle. In this way the telescope can be used as a *tacheometer* without stadia hairs. It may also enable the telescope to be used as a *grading instrument*.

gradient post [rly] A short post set beside a railway at each change of gradi-ent. It has an arm at each side on which is painted the length of track for one unit rise or fall, or some other simple indication of gradient.

gradient speed The theoretical wind

speed at 600 m elevation. It can be calculated by meteorologists from isobars. On land it has only one third of this value because of the friction from obstructions and at sea about two thirds of it.

grading (1) Shaping the ground surface, usually by *earth-moving* plant such as *graders*.

(2) The percentage by weight of different grain sizes in a sample of soil or aggregate, expressed on a *grading curve*.

(3) The purposeful modification of the proportions of the different grain sizes in an aggregate for concrete or in a soil for an earth dam or other structure, so as to produce the densest or most stable material. See **gap-graded aggregate**.

grading curve [s.m.] A curve on which the grain size of a sample is plotted on a horizontal, logarithmic scale, and percentages are plotted on a vertical, arithmetic scale. Any point in the curve shows what percentage by weight of particles in the sample is smaller in size than the given point. It is therefore often called a grain-size accumulation curve. See **classification of soils, coefficient of uniformity, effective size, graded filter, mechanical analysis, fineness modulus**.

grading instrument, gradiometer [sur.] A surveying level with a telescope which can be raised or lowered to set out precisely a required gradient. Cf. **gradienter**.

gradiometer See **grading instrument**.

graduate member An *engineering institution* member who has an engineering degree and works for a firm approved by his or her institution under training as a *chartered engineer*.

graduation of tapes [sur.] Most accurate tapes or steel bands of 100 ft or more were graduated only at every foot. To obtain the precise length it was therefore necessary to apply a *reglette* to the foot mark nearest to the point whose distance was required. The other end of the band was set at a precise foot or read similarly with another scale.

grafting tool, clay spade, graft A narrow stiff spade for digging hard clay, often used on the London blue clay. It may be driven in by the foot or by compressed air. In the latter case it may also be called a clay digger or spader and weighs about 10 kg.

Graham's law of diffusion [min.] The rates of diffusion of different gases at the same temperature and pressure are inversely proportional to the square roots of their densities (molecular weights). See **diffusion**.

grain See **rift**, also (*B*).

grain-size classification [s.m.] Any *classification of soils* based on grain size. See **grading curve**.

Grand Canal Completed under Kublai Khan, the first Mongol emperor of China, around AD 1290 but begun much earlier, this main Chinese north–south waterway of canals and *navigations* has been modernized to take ships of 600 and in places 2000 tons. It is 1600 km (1000 miles) long and is probably the world's longest manmade waterway. Its northern half, the world's first *summit canal*, was dug to provide taxes in the form of grain for Kublai Khan's new capital, Peking (Beijing).

grandstand Grandstand roofs may be difficult to design because of their long overhangs but their *prestressed concrete* roofs can be elegant.

granite A rock formed of relatively large, interlocking crystals (indicating that the fluid rock cooled and crystallized slowly) of *quartz*, felspars and mica. The crystals are often half an inch in size or larger. Its *crushing strength* can reach 140 MPa (20,000 psi) (Aberdeen), which is stronger than blue bricks. Since it is not easy to achieve this strength with concretes it is conceivable that *prestressed* structures of the future may use granite. The relative density is from 2.4 to 2.8. It is a typical *igneous intrusion*. Its fire resistance is low.

granolithic screed, grano, high-strength concrete topping A 1:1:2 *screed* of cement, sand and granite chips laid over a concrete floor to give a smooth, hard-wearing surface, usually 50 mm (2 in.) thick, but of 15 mm absolute minimum thickness if laid as a *monolithic screed* on *green concrete*, or 25 mm on clean, hard concrete. Emery or Carborundum powder may be mixed in to make a non-slip surface. Bays should not be wider than 4.5 m (15 ft) nor longer than 10 m (33 ft) if reinforced or 6 m (20 ft) unreinforced. See also **jointless flooring** (*B*); cf. **unbonded screed**.

granular active carbon See **GAC**.

granular stabilization [s.m.] *Soil stabilization.*

granulator A *breaker* for making small aggregate from large stone.

graph [d.o] A line drawn, usually on squared paper, to show the variation of one quantity with another; for instance, the increase of concrete strength with age or the temperature increase of the rock in a tunnel as the depth of the tunnel increases. A vertical measurement to a graph from the *origin* is an ordinate, a horizontal one an abscissa. Both are *coordinates*.

graphics package A collection of programs for *computer graphics*, sometimes including the *microcomputer*, *VDUs* and accessories.

grass A *revetment* for river banks, obtainable by turfing or sowing grass seed. *See* **maritime plants**, **reinforced grass**.

graticule (1) [sur.] The *reticule* of a surveying telescope.
(2) [d.o.] A squared grid.

grating A wooden *grillage* foundation. Wooden foundations must be well below the permanent water level to prevent them rotting.

gravel (1) Untreated or only slightly washed, rounded, natural building aggregate, larger than 5 mm (for concrete 10 mm).
(2) [s.m.] Granular material, smaller than 60 mm, which remains on a 2 mm square mesh.

gravel pump A *centrifugal pump* with renewable shoes on the impellers and a renewable lining, used for raising gravel loosened hydraulically. Sometimes the linings to pump and pipes are of rubber, which lasts longer than steel or iron. See **rubber linings**.

graving dock A *dry dock*, originally a dock where ships' bottoms were scraped and smeared with graves (the dregs of tallow).

gravitational water (1) [s.m.] Groundwater moving downwards in the unsaturated ground above the water table.
(2) *Irrigation* water that is not pumped.

gravity The force which attracts matter towards the centre of the earth and gives it its weight. The acceleration due to gravity is 9.81 m/s each second at sea level, but decreases slightly as the distance from the centre of the earth increases.

gravity-arch dam A *dam* which resists the thrust of water both as an *arch dam* and by its own weight.

gravity axis US term for *centroid*.

gravity correction [sur.] A *tape correction* to a band standardized at one place and used at another place of different *gravity*. This correction is made only for the most exact work in which the tension to the tape is applied by weights on a cord running over a pulley.

gravity current See **density current**.

gravity dam A *dam* which is prevented from overturning by its weight alone. If it is high it is very heavy and expensive, therefore other types are preferred where possible. See **arch/gravity-arch dam, gravity retaining wall**.

gravity main The pipeline in which water flows downhill from an *impounding reservoir* to a *waterworks*. Its diameter is fixed by the designer, who calculates the friction loss of the maximum flow through it. This friction loss in metres must not exceed the difference in level between the bottom water level of the impounding reservoir and the top water level at the waterworks. See **break-pressure tank**.

gravity platform A heavy concrete oil-tank structure placed directly on the seabed without supporting piles. Although cheaper than steel, this is not of vital importance because the structure is only about 10% of the cost of an *offshore platform*. The superstructure is often steel. The oil tanks, empty, provide buoyancy for floating the platform to its site. All gravity platforms up to 1988 were on firm seabed, either dense sand or very stiff clay, unprepared except for the removal of boulders.

The first gravity platform was installed in 1973 in the Ekofisk field of the North Sea to store three days of oil flow (a million barrels or 160,000 m³) during bad weather before the pipeline to the shore had been laid when high waves prevented the loading of the tanker. It was followed by many others, some used also as production platforms.

A gravity platform is built in several stages of which only the first is on land in dry dock where the base and walls are cast. These can be floated out to the relatively deep-water construction site where the tanks are roofed. If the towers are to be of concrete they are *slipformed* and finally the concrete deck is cast. See also **dowel, skirt**.

gravity retaining wall A *retaining wall* which, like a gravity dam, is prevented from overturning by its own weight alone, not by the weight of any soil it carries. It must therefore be designed to take no tension, i.e. the *line of thrust* must pass through the middle third of the wall. The design is usually massive and, for deep walls, extravagant compared with reinforced-concrete walls. Gravity walls are of masonry, brick or mass concrete, or they may be *crib dams*. See **flexible wall**.

gravity scheme [hyd.] A water supply scheme in which all flow (or most of it) into or out of the *impounding reservoir* is by gravity and no pumping is therefore needed.

gravity water (1) [s.m.] A term used in the USA for *gravitational water*.

(2) [hyd.] A water supply from a *gravity scheme*.

GRC *Glassfibre-reinforced cement* or *concrete*.

grease trap A large household *gulley* for kitchen waste fitted with a metal tray in the bottom, fixed to a handle reaching above water level. The grease carried into the gulley by the sink waste congeals and floats to the surface of the water in the trap. The rest of the

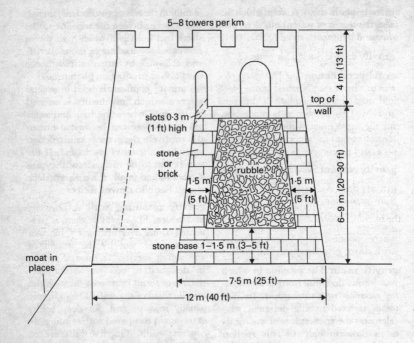

5–8 towers per km

4 m (13 ft)

top of wall

slots 0·3 m (1 ft) high

stone or brick

rubble

6–9 m (20–30 ft)

1·5 m (5 ft)

1·5 m (5 ft)

stone base 1–1·5 m (3–5 ft)

moat in places

7·5 m (25 ft)

12 m (40 ft)

Great Wall of China (section), showing tower in elevation.

waste water flows out at a low level. If the tray is pulled out and emptied daily, the drains are kept free of grease and stoppages of drains are unlikely.

Great Belt (**Storaebelt** in Danish) The busy 18 km wide stretch of water separating Zeeland (with the capital Copenhagen) from most of Denmark to the west. A £1500 million project to cross the Belt with road bridge and rail tunnels began in 1989, with the 8 km long twin rail tunnels of 7.7 m (25 ft) dia. due to hole through in 1991 and to have trains running in 1993. The road bridge is expected by 1996.

Greathead shield A protection to workmen tunnelling in soft ground, first used successfully in London by Greathead in 1879, and later for driving the circular London *tube railway* tunnels. See **shield**.

Great Salt Lake See **artificial recharge**.

Great Wall of China (see diagram) Apart from the prehistoric marks in Peruvian deserts, the only manmade object on earth visible from a satellite. It runs from the Gulf of Chihli on the Yellow Sea 2400 km (1500 miles) westward into central Asia. Begun around 400 BC, the various lengths were joined in 214 BC by the first emperor of all China to protect the country from the Hun cavalry to the north. Generally 9 m (30 ft) high, it has towers at intervals, 12 m (40 ft) high and 12 m square. Faced partly with stone or brick, the wall usually kept out the horsemen.

green concrete Fresh, set but still soft concrete. See (*B*).

greenhouse effect Carbon dioxide (CO_2) and other gases reflect back to the earth some of the heat radiation that otherwise would be lost to space. They tend to warm up the earth. The average concentration of CO_2 has risen 25% in the last century mainly from the burning of coal and oil. The atmosphere contains only 0.03% of CO_2 but scientists believe that if this were to double, as seems likely before 2050, the earth's average surface temperature would rise by 3°C, possibly flooding coastal cities because of the melting of the polar icecaps.

Methane also has a greenhouse effect, and from analyses of air trapped in dated ice cores its concentration also may have doubled in the last 350 years. Other important 'greenhouse gases' are *ozone*, nitrous oxide (a precursor of ozone in the atmosphere), and two *chlorofluorocarbons* (halocarbons or fluorinated hydrocarbons) used in spray cans, $CFCl_3$ (also called Freon 11) and CF_2Cl_2 (Freon 12). These should be banned.

Though it is difficult to be certain that the coast is not falling at the measurement points, scientists believe the oceans have been rising at 1.2 mm yearly since 1925. See **subsidence**.

green pellets See **briquette**.

GRG *Glassfibre-reinforced gypsum* (*B*).

gribble A crustacean *marine borer*.

grid (1) An open frame of wooden beams resting on the foreshore where a vessel can float in, moor, and be repaired lying on the grid when the tide is low.

(2) [d.o.] Any rectangular layout of straight lines, generally used in locating points on a plan. See *grid plan* (*B*).

(3) [stru.] In plane frames the members are simple and repetitive, therefore cheap. These properties have been copied in the space frames known as space grids or space decks, but the spans are so large that the structures are extremely expensive. Many types of grid are used as space frames. A two-way grid has two sets of trusses of the same depth throughout, intersecting each other at right angles. A three-way grid has three sets intersecting each other at 120°, forming a triangular mesh. Hexagonal meshes have also been built. See **double-layer grid**.

grid bearing [sur.] The angle between a line in a *grid*, generally a north–south line, and a required direction.

grillage foundation, grillage A *foundation* for concentrated loads such as columns, consisting usually of two layers of rolled steel joists (occasionally timbers, see **grating**) on top of and at right angles to each other. The lowest layer is placed on the concrete foundation, the next laid across it. Steel joists are invariably covered with concrete as a protection from rust. By this means an intense column load can be spread over an area large enough to carry it.

grinding (1) [min.] The final size reduction of ore from about 18 mm down to about 0.06 mm (slime) depending on the size of the mineral particles to be extracted from the ore. See **ball mill**, **breaker**.

(2) [mech.] The removal of metal by a *grinder* (*B*).

grip (1) **catch drain, intercepting channel, drain** A small channel cut into the ground on the uphill side of an excavation to lead rainwater clear of it. See **catchwater**.

(2) A shallow channel dug into the *verge* of a road to lead rainwater from the road to the ditch or drain.

(3) See **bond** and below.

grip length, [stru.]**, bond l. (embed-ment l.** *ACI*) The length of straight reinforcing bar, expressed in bar diameters, required to anchor the bar in concrete. For round bars it is taken in Britain as equal to the stress in the bar divided by four times the permissible *bond stress*. See **transmission length**.

grippers Powerful jacks on a *tunnelling machine* in hard ground, which provide the machine with the resistance to push it forward with its horizontal jacks. In soft ground the machine pushes on the *segmental rings* of the *tunnel lining*. See also **bladed shield**.

grit blasting [mech.] *Sand blasting* with grit.

grit chamber [sewage] A *detritus tank*.

gritter (1) A towed or self-propelled implement which spreads chippings over the surface of a road in *surface dressing*.

(2) **winter gritter** A similar machine for spreading chippings, grit or salt over a frosty road.

gritting *Blinding* on a road surface.

gritting material Small chips for *surface dressing* or for making a temporary non-skid layer on a road.

grizzly, scalping screen A coarse screen usually built of parallel sloping steel rails with a gap of at least 2–5 cm (1–2 in.) between them to separate the small stones.

groin US spelling of *groyne*. See also (*B*).

grommet or **grummet** (1) A circular

washer made of hemp and red lead, inserted on to a fixing bolt on the flange of cast iron *tubbing* or tunnel lining to make it watertight. A grommet is also used for making a watertight packing between a steel pipe in a hole bored in rock (for *grouting*) and the rock itself.

(2) A coil of rope placed on top of a *dolly*.

groover Equipment for making grooves in the surface of a concrete slab during construction to control the location of cracks (*ACI*).

gross duty of water, diversion requirement [hyd.] The irrigation water diverted to the intake of a canal system, usually expressed in depth on the irrigable area concerned.

gross error, mistake [sur.] An error which is easily detected because it is large, in proportion to the value being measured. This error would be detected by making three measurements. Two of the measurements would show that one measurement was incorrect and could be rejected. The value used would then be the average of the two more accurate measurements.

gross loading intensity [stru.] The total load per unit area at the base of a foundation. In order to calculate the likely *subsidence* of a structure the *net loading intensity* has to be calculated. This is the gross loading intensity less the pressure of the earth that has been removed from the dig. It is possible for the net loading intensity to be less than the gross, in which case there will be no settlement. The earth removed is then heavier than the structure put in its place.

gross ton See **long ton**.

ground anchor *Roof bolts* in rock or an *anchor pile* in soft ground to hold back a *sheet-pile wall* or the timbered face of a deep excavation, etc.

ground bashing [s.m.] *Dynamic consolidation*.

ground beam (grade b. USA) [stru.] A reinforced-concrete beam at or near ground level which acts as a foundation for the walls or floors of the superstructure. It may span between foundation piers or piles or be itself a *strip foundation*.

ground control [sur.] In *photogrammetry* the marking or identification of points on the ground so that they can be recognized in the photograph. In addition some surveying work must be done to verify or measure the positions and altitudes of these points. *Radar* can now *fix* the aircraft position at the instant at which each photograph is taken, and this is another very important sort of ground control. See **stereoscopic pair**.

ground duct, creep trench A concrete channel built into the ground to carry pipes, cables, etc. between buildings. See also **duct** (*B*).

ground engineering, geotechnical processes [s.m.] Improvement of the ground from the viewpoint of the structural engineer. With *site exploration* it includes all the processes of *ground improvement*.

ground frame See **top frame**.

ground granulated blast-furnace slag (ggbfs) *Blast-furnace slag* which has been granulated and then crushed for mixing with Portland cement. It heats up less than *ordinary, rapid-hardening* or sulphate-resisting cements and is slower to harden, but *alkali–silica reaction* with it is unlikely. It generally replaces cement weight for weight in proportions from 30 to 50%. Like *fly-ash* it is water-reducing and improves *workability* and sulphate resistance. The first BS for ggbfs cements was

issued in 1923; they are very widely used on the Continent.

ground improvement [s.m.] (see diagram, p. 202) *Artificial cementing, compaction, deep blasting, dynamic consolidation, electro-osmosis, freezing, groundwater lowering, injection, sand piles, stone piles* and *vibroflotation* are ways of strengthening or dewatering ground, sometimes to carry heavy foundations.

ground investigation That part of *site investigation* which concerns the soils of a construction site and their properties, mainly their strength for the purpose of foundation design. The most costly part of this work is the sinking of boreholes, test pits, shafts or headings under the proposed structure to a depth equal to about 1.5 times the building width unless *bedrock* is found above this level. Normally in the UK the boring has a 15 cm (6 in.) casing so that *undisturbed samples* of 10 cm (4 in.) dia. can be taken easily. The cost of ground exploration may be only 0.5% of a large job but for a small job even 5% may be too little. In any case ground exploration results are only provisional until more information is obtained from digging up the site.

ground prop A *puncheon* between the lowest *frame* and a foot block on the formation level of a timbered excavation. It carries the weight of the timbering.

ground truth [sur.] Field studies made to check the accuracy of *remote sensing*.

groundwater [s.m.] Water contained in the soil or rocks below the *water table*; if this is lowered too much, the ground may settle disastrously. Venice has suffered *subsidence* at about 5 mm a year because of excessive withdrawal of sweet water from wells and fountains. To prevent subsidence and in

(max. grain size, mm)	CLAY	SILT			SAND			GRAVEL		(fine gravel is coarse sand)
		fine	medium	coarse	fine	medium	coarse	medium	coarse	
	0·002	0·006	0·02	0·06	0·2	1·0	5	10	100	

Ground improvement methods by applicable soil particle size range:
- cement, etc. grouts
- resin or chemical grouts
- ground freezing
- compressed air
- wellpointing
- electro-osmosis

Approximate soil particle sizes appropriate to ground improvement methods. (Pipe Jacking Association.)

the hope of raising the city by 30 to 50 mm yearly, ICOS proposed in 1973 to sink a *diaphragm wall* 120 m deep and 60 cm wide in the lagoon around it, inside which the water level could be raised by pumping fresh water into recharging shafts within the watertight ring. All the diaphragm wall work would be done from barges in the lagoon. Such a deep diaphragm wall has never been attempted, the previous maximum depth being about 30 m.

The diaphragm wall was never begun but in 1988 an effort was made to close the three entrances to the lagoon against abnormally high tides by sinking concrete structures across them. Built into these vast structures were hinged, hollow flap-gates normally filled with water and resting on the bed of the lagoon. When a high tide approaches, the gates will be inflated full of air and will rise above the sea in 11 minutes to keep out the high tide. It is hoped that the flaps will all be in place in 1995. Normally the structures are invisible, being below water. But this barrier against high tides will not solve Venice's problems of a falling population, halved from 1966 to 1988, and pollution.

groundwater dam An underground dam holding back polluted groundwater, e.g. to prevent it polluting drinking-water. Around Chernobyl about 130 such dams have been built since the disaster of April–May 1986. Pollution of the groundwater below John F. Kennedy Airport by the fuel of jet aircraft has been held back by groundwater dams. An aquifer may be penetrated and sealed off with a wall of clay or a *slurry wall* or a *grout curtain*.

groundwater lowering [s.m.] (see diagram) Lowering the level of *groundwater* to ensure a dry excavation in sand or gravel or to enable the sides of the excavation to stand up. Groundwater lowering in this sense is always carried out from outside the excavation either by *wellpoints* or from *filter wells*. Filter wells have the advantage for deep lowering that the pump can be submerged in the bottom of the well. There is therefore no limit to their depth. The limit to the depth of a

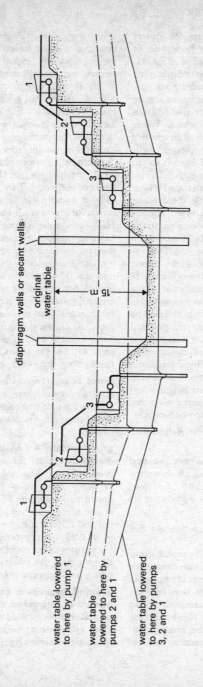

diaphragm walls or secant walls

original
water table

15 m

water table lowered
to here by pump 1

water table
lowered to here by
pumps 2 and 1

water table lowered
to here by pumps
3, 2 and 1

Groundwater lowering. A diaphragm wall or secant wall reduces the excavation and pumping to a fraction of that needed by groundwater lowering with pumps. In addition the wall can go down to 30 m or even deeper. But for heavily water-bearing ground the walls need anchorages (not shown), and the foot of the diaphragm wall should be sealed by entry into an impermeable soil such as clay.

wellpoint is about 4.5 m; for greater depths than this, two or more series of wellpoints must be installed in succession. Filter wells and wellpoints must be provided with a *graded filter* surrounding the pump suction to ensure that fine material is not removed from the ground with the water. Water cannot be lowered in soils of an *effective size* below 0.05 mm although wellpoints may still be used in such soil in the so-called vacuum method. In this method although the water level is lowered not at all or very little, the strength of the silt is greatly increased; the increase in strength is proportional to the vacuum and may be about 0.05 MPa (7 psi). All groundwater-lowering installations, except those in the purest coarse gravel, tend to remove fine material from the ground with the water they pump out. The water should therefore, before discharge from the site, be led to a tank where any soil pumped out can be seen. If soil is pumped out the ground will settle, and damage may be caused to buried cables, pipelines, drains or other structures. Such *loss of ground* can be prevented by installing properly *graded filters* at the *wellpoints* or *filter wells*. See **electro-osmosis**.

groundwater recharge *Artificial* or natural *recharge*.

ground wire, alignment w., screed w. A strong steel wire which shows the line and sometimes also the level needed for a tunnel or a *slipform paving train*.

grout [s.m.] (1) To fill with grout. See **artificial cementing**.

(2) Fluid or semi-fluid cement *slurry* or a slurry made with other materials for pouring into the joints of brickwork or masonry or for injection into the ground or prestressing ducts. Grouting of ducts improves the bond and may reduce corrosion of the tendons but it prevents their inspection and re-

tensioning or renewal so is not always desirable. Grouts injected for *ground improvement* can include slurries of cement alone or with sand, *fly-ash*, clay or alum, or bentonite alone or with silicate; the *Joosten process* and other two-shot processes are effective and include epoxy resin or polyester resin injections, acrylamides, chrome lignin (calcium lignosulphonate with sodium dichromate), urea formaldehyde, resorcinol formaldehyde, etc. *Ordinary Portland* cement particles are often as large as 100 microns, limiting its use to grouts for coarse sands and fine gravels. *High-early-strength* cements have particles up to about 20 microns so can be used for grouting finer soils. *Bentonites* are also fine, usually smaller than 10 microns. The main division of grouts is into two types: (i) particulates, of cement, fly-ash, bentonite, etc., ordinarily injected first to close the coarse gaps; (ii) chemical grouts, very much finer, injected last, to close the finest gaps. (2) See **jet grouting**, **stage g.**, **thickening-time test**.

grout box A conical *expanded-metal* box, cast into concrete, with an anchor plate at the foot (the narrow end of the cone) through which an *anchor bolt* passes.

grout curtain A row of holes drilled downwards under the *cut-off wall* beneath a dam. The holes are drilled vertically down, also upstream and downstream, and are filled with grout at one pressure. They are then drilled out and filled at a higher pressure and so on until the engineer is satisfied that the fissures in the rocks are all filled or that the rock can take no higher pressure without danger of breaking up.

grouted aggregate concrete, preplaced c., intrusion c., prepacked c., Colcrete A method of placing concrete either under water or in large quantities. Coarse aggregate, of 25 mm dia. or larger, is first placed and compacted. The cement and sand, mixed in a special high-shear mixer (colloidal grout), is injected from the bottom of the aggregate through *trémie* pipes previously placed there. Only one third of the concrete passes through the mixer, since coarse aggregate amounts to about two thirds of any concrete. A typical mortar injected is in the ratio by weight 2:1:3–4 (cement:fly-ash:fine sand). It is economical in cement, 120–50 kg/m³ being adequate, but must be fluid for pumping so is not very strong, though it has lower shrinkage, lower permeability and higher freeze-thaw resistance than ordinary concrete.

grouted macadam A road built with coarse aggregate in which the voids are filled by pouring in bituminous grout or cement grout. See **Colcrete**.

grouting *Artificial cementing*. See also **grouted aggregate concrete**.

grouting machine, grout pan A *boojee pump*.

grouting method of shaft sinking, cementation [min.] In fissured water-bearing rock a ring of holes is drilled parallel to the shaft and just outside it for the full depth of the expected water. Grout is pumped into each hole until the engineer is satisfied that the cavities are filled. (See **grout curtain**.) Where fissures are too large to be sealed by cement alone, other materials have been used, some of which swell in water, such as cottonseed hulls, chopped straw, *bentonite* clay, sawdust or *bituminous emulsion*.

groutnick See **joggle**.

grout vents Vertical tubes connected to the upper part of a *cable duct* in concrete to release the air during injection of *grout*. The grouting of *tendons* protects them from rusting and provides bond with the concrete, but is not desirable if *tendons* have to be inspected or re-stressed. The entire void should be filled with fluid *grout* of low shrinkage. *Ordinary Portland* cement or Portland blast-furnace cement should be satisfactory. Any *admixture* should contain no chloride, nitrate, sulphate or sulphide.

groyne (jetty, groin USA) A wall built out from a river bank or sea shore to increase *scour* or raise a beach level. It may be built of wooden piles, stone, concrete, etc. See **spur**.

GRP *Glassfibre-reinforced polyester* or *plastic*.

GRPF *Glassfibre-reinforced plastic ferrocement*.

grub axe A tool with an adze-like blade for pulling up roots and an axe blade at the other end of the head.

grubbing See **clearing**.

grummet A *grommet*.

guard lock [hyd.] A lock which separates a *dock* from tidal water.

guard post A *bollard*.

guard rail A *check rail* or a handrail.

guided drilling See **directional drilling**.

guide frame See **frame**.

guide pile In an excavation supported by sheet piles, a heavy vertical square timber which is driven close to them and carries the horizontal members (walings) which first guide and later support the sheet piles. It is strutted to a similar timber on the far side of the dig, to a *dumpling* or to a raking shore. It carries the full *earth pressure* from the walings. See **king pile**.

guide rail [rly] A *check rail*.

guide runner A *runner* driven ahead of others to guide them.

guide trench A *slurry trench*.

guideway A transport structure, usually elevated, whether for *maglev*, an automatic railway, overhead monorail, belt conveyor, hovercraft, etc.

gullet (1) A narrow trench dug to formation level in an earth or rock cutting, wide enough to take a track for the wagons which remove the soil. The trench is widened as convenient to the full width.
(2) The gap between two teeth on a saw blade.

gulley, gully (1) A pit in the gutter by the side of a road. It is covered with a grating and water drains away from it through a trap, but silt can collect in the pit. The silt is periodically removed by a *gulley sucker*.
(2) **yard gulley** A small grating and inlet to a drain to receive rainwater and waste water from sinks, baths or basins. See **grease trap**.

gulley sucker, sludge gulper A heavy lorry carrying a large tank with a pump for sucking silt out of road *gulleys* and forcing it into the tank.

gulley trap, yard t. [sewage] A water seal provided in all gulleys to keep foul gases in the drains. It may get unsealed in very hot weather if the water evaporates. The remedy is to pour water into it.

gully See **gulley**.

guncotton *Nitrocellulose*.

gunite, pneumatic mortar, shotcrete, sprayed concrete Mortar sprayed on to a surface by compressed air to form a dense, strong concrete, sometimes reinforced, in tunnelling, with wire mesh and roof bolts. Gunite has aggregate smaller than 10 mm, *shotcrete* uses aggregate larger than 10 mm. Fine gunite (0–4 mm) should have 400–600 kg cement/m^3 sand; with a size of 0–8 mm the cement content falls to 350–450 kg/m^3. It is used for repairing reinforced concrete and for covering the walls of mine or civil engineering tunnels to make them safe. 'Dry' mix is gunite to which most of the water is added at the nozzle. 'Wet' mix gunite is mixed wet and conveyed wet to the nozzle. Dry mix needs one extra hose to the nozzle, the water hose. Rebound (lost gunite which bounces off) can reach 30% when spraying overhead but should not exceed 20% from walls. The rebound is mainly coarse material. Since an *accelerator* is included, the rebound is coated with partly set cement and should not be reused. In the USA polymers added to the mix have reduced rebound. See **NATM**.

guniting team, shotcreting t. Apart from the foreman and any labourers needed, the team needs two people at the nozzle and another two at the mixer.

gunmetal [mech.] An alloy of copper, tin, lead and zinc.

gunning Applying *shotcrete* or *gunite* (*ACI*).

Gunter's chain [sur.] A surveyor's 66 ft chain for land measurement in Britain. One chain is one tenth of a furlong and 10 sq. chains (4840 yd^2) are one acre.

gusset, gusset plate [stru.] A piece of steel plate, usually roughly rectangular or triangular, which connects the members of a *truss*.

gusset plate See **gusset**.

GUTS [stru.] Guaranteed ultimate tensile strength of a tendon, the strength of the weakest tendon, therefore a value below which none of the tensile tests falls.

gutter [hyd.] (1) A paved channel beside a street, to lead rainwater away.
(2) A trench beside a canal, lined with *clay puddle*.

guy, g. rope A rope which stays a mast, *shear legs*, *derrick* or other temporary structure.

guy derrick, guyed d. See **derrick** (3).

guyed mast See **derrick** (1).

guyed tower, g. mast A *tower* held upright by *guy ropes*. Many guyed television or radio towers exist, but the first offshore guyed tower was built in 1983, in the Lena project in the Gulf of Mexico, in 305 m (1000 ft) of water. Its permanent buoyancy tanks in the upper part improve stability, make installation easier and reduce the foundation loads. But it still needs a piled foundation. It was also the first offshore structure to be launched sideways off a barge in one piece, saving 2700 tons of steel that would have been needed to strengthen the tower for an end launch.

The 20 guy ropes are of 127 mm (5 in.) dia., tied to anchor piles 950 m (3100 ft) away, and sheathed in high-density polythene. One third up the guys from the piles are 180 ton hinged-steel clump weights, normally resting on the seabed. But when the tower sways in a storm they are lifted off the seabed and help to bring it back upright.

Twelve buoyancy tanks of 6 m (20 ft) dia. and 35 m (115 ft) long are fixed at three levels between 23 and 130 m below sea level. They offset about 70% of the deck weight.

The tower is 36.5 m (120 ft) square throughout its height and is built of conventional tubular steel, except for the highly stressed level to which the 20 guy lines are connected.

The Lena project proved that a guyed tower is highly competitive in deep waters from 300 to 750 m (1000–2500 ft), but over $50 million had to be spent on research, including the modification of the barge for side-launching, as well as building the equipment to place the guy lines. For depths of 450 m (1500 ft) or more, guyed towers might need no clump weights, and possibly no guys. But the savings might not cover the cost of the extra buoyancy tanks. See also **derrick** (1).

gypsum concrete Concrete in which the binder is partly dehydrated plaster (calcium sulphate ($CaSO_4$)) (*ACI*).

gyratory breaker, g. crusher [min.] A fixed, crushing surface in the shape of a hollow erect cone, within which a solid erect cone gyrates on an eccentric bearing. This is a very widely used machine, for the first breaking of mineral or stone (*primary breaking*). It has a reduction ratio of about 6 and breaks down to a size of about 50 mm like the *jaw breaker*.

gyrocompass [sur.] Part of an *inertial surveying system*.

H

HAC *High-alumina cement*

haematite (hematite USA) Iron oxide (Fe_2O_3), an iron ore used in a *biological shield* as a dense aggregate. It varies in *relative density* from 3.8 to 5.1.

half-cell potential A *semi-destructive test* for finding whether the steel in concrete is corroding, by connecting it through a high-impedance voltmeter to a copper electrode in a saturated solution (half-cell) of copper sulphate. The voltage measurement shows the seriousness of the corrosion. One way of making electrical contact with the steel is by breaking out the concrete cover, drilling a small hole, and screwing a self-tapping screw into the steel (*ASTM*).

half-hour rating [elec.] See **rating**.

half-joist [stru.] A joist cut in two along the web, to form a T-section often used in welded steelwork. See also **castellated beam**.

half-lattice girder [stru.] A *Warren girder*.

half-silvered mirror [sur.] A mirror of which half the area is silvered, the other half clear. It can bring a mirror image into line with a direct image so is very conveniently used in many instruments, modern or ancient, including the *optical square*, *sextant*, *prismatic coincidence bubble*, etc.

half-sized aggregate See **single-sized aggregate**.

half-socket pipe A *subsoil drain* socketed in the lower half only. If of concrete, the upper half of the barrel may be made porous and the lower half impervious.

half-tide cofferdam A cofferdam in the sea or an estuary which is not built high enough to exclude the water at high tide and therefore needs dewatering after every full tide. See **full-tide cofferdam**.

half-track tractor A *tractor* with wheels in front and *crawler tracks* in the rear.

hammer See **hydraulic hammer**, **pile hammer**, **water hammer**, also (*B*).

hammer drill [min.] The usual compressed-air-operated *rock drill* which has superseded the piston drill.

hammer tunnelling The use of a *hydraulic hammer* (heavy *concrete breaker*), powered by a hydraulic *excavator* or *loader* for *breaking ground* instead of the usual drilling and *blasting*. In 1988 five large Italian tunnels were being driven by this means, one of them nearly 100 m² (1000 ft²) in cross-section. Advantages claimed are that the rock is less shattered, the tunnel is safer, and the method is less noisy than blasting. The same excavator can be used for loading broken rock and breaking ground. The hammer can be changed for a digging bucket in a few minutes, even though a heavy hammer weighs 3.5 tonnes.

hand boring Holes for *site investigation* in loose soils can be bored inexpensively to a depth of about 5 m (16 ft), at diameters from 0.04 to 0.6 m (1½–24 in.) by *shell-and-auger* operated by three people with a small derrick. The smallest holes can be bored without a derrick using a *posthole auger* turned by two operators.

hand distributor A *hand sprayer*.

hand finisher A tool such as a *screed* rail for forming a surface of compacted concrete to the right level and shape. It may carry a *vibrator*.

hand lead, sounding l. [sur.] A lead weight for attaching to a *lead line* of 100 fathoms or less in *hydrography*.

hand level [sur.] A hand-held instrument such as the *Abney level* having a spirit level which provides a level line of sight. Contours can be made with these instruments at distances of up to 120 m from a point of known level.

handling plant [mech.] See **materials handling**.

hand rammer See **punner**.

hand sprayer, h. distribution A hand-directed spray for spreading road binder, in which the pressure is developed by hand pump or power-operated pump. See **spray lance**.

hangar A shed of large span for sheltering aircraft.

hanger A *suspender* in a *suspension bridge* or elsewhere.

hanging leaders *Leaders* hung from a crane jib.

harbour Sheltered water where ships can lie and, in some harbours, load or unload. It may be natural or artificially sheltered (by breakwaters). See **harbour of refuge**.

harbour models [hyd.] Before building a harbour a scale model of the harbour is built, to such a scale that waves of measurable height can be generated. Harbour models are built to horizontal scales from 1 in 50 to 1 in 180, with a vertical scale that may be five times as large, resulting in waves outside of about 18 mm in height. The height of the waves in the model generally gives to scale an accurate measure of those at the harbour. Models are also used for solving problems of silting and scour. Harbour designs are frequently modified by the information given in the *model analysis*. Their main drawback is the difficulty in obtaining any true information about *silting* or *scour*. The particle size of the *silt* in the harbour cannot be reduced to scale in the model or it would become a different material (*clay*) with different settling properties and might not settle at all.

harbour of refuge A *harbour* without loading facilities, provided at an inhospitable coastline merely to allow shipping to shelter during storms. Every harbour must serve as a harbour of refuge, so the term is applied only to those that serve no other purpose.

hardcore Hard lumps of stone, brick, furnace slag, old concrete, etc. suitable for filling soft ground in a foundation or under a road, etc.

hardenability In a steel, the degree of hardness, and usually loss of *ductility* caused by *quenching* or other rapid cooling, e.g. from welding or cutting. It depends mainly on the carbon content but also on the mass and cross-section of the steel and the severity of the quench. The faster the cooling, the harder the steel is.

hardener (1) Chemicals including fluosilicates or sodium silicate, applied to a concrete floor to reduce wear or *dusting* (*ACI*).

(2) **accelerator, polymerizing agent** A substance mixed with a resin to make it set hard.

hardening Hardening of metal usually means that the tensile strength improves but the toughness gets worse: the metal gets brittle. This happens with copper and steels. Concrete hardens with age, getting stronger especially in the first few weeks. Steel and some non-ferrous metal alloys may be hardened by *heat treatment*. Steels with more than about 0.3% carbon are *quenched* in water or oil from above 800°C to acquire maximum hardness. To remove brittleness the steel is softened slightly by *tempering*. An increase in carbon content or alloying elements normally increases the hardness. Non-ferrous alloys like *duralumin* are softened by quenching from 450° to 500°C and are then 'aged' at normal or high temperatures for optimum hardness. *Annealing* softens fully cold-worked or heat-treated metals.

hard facing A hard alloy such as *Stellite, tungsten carbide*, etc. welded on to metal, usually in the form of an abrasion-resistant cutting edge to an oilwell drilling tool, excavator bucket, *tunnelling machine* cutter, etc.

hardness (1) The hardness of a metal is very roughly proportional to its *tensile strength* and is measured by such indenting tests as the *Brinell, Rockwell* and *Vickers hardness tests*. Scratch tests also exist. The hardness numbers for steels, obtained from these tests, range from 100 to 1000.
(2) The surface hardness of compacted concrete is related to the strength of the material near the surface and is measured by the *rebound hammer* or by indentation testing. The method cannot be used for concrete that is honeycombed whether unintentionally or purposely as in concrete blocks or no-fines concrete. Wet surfaces are at least 20% weaker than dry ones. Hardness testing is, however, a good measure of the uniformity of concrete pieces cast in similar conditions. See **non-destructive testing**.

hardpan [s.m.] (1) The lower part of the topsoil which has been cemented by iron or calcium salts leached from the upper part. See **B-horizon**.
(2) In the USA a glacial drift cemented by the weathering of overlying rocks. It may or may not contain boulders, sand or clay.

hard standing Any hard surface suitable for parking vehicles.

hardware Computers and their associated machines, in contrast with their *programs* (software).

Hardy Cross method [stru.] This usually refers to the method of *moment distribution* in continuous beams described by Professor Hardy Cross in 1936. Another valuable short cut due to him is the *column analogy*. His method of successive approximations can be used also for calculating the flows in pipe networks.

harmonization By 'harmonization' the European Community understands equal access for all countries to all markets. Favouring one's own country is illegal.

hatching [d.o.] Drawing parallel lines in sections of buildings or machines to distinguish between different materials. New brickwork is usually shown in Britain by pairs of lines at 45° to the horizontal drawn across the sectioned walls.

haulage rope A *traction rope* in an *aerial ropeway* or *cableway*, etc.

hauler US term for a *dump truck*.

hauling plant For short hauls of 30 m or less, *dozers, draglines*, or other *excavators* or *loaders* are used; for intermediate hauls of some hundreds of yards, *bowl scrapers* are best. For special conditions, *slackline cableways* or *hydraulicking* may be suitable. Tipping-lorries are good for long road journeys but, off the road, *dump trucks*. A *belt conveyor* instead of a *haul road* may be cheaper.

haul road A hard road built on an earthmoving site mainly to move earth. If built and maintained by a *grader*, the cost of the grader is generally more than paid for by the savings in time and the low repair and low tyre costs of the earthmoving vehicles. Cf. **belt conveyor**.

haunch (1) The part of an arch near the springing, roughly one quarter of the span.

(2) In the USA the haunch may mean a complete half-arch from springing to crown.

(3) **flank** The outermost strip of a road.

(4) A deepening of a beam at its support.

(5) **haunching** A fillet of concrete behind or beside other work, such as road kerbs or buried drains. UK building regulations require haunching up to half the height of a buried drain or sewer, whether the pipe is inside or outside a building.

hawk See (*B*)

haydite *Lightweight aggregate* used in the USA for many years. Ships were built of it in 1914–18.

Hazen's law for clean sands [s.m] Since the *permeabilities* of soils vary from 10^{-7} cm/s for clays to 1 mm/s for coarse sand, it is helpful to have an approximation to the permeability based on the particle size. Hazen's law, based on the *effective size* (mm), does this. It states that the permeability in cm/s is approximately equal to the D_{10} size squared.

head [hyd.] The potential energy per unit weight of fluid above a certain point, i.e. the height, usually in metres, of the free-water level above this point. Theoretically the velocity head and pressure head should also be included, but in water problems these are often small by comparison.

head bay [hyd.] The part of a canal lock upstream of the lock gates.

head board A horizontal board in the roof of a heading which touches the earth above and is carried by *head trees* at each side. See also (*B*).

header [mech.] Generally any conduit or pipe which distributes fluid to or extracts fluid from other conduits or pipes, e.g. a pipe through which a pump draws water from *wellpoints*, or the part of a boiler to which the boiler tubes are connected. See also (*B*).

head gate [hyd.] The upstream gate of a lock or a conduit. See **tail gate**.

heading Any small, hand-driven tunnel, timbered in soft ground, e.g. short tunnels under city roads. Long headings have been more or less superseded by various methods of *microtunnelling*, *pipe jacking*, etc. But they are still used for excavating small sections of a long tunnel in treacherous ground to make the roof safe before the bulk excavation. A *pilot tunnel* or heading may be driven in the UK as small as 1.8 m (6 ft) high and 1.2 m (4 ft) wide, but Iranian *khanats* are much smaller. Headings have many uses in mining and in acquiring underground water through *adits*.

head race [hyd.] The channel bringing water to a turbine or water wheel from the *forebay*.

head tree A horizontal timber or steel joist carried on posts such as *dead shores* (*B*).

head wall A retaining wall at the end of a culvert or drain.

headwater [hyd.] The water upstream, or the source of a stream.

headway The amount of time that elapses between vehicles in the same lane on a road.

headworks (intake heading USA) [hyd.] In irrigation, the structure at the head of a channel for diverting water into it.

heat-affected zone (HAZ) The metal next to a weld, which has not melted but is changed, sometimes harmed, by the heat of welding or thermal cutting.

heating value See **calorific value**.

heat of hydration Heat given off by the chemical action of the wetting of *quicklime* (*B*) or any sort of *hydraulic cement*. The second is slower and less evident than the first, but may be as large. When a large mass of concrete is cast, 'thermal cracking' may result during cooling. According to the *ACI* Guide, Mass Concrete, 1987, the most effective way to reduce the heat of hydration is to reduce the cement content either directly or by substituting a *pozzolan* for up to 30% of it. *Low-heat cement* may help. See **pre-cooling of concrete**.

heat pipe Automatic heat transfer along a sealed tube which is empty except for a wick from end to end and a small quantity of liquid with a boiling point appropriate to the temperature of use. The wick sucks the liquid up to the cold top from the relative warmth of the bottom end. The drop in vapour pressure because of the cold at the top end causes the liquid to vaporize and then condense on the cold walls of the tube, releasing heat. The liquid drops to the bottom in a sort of perpetual motion. On the *Alaska pipeline* the soil below the supports is thus kept frozen by transfer of heat from the bottom to the top of the supports.

heat treatment Any controlled heating and cooling of steel or other metals to improve their properties. Full heat treatment involves *quenching* and is usually followed by *tempering*. See also **normalizing**.

heave (1) *Elastic*, immediate lifting of the base of an excavation in clay. Swelling is the long-term expansion. Cf. **bulking**.
(2) Because of the *rising water table under cities*, the lowest basement floors of several buildings with deep foundations in London, including the British Library at St Pancras station, have been built as *suspended floors*, 60 cm (2 ft) clear of the clay below them so as to allow it to expand upwards if need be. See below and **frost heave**.

heaving shale [min.] When a shale at depth (and therefore dry) is penetrated by a tunnel or by an oilwell, the moisture in the air or in the mud of the *drilling fluid* enters the shale and may cause it to expand and lumps to break off. In tunnelling this is troublesome but not disastrous before the tunnel is lined. If it occurs after the lining is built, much costly lining may be destroyed. One solution is to seal the shale with *gunite* as soon as it is exposed. In oilwell drilling, shales can be prevented from heaving by using special *drilling fluids* containing no water (oil-based muds) or water-based muds containing dissolved sodium silicate. Before these techniques were learnt many costly miles of well had to be abandoned.

heavy soil [s.m.] A soil which is largely clay (therefore damper than a sand and for this reason heavier).

heavyweight concrete, high-density c. See **biological shield**.

hecto- A prefix denoting one hundred times.

heel (1) The part of the base of a dam or *retaining wall* which is on the water side of a dam or on the earth side of a retaining wall. See **toe**.

(2) [rly] The unplaned (hinged) end of a *switch tongue*.

heel post, quoin p. [hyd.] The corner post of a lock gate, carried on the *hollow quoin*.

height of instrument method [sur.] The *collimation method*.

held water, capillary w. [s.m.] **(free w., vadose w.** USA) Water held in the soil by surface tension above the *standing-water level* against the action of gravity. The term is sometimes, in Britain, reserved for adsorbed water which can be driven off only by baking for 24 hours at 105°C. See **fringe water, hygroscopic moisture, specific retention**.

heli-arc welding *MIG welding* using helium as the shielding gas, relatively common in the USA, where helium is available.

helical conveyor, c. screw/worm [mech.] A conveyor for small coal, grain and similar material which consists mainly of a horizontal shaft with helical paddles or ribbons, rotating on its centre line within a stationary tube filled with the material. It is compact and cheap but has a low capacity and a power consumption about 20 times that of a belt conveyor. It is therefore suitable only for small outputs or for condensing and expelling air from very fine material, like cement, before delivering it to a belt conveyor. In the screw-tube conveyor the helical ribbon is fixed to the tube, which itself rotates, and there is no central shaft.

helical reinforcement Steel rod reinforcement bent into a spiral curve, sometimes used as a link in a column.

heliograph An instrument at a survey station which reflects the sunlight to make itself visible. It may include a flashing device for signalling in Morse or other code. In sunny countries it is indispensable for sighting a distant survey point, as far away as 100 km.

helipad A small, often circular landing area on a ship, *offshore platform*, tall building, etc. It has a diameter of at least $1.3R$ where R is the rotor diameter of the largest helicopter to use it. Around the helipad is an obstacle-free perimeter ring at least 5 m (16 ft) wide. Below the helipad level the perimeter ring is surrounded by a nearly horizontal safety net 1.5 m (5 ft) or more wide, able to carry 200 kg per metre length. The helipad floor is non-slip and surrounded by a 10 cm (4 in.) upstand.

helium [min.] A gas found in some uranium mines and used in the USA for lighter-than-air craft. The richest wells are reserved for the US government. Helium is also used as a breathing atmosphere mixed with oxygen in deep diving or *caissons*, since it is less soluble in the blood than nitrogen, and diffuses out of the blood more quickly, being less dense. See **caisson disease, Hydreliox**.

helium diving bell A *diving bell* in which the divers breathe a *helium*–oxygen mixture instead of the nitrogen–oxygen mixture of ordinary air.

helmet See **pile helmet**.

hematite See **haematite**.

hemihydrate A molecule combined with half a molecule of water. Thus *plaster of Paris* (B) is calcium sulphate hemihydrate. ($CaSO_4 . \frac{1}{2}H_2O$).

hemp Vegetable fibres used for making *fibre rope* or the absorbent oil-filled cores of steel ropes.

herringbone drain A *chevron drain*.

hessian See (*B*).

heuristic By trial and error, not by science.

high-alumina cement (HAC), aluminous c., Ciment Fondu, fondu A *refractory* cement with more alumina than *Portland*. It hardens enough for normal loads in 24 hours but heats up during hardening, so it can be cast only in thin layers. Because of failures of precast concrete roof beams made with this cement, such critical uses are no longer common, but it resists acids and sulphates well and can sometimes be used below ground, particularly for temporary work. See **calcium aluminate**.

high-bay warehousing system, high-storage-racking s. Modern high-density warehouses may have storage racks 14 m (46 ft) high with aisles between them only 1.3 m wide. Any unevenness in the floor is therefore multiplied by 14 at the top, so the floor must be very flat and level, without roughness at joints. *Long-strip flooring* in a *prestressed-concrete pavement* makes this possible.

high-bond bars *Reinforcement* bars with surface deformations such as raised patterns, which improve the bond strength.

high-carbon steel [mech.] *Carbon steel* containing over 0.5% and up to 1.5% carbon, such as spring steels which are stronger and more easily tempered but less ductile than *mild steels*. Low-carbon steel has 0.04–0.25% C and medium-carbon steel has 0.25–0.5% C. The hardness of high-carbon steel is raised and its *ductility* lowered by heating and *quenching*. See **high-tensile steel**.

high-density concrete See *biological shield*.

high-early-strength cement *Rapid-hardening cement*.

high explosive [min.] An explosive which contains at least one chemical compound that, when fired, decomposes at high speed, a process called *detonation*. Most explosives are now of this type and are thus the opposite of low explosive like black powder or liquid-oxygen explosives, in which at least two materials combine with a rending or heaving slow explosion, very different from the shattering detonation of high explosives.

high-pressure steam-curing See **autoclaving**.

high-range water-reducer A *superplasticizer*.

high-speed compactor (see diagram) A soil-compaction *roller* with four octagonal wheels. The ridges on the surface of the front wheels correspond to the troughs on the rear wheels. It travels at 10–16 kph (6–10 mph) and can be used over many different soil types.

high-storage-racking system See **high-bay warehousing**.

high-strength concrete Norwegian researchers have achieved 125 MPa (18,000 psi), but in 1988 expensive high-strength concrete saved money at *Seattle* and elsewhere. In Chicago, deck concrete at 62 MPa (9000 psi) allowed a thin deck slab to be used at 311 South Wacker Drive in a building 292 m (960 ft) high. Also formwork was stripped quickly, and steel was saved.

high-strength concrete topping A *granolithic screed*.

high-strength friction-grip bolts (HSFG bolts) High-tensile, heat-treated, quenched and tempered steel bolts and nuts with quenched and tempered steel washers, tightened accu-

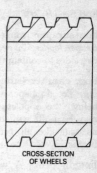

CROSS-SECTION
OF WHEELS

High-speed compactor. (F. C. Harris, *Construction Plant*, Granada, 1981.)

rately either by a torque spanner or the use of *load-indicating washers*. The clamping force transfers loads by friction between the connected members and not by *shear*, *bearing pressure*, etc. These bolts are inserted, not driven, into a *clearing hole*. See also **interference-body bolt**.

high-tensile bar *Alloy-steel* bar for *prestressing* and other purposes is used in diameters from 20 to 40 mm, usually in straight lengths with threaded ends for *post-tensioning*. The *ultimate strength* is at least 835 MPa (120,000 psi).

high-tensile steel Steel for bridges, buildings and similar purposes is made in Britain with a yield point up to 350 N/mm², as compared with *mild steels* 230 N/mm². It may contain up to 0.3% carbon, up to 0.5% copper and up to 1.5% manganese. See **high-carbon steel** and above.

highway A road where traffic has the right to pass and to which owners of abutting property have access. Cf. **freeway**.

highway drain Motorway drains

are often polluted by oil, antifreeze, rubber, cadmium from tyres, salt and accidental spillage from lorries. Purification has been the responsibility of the Department of Transport, an unfortunate mistake since it has no purification facilities, so the local *water service plc* suffers the resulting pollution.

hindered settling [s.m.] The condition in which closely packed particles sink in a liquid. They settle more slowly than when one particle sinks freely alone and the latter is the case to which *Stokes's law* truly applies.

hinge [stru.] A point in a structure at which a member can rotate slightly, sufficiently to eliminate all bending moment in the members at the joint. At a hinge in an arch or other thick concrete member the concrete must thin out to reduce cracking and allow bending. In one type the top steel crosses over at the hinge to become the bottom steel on the other side, and vice versa for the bottom steel. Cf. **fixed end**. See (B).

histogram [stat.] A *frequency diagram*.

hod See (*B*).

hoe scraper [min.] A bucket for a *scraper loader*, shaped like a hoe with two sides to contain the ore.

hog, camber [stru.] Upward bending, i.e. a shape which is concave below, the opposite of sag. A beam may be built with hog to counteract its sag, like all prestressed-concrete beams.

hoggin [s.m.] A well-graded gravel containing enough clay binder to be used in its natural form for making roads or paths.

hogging moment, negative m., support m. [stru.] A *bending moment* which tends to cause *hog* such as occurs at the support of a *continuous beam*. Cf. **sagging moment**.

hoist (1) [mech.] A drum driven by a prime mover or motor. The drum winds or unwinds a steel rope, which passes over a hoisting sheave and raises or lowers the load. See **platform hoist**.

(2) A set of *lifting tackle*.

hoist controller [elec.] Plant which controls the speed of an electric motor for a hoist or winding engine.

holdfast A *dead man* for a *guy rope*.

holding-down bolt An *anchor bolt*.

holding time [sewage] *Detention time*.

hollow-block floor See **hollow-tile floor**.

hollow clay tile See (*B*) and **hollow-tile floor**.

hollow-core beams See **pretensioned floor beams**.

hollow dam A reinforced-concrete, plain-concrete or masonry dam in which the thrust of the water is taken on a sloping slab or vault carried by a row of regularly spaced buttresses. See **multiple-arch dam**.

hollow quoin [hyd.] The masonry socket carrying the *heel post* of a *lock gate*.

hollow sections [stru.] See **tubular sections**.

hollow tile, pot Concrete or burnt-clay hollow blocks. Flooring blocks in the UK measure 0.3 × 0.3 m and are from 7.5 to 30 cm deep. Similar blocks are used for building outside walls (rendered outside, plastered inside).

hollow-tile floor, hollow-block f., pot f., ribbed f. A reinforced-concrete floor in which the load-bearing part consists of *tee-beams* reinforced with steel bars in the bottom of the span, usually bent up over the support. Between each pair of ribs with this reinforcement, a row of burnt-clay *hollow tiles* is placed, generally 0.3 m wide. These are laid on the shuttering and are later plastered. They reduce the dead load of the concrete by an amount which varies between 100 and 250 kg/m² according to the thickness of the pot. These floors are therefore very suitable for spans of 6 m or more, but not if many holes are to be cut through them in positions which are unknown at the design stage. The concrete in hollow-tile floors is not vibrated. See **plate floor**.

hollow-web girder [stru.] A *box girder*.

honeycombing (rock pocket USA**)** Local roughness of the face of a concrete wall caused by the concrete having *segregated* so badly that there is very little sand to fill the gaps between the stones at this point. Such concrete is weak and should be cut out and rebuilt if the wall is heavily loaded.

Honigmann method of shaft sinking [min.] Sinking a shaft with a *trepan* and a central *air-lift pump* as in the *forced drop-shaft method*.

Hooghly Bridge, Calcutta [stru.] See **cable-stayed bridge**.

Hooke's law The elastic deflection of a material is proportional to the load on it up to the *elastic limit*. This means that *stress* is proportional to *strain*, a basic principle of engineering. See **modulus of elasticity**.

hook gauge [hyd.] A pointed hook fixed to a vernier moving along a vertical graduated staff. The hook is lowered into the water and raised until the surface is just pierced. It is an accurate laboratory instrument for measuring water level, and a skilled observer can measure a change in level of one tenth of a millimetre. See **point gauge**.

hooping Curved reinforcement such as the steel in a circular concrete tank which resists *ring tension*.

hoop stress, hoop t. [stru.] See **ring tension**.

hopper A *bunker*.

hopper dredger [hyd.] A dredger which can act as its own hopper and can thus transport the mud which it picks up to dump it at the desired place.

hoppit [min.] A *kibble*.

horizon [sur.] At any point, a plane at right angles to a plumb line. See **artificial horizon, sensible horizon**.

horizon glass [sur.] Part of the *sextant*, a glass which is half silvered and half clear.

horizontal circle of a theodolite or transit [sur.] The circular, graduated *plate* under a *theodolite* telescope, by which horizontal angles can be accurately measured.

horizontal control [sur.] Connecting a survey with accurate *trilateration*, *traverse*, *satellite-Doppler* or other points.

horizontal curve A *curve* in plan.

horizontal drilling *Augers* can be used horizontally, e.g. for mining a coal seam that crops out on a hillside. But recently petroleum engineers at depths as great as 3500 m (12,000 ft) have placed horizontal lengths of borehole to increase the output from a well. The hole is diverted by 1.5–2°/30 m (100 ft) in two or more lengths, until the hole becomes horizontal, by *directional drilling*.

horsepower [mech.] In Britain and USA originally 550 ft-lb/s and on the European continent 75 kg-m/s. Since 550 ft-lb = 76 kg-m, not 75, the hp is now avoided by British engineers and for precision they prefer the kilowatt. 1 UK hp = 746 W.

horsepower-hour [mech.] The work done when one horsepower is spent for one hour. In precise statements, kilowatt-hours are now preferred in the UK so as to avoid the confusion that is possible between continental and UK hp.

hose coupling [mech.] A joint from hose to metal pipe or to another hose.

hot-dip coating [mech.] Dipping metal parts in molten tin or zinc to give them a protective coating. See **galvanize**.

hot miller [mech.] A compressed-air tool with cutting wheels which mill the hot cutting edges of rock drill bits.

hot rocks See **geothermal energy**.

hot-rolled bars Several types of bar *reinforcement* are hot rolled, including the cheapest – plain *mild steel* bar with a yield point of 250 MPa (36,000 psi) as well as deformed mild steel and deformed high-yield steel with a yield point of 410 MPa (60,000 psi).

hot rolling [mech.] The method by which *rolled-steel sections* are made in a rolling mill, by passing hot steel bars through pairs of massive steel rolls. See **hot working**.

hot shortness, red s. [mech.] Brittleness when working hot metal, caused in steel by a low manganese and a high sulphur content.

hot working [mech.] The shaping of metal parts by *extrusion*, smith *forging*, *hot rolling* and so on at temperatures (for steel around red heat) high enough to prevent the *hardening* and brittleness which may be caused by *cold working*.

house drain A drain which takes all waste or sewage from a house, in the USA also called a collection line. In Britain it extends from the house to the *water authority*'s sewer. Before each house drain joins the *sewer* it must be provided with an *inspection manhole* and a *trap* (*B*).

Howe truss [stru.] A roof truss used in spans up to 24 m. As used in the USA it has steel verticals and timber (or timber and steel) horizontal members and sloping members, but it was also entirely of steel.

Hoyer method of prestressing See **pretensioning**.

H-Section [stru.] A *rolled-steel joist* of depth not greater than 1.2 times its flange width. Cf. **I-section**.

HSFG bolt *High-strength friction-grip bolt*.

hub A *survey point*.

huckbolt A proprietary, threaded fixing for thick steel sheet or thin plate, inserted into a predrilled hole. The shank is notched, enabling the end to be pulled off by a special tool when the nut has been screwed on. The fixing is claimed to be vibration-proof. *Pop rivets* are similar but not threaded.

Humber Bridge (1981) The longest UK *suspension bridge*, resembling the *Severn Bridge* but of 1410 m main span.

humidity of air, absolute h. The weight of moisture present in unit volume of air, to be contrasted with *relative humidity*, which is a percentage or ratio.

humus (1) [s.m.] Dark-brown fertile material in topsoil, largely rotting vegetation, a useless foundation material.
(2) The solids that flow out in the *effluent* from *trickling filters*.

humus tank [sewage] A sedimentation tank in which the so-called *humus* from *trickling filters* settles out. Humus sludge may be dewatered and disposed of either directly or together with other *sludges*.

hurdle work, wattle w. Osiers (*laths*) interlaced with vertical sticks to make a low fence on a river bank which encourages silting or discourages scour.

hurricane A wind of 120 kph. (*Beaufort*'s number 12.)

hybrid construction [stru.] Building with materials of different properties in the same structure. For a steel structure it may mean merely steels with different *yield points*.

hydrate Any chemical compound containing water – important in building because all *hydraulic cements* and *plasters* (*B*) are hydrates when they have hardened after wetting (*hydration*).

hydrated lime See (*B*).

hydration Chemical combination with water. Every *hydraulic cement* hydrates as it stiffens, and the hydration warms it. See **heat of hydration**.

hydraulic Relating to the flow of fluids, particularly water. See **hydraulics** and (*B*).

hydraulic breaker See **hydraulic hammer, jackhammer**.

hydraulic cement *Cements* like *Portland cement* that need water for hardening, and harden under water, See also (*B*).

hydraulic dredger A *suction* or *suction-cutter* or *draghead dredger*.

hydraulic ejector, eductor, elephant's trunk, silt ejector A pipe that removes water containing sand, mud or small gravel from the working chamber of a *pneumatic caisson* or the base of a *bentonite*-filled pit such as a *diaphragm wall* before concreting. It works like an *air-lift pump* but a jet of water, not air, is injected into the foot of the pipe. See **underwater excavation** (diagram).

hydraulic elements [hyd.] The depth, cross-sectional area, *wetted perimeter*, velocity and so on of water flowing in a channel.

hydraulic elevator A *hydraulic ejector*.

hydraulic engineering [mech.] The design and manufacture of pumping plant, reservoir valves, penstocks, pipelines and so on.

hydraulic excavation, hydraulicking Excavation by *giants* delivering a jet of water at high velocity against an earth or gravel bank and breaking it up. The flowing water carrying the mud and soil is led to *flumes* and from

there to the embankment or treatment plant as required. The gradient for carrying fine material in a flume must be at least 2% and for coarse material 6–8% See **hydraulic fill dam**.

hydraulic excavator A usually tracked *excavator* in which the *jib*, digging *bucket* or *grab* are controlled by *hydraulic power*, See **hydraulic grab**.

hydraulic fill [s.m., min.] Embankment material carried by water in *flumes* or pipes, sometimes after *hydraulic excavation*. Hydraulic fill can be used in mines to fill voids left by the removal of mineral. See **rubber linings**.

hydraulic fill dam An embankment or *dam* built up from water-borne clay, sand and gravel carried through a pipeline or flume. The discharges from the pipe are arranged at both edges of the dam flowing inwards towards the centre. By this means the heaviest material settles near the edges and the finest, most impervious material settles at the centre, forming what amounts to a clay *cut-off wall*. The water is removed at the centre over the clay core, sometimes by a *floating pipeline*. Hydraulic fill is in suitable conditions the cheapest method of transporting fill and building dams.

hydraulic fluid, h. oil The fluid circulating in a *hydraulic power pack*. It transmits power to the moving parts of an *excavator*, *drill*, *hydraulic hammer* or other equipment. Hydraulic fluids are often based on water, some of them 95% water, without fire hazard and not very expensive.

They should be lubricants, unaffected either by temperature change or the small quantity of water which always enters. They should not go sticky with age, which would result in jammed valves. Their viscosity should be a suitable compromise. A thick fluid is needed for good lubrication and tight seals, but a flowing one is needed for easy pumping and flow.

hydraulic friction [hyd.] The resistance to flow caused by roughness or obstructions in a pipe or channel. See **loss of head**.

hydraulic grab (see diagram) An *excavator* with a hydraulically controlled *grab bucket*. Because of the need for flexible hydraulic pressure pipes to

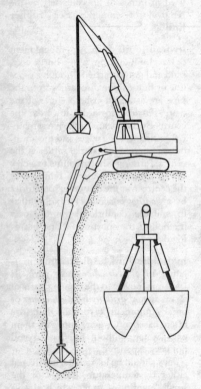

Hydraulic grab showing maximum depth. (F. C. Harris, *Construction Plant*, Granada, 1981.)

reach the grab along the line of the hoisting rope, its depth of reach is short compared with a rope-operated grab.

hydraulic gradient (1) [hyd.] (**h. grade line USA**) An imaginary curve along a flowing pipeline which shows the levels to which the water could rise in open pipes leading up from it. It can be expressed as a ratio, the slope of the curve, or as a fractional drop in m/km. It is also the slope of the surface of water flowing uniformly in an open conduit.

(2) [s.m.] The difference in the water level between two points, divided by the length of the shortest soil path between them. See **Darcy's law**.

hydraulic hammer, impact h., rock breaker A hydraulically operated tool mounted on the jib or bucket of an *excavator* or *loader*. Though in principle it resembles the hand-held, compressed-air-driven *concrete breaker* or pneumatic pick, it is many times more powerful, and with an 80 mm (3 in.) thick point is far too heavy for one person to handle. In 1988 the largest hydraulic hammer weighed 3.5 tons. It is used for tunnel driving (even in granite) instead of blasting, or for *scaling* rock after blasting while the operator shelters in the cabin. The hammer can be changed for a digging bucket, rock drill or *roof bolting* tool in a few minutes. Underwater hydraulic hammers are used for driving *piles* at *offshore platforms*.

hydraulic jack [mech.] A ram which works on the principle of the hydrostatic press, used in civil engineering for loading up and testing driven or bored piles or other structures, or for *prestressing* concrete beams and slabs. It is generally oil-filled and loaded up by a hand pump.

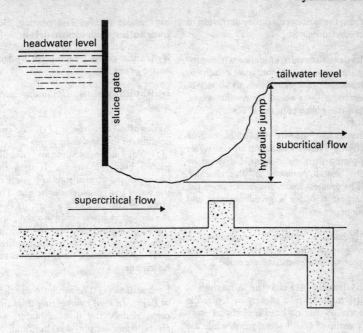

Sluice gate, with supercritical and subcritical flow, and hydraulic jump.

hydraulic jump [hyd.] (see diagram) In an *open channel* of uniform slope the flow is in equilibrium at a single depth (the normal depth) at which the gravitational pull just balances the friction of the bed and walls. When, because of obstacles, a fast flow meets a slower flow downstream, there is a sudden change of depth called the hydraulic jump as the stream gradually recovers the 'normal depth'. The jump can be artificially induced in a measuring flume, or at an outlet from a sluice gate, and involves violent eddying, turbulence and loss of energy.

hydraulicking [min.] Earth-moving by flowing water, *hydraulic excavation*.

hydraulic lift, electro-h. l. [mech.]

A lift operated by hydraulic ram. The hydraulic pressure for it is obtained from an electrically driven pump nearby. Compared with rope-hoisted lifts, hydraulic lifts have many advantages; they need no overhead hoist and the siting of the pump and control room is not critical. They are smooth and accurate in operation, need little maintenance (no adjustment of the rope) and have no counter weights so the lift shaft is smaller. They can be installed in an existing building without strengthening it because all the load is carried on the ram and this transfers it to the ground through a borehole in the cellar. Boreholes are feasible for low buildings. For tall ones the ram can be installed at the side of the lift shaft with the cage load carried by rope from the ram over a pulley at the top of the building.

hydraulic main A pipeline which supplies water under a pressure of tens or hundreds of megapascals to subscribers who thus obtain *hydraulic power*, but it is obsolete in the UK.

hydraulic mean depth, h. radius [hyd.] The cross-section of the water flowing through a channel divided by the *wetted perimeter* of the channel. The greatest hydraulic radius gives the largest flow for a given channel and slope.

hydraulic mole A *microtunnelling machine* worked by hydraulics, quieter than the *pneumatic impact mole*.

hydraulic pile driving A method of driving *sheet piles* silently by hydraulic force. A driving head is clamped on to a group of about seven piles; while the jack forces each pile down it reacts against the other six.

hydraulic power, h.p. pack Hydraulic power systems are often more convenient than compressed air since most hydraulically controlled self-propelled units (*excavators*, *loaders*) can act as power packs for an otherwise disconnected hydraulic unit such as a diamond drill, *vibrating plate* soil compactor, *hydraulic hammer*, etc. A Land-Rover has provided the hydraulic power for drilling a 200 mm (8 in.) dia, taphole through the brickwork of the blast furnaces of British Steel. But separate hydraulic power packs also exist, driven by an electric motor or *prime mover*. In the UK *hydraulic mains* have been made obsolete by reliable electric motors and electric power. The flow rate of *hydraulic fluid* through its pipes should be below 3 m/s (10 ft/s) to avoid *turbulence*, and the result-

ing noise, friction and heating. See **hydraulics**, **hydrostatic drive**.

hydraulic press [mech.] See **hydrostatic press**.

hydraulic radius *Hydraulic mean depth*.

hydraulic ram [mech.] (1) The *ram* or plunger in a *hydraulic press*.
(2) An automatic pump, usually placed beside a stream. It uses the energy of the stream to pump some of the water to a considerable height. The stream is diverted into a pipe, its flow is checked at intervals by a valve, and the resulting increased pressure in the pipe drives the water to the required level up a branch-pipe. See **water hammer**.

hydraulics (1) The study of the flow of fluids. In civil engineering this concerns mainly the flow of water, i.e. rivers, water supply, dam building, irrigation and drainage; and in mining the flow of viscous *drilling fluids*, ore pulps and *hydraulic filling*. Cf. **hydrostatics**.
(2) *Hydraulic power* now operates the jaws of *excavator* or *loader* buckets and other powerful *contractors' plant*. The steel wire ropes that originally transmitted motion in the 1950s have been almost completely replaced by hydraulic rams operating at 250 atm (25 MPa (3600 psi)) or more, except on large *shovels* and *draglines*. Ropes are still a convenient way of working these very large machines. Hydraulically operated tools have the advantage that they do not freeze up when working at full power, which can happen with a large compressed-air tool.

hydraulic system Mechanical power transmission – *hydraulic power* by pumps pushing *hydraulic fluid* through pipes to drive motors, rams, etc.

hydraulic tensor [rly] A *tensor*.

hydraulic test (1) [mech.] A test for pipes, boilers, pressure vessels, etc., which are filled with water to the design pressure or slightly above it.

(2) **water t.** A test for new *drains*. They are stopped at the lower end and filled with water for an hour to a maximum head of 2 m. If no fall of level occurs during the hour the drain is accepted. The test is a severe one. Cf. **rocket tester** (*B*).

Hydreliox A mixture of hydrogen, *helium* and oxygen breathed by divers and recommended by the French firm Comex for depths of 500 m or more. Comex uses helium and oxygen down to about 400 m but below 250 m the helium is gradually replaced by hydrogen. Depths of 1000 m are under consideration. The danger in breathing this explosive mixture has not been overlooked but is not thought to be serious in view of the precautions taken.

hydrodynamics The branch of *hydraulics* which deals with flow over weirs and through openings, pipes and channels.

hydroelectric power station A power station at which electricity is generated by the energy of falling water, causing turbines to turn and drive generators.

hydroelectric scheme A project or completed structure for generating water power, including the building of a dam, the diversion tunnels or channels, spillways, power station, intake structures, penstocks and any roads, bridges, houses, irrigation works, villages and cement works which may be required.

hydrofracture [s.m] Breakage or disturbance of ground, near an injection

point, by a grouting pressure that exceeds the *overburden pressure* and consequently can lift individual rock masses. It may sometimes purposely be used to increase the permeability of the ground. See **artificial cementing**.

hydrofraise A *rock mill*.

hydrogeology *Hydrology*, but with emphasis on geology, chemistry and water exploration.

hydrograph [hyd.] A graph showing, against time as abscissa, the level, flow rate or velocity of water in a channel, indispensable for the planning of a water supply or hydro-electric or drainage scheme. See **unit hydrograph**.

hydrographer [hyd.] A person who records measurements of water level, flow, rainfall, run-off, and so on.

hydrography Surveying and charting oceans, seas and other waterways.

hydrological cycle, water c. The circulation of water from hills and streams down to lakes or seas, through evaporation up into clouds, then as rain or snow on to the hills again and back to the streams. Some of the rain enters the earth as *infiltration* to recharge the groundwater that flows out elsewhere as springs, often in a stream or lake bed, or the sea.

Freshwater amounts to only 2.6% of the world's water and 98% of this is in the polar icecaps. Most of the remaining 2% is groundwater.

hydrology The study of water; for civil engineers mainly what lies on or under the earth's surface, including water discovery, storage and flow.

hydrometer An instrument floating in water or other fluid, which by its level of submergence measures *relative density*, shown by graduations on the stem. It is used in the *wet analysis* of soils.

hydrometry (1) The use of *hydrometers*.

(2) The measurement and analysis of the flow of water (ASCE).

hydrophobic cement *Water repellent cement*.

hydro-pneumatic accumulator See **accumulator** (diagram).

hydroseeding Propelling seed, on to an area to be grassed, by a jet of water containing grass seeds with substances to encourage germination and growth.

hydroshield A *slurry shield*.

hydrostatic catenary [stru.] The curve taken by an inextensible cord which is pulled by a load proportional to the distance of the cord below its supports. It is the shape taken by a *flume* which bridges a gap.

hydrostatic drive A *hydraulic power* supply, producing torque and rotary motion where needed. It is not hydrostatic but is so called because the flow rate of the *hydraulic fluid* is almost constant. Stepless road speed variation is possible, the starting torque almost equals the full-speed running torque, and moving parts can be stalled yet maintain full torque. With all these advantages it is natural to use it for controlling cranes or earthmoving equipment or for steering.

As a drive method it needs no clutch, being infinitely variable. One joystick controls forward, reverse and speed. A 45 ton *bulldozer* needs no *torque converter*, no power-shift transmission, no steering clutches or brakes. Its three joysticks control: (i) travel, (ii) raising, lowering or tilting the *mouldboard*, (iii) the *ripper*. Perhaps the main disadvantage is that the engine must be appreci-

ably more powerful, but at least one maker thinks this does not matter.

hydrostatic excess pressure [s.m.] The pressure per unit area of soil which exists in the pore water at any time in excess of the hydrostatic pressure, due to applied loads. The hydrostatic excess pressure is equal to the *total pressure* minus (*effective pressure* plus the *neutral pressure*). See **pore-water pressure**.

hydrostatic joint A spigot-and-socket joint in a large water main formed by forcing lead into the socket with a hydraulic ram.

hydrostatic press, hydraulic p. [hyd.] A large ram whose surface is acted on by a fluid in contact with a small ram. The small ram is moved to and fro to increase the liquid pressure, thus producing a large force on the large ram. The original example of this was Bramah's press (1796); modern industrial examples of it are used for many purposes, including clamping plywood during gluing.

hydrostatic pressure The pressure at any point in a liquid at rest is the hydrostatic pressure. This is equal to the depth of the liquid multiplied by its density.

hydrostatic pressure ratio In calculations of *active earth pressure*, the ratio between the pressure on a vertical plane due to the soil and that which would exist at the same point in a liquid of the same density as the soil. It is Rankine's coefficient of *active earth pressure*.

hydrostatics [hyd.] That part of *hydraulics* which concerns the pressures in fluids at rest.

hydrostatic test A *hydraulic test*.

hydro-stressor [rly] A *tensor*.

hydrothermal power *Pumped storage*

combined with steam generation of electricity.

hydroxylated polymers, polyhydroxylated p. *Admixtures* to concrete that were developed in the 1970s and are claimed to improve workability, surface finish, homogeneity, cohesion, pumpability and durability in the concrete.

hygrometer A device for estimating the *relative humidity* of the air. The commonest hygrometer is called the wet-and-dry bulb thermometer. It consists of two thermometers, the wet-bulb thermometer whose bulb is covered with a wet muslin bag dipping in water, and an identical dry-bulb thermometer beside it with no muslin bag. The temperature of the wet bulb is usually lower than the dry bulb because it is cooled by the evaporation of the water. At 100% relative humidity no evaporation is possible and this is the only occasion when the wet-bulb and dry-bulb temperatures are the same. The humidity of the air can be determined by noting the wet-bulb and dry-bulb temperatures and reading from a table or graph the relative or absolute humidity for any given dry-bulb temperature with the observed wet-bulb depression.

hygrometry The measurement of the *humidity* of the air.

hygroscopic coefficient [s.m.] The moisture, in percentage of its dry weight, that a dry soil will absorb in saturated air at a given temperature (ASCE).

hygroscopic moisture [s.m.] The moisture in an air-dried soil. It evaporates if the soil is dried at 105°C.

hypar, hyperbolic paraboloid roof A roof of unusual shape often used in recent years for building thin shells in concrete or timber. Although the surface is doubly curved, every point on it is at an intersection of two straight lines on the surface. Therefore the shuttering for it (or the timber roof itself) can be conveniently built from straight strips of wood.

hyperbaric 'High-pressure' is the sense in the applications listed below.

(1) Hammelmann's Hyperbaric (trade name) pumps can reach pressures of 250 MPa (3600 psi). A water jet from them can abrade weak concrete or wash out stalactites in sewers.

(2) A *medical lock*.

(3) Hyperbaric welding is *arc welding* in a dry underwater pressure chamber where welds of high quality can be made that are impossible by *wet welding*.

hyperstatic frame See **redundant frame**, **statically-indeterminate frame**.

hypsometer [sur.] An instrument in which water is boiled and the boiling temperature is measured, either for determining the air pressure as a measure of altitude or for *calibrating* the thermometer.

hysteresis The loop formed in the stress–strain curve when a specimen is strained beyond the elastic limit in alternating cycles of tension and compression. This is one sort of hysteresis but many magnetic or other sorts exist, since hysteresis means 'lag'.

I

I [stru.] See **moment of inertia**.

I-beam [stru.] A rolled-steel joist, generally implying one which is tall and narrow.

ice See **flake ice**, **polar ice** and below.

ice apron, i. breaker A ramp on the upstream side of a bridge pier which slopes up from well below water level. Ice carried down by the stream is lifted and harmlessly broken by this simple device.

ICE Conditions of Contract Printed *conditions of contract* published by the Institution of Civil Engineers.

ice lens A horizontal block of ice in *permafrost*. Cf. **ice wedge**.

ice warning See **infrared thermography**.

ice wedge A vertical block of ice in *permafrost*. Cf. **ice lens**.

idler [mech.] (1) A wheel interposed between two others in a gear train. It does not alter the relative speeds of the wheels but enables them to turn in the same direction instead of in opposite directions as they would if they meshed together.

(2) A broad pulley carrying the weight of a *belt conveyor* and its load. Idler pulleys are symmetrically placed about the belt centre line in sets of three, sometimes five, at an angle to each other so as to bend the belt into a *troughed belt*.

igneous intrusion [min.] Rock forced molten from the centre of the earth to form thick masses or thin flat deposits (sills) covered by other rocks. They cool and crystallize relatively slowly and therefore have crystals large enough to be seen by the naked eye, unlike volcanic (extruded) rocks which have very few visible crystals. Many valuable ores are found near igneous intrusions like *granite*.

ignition powder [mech.] A mixture usually of powdered aluminium and oxidizing material which starts the reaction in *thermit* welding.

IKBS (intelligent knowledge-based system) An *expert system*.

Imhoff tank [sewage] A deep two-storeyed tank in which sewage ferments to form *methane* and the sludge settles to be drawn off.

immersed tube (see diagrams, pp. 227–8) An underwater tunnel made by sinking precast concrete boxes (tube elements) into a channel dredged for them and joining them up under water. Concrete tubes are often provided with a steel watertight 'skin', while 'steel' ones have to be loaded with a concrete base to overcome their buoyancy. The precasting dock, if specially excavated for the job, is a large part of the expense. Each element is closed by *diaphragms* at the ends, floated out, accurately sunk and rammed to fit its neighbour. Immersed tubes can be less deep than conventional tunnels, which need much more than the few metres of cover that protect an immersed tube from ships' anchors. The first immersed tube for vehicles was 800 m long, built in 1910, and contained two railway tunnels between Michigan, USA, and Ontario, Canada, under the Detroit River.

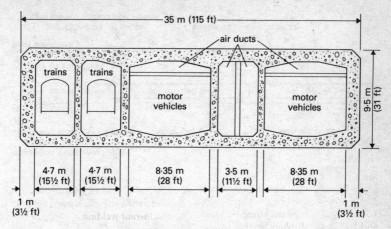

Immersed tube, Hong Kong. Typical cross-section of the 15 concrete immersed tube units, from 122 to 128 m long, laid to cross Hong Kong harbour in 1987–8, and weighing from 40,000 to 42,000 tons each. Contractor: Kumagai Gumi; consulting engineers Freeman Fox and also Maunsell and Partners. (*Construction News*, 22 September 1988.)

The first British immersed tube, begun early in 1987 under the *Conwy River* at Conwy, North Wales, was expected to cost £100 million. A tunnel was chosen for aesthetic reasons. In 1988 the concept of a toll-paying motorway in an immersed tube along the bed of the Thames from Chiswick to the City 25 km away was mentioned as a possible relief to east–west traffic congestion in London, but costing nearly as much as the *Channel Tunnel*. Another concept, but in 1850, was an immersed steel tube across the English Channel proposed by a Frenchman. It would have had ventilation towers above the sea at intervals.

Possibly the world's largest immersed tube, across Hong Kong harbour, was completed in 1988. It has two rail tunnels, two road tunnels and a service tunnel divided in two to provide two ventilation ducts. To reduce the heat of hydration of the heavy concrete units, 20% of the cement was replaced by fly-ash, improving the sulphate resistance of the concrete. Visible cracks in the concrete were grouted with epoxy resin, and the sides and roof of the units were sprayed all over with an epoxy rubber membrane. See **Bank of China**, **Chesapeake Bay bridge-tunnel**, **jack bag**.

immersion vibrator See **internal vibrator**.

impact [stru.] The collision of bodies. From the viewpoint of the structural engineer, even the small weight of a person walking gently over a floor produces an impact effect on it which must be allowed for by adding something to the weight of the person to

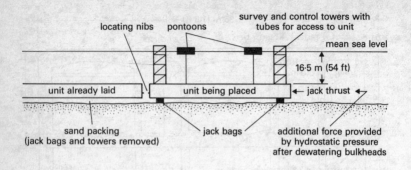

locating nibs   pontoons   survey and control towers with tubes for access to unit

mean sea level

16·5 m (54 ft)

unit already laid   unit being placed   ← jack thrust →

sand packing
(jack bags and towers removed)   jack bags   additional force provided
by hydrostatic pressure
after dewatering bulkheads

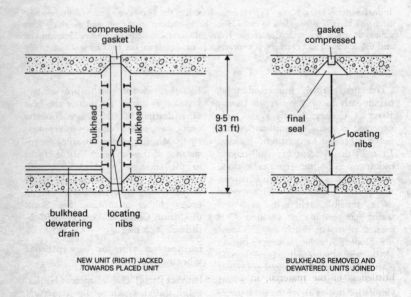

compressible
gasket

gasket
compressed

bulkhead

bulkhead

9·5 m
(31 ft)

final
seal

locating
nibs

bulkhead
dewatering
drain   locating
nibs

NEW UNIT (RIGHT) JACKED
TOWARDS PLACED UNIT

BULKHEADS REMOVED AND
DEWATERED. UNITS JOINED

Hong Kong harbour immersed tube, 1988: connection details. (*Construction News*, 22 September 1988.)

calculate his or her full effect on the floor. See below.

impact factor [stru.] A factor between 1 and 2 by which the weight of a moving load is multiplied to give its full effect on a bridge or floor. For a long span the factor for the same load is smaller than for a short span. The greatest impact factor for a short single-line road bridge is 1.6. For a bridge with more than one traffic line the biggest factor is 1.5.

impact hammer A *rebound hammer* or a *hydraulic hammer*.

impact mole See **pneumatic impact mole**.

impact ripper A *ripper* combined with a *hydraulic hammer* to strengthen its impact on hard rock.

impact spanner, i. wrench, power wrench A compressed-air-operated spanner which tightens *high-strength friction-grip bolts* until it reaches the correct *torque* for the nut being tightened. These impact spanners must be recalibrated every shift and for every change of bolt diameter. A torque spanner is a hand-operated spanner for doing the same work, but for reasons of labour cost it should be used only for tightening a small number of bolts where no air supply is available. The *Torshear bolt* is tightened by a specialized impact spanner.

impact test [mech.] A test on a notched bar which is broken by a pendulum or other striker while the energy absorbed by the broken specimen is recorded. It is a measure of the brittleness of the material, in particular its sensitivity to the *notch effect*. Common tests are the *Izod* and the *Charpy* tests.

Impact test pieces are cut with their length parallel to the direction of rolling, and the notch is cut across the thickness of the plate. Impact tests are not usually demanded for steel thinner than 6 mm. Thin steel is more reliable than thick steel because it has been rolled more. See also **mechanical properties**.

impact wrench See **impact spanner**.

impeller [mech.] The rotating curved blades of a centrifugal (or rotary) pump, blower, compressor or fan.

impending slough Description of the wettest *shotcrete* that can be placed without flowing down (*ACI*).

imperfect frame [stru.] A frame which has fewer members than are needed to make it stable. Some writers also consider that a *redundant frame* is imperfect.

impermeability factor, run-off coefficient (coefficient of imperviousness USA) [hyd.] The ratio of the amount of rain which runs off a surface to that which falls on it. It is therefore a factor from which the *run-off* can be calculated. It is as follows:

watertight roof surfaces	0.70–0.95
cobblestones	0.40–0.50
macadam road	0.25–0.60
gravel road	0.15–0.30
parks	0.05–0.30
woodland	0.01–0.20

The lower factor applies at the start of the rainstorm. See **Lloyd Davies formula**.

impervious [s.m.] A description of relatively waterproof soils such as clays through which water percolates at about one millionth of the speed with which it passes through gravel.

imposed load [stru.] **Live load**.

impounding reservoir, storage r. A large reservoir in which water is stored from the wet to the dry season, as opposed to the smaller *service reservoir*.

impregnation The soaking of wood with creosote, zinc chloride, mercuric chloride or other *preservatives*.

impressed current system *Cathodic protection* in which a permanent voltage to an inert *anode* prevents it dissolving.

improved Venturi flume [hyd.] The *Parshall measuring flume*.

impulse turbine [mech.] A steam or water turbine (such as a Pelton wheel) in which the driving energy is provided by the speed of the fluid rather than by its change in pressure. Cf. **reaction turbine**.

incendive spark [min.] A spark which may ignite a fire or explosion. No incendive spark can arise from an *intrinsically safe* circuit.

incineration Disposal of domestic refuse by burning it, resulting in a reduction to half its previous bulk but often with the release of poisonous gases and dusts. Consequently in Germany most doctors oppose it.

incinerator A furnace where rubbish is burnt.

incise To put regularly spaced shallow wounds into a log so that it can absorb preservative such as creosote. See also (*B*).

incline [rly] A length of track laid at a uniform slope.

inclined cableway A monocable *cableway* in which the track cable has a slope along its full length (about 1 in 4) steep enough to allow the carrier to run down under its own weight.

inclined gauge [hyd.] A sloping staff graduated to read vertical heights (or depths) above a certain datum.

inclinometer [sur.] A *clinometer* or *dip needle*.

incorporated engineer (In.Eng.) The title granted in 1988 to *technician engineers*.

incremental launching (see diagram) Launching from one bank may be the only way of bridging a gap in battle conditions. In peace time, especially for a *segmental bridge* of many equal spans, it can be economical. The complete bridge is built one span or half-span at a time on the bank and launched on the prepared piers as each unit is completed. To reduce the launching stresses from overhanging weight, a lightweight metal nose (span or half-span) may be attached purely for launching and discarded when the last gap is closed.

The elegant 890 m long Dornoch Firth road bridge in the north of Scotland, completed in 1991, shortens the road by some 25 km. The end spans are of 31.75 m, the 19 main spans of 43.5 m. The 20 piers are carried on pairs of 2.1 m dia. steel raking piles 16–24 m long, excavated inside and then filled with concrete and reinforcement. The box-girder units of the bridge, 18.9 m wide, 2.6 m ($8\frac{1}{2}$ ft) deep, were precast in half-spans of 21.75 m and connected by pre-stressing.

indented bolt An *anchor bolt* with indentations forged on it to increase its grip in concrete or grout.

indenting roller, branding iron, crimper A roller with a pattern cast on its surface which it impresses into fresh concrete or hot asphalt when pushed over it, making a non-slip texture.

index glass [sur.] The movable reflecting glass of a *sextant*.

index of liquidity, liquidity i. [s.m.] It is equal to the:

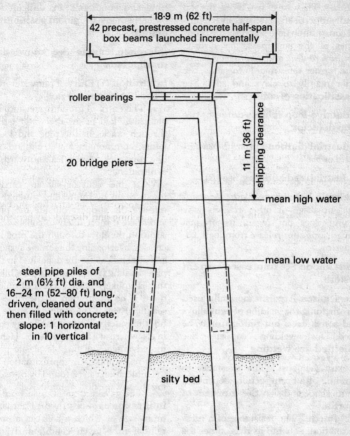

Incrementally launched bridge, 890 m long, over the Dornoch Firth, completed in 1991. 19 spans of 43.5 m (143 ft) and two end-spans of 31.75 m (107 ft). Contractor: joint venture between Christiani and Nielsen and Morrison Civil Engineering.

$$\frac{\text{(water content of sample) minus}}{\text{index of plasticity.}}$$
(water content at plastic limit)

This figure is the reverse of the *consistency index* and gives a value of 100% for a clay at the *liquid limit* and 0 for a clay at the *plastic limit*.

index of plasticity, plasticity i.

[s.m.] The range of water contents for which a clay is plastic; it is the difference between the water contents of a clay at the *liquid* and the *plastic limits*. A clay with high plasticity.index is said to be very plastic. See **toughness index**.

index properties [s.m.] Those properties which distinguish one soil from another. They are of two types, the soil grain properties (size, shape, and chemical constitution) and secondly the properties of the soil mass, *dry density*, *moisture content* and *consistency limits*. The former refer mainly to sands, the latter mainly to clays and silts. See **classification of soils**.

inductive-loop vehicle detector See **loop detector**.

industrial diamond [min.] *Black diamond* or *bort*.

industrialized building See (*B*).

inelastic behaviour *Plastic deformation*, breakage or other reasons why a unit does not regain its original dimensions after release from load. Cf. **elastic**.

inert anode See **impressed current system**.

inert gases Argon is normally used for shielding the welding of light alloys and some steels but nitrogen may be used for welding copper. See **shielded-arc welding**.

inertia (1) [stru.] The resistance to bending of a beam section, dependent on its shape and size. See **moment of inertia**.

(2) [mech.] The resistance of a mass to rotation, equal to its mass times the distance from its centre of rotation squared.

inertial surveying system [sur.] A surveying system based on a gyrocompass, carried on a land vehicle or an aircraft. This includes three gyroscopes with three accelerometers, one each for the *x*, *y* and vertical directions, as well as a microcomputer to calculate the displacement of the system from a known point to the new points whose coordinates are required. Because the system is expensive it is used when many points must be surveyed in a short

time and conditions are difficult for other methods. Cf. **global positioning system**.

inference engine See **knowledge manipulator**.

infiltration (1) Entry of rainwater into the soil, as opposed to *run-off*.

(2) The entry of groundwater into an old *sewer*, laid below the water table, through cracks in the pipe and loose joints, common in a wet climate. In dry countries, sewers leak outwards – exfiltration.

When fine sand or silt are carried into a drain by infiltrating water they remove the support outside the pipe and make it more likely to collapse. In addition the flow through the pipe increases, overloading the sewage system, and silt and calcite are deposited in the pipe. Sealing from inside is a possibility that is unlikely to be successful unless there is no infiltration at the time of sealing. Even if the sealing is successful, the water will rise up and may seep in elsewhere at a higher pessure. Hence the advantages of flexible plastic pipe with few joints. An approximation to the infiltration can be found by measuring the sewage flow at night.

infiltration capacity [hyd.] The maximum rate at which a soil can absorb rainfall in a given condition. It decreases as the rainfall continues, enabling the clay particles to swell and block the soil pores.

inflatable structure A structure that receives its shape by filling with fluid, usually air or water. Balloons were the first inflatable structures and they are still used as the *formwork* inside *shotcrete* domes, *SHEDs*, *Ductubes* or *air houses* (*B*). Balloons were followed by tyres, first for bicycles, then for cars. Inflatable tubular plugs of *polythene* or other plastic can block water pipes or sewers up to 4 m (15 ft) dia. Inflatable dinghies are two-skinned structures. Other structures or tools that may be inflated with air or liquid are *Dracones*, *Fabridams* and *flat jacks*. See **air-supported structure**, **duct former**.

inflexion, inflection [stru.] *Contraflexure*.

influence line [stru.] A graph used for examining the effects of different loads on beams. The curve extends the full span (or half span) of the beam. Its ordinates show, for any point of particular interest on the beam, such as the support, or midspan, either the *shears* or the *bending moments* caused by unit load at any position.

influent stream [hyd.] A stream that loses water to the ground, thus recharging the *groundwater*, unlike an *effluent stream*.

infrared distancer [sur.] An *EDM* instrument with an infra-red *carrier wave*.

infrared photography [air sur.] Photography on special film which is more sensitive to heat radiation (infra-red rays) than to light. It can therefore be effectively used in misty weather (though not in dense fog) for *photogrammetry*. Infra-red colour photography is increasingly used for environmental investigations into vegetation under stress.

infrared thermography, thermal mapping Use of an infra-red camera or more usually an electronic sensor or scanner to map temperature variations of bridges or other decks or the ground, or to show up *delamination*. Most scanners can discriminate temperature differences as small as 0.2°C. The sensor can be mounted on a tripod, in a vehicle, or on a helicopter or other aircraft, and can record data on film or tape. Geological, groundwater, energy conservation, road icing, fog prediction and other investigations are possible. Devon County in 1989 advertised a contract for thermal mapping of 500 km of Devon roads.

infrastructure Roads and other means of transport, communications (telephone, radio, postal services), water supply, sewerage, sewage treatment, and electricity, gas and other fuel supplies. Much of civil engineering is the building of infrastructure.

In 1722 when Daniel Defoe wrote that a 'lady of very good quality' had to go to church in a coach pulled by six oxen because the mud was too deep for horses, the difficulty was the absence of roads. Some 50 years later turnpike (toll) roads were being built and until about 1825 railways could be used by anyone willing to pay the railway fee and possessing a suitable wagon and a horse or mule. In 1823 an Amending Act for the Stockton–Darlington Railway allowed stationary steam engines to be installed at steep places to provide rope haulage uphill. They were never put in because this very railway in 1825 became the first locomotive-drawn railway for passengers and goods. Created by George Stephenson from 1823 it was operated by him from 1825. From then on railways were operated by companies with locomotives, not by people with their own wagon and horse.

At present traffic congestion on roads and even on commuter railways, underground and surface, has been increasing so severely that it is one of the most important political issues. Traffic in London has been growing at 1% yearly since 1976 and central London traffic is getting slower every year. In 1987 road speed in the working day (7 am to 7 pm) was only 18 kph (11 mph). The roads were full. In early December 1987 the traffic of central London became blocked for some ten hours.

Luckily this *cascading failure* did not end with drivers abandoning their vehicles.

In November 1988 British Rail invited tenders from large contractors to finance, build, own and maintain a high-speed rail link from Folkestone, possibly via Gatwick and Heathrow, to take the Channel Tunnel traffic. The line might be operating by 1995, two years after the opening of the tunnel. It was expected to cost £1500 million. See **carbon monoxide sensor**, **road pricing**, **traffic restraint**.

ingot [mech.] A metal bar which, if of steel, may be up to 50 cm square in cross-section and about 2 m long, cast into this shape for *hot working*, usually *cogging*. Modern steel makers are now using continuous-casting machines in which ingots of any length can be cast at a speed of 300 m/h (for a 5 × 5 cm ingot, 6 tons). Non-ferrous metals are cast into ingots which are generally smaller. Cf. **pig**.

inherent settlement [stru.] The sinking of a foundation due only to the loads which it puts on the soil below it and not to the loads on any nearby foundations. In city sites where the foundations are on clay, as in London, all foundations suffer both inherent and *interference settlement*.

initial setting time The time required before a concrete mix can carry a small load without sinking like a mud. This is after about one hour in warm weather with wet concrete. However, with ordinary *vibrated concrete* or *vacuum concrete*, no impression with the full weight of the foot can be made immediately after vibration or vacuum treatment. Nevertheless such concretes do have an initial set which increases their strength in the same way as that of wet concretes. They have initially a much higher strength owing to their lower water content. But for all cements and concretes the strength at initial set is less than at final set.

initial surface absorption test A test of water absorption by a concrete surface, aimed at proving the quality of concrete roofing tiles (BS 1881).

injection [min.] See **artificial cementing**, **grout**, **grouting method of shaft sinking**.

injection well [hyd.] A *disposal well* or one for *artificial recharge*.

inlet, street i. US term for a *gulley*.

innings Land regained from the sea or from a marsh.

insert An object cast into concrete for any purpose such as carrying load or joining one concrete part to another.

in situ See **cast-in-situ**, and below.

in-situ concrete piles Concrete *bored piles* poured in place in holes bored or driven in the ground, as opposed to precast piles which are always fixed in the ground by driving, jacking or *jetting*.

Insituform See **sock lining**.

in-situ soil tests [s.m.] Tests made on soil in a borehole, tunnel or trial pit. For example, static or *dynamic penetration tests*, *vane tests*, field *permeability* tests made by pumping from boreholes, measurement of soil density in place, and vertical or lateral loading tests.

insoluble anode An inert anode in *cathodic protection*.

inspection chamber (1) **manhole** A shaft down to a sewer or duct, enabling it to be inspected and if necessary unblocked by *rodding*. Manholes are conveniently built of precast concrete units with *step irons* cast in, or of brickwork. Minimum dimensions are 0.6 × 0.9 m (2 × 3 ft) throughout. See **deep man-**

hole, **manhole spacing**, **respirator**, **shallow manhole**, **side-entrance manhole**.

(2) A shallow pit of small diameter, e.g. 450 mm (18 in.), through which a sewer or drain can be inspected and rodded.

inspector, i. of works An experienced tradesman (*B*), ganger or foreman, employed under the supervision of a clerk of works or a resident engineer to report on progress and to control a contractor on a section of civil engineering work. Cf. **building inspector** (*B*).

Institute of Water Pollution Control See **Institution of Water and Environmental Management**.

Institution of Concrete Technology An *engineering institution* founded in 1970 by the first group of successful graduates from the Advanced Concrete Technology course (now held at Imperial College). It is an examining body with an international membership.

Institution of Public Health Engineers See **Institution of Water and Environmental Management**.

Institution of Water and Environmental Management (IWEM) Since July 1987 the main *engineering institution* for water. It came into existence to unite the publications and activities of three other institutions: the Institution of Public Health Engineers, the Institute of Water Pollution Control and the Institution of Water Engineers and Scientists. This strong body should be able to defend the water industry against any predator.

Institution of Water Engineers and Scientists See above.

instrument [sur.] A measuring appliance which needs unusual skill, care or training to be effectively used, e.g. a *plotting instrument*, *EDM instrument*, *planimeter*, *theodolite*, *sextant*, etc.

instrumental shaft plumbing [min. sur.] Looking down (or up) a shaft through a *theodolite* telescope to obtain the *bearing* of a line consisting of two points at the foot of the shaft, using a *set-up* at its top (or bottom). See **optical plummet**, **shaft plumbing**, **Weisbach triangle**.

insulating concrete *Aerated concrete*.

intact clay [s.m.] A *clay* with no visible fissures. Cf. **fissured clay**.

intake heading US term for *headworks*.

integrally stiffened plating [stru.] Extruded aluminium sheet shaped like an L upside down ('⌐') and used for decking.

integrated circuit, IC, chip, microchip A piece of silicon about 6 mm ($\frac{1}{4}$ in.) square with thousands of miniaturized circuits etched into it. Many of these tiny circuits include *transistors* which perform rapid electronic switching in computers or similar devices. See **gallium arsenide**.

integrating meter A *meter* which records the total of fluid or electricity which has passed it. It is usually simply called a meter. See also **flow meter**.

intensity of rainfall The rainfall in mm per unit of time.

intensity of stress [mech.] A US term for *stress*.

interactive computing Helpful activity by the computer can take many forms – access to databases including encyclopaedias, electronic mail or teleshopping, electronic data interchange, or even distance learning or the remote monitoring of health. Perhaps the most familiar are *interactive graphics* or the interactive compact disc player.

interactive graphics Work with *computer graphics* using *light pens*, joysticks or, in the most expensive devices, even speech.

intercept [sur.] The length of a staff which is seen between the two *stadia hairs* of a telescope in *stadia work*.

intercepting channel, i. drain A *grip*.

interception [hyd.] US term for *fly-off*.

interceptor A *sewer* connecting individual drains which formerly went direct to a river or the sea. It takes their flows to the *sewage treatment* plant.

interchange See **grade-separated junction**.

intercooler [mech.] A device for cooling air or gas between one stage of compression and the next in a blower or compressor. Cf. **aftercooler**.

interface strength See **bond**.

interfacing In *microcomputing*, process control, etc., making the right connections between an electronic system, often including a *transducer*, and the activity it is measuring or controlling. It involves the design of circuits which match, have the same timing and do not suffer too much from *electronic noise*.

interference-body bolt Bolts made by the Bethlehem Steel Co. in the USA and first publicized by them in 1963, which were in their way as revolutionary as *high-strength friction-grip bolts*. Like these, they are used in a clearance hole, but unlike them they are driven in and they completely fill the hole. They thus have high bearing strength and shear strength, which are claimed to be better than those of rivets. They carry V-shaped corrugations on the unthreaded part of the shank, and these are deformed during driving. The head is cup-shaped and cannot be turned, but this is unimportant since the shank is fully gripped by the slightly crushed corrugations pressing against the walls of the hole.

interference settlement [stru.] The sinking of a foundation due to loads on foundations near it and the natural extension of their *settlement craters* beyond their own boundaries. Cf. **inherent settlement**.

interflow [hyd.] Flow of groundwater from one *aquifer*, often at a *perched water table*, to another at a lower level.

intergranular pressure [s.m.] *Effective pressure*.

interheater [mech.] A *reheater*.

interior span In a *continuous beam* or slab, a span which at both supports has *continuity* with neighbouring spans; cf. **end span**.

interjack station, intermediate jacking point In *pipe jacking*, a point between a *launch pit* and a *reception pit* at which a set of jacks is inserted. This is done where the length to be jacked is likely to be too long for a single push to be possible, or where difficult ground is anticipated ahead. It avoids the heavy expense of digging a new launch pit.

interlock A *clutch* in steel-sheet piling.

interlocking piles *Steel sheet piling*.
  Occasionally concrete or timber piles have *tongue-and-groove joints* (*B*), but this *clutch* is much less likely to be waterproof than with steel sheet piling.

intermediate sight [s.m.] In levelling, a staff reading which is neither a *back sight* nor a *fore sight*.

internal-combustion engine [mech] An engine such as a petrol, gas or *Diesel engine* in which the burning of a gas or vapour provides the energy which turns the wheels round. Generally the burning takes place in a cylinder in which a piston is driven down by the increased gas pressure formed by burning, but jet engines and turbojets have neither piston nor cylinder.

internal flow [hyd.] See **density current**.

internal fracture testing of concrete A *semi-destructive test* devised by the Building Research Establishment to minimize damage to the structure under test. A hole is drilled of only 6 mm ($\frac{1}{4}$ in.) dia. and 35 mm ($1\frac{1}{2}$ in.) deep. In this modified *pull-out test*, the bolt inserted into the hole is expanded gently until a torque meter shows that the maximum has been passed. Using the averaged readings a value of compressive strength can be read from a calibration curve.

internal friction [s.m.] See **angle of internal friction**.

internally focusing telescope [sur.] An *anallactic telescope*.

internal vibrator, immersion v., poker v. (spud v. USA) A cylinder of 3–8 cm dia. containing a vibrating mechanism. It is inserted into wet concrete so as to compact it. See **vibrator**.

international conditions of contract, FIDIC conditions of contract Printed *conditions of contract* based on the fourth edition of the ICE Conditions of Contract and published by FIDIC, the international federation of consultants.

interpolation Inferring the position of a point between two known points on a graph by assuming that the variation between them is smooth. Usually the assumption is that the variation is linear, a straight-line variation. See **extrapolate**.

intersection [sur.] A method of using the *plane table* or *theodolite* at each end of a *base line* alternately. The points to be mapped are sighted at each set-up and located on the plan by the intersection of corresponding pairs of sight lines or the computed intersection of observed *bearings*.

intersection angle [sur.] A *deflection angle*.

intersection point [sur.] The point at which two straights or tangents to a road or railway curve meet when produced. See **tangent distance**.

interstate US word for *motorway*.

interstitial water *Groundwater*.

intrinsic safety [min.] Coal mine equipments such as telephones and signals, which use only small electrical power, are designed to be intrinsically safe – the electrical circuit is purposely restricted to a power output so small that it can never ignite gas or dust and so needs no heavy, expensive *flameproof enclosure*.

intruder alarms Those that are mains-wired have the advantage that they can be continuously monitored. Unwired devices have to contain a battery, and are either radio sets or ultrasonic. They can be carried on the person and are much cheaper to install than wired devices.

intrusion [hyd.] See **encroachment**.

intrusive rock [min.] See **igneous intrusion**.

invar [sur.] An alloy of one third nickel, two thirds iron and other elements, used for making measuring *tapes* because of its very low *coefficient of expansion* which varies with different tapes, and is even occasionally negative, but is usually less than 0.0000014/°C. Invar tapes vary in length with time and therefore need fairly frequent *standardization*.

invert (see diagram) The lowest visible surface, the floor, of a *culvert*, *drain*, *sewer*, channel or tunnel. Cf. **crown**.

inverted siphon [hyd.] See **sag pipe**.

invert level The level of the lowest part of the invert; the level by which the elevation and slope of a drain, sewer or channel is defined, since however much the diameter changes, the invert is usually kept smooth.

ion [elec.] That part of an electrolyte which moves to one or other *electrode* in *electrolysis*. Cations move to the cathode, anions move to the anode.

ion exchange [hyd.] Chemical exchange of a dissolved substance, usually harmful to water, for another that is less harmful. Thus the zeolites used in *base exchange* water softening take in calcium and magnesium in exchange for the sodium they give out. The sodium salts thus introduced to the water do not form curds with soap, but soften the water. Artificial resins used in ion exchange in power stations reduce the solids content of water below 1 ppm in *demineralized water*.

ion-exchange resins 'Synthetic' *zeolites* used in *base exchange*. They can remove the cadmium, copper and chromium salts in the rinse water from metal plating, and can be regenerated by back-washing with caustic soda. The capital investment is high but soon pays for itself by smaller losses of plating metal, smaller water consumption and less boiler maintenance. The system was first introduced to the UK in the late 1950s at Morris Radiators, Oxford, to reduce toxic discharges to sewers.

Irish bridge A paved ford, a type of bridge used in UK mountain districts.

One or more pipes surrounded with concrete form a *culvert* which may submerge in a flood yet remain fordable.

iron (1) [mech.] *Pig* iron or *cast iron*.
(2) [mech.] *Wrought iron*, the purest form of iron, much purer than cast iron or steel and with none of the brittleness of cast iron. It is unfortunate that these two very different materials should be similarly named.
(3) A smoothing iron for sealing and smoothing an *asphalt* surface.

iron pan [s.m.] *Hardpan*, sometimes impervious, cemented by iron oxides.

iron paving A non-skid surface of studded cast-iron blocks.

irrigable area The amount of arable land which is low enough to be irrigated. The area includes houses, roads and other areas never irrigated.

irrigating head The flow of water used for a particular tract of land or a particular irrigation ditch, or that which rotates among a group of irrigators.

irrigation The distribution of water to land for plants. It may involve digging canals and building civil engineering structures such as *dams*, *aqueducts* or *bridges*, or at least ploughing furrows and perhaps spraying by pump. In the USA and Australia it is an important method of *effluent* disposal because those countries are warmer and less densely populated than the UK, where it is obsolescent. Most of the *pollutants* are removed from the water by vegetation at *sewage farms* or by trees and grass in forests, parks and golf courses. Land must be allowed to rest and re-aerate after irrigation with effluent or sludge.

irrigation requirement [hyd.] The quantity of water, including wastes but excluding rain, that is needed to grow a crop.

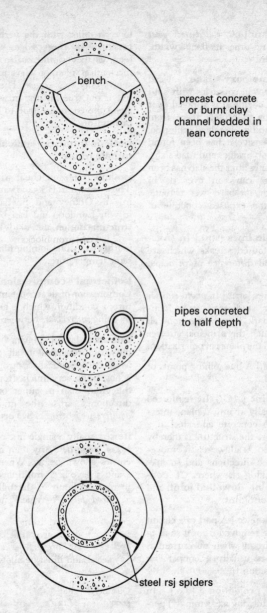

bench

precast concrete
or burnt clay
channel bedded in
lean concrete

pipes concreted
to half depth

steel rsj spiders

Ways of ensuring that inverts of drains, set in a pipe that is too large, may be adjusted to the right slope.

I-section [stru.] A *rolled-steel joist* deeper than 1.2 times its flange width. Cf. **H-section**.

isobutyl-trimethoxy-silane Used since the 1970s in countries with hard winters, this chemical is sprayed over concrete slabs to protect them from water containing dissolved chloride. Though expensive it has been found worth while on bridges and road slabs. Only 20 minutes after the slab has been sprayed, cars can run over it. In Iceland, with its hard wet climate, it protects the exposed concrete of housing.

isochromatic lines [stru.] In *photo-elasticity*, coloured streaks which are lines of equal difference of principal stress.

isoclinic lines [stru.] In *photo-elasticity* or other studies of stress concentration, dark lines which join all points at which the principal stresses are parallel to the planes of polarization.

isohyet [hyd.] A line joining points of equal rainfall.

isolation joint (*ACI*) In reinforced concrete usually a vertical plane interrupting both concrete and steel at a location where the structure is thereby least disturbed. It allows relative movement in three directions and so may prevent cracking elsewhere. Cf. **construction joint**, **dowelled joint**; and see **movement joint**.

isolator link [elec.] A part of a circuit which can be removed from it so as to break the circuit when no current is flowing. It is usually a copper bar bolted in position.

isometric projection [d.o.] In mechanical drawing a (seemingly) perspective view of an object in which, for a cube as an example, the nearest three edges would be drawn at 120° to each other with the vertical edges vertical. All three edges would be seen, as they would not in the usual mechanical plan or elevation, which shows only two edges. The perspective is not true but the measurements along the edges are true to scale. See **projection**.

isostatic frame See **statically determinate frame**.

isotach A line of equal wind speed. Those shown in BS 5400 vary from 36 m/s in Orkney and Shetland to 26 m/s in London, and can be used in structural design.

isotherm A line joining places at the same temperature.

isothermal compression [mech.] Compression of air at constant temperature, a condition which is approached by using *intercoolers* between stages.

isotropic [stru.] Having the same physical properties in all directions. Metals are practically isotropic. Wood is not, being very much stronger along the grain than in either of the two directions across the grain. Some *aquifers* are isotropic. See **orthotropic**.

item Usually a single line of text in a *bill of quantities*. Any item has the six entries shown below. When columns 4 and 5 have been multiplied together to make column 6, the bill becomes a *priced bill*. A typical item could be:

1. Item No.	2. Description	3. Unit
254	Excavation of foundations	cu. m

4. Quantity	5. Rate	6. Amount
5000	£12	£60,000

IWEM *Institution of Water and Environmental Management.*

Izod test [mech.] An *impact test* in which a notched bar is broken by a blow from a pendulum. The height to which the pendulum rises is recorded after the test piece is broken. From this height, and the height from which the pendulum was released, the energy absorbed in breaking the specimen can be deduced.

J

jack [mech.] A piece of equipment for raising heavy loads from below. The smallest jacks are operated by screw and called screw jacks; larger jacks work by a hydraulic ram. They are used for *prestressing* and other purposes. See **flat jack**, **kentledge**, **pipe jacking**.

jack arch (1) (**floor a.** USA) A brick or concrete arch of about 1 m span springing from the bottom flange of a rolled-steel joist or rail. Jack arches were much used before reinforced concrete was developed and are still used to a limited extent in short-span bridge decks or heavy floors, but they have generally been replaced by the *filler-beam floor* in which the soffit is flat.

(2) US term for a *Welsh arch* (B) or *flat arch* (B).

jack bag In the placing of the 15 *immersed tube* units under Hong Kong Eastern Harbour in 1987–8, a cylindrical, rubber, water-filled bag was placed under each end of the unit, enabling it to be set to precise level while its sand foundation was pumped in below it. The units are 9.5 m (31 ft) high, 35 m (115 ft) wide and 122–8 m (400–420 ft) long and weigh 40,000–42,000 tons each. The water in the jack bag is pumped in or out to adjust the level of the unit.

jackblock method Building a multi-storey block by an idea developed from *lift-slab construction*. The first building of this type was a 17-storey block of flats for old people at Barras Heath, Coventry, completed in 1963. One real advantage of the method was that its completion was not delayed by the extremely hard frosts from January to March 1963. As in lift-slab, the floors are cast at ground level, and as soon as each floor has been cast, matured, and stressed up by its prestressing cables (with rapid-hardening cement, after about four days), it is jacked up one storey. The ground-floor slab is an exception to this rule, because it is cast first, remains in position throughout its life, and acts as casting bed for the other floors and roof. The roof slab is cast next. As soon as the roof slab has been raised one storey, the top floor is cast, its walls are built, and with the roof it is jacked up one storey. The main difference from lift-slab is that the supporting walls are built at the same time as the floors and are jacked up with them. With lift-slab, only the floors are lifted; the columns are already in place to their full height. The finishing trades work at about two storeys above ground, the wet trades at one storey below this, and the basic concreting is at ground level. The contractor claims that this converts the building site into a factory which is relatively easy to keep warm and well-organized; the work flows efficiently because the building moves upwards, like a conveyor, and completion is fast and efficient. The name originates from the precision-made precast concrete blocks, equal in height to one stroke of the jacking rams (about 20 cm) which are laid dry for the full height of the building. With their enclosing vertical strips of high-alumina cement concrete, the jackblocks form the supporting walls.

jacked pile A *pile* forced into the

ground by jacking against the building above it in underpinning, usually installed in short lengths. See **pretesting**.

jacker-packer [rly] A small rail-borne machine which can lift track and roughly pack *ballast* beneath *sleepers*.

jacket platform A tubular steel *space frame* carrying an *offshore platform* and resting on piles on the seabed. The first jacket was used in 6 m (20 ft) of water in the Gulf of Mexico off Louisiana in 1947. Since then some 10,000 have been erected, mainly in shallow water. Originally the foundation piles were driven through the tubular legs and grouted to them, hence the name 'jacket'. Self-floating jackets have legs of large diameter to float themselves out before up-ending. The Brent 'A' legs were of 8 m (26 ft) dia. and needed very careful design to avoid *buckling*.

jackhammer See **concrete breaker**.

jack roll A windlass used for hand-hoisting from pits. A bucket is hung from each end of the rope so that an empty bucket goes down as every full one comes up.

jack-up rig, j.-u. platform An *offshore platform* carried on at least three, sometimes as many as 14 welded tubular legs. Each leg is rigidly connected to its foundation plate (spudcan), which may be as much as 14 m (45 ft) across. In 1985 legs were up to 150 m (500 ft) long and could be built as lattice towers. The legs are floated to their site horizontally, then sunk upright to their final position. When all the legs have been sunk, the first platform is built over them. It carries a tank to be filled with seawater, ensuring that the legs sink into the bed more than they

would under working load. Some legs have sunk 36 m (110 ft), but not in the North Sea. When the legs have sunk enough the seawater is released, the platform is built higher if necessary and completed. The height above sea level needs to be about 20 m (65 ft). These rigs can be moved from one site to another. See also **mat-supported jack-up platform**.

jaw breaker, Blake b., j. crusher [min.] A machine for breaking ore or rock. The jaw opening in different machines varies from the laboratory size, 18 × 25 cm, to the mammoth size, 2 × 3 m. Jaw breakers have roughly the same duty as *gyratory crushers*.

jet cement, regulated-set c. (Schnellzement Germany) An extremely *rapid-hardening cement* developed in the USA which sets in 1–30 min depending on its content of calcium fluoro-aluminate and the *retarder*, often citric acid. Using 330 kg/m^3 at 5% calcium fluoro-aluminate, the concrete strength reaches 6 MPa in 1 hour; with 50% it reaches 20 MPa (2900 psi) in 1 hour. Later increases of strength resemble *ordinary Portland cement*, but at 15°C there is little strength gain from one to three days. The cement is used in the precasting industry.

jet drilling [min.] A US method of rock drilling by fire (1946). Fuel is injected through a water-cooled pipe with oxygen and burnt within the hole. Drilling speeds are usually 5 m/h but may reach 11 m/h. About 65 m^3 of oxygen are burnt per metre of hole, the usual diameter being 18 cm. It is very suitable for quarrying. Holes can be *chambered* without explosives. See **fire setting**, **thermic boring**.

jet grouting, high-pressure soil grouting The creation of soil–cement *forepoles* 12 m (40 ft) long or more, very effective in the weak ground over the Vienna metro. Cement *grout* is injected at high pressure through a 50 mm (2 in.) dia. drilling tube while it is being withdrawn and rotated. It has also been used in vertical down holes to protect a shaft to be sunk, and in California to stabilize a hillside against slips. The diameter of the concrete cylinder created varies up to 0.6 m (2 ft), depending on the soil. In Vienna the jet piles were given a 0.4 m (16 in.) *shotcrete* lining after excavation of the tunnel.

Milan metro, due to open in 1992, was being jet grouted in 1988 with 56 mm (2¼ in.) holes at 350 mm (14 in.) centres.

jetting Sinking *piles*, *wellpoints*, etc. into sandy soils where a *pile hammer* is impracticable because it might damage the piles or neighbouring structures. A concrete pile is cast with jetting tubes in it. A timber pile has jet pipes fixed to opposite sides. Large quantities of water are needed with or without compressed air but jetting is often effective after piles have been driven to refusal. Air has been used alone. The water jet may break up the ground so must be used with caution (apart from dangers of flooding). It is not usable in clays but can be a help in *microtunnelling* when making a hole for a pipe or cable.

High-pressure water jets are good for cleaning small sewers but for diameters above 600 mm (2 ft) they are less effective. See also **hyperbaric, supersonic air knife, water-jet assisted-boomheader**.

jetty, landing stage (1) A *berth*, usually one projecting out from the shore line. See **open jetty** (diagram); cf. **wharf**.

(2) A *groyne* to check scour or encourage silting.

jetty cylinder See **screw pile**.

jib [mech.] (1) **boom** The lifting arm

of a *crane* or *derrick*. At the outer end is a pulley over which the hoisting rope passes.

(2) The cutting arm of a *boomheader*, *chain saw*, coalcutter, etc.

jib crane [mech.] A crane with a jib, as opposed to *overhead travelling cranes*, *transporter cranes*, etc., which usually have none.

jig [mech.] A template shaped to guide a cutting tool, such as a drill, quickly and accurately to the right places. See **manipulator**.

jig back (reversible tramway USA), **to-and-fro aerial ropeway** [min.] An *aerial ropeway* with one or two track cables and only one carrier on each track cable. A single traction rope is connected to both carriers, which travel in opposite directions on the two track cables, reversing at the end of each run. The intermittent action limits the capacity to about 25 trips per hour or 100 tons maximum.

jiggle bars A *rumble strip*.

jim crow [rly] A hand-operated rail bender with a heavy *buttress screw thread*. See **railway tools** (diagram).

joggle (1) **groutnick** In block work or masonry walls, a recess on one block which fits a projection on another block, or which forms, with a similar recess on the other block, a cavity into which mortar is poured, making a *cement joggle*.

It may also be a trapezoidal shear key cast in a floor slab within its thickness at a *construction joint*.

(2)[mech.] The slight sharp bending of a steel angle needed to make it fit as a web *stiffener* over the angles at the top and bottom of a built-up girder and to fit tightly on to the web plate. This bend is usually made cold in a joggling press but for a heavy bar may need to be done by *forging*.

joint (1) A discontinuity in rock, where it breaks easily. See **rift**.

(2) In steel sheet piling, a *clutch*.

(3) In the USA one pipe length. See **double joint**.

(4) See **movement joint**.

joint box [elec.] A cast-iron box which is built up around a joint formed between the end of one cable and the beginning of the next. The wire armouring or lead sheath is gripped by bolted glands outside the box. The box is often filled with insulating compound after the joint is made.

Joint Contracts Tribunal (JCT) Not a tribunal but a voluntary body supported by many organizations in the building industry, including the AMA, NFBTE, RIBA, RICS, the Committee of Associations of Specialist Engineering Contractors and the Scottish Building Contract Committee. The JCT compiles, publishes and maintains various versions of its standard form of building contract and *tender* documents. Its decisions are unanimous.

joint filler Compressible strip material used as a spacer between precast or in-situ concrete units, permitting them to expand without developing serious compressive stress in the concrete. Bitumen, bituminized felt, granulated cork or loose cork are used, some of them with a **sealant**.

joint-sealing material See **sealing compound** (*B*), **sealer** (*B*) and above; also **sealant**.

joint venture Collaboration between one or more *contractors* or *consultants* on a large project such as the *Channel Tunnel*.

joist In Britain, a wood, steel or precast concrete beam directly supporting a floor, usually a wooden joist. Steel joists are usually distinguished by calling them RSJs or *rolled-steel joists*.

Joosten process [s.m.] Injection of two separate solutions through pipes driven or *jetted* into the soil. The solutions injected are calcium chloride after sodium silicate. The two solutions meet in the ground and react to form a gel which strengthens the soil and makes it watertight. The method is suitable only for sands and gravels, but in them it forms a solid cylinder of about 0.6 m dia. round each injection tube. (First used about 1920.) See **artificial cementing**.

joule The SI unit of work, energy or heat. It is therefore the ideal unit for expressing the *mechanical equivalent of heat*. One joule = 1 newton-metre and one megajoule (MJ) = one mega-newton-metre (MN-m). One joule/second = one watt and one megajoule/second (MJ/s) = one megawatt (MW).

jubilee wagon A *tipping wagon*.

jumbo [min.] A *drill carriage*.

jumper A heavy steel bar with a chisel point used for boring holes in rock or soft ground by raising it, twisting it, and dropping it repeatedly on the same point. For making horizontal holes it is usually struck with a hammer, with two people working together.

jumping up [mech.] *Upsetting*.

jump join [mech.] A forge weld made by *upsetting* the ends of two bars before welding them.

junction point [sur.] A point on a curve where the circular part joins a non-circular part, either a straight or a transition curve.

jute fibre The fibres of plants grown in India and America used for making *fibre rope*, hessian canvas, reinforcement for plaster and so on.

K

kanat [hyd.] See **khanat**.

Kansai International Airport An airport being built on an *artificial island* in Osaka Bay, Japan, to cost about as much as the *Channel Tunnel* and to open at about the same time – spring 1993. The size is 4.4 × 1.25 km, in water 18 m (60 ft) deep, requiring 150 million m³ of fill from land near by. But the seabed is extremely soft; the 20 m of soft clay is expected to settle 6 m (20 ft) initially and a further 3 m (10 ft) by the year 2050. Consequently the settlement of the clay must be accelerated by *vertical sand drains* inserted by barges driving 14 steel casings at a time about 20 m (70 ft) into the seabed. Sand is poured into the casings as they are withdrawn. By 1988 more than a million had been inserted. A 3.8 km (2.4 mile) long double-decked steel viaduct to the island will carry a six-lane road and two rail tracks. (*Construction Today*, July–Aug. 1988.)

Kaplan turbine [hyd.] A *water turbine* of propeller type. To increase efficiency, the pitch of its blades can be automatically varied with the load.

kathode See **cathode**.

keel A downward projection from the underside of the foundation of a *retaining wall* to prevent it from being pushed forward by the earth behind it. See **retaining wall** (diagram).

keel blocks *Docking blocks*.

kelly [min.] The topmost drilling pipe of a *rotary drilling* rig, carrying the water swivel.

kelly bar The square sliding bar on a

rotary drilling rig which transmits the *torque* to the drill pipe. At its top it carries the water swivel providing the *drilling fluid*.

Kennedy's critical velocity See **critical velocity**.

kentledge, cantledge Loading to give weight and thus stability to a crane, to provide a reaction over a *jack*, to push down a plate in the *plate bearing test*, or to test a caisson or a bearing pile. It may be scrap metal, large stones, tanks filled with water, or any other convenient material.

kerb (curb USA) A hard stone like granite or good-quality precast concrete used for bordering a road and limiting the footway. Usually it raises the footway above the road level by about 15 cm.

kerb raise, k. race A narrow ridge of semi-dry concrete, usually 15 cm (6 in.) or more thick, on which *kerbs* are bedded.

kerf Any saw-cut.

kern [stru.] The *core* of a section.

key (1) *Mechanical bond* or that type of it achieved by an irregular or serrated surface in any construction joint. See also (*B*).

(2) [rly] A steel spring or a hardwood wedge driven between a *chair* and the rail to hold the rail firmly.

k factor, k principle A factor calculated by the Government's financial advisers for each of the ten *water authorities* and 29 companies before *privatization*, to enable financial analysts to estimate its profitability. Since the k factor is fixed for the next ten years, it is certain to be wrong because future interest rates and other variables are unknown. Water charges are fixed by a formula related to inflation plus or minus k.

khanat, kanat, ganat, ghanat, infiltration gallery (galeria Mexico) [hyd.] (see diagram, p. 248). A long, ancient tunnel or network of tunnels dug to collect groundwater, usually from alluvial deposits, certainly used 2500 years ago in Persia, possibly 1000 years earlier in Armenia. They are also in use in Afghanistan, Egypt, Mexico and Chile. About 25,000 are said to be in use in Iran. A khanat is an *adit* or several adits branching off each other, dug by sinking vertical shafts at a spacing of around 100 m to remove the rock from the tunnel and provide ventilation for the tunnellers. The adits are about 1 m high and 0.8 m wide and slope at 0.5–5 m/km. The shafts may serve as wells when the khanat is completed. If there is enough water it is distributed, when the adit reaches the surface, through irrigation canals to the farmers near by.

kibble, bowk, hoppit, sinking bucket, skip [min.] A large bucket used in shaft-sinking for hoisting rock, water, people and tools. It may be from 60 to 600 l capacity.

kicker, starter stub A concrete plinth at least 7 cm high above the concrete floor, forming the start of a concrete wall or column. It is accurately set and sometimes cast with the floor slab. It thus forms a strong and very convenient way of clamping the feet of the wall shutters.

kicking piece A short timber spiked to a waling to take the thrust from a raking strut.

killed steel Steel, usually with at least 0.25% carbon, that has been fully deoxidized (killed) before casting, by small additions of aluminium, ferrosilicon, manganese or silicon. No bubbling is caused by any reaction between carbon and oxygen during solidification and the ingots are homogeneous. Cf. **rimmed steel**.

Kill Van Kull Bridge (now called Bayonne Bridge) A steel, two-hinged trussed-arch bridge near New York of 510 m clear span built in 1931, the longest of its type in the world. Unlike *Sydney Harbour Bridge* it was built with the help of intermediate supports.

kilo- A prefix meaning 1000 times, thus a kilogram is 1000 grams.

kilowatt hour (kwh), Board of Trade unit, kilowatt unit [elec.] The commercial unit of electrical energy, the unit of payment. It is equivalent to 1.34 UK horsepower-hours.

Kind–Chaudron method [min.] An obsolescent method of shaft sinking in waterlogged ground, using a *trepan*. When the full depth is reached, the complete cast-iron *tubbing* is lowered in and *dewatering* begins.

kinematic similarity A *dimensional analysis* of a *harbour model* and prototype indicates kinematic similarity if the flow rates of objects travelling over geometrically similar tracks are to the correct time scale.

kinetic energy [hyd.] The *energy* of a moving body due to its mass and motion, equal to $\dfrac{Wv^2}{2g}$ kg-m where W is the mass in kg, v its speed in m/s, and g is 9.81 m/s each second.

kinetic head [hyd.] See **velocity head**.

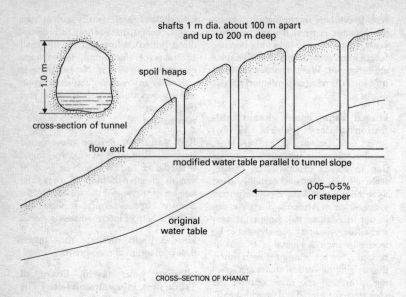

shafts 1 m dia. about 100 m apart
and up to 200 m deep

spoil heaps

cross-section of tunnel

flow exit

modified water table parallel to tunnel slope

0·05–0·5%
or steeper

original
water table

CROSS–SECTION OF KHANAT

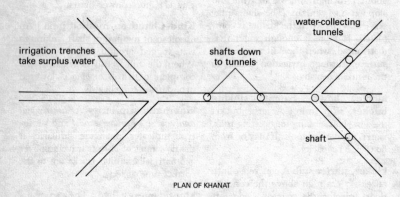

irrigation trenches
take surplus water

shafts down
to tunnels

water-collecting
tunnels

shaft

PLAN OF KHANAT

The khanat is one of the most useful and ancient of man's creations. It cannot be called a structure because it is entirely dug and not built. In Iran they reach to 70 km in length.

king pile (1) In a wide, strutted sheet-pile excavation, long piles driven at the strut spacing in the centre of the trench before it is dug. King piles thus halve the length of the main timbers from side to side of the excavation, since the main timbers from each side bear on the king piles.

(2) A *guide pile*.

king tower See **crane tower**.

kip, kilopound [stru.] 1000 lb, a convenient unit for structural calculations in the USA (4448 N = 1 kip = 454 kg).

knee brace [stru.] A brace between a column and a roof truss to make a building frame stable under wind load.

kneeler [hyd.] *Pitching* laid to protect a bank from *scour* is usually divided into panels to localize damage. The boundaries of the panels are deep stones or reinforced-concrete beams sunk deep into the bank. These are kneelers.

knowledge base In an *expert system* the accumulation of facts, theories, rules of thumb, etc. put into it by the *knowledge engineer*. For a water source it would include information on such properties as turbidity, contents of iron, manganese, *algae, suspended solids*, dissolved oxygen, etc., as well as colour, and seasonal or other variations in quality.

knowledge engineer A programmer of wide experience who has the most demanding task in the writing of an *expert system* – the extraction, articulation and computerization of the expert's knowledge. He or she must talk to experts to discover their knowledge and find how it can be used. The expert and the knowledge engineer continually change their attitudes as they work their way through the subject.

knowledge manipulator, inference engine That part of an *expert system* that makes it work, its *program*.

Kuwait Kuwait's *pipe renewal* programme before the 1990 invasion involved some 1000 km of sewers failing after only ten years' service, 95% of them asbestos cement and very vulnerable to acids. The sewage temperature at Kuwait often reaches 38° C (100° F), and water is expensive, so the sewage is not dilute. Moreover, the sewers have very little fall so the sewage stagnates. This combination results in strongly acid sewage corroding the sewers. Some 2500 km of modern gravity sewers are affected as well as 650 km of pumped sewers, the pumping stations, and treatment works.

In any hot country, sewers with a gentle gradient are likely to suffer similarly. In Cairo, *pultruded* ladders left in sewer manholes have failed in six weeks. See **pipe linings, septic sewage**.

kWh See **kilowatt hour**.

L

laced column [stru.] A column built up with *lacing* from several members.

lacing (1) [stru.] Light metal members fixed diagonally to two channels or four angle sections to form a composite strut or beam (laced or lattice strut or girder). Cf. **batten plate**.

(2) [stru.] Light wooden bars fixed diagonally to the wooden posts at each corner of a laced timber column to make one strut.

(3) [stru.] The *distribution steel* of a reinforced-concrete slab.

ladder The mud buckets of a *bucket-ladder dredger*.

lagging *Sheeting*.

lagging rail [rly] A rail bolted to sleepers to support weak track.

laitance A weak, milky-looking layer on a concrete surface, which should be cut away and replaced by cement *slurry* before another *lift* is cast on top of it. It can cause *dusting* on a floor. ACI considers that the laitance is brought up by *bleeding*, excess water and overworking of the surface.

Lake Washington Bridge See **pontoon bridge**.

Lally column [stru.] Trade name for a hollow, *cold-rolled* steel column, sometimes filled with concrete (USA).

lamella roof [stru.] A large-span vault or dome built of concrete, wooden or metal members joined by bolts or other means, connected in a diamond pattern. Since trusses are not used, the space below can be well lit by roof lights and has a feeling of volume. The system was patented by a German engineer in 1925. Steel lamellas can span 425 m.

lamellar tearing Failure of hot-rolled steel along planes made by slag inclusions.

laminar flow [hyd.] *Streamline flow*.

laminar velocity [hyd.] That speed in a particular channel for a certain liquid below which *streamline flow* always occurs and above which the flow may be streamline or turbulent.

laminate See (*B*).

lamination In *rolled steel*, a weakness caused by thin sheets of slag or other impurity separating the layers of steel. See **delamination**.

lamphole A 225 mm (9 in.) vertical pipe down to a sewer, protected by the frame and cover of an *inspection chamber*. A lamp can be lowered down it on a string into the sewer to show an observer at the next manhole whether the sewer is blocked or damaged.

land accretion, l. reclamation Gaining land from the sea or from a marsh. The cheapest method is by planting reeds or other *maritime plants* to encourage the deposition of silt, a method used in Holland many centuries ago. Other methods are the dumping of dredged mud or other material as an enclosure and pumping out.

land drain A *field drain*.

land drainage See *water authority*.

landfill Low-lying land can be reclaimed by dumping soil on it but the term usually implies 'controlled tipping' of solid refuse, consolidated by earthmoving plant and sealed each night with 15 cm (6 in.) of earth to deter rats, gulls, etc. 'Controlled tipping' is not in any way offensive, and the tip does not leak *pollutants*. Cf. **incineration**; see below.

landfill gas Gas formed by *anaerobic* decomposition of vegetation, usually a mixture of *methane*, carbon dioxide, nitrogen, water, hydrocarbons and sometimes hydrogen sulphide, (H_2S). H_2S is very poisonous but its stink is detectable even at the very low content of 0.025 ppm. At more than 50 ppm, curiously enough, the smell may be unnoticeable. Landfill gas can be expected anywhere above refuse disposal sites, coal mines, oilwells, marshland, old gasworks, etc.

In 1986 a methane explosion at Loscoe, Derbyshire, destroyed a house and injured two people. The methane had moved sideways (lateral migration) from a refuse tip. In 1988 there were 3000 landfill sites in the UK; 60% of the active sites were generating enough gas to be worrying, as were 75% of those closed since 1978. In the USA, 2 mm thick polythene sheet has been used to seal 4-hectare (10-acre) landfill sites all round, including a sheet welded over the top, covered with gravel, topsoil and grass. Some 12 months after the final sealing, holes are drilled to draw off the gas as a fuel.

landing stage A *wharf* or *jetty*, occasionally floating.

land reclamation See **land accretion**.

landslip, landslide A sliding down of the soil on a slope because of an increase of load (due to rain, new building, etc.), or a removal of support at the foot due to cutting a railway, road or canal. Clays are particularly liable to slips. See also **detritus slide, flow slide, rotational slide, shear slide, sidelong ground**.

land surveyor, topographical s. [sur.] A person who measures land and buildings for mapping. His or her professional qualification may be membership of the Royal Institution of Chartered Surveyors and possibly also a university degree. Some engineers (structural, civil, mechanical) can act as surveyors since *setting out* requires a knowledge of surveying. Outside the UK, land surveyors are responsible for *cadastral mapping*. See **cartographer**, **photogrammetry**.

land tie A tie-rod holding a *sheet-pile* or other retaining wall to a buried *dead man* or *stay pile*. See **barge bed** (illustration).

land treatment [sewage] Any method of treating *sewage* (usually *effluent* but sometimes *sludge*) by letting it flow over the ground. The three main methods are used intermittently, with 2–14 days of effluent flow followed by 4–14 days of rest, during which the groundwater level drops and air is sucked into the ground, oxygenating it. The methods are: (i) *irrigation*, (ii) *rapid infiltration*, (iii) *overland flow*. Though obsolescent in the UK, land treatment is still in use for small plants or small quantities of sludge. But it is well suited to warm countries with low population densities like the USA or Australia.

lane rental A style of *contract* from the UK Department of Transport which has dramatically shortened completion times for motorway repairs. A large daily bonus is earned for early completion and a corresponding penalty is paid for each day of delay. In *tendering*, each *contractor* states completion times as well as prices. The contract is awarded to the contractor who offers the best time–cost combination. A saving in time can be converted into cost by multiplying it by the lane rental, around £20,000 per day.

Several varieties of this contract have been used since 1984, all with faster times and with no appreciable increase in cost but with additional pressure on supervisory staff, which can be enjoyable.

Lang lay, Lang's l. [mech.] A wire-rope construction in which the strands are twisted in the same direction as the wires in them. This sort of rope can only be used for hoisting a guided load because it spins dangerously and loses its twist when hoisting a free bucket, but it wears less than a rope of *ordinary lay*.

language, computer l., programming l. For the computer itself there is only one language – machine code. But with its translator *program*, the computer translates programs from other languages into machine code. Languages are of two main types, high-level and low-level languages. Machine code, a mere succession of 'current on' and 'current off', is the lowest level of language. BASIC is a high-level language. Unfortunately each language has hundreds of dialects, so one type of BASIC may not be acceptable to another; similarly for *FORTRAN*.

lap The length by which one reinforcing bar must overlap another bar which takes its place. It is equal to the *grip length*.

lap joint A simple joint between overlapping metal plates, welded, bolted, riveted or even glued together. See also (*B*).

large calorie A kilogram *calorie* (*B*) 1 kg cal = 3.97 Btu = 4.19 kJ.

laser [sur.] Lasers were first used around 1960 and at that time consisted of infra-red (invisible) light only. They were subsequently made visible for use by surveyors. The beam is intense, more sharply defined than is possible with ordinary light, and can be maintained over distances above 1 km, so it is used for aligning tunnels during driving and for plumbing tall buildings during construction.

laser cutting A carbon dioxide laser beam combined with a gas jet, which for steels may be oxygen or for non-metals air or an inert gas, can cut fast and accurately. Pulsed solid-state lasers are used for drilling.

laser level [sur.] See **beacon laser**.

laser sensor [sur.] A *sensor* which among other uses can be moved up and down a levelling staff until it detects the horizontal beam from a *beacon laser*. Laser levelling can use longer sights than is otherwise possible, and only one person is needed. Several lasers built into a road *profilometer* can measure the road surface within 1 mm. See **three-light display**.

laser welding The use of a focused laser beam to fuse and weld metal. The concentrated, high-energy beam penetrates deeply, giving welds free from distortion. At a power up to 5 kW it is used for welding car or lorry parts, but 20 kW lasers are available.

latent damage Damage or faulty construction which is not seen until after completion of a structure.

lateral [hyd.] A small *irrigation* channel, also a branch sewer.

lateral canal A canal with continuous fall along a river, e.g. *canal latéral du Rhône*. Cf. **summit canal**.

lateral-force design [stru.] (see diagram) A method of designing a building in an earthquake region, so that it shall carry safely a horizontal force (lateral force) in any direction equal to a proportion of its deadweight (plus some live load). See **reinforced grouted brick masonry**.

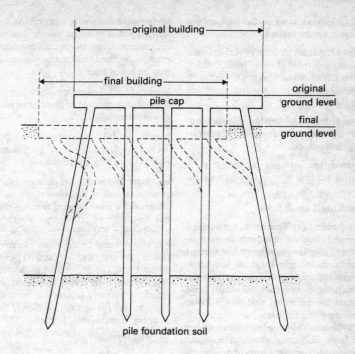

original building

final building

pile cap

original ground level

final ground level

pile foundation soil

Earthquake effect: building moved sideways and lowered several metres, showing the need for lateral-force design. (D. J. Dowrick, *Earthquake-Resistant Design for Engineers and Architects*, Wiley, 1987.)

lateral migration See **landfill gas**.

lateral support [stru.] Horizontal propping to a column or wall or pier across its least dimension. A wall or column with lateral support at close intervals to prevent buckling is regarded as a *short column* and is thus allowed to carry much higher loads than a *long column*.

laterite A red soil widespread in wet, well-drained tropics or subtropics. It has been leached of silica, but usually contains aluminium and iron, sometimes even their ores or those of manganese or nickel.

latex cement See **cement–rubber latex** (*B*).

lath [hyd.] The lengthwise, horizontal osiers of *hurdle work*.

latitude [sur.] The distance in metres north or south of a reference line which runs east and west. In *geodetic surveying* the distance in metres or degrees from the equator. Cf. **departure**.

lattice Description of an open girder, beam, column, etc., built up from members joined by intersecting diagonal bars of wood, steel or light alloy. See **lacing**, **space lattice**.

launch gantry A long movable beam above a bridge being built, which moves forward as the bridge is built, placing sections of the spans. See **Pont de Ré** (diagram).

launching See **bridge launching**, **guyed tower**, **incremental launching**, **launch pit**.

launch pit, jacking p., thrust p. A pit for starting *pipe jacking* or a *microtunnel*.

launder A small *flume*.

lay [mech.] (1) The lay of a wire rope is its method of twist, which may be either *Lang lay* or *ordinary lay*, further classified as right-hand or left-hand. Right-hand lay is the standard lay, in which the *strands* (not the wires) bend round to the right, that is clockwise looking along the rope.

(2) Numerically the lay is that number of helix diameters in which a strand makes one complete turn of 360° in its helix. A lay of 20 means that each strand turns round the rope in 20 rope diameters.

lay barge A vessel which lays a submerged pipeline. On it the pipes are welded together, corrosion protected, weighted with concrete, and progressively lowered from the *stinger* to the sea bed, as the barge pulls itself forward by its anchors. It is followed by the *bury barge*.

lay-by A part of a road or railway out of the traffic lanes where vehicles may wait.

layered flow [hyd.] See **density current**.

layered map [sur.] A *contour* map in which the areas enclosed by the different contours are given different colours.

laying-and-finishing machine A self-propelled machine which receives road material, spreads it, and compacts it into a finished road surfacing which may be rolled later.

layout [d.o.] A drawing showing the general arrangement of plant or of proposed construction.

leach [s.m.] To remove salts or other substances from anything by passing water through it. See **leachate**.

leachate Water that has dissolved salts from materials it has passed through. The word is often applied to water that has flowed from refuse tips, which is usually more polluting than sewage. It has been treated by *digestion*. Six cavities in rock salt 1650 m (5400 ft) below ground and 150 m (500 ft) high were leached between 1970 and 1980 by seawater pumped in, each to provide 230,000 m³ for storage of North Sea gas near Humberside, England.

lead [rly] *Points*.

leaders Steel channel-sections or tube-sections which guide the *pile hammer* and *pile* during driving. They may form part of a *pile frame* or be hung from the jib of a *mobile crane* or stand on the ground supported by guy ropes. See **false/hanging leaders**.

lead line, sounding line [sur.] A strong cord marked at metre intervals for taking soundings in *hydrography*. It is weighted with a lump of lead, which may be concave underneath and greased to pick up a sample of the bed.

lead sheath [elec.] A lead tube covering a power cable or cable for electrical communication. See **extrusion**; also (*B*).

leaf See (*B*).

leaky feeder radio Underground radio communication is usually impossible but researchers of British Coal found how to do it in 1970. An insulated cable from the surface, installed along the underground roadway that needs communications, carries oscillations (a radio *carrier wave*) which can be picked up by a receiver or modified by a transmitter underground. In this way speech or data can travel along the cable. The London tube railways began to install the system urgently in 1987, after two fires in that year, one of which killed 31 people.

lean concrete Concrete containing little cement and usually little water. *Roller-compacted concrete* might contain only 100–120 kg cement/m³ of concrete. See **flocculating agent**.

leaning-wheel grader See **grader**.

leasing See **contractors' plant**.

least count [sur.] The smallest direct reading made with a measuring instrument.

least squares adjustment [sur.] Simultaneously adjusting the survey observations in a network by computer so that all geometrical conditions are satisfied and the sum of the squares of the differences between observed and adjusted values is a minimum. Inexpensive *microcomputers* and programs are increasingly used for such work in civil engineering.

leat [min.] A channel dug along a contour to carry a stream of water (generally a mill stream) for power purposes.

Leca Light expanded clay aggregate, a *lightweight aggregate*.

ledger A horizontal timber in *formwork* or other wooden structures. See (*B*).

ledge rock, ledge US terms for *bedrock*.

leech A *limpet*.

Leipzig market halls [stru.] Circular buildings of 75 m span which were remarkable for their concrete dome roofs, the first large *shells* built according to the *membrane theory*. Like all such shells they were remarkably light, having a weight per square metre of floor area about one tenth that of St Peter's, Rome, which has a span of 40 m.

Lena project See **guyed tower**.

lenticular Description of the shape of a double-convex lens, often given to an *ice lens*.

letting down [mech.] Reducing the hardness and brittleness of a quenched steel by *tempering* it.

levée An embankment, usually parallel to a river, to prevent flooding of low-lying land or to store the water. It may have an impervious core. A natural levée is a ridge of sand or gravel left by a flood. Most levées are artificial.

level (1) [sur.] A *dumpy level* or similar instrument with a telescope and bubble tube which enable the surveyor to take level sights over considerable distances, shots of 30 m being normal practice. It is used with a *levelling staff*. See **reciprocal levelling**, **automatic level**.
 (2) [sur.] To use a level of any sort for measuring differences in altitude.
 (3) [sur.] The elevation of a point, corresponding in Britain to *grade* in the USA.
 (4) To form a horizontal earth surface.
 (5) [hyd.] A drainage canal in the English fen country.

level book [sur.] A field book with special vertical rulings to record the staff readings taken with a *dumpy*, tilting or *automatic level*.

levelling rod See **levelling staff**.

levelling screw [sur.] A *foot screw*.

levelling staff, l. rod [sur.] A staff from 2 to 5 m long carried by a staff man or woman in *level* or *stadia work*. For most work it is graduated in metres and centimetres and can easily be read to 5 mm. A staff is either *self-reading* or a *target rod* and is almost always telescopic so as to be carried easily.

level-luffing crane [mech.] A *crane* in which, during any alteration of radius (luffing or *derricking*), an automatic device causes the load to move horizontally. The crane therefore operates more quickly and with a smaller power output than simpler cranes.

level of control [stat.] A measure of the producer's mastery over his or her production processes. In concrete it is often measured by the cube strength and its *standard deviation*. See **statistical uniformity**.

level recorder [hyd.] A pressure-operated or float-operated instrument which records continuously the level of the water in a channel.

level trier, bubble t. [sur.]. An instrument which measures the slope corresponding to a noted number of graduations through which the bubble has moved.

level tube, bubble, bubble t., vial A glass or other transparent, graduated tube filled with liquid and a small air bubble. The upper surface is barrel-shaped so the bubble centres itself when the tube is horizontal. The bubble tube is an indispensable part of all levels, *dumpy* or *spirit* (*B*), and of the *theodolite*.

lever [mech.] A rigid rod which rotates about a fixed point called the fulcrum. A force applied at the distant end is multiplied at the near end in proportion to the ratio of the distances of the ends from the fulcrum.

lever arm [stru.] The arm of a *bending moment*, that is the bending moment divided by the force producing the moment.

Levy facing [hyd.] A watertight facing to the water face of a French dam. The facing is built on to a row of small arches in a horizontal plane which form vertical openings to the full height of the dam. The arches are drained at the foot. Any leakages which occur through the facing can thus not produce dangerous hydrostatic pressures within the masonry of the dam, since the water within the Levy facing can never be at a pressure higher than atmospheric.

Lewis bolt, rag b. (see diagram) A steel *anchor bolt* with an enlarged indented conical base for anchoring it into the concrete. It is fixed by casting it into the concrete or by leaving a hole and grouting it in later with cement mortar or other *grout*.

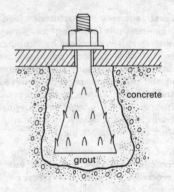

concrete

grout

Lewis bolt.

LHD (load–haul–dump) A self-propelled, usually rubber-tyred mine wagon equipped with a *loader*.

Liability See **defects liability period, responsibility for construction**

life linesman A diver's helper, who stays at the diver's lifeline as long as the diver is under water. He or she lowers the diver into the water and pulls him or her up as signalled by the diver on the lifeline (a stout cord).

life-safety design [stru.] *Structural design* that provides a safe refuge for the occupants in an *earthquake*. Modern tall buildings are usually much less safe than old, small, low ones, but it is possible to design them for life safety. (*Civil Engineering* (ASCE), July 1988.)

lift (1) The vertical height or depth of concrete or other material placed and compacted in one stage; for concrete usually about 1 m but occasionally a storey height (3 m). Deeper lifts can be achieved with a *trémie*, which prevents *segregation*. *Formwork* must be strong to resist the thrust of the wet concrete if it is only rodded but even more so if it is vibrated. A 'lift' or 'tier' can also be a storey height of scaffolding.
(2) [hyd.] The distance through which a vessel rises or falls in passing through a lock.
(3) [hyd.] A power-operated hoist for raising or lowering vessels from one reach to the next without using a lock.

lift bridge A bridge in which both ends of the deck are lifted at the same time to let ships pass. Cf. **bascule bridge**.

lifter hole [min.] A blast hole in tunnelling drilled near the floor and fired after the *cut holes* and *relief holes*.

lift gate [hyd.] A lock gate which opens by rising vertically.

lifting block [mech.] An arrangement of pulleys to enable heavy weights to be lifted, e.g. a *differential pulley block*.

lifting magnet An *electromagnet* that hangs from a crane hook and is used for lifting iron or steel. A further advantage is that it lifts only magnetic material and thus separates it from non-magnetic material (non-ferrous metal).

lifting tackle [mech.] Lifting blocks, hooks, ropes, chains, slings, eye-bolts, bulldog grips, *Bordeaux connections*, and hand-operated pulleys or hoists for raising or lowering heavy weights. Power-operated lifting tackle is usually a *crane* or a *derrick*.

lift joint The surface at which two successive lifts of concrete meet (*ACI*). Though usually visible, it is not always a blemish.

lift pump [mech.] A suction pump. See **suction head**.

lift-slab construction Pouring reinforced-concrete floor slabs and roof slab on the ground floor, one on top of the other, separated by sheets of building paper. Before the slabs are poured, the columns which will eventually carry them are erected to their full height. After hardening, the roof slab is raised into position by lifting jacks. The lower slabs are then jacked into position after the roof. One method of fixing slabs to steel columns has been the welding of the columns to metal lugs cast into the slab. The method was first used in the USA in 1948, and though proposed for flats at Saint-Ouen, was banned in France. See **core wall, jackblock method**.

light alloys Alloys of aluminium have been used in building since about 1950 and have a relative density of 2.7 compared with 7.8 for steel, but they are more expensive. See **bascule bridge, duralumin**.

lighthouse A structure having a recognizable flashing light which guides ships or warns them of danger. The famous beacon or lighthouse 134 m high on the island of Pharos near Alexandria was built about 285 BC.

light pen A popular *computer graphics* device containing a photo-electric cell to read bar codes, or to show the computer a position on the screen that demands computer action.

light railway [rly] A railway with light traffic; because of this it is subject to less demanding regulations than most railways. It may be a narrow-gauge railway.

lightweight aggregate *Foamed slag* (*B*), *perlite* (*B*), *vermiculite* (*B*) and other lightweight aggregates have been used in plasters and screeds for many years. To the structural engineer, however, the most interesting lightweight aggregates are those which can be used in *reinforced* or *prestressed concrete*, e.g., *expanded clay* (*B*), *sintered fly-ash* and foamed slag. They make an insulating as well as a lightweight structure.

Several synthetic lightweight aggregates are available in the UK – Aglite, Leca and Lytag – and although more expensive than dense aggregates and slightly more difficult to use because of their high water absorption, they can sometimes reduce the cost of a structure. The decks of two London bridges, Southwark and Blackfriars, were replaced in 1988–9 with lightweight aggregate concrete. The weight reduction saved money by eliminating the need to rebuild the piers. Another structure, the rebuilt offices over Moorgate Underground station in the City of London, took on an extra floor without overloading the foundation because its new floors, of lightweight aggregate concrete, saved so much weight.

lightweight concretes *No-fines concretes* (*B*), *aerated concretes* and concretes made of *lightweight aggregate*. Sometimes two of these types are combined. For example, a no-fines concrete made of coarse foamed slag, clinker or other lightweight aggregate is not only more insulating and lighter in weight than a no-fines wall of dense stone and the same thickness, but is nailable, which is very convenient. Aerated concretes are the most insulating and lightweight, having a conductivity (*K*-value) of about 0.1 W/m °C at a density of 500 kg/m³. Loadbearing roof slabs at this density have a wet *crushing strength* of 1.9 MPa (275 psi) and an *E*-value of 1.4 GPa (200,000 psi).

Concretes made of lightweight aggregates reach much higher crushing strengths than aerated concretes but their insulation value is lower. Usually they are only 25% lighter than dense concrete. A typical *K*-value for lightweight aggregate concrete would be 0.3 W/m °C at a density of 1440 kg/m³, with a 28-day crushing strength of 20 MPa (2900 psi).

limb, lower plate [sur.] That part of a *theodolite* which pivots on the *tribrach*, and is graduated from 0° to 360°. Both the upper *plate* and the lower plate may be called limbs, but at present 'limb' usually means lower plate. In astronomical observations, the limb of the sun is its edge.

lime See (*B*).

limestone composite cement See **fillerized cement**.

limiting gradient, ruling g. [rly] The maximum gradient on a route. This gradient dictates the size of the trains that can use the route, unless a second locomotive is used, which is unusual.

limiting span See **long span**.

limit of liquidity, l. of plasticity [s.m.] See **liquid limit, plastic limit**.

limit of proportionality [mech.] The

point on a *stress–strain curve* at which the strain ceases to be proportional to the stress and begins to increase more rapidly than before. For mild steel this point is very near the *yield point* and the *elastic limit*, and commercially all three are generally taken to be the same.

limit states [stru.] Conditions in which a structure would become unfit for use. BS 5950 lists eight limit states, four of them 'ultimate' and four of them 'serviceability limit states'. The ultimate limit states are: yielding or *buckling*; overturning or sway; *brittle fracture; fatigue failure*. The serviceability limit states are: deflection; vibration, e.g. from wind; repairable damage due to fatigue; corrosion or general durability. *Eurocodes* insist on structural design using limit states, not *permissible stresses*. See also **plastic design**.

limit switch [elec.] A control and safety device for an electrically driven lift, winding engine or hoist. Limit switches come into operation at the end of the journey, and prevent the cage overwinding or underwinding. See **proximity switch**.

limonite Brown iron oxide, sometimes used as dense aggregate for a *biological shield*.

limpet, leech, limpet dam [hyd.] A sturdy box, open at the top, shaped to fit a dock wall. It is lowered by crane into the water, placed in contact with the section of wall to be repaired, and pumped out. The water pressure forces it into close contact with the wall at the edges and holds it in place. Access is through the open top. Most large harbours possess a limpet, since it is very much more convenient, easier, and quicker to work in a limpet which is open to the air than in a diving suit or a *diving bell*.

linear construction, l. procedure The conventional but slow process of promotion, *feasibility study*, *design*, *drawing*, *tendering* and finally negotiation for a complete *contract* before any building begins. This is the opposite of the *management contract* which may be *fast track*, sometimes called phased construction, in which different parts of the job may be in any of these stages, 'out of phase' with the remainder.

linearity A description of a variable. If it has straight-line variation (varying like a straight line in a graph), it is linear. A non-linear variable is any variable that cannot be expressed as a straight line. Below the *elastic limit*, stress and strain are linear variables. Any measuring *transducer* can much more easily measure a linear variable than a non-linear one.

line drilling [min.] *Broaching*.

line drop [elec.] The loss in electrical pressure (voltage) due to the resistance of the conductors (the line) between the power station and the consumer.

line of least resistance [min.] In blasting, the shortest distance between the centre of the explosive charge and the nearest *free face*. It may be slightly shorter than the *burden*.

line of thrust [stru.] The locus of the points through which the resultant force in an arch or retaining wall passes. In an unreinforced arch or wall, it should be within the *middle third* throughout. See **gravity retaining wall**, **Eddy's theorem**.

liner, stretcher In timbering, a board cut to the length available between opposite members of a frame and spiked between them to lock them in place. See **sill**, also **pipe lining**.

lining (1) [hyd.] A layer of clay, concrete, brick, stone or wood on the bed of a canal to reduce scour, friction and leakage.

(2) [min.] Any protective support within a tunnel or shaft. See also **form lining, pipe lining**.

link (1) **binder, stirrup, secondary reinforcement, transverse r.** Thin steel rods bent to the shape of their beam, column or pile. They are tied round the main steel and provide strength in *shear*. For circular piles or columns, links may be wound on as a continuous helix (helical reinforcement).

(2) One hundredth of an engineer's chain, in which a link is 12 in., or of a Gunter's chain, in which it is 7.92 in. See **tally**.

(3) [elec.] An *isolator*.

Linville truss See **Pratt truss**.

lip block, l. piece, lipping A piece of wood spiked over a strut in trench timbering. It overhangs the waling which the strut supports, so that if the sides of the trench fall away the strut will not drop on the workers below.

liquefaction [s.m.] A saturated sand which has a *voids ratio* higher than the *critical voids ratio*, i.e. a loose sand, is liable to liquefaction when it is shocked by earth tremors, by more water flowing into it or by sudden loading. The sand becomes temporarily quick and a *flow slide* is caused.

liquidity index [s.m.] See **index of liquidity**.

liquid limit [s.m.] The *moisture content* at the point between the liquid and the plastic states of a clay. A mixture of soil and water is placed in a standard cup and divided by a standard grooving-tool along a diameter of the cup. A curve is plotted of moisture content against the number of blows on the cup which make the two divided parts rejoin for 12 mm at the bottom of the cup. The curve should be smooth (normally a straight line) from 10 blows to 50 blows, and the water content at 25 blows is the liquid limit. See **consistency limits, flow curve**.

liquid-membrane curing compound A *sealant* (3).

liquid nitrogen method of freezing ground [s.m.] Liquid nitrogen, a byproduct of the manufacture of oxygen, boils at $-196°C$ and can be applied to tubes for *freezing* ground through 5 cm pipes at about 1 atm pressure; but it is expensive since normally it is not recirculated. The discharge point, where the used nitrogen is released, must be well ventilated, since being devoid of oxygen it could suffocate people.

littoral drift The moving of beach material along a coast line by the sea. The waves are one cause since they move and break heavy stones, but transport of fine material over large distances is by marine currents which hold sand in suspension.

live load, superload [stru.] A *load* which may be removed or replaced on a structure, not necessarily a dynamic load and including neither *wind force* nor *earthquake* loads. See **moving load**.

Liverpool datum [sur.] A levelling datum used in England, based on a fortnight's observations of the mean sea level at Liverpool in 1844. This datum is still shown on old maps but has been replaced on newer maps by *Newlyn* (Cornwall) datum.

lixiviate To *leach*.

Lloyd-Davies formula, rational method for the design of sewers An old way of estimating *run-off* by the simple formula: $Q = 0.167ARC$ in which Q is the rate of run-off in cubic metres per minute, A is the catchment area in hectares, R is the rainfall in mm per hour and C is the *impermeability factor*.

load (1) [stru.] The weight carried by a structure, its *dead load* and *live load*. (2) [mech.] The work being done by an engine or motor.

loadbearing wall [stru.] A wall which carries any load in addition to its own weight and the *wind force* on it. Partitions and panel walls are generally not loadbearing.

loaded filter, reversed f., or weighted f. [s.m.] A *graded filter* at the foot of an earth dam or elsewhere, which stabilizes the toe of the dam by its weight and permeability, since no water can exist in it under pressure.

loader, wheeled l., front-end l., loading shovel, tractor shovel A self-propelled, wheeled (occasionally tracked) machine which may steer on all four wheels or be front-wheel or rear-wheel steered. In front a loading bucket, *four-in-one bucket*, *pallet fork* or even a *bulldozer* can be fitted. Usually the engine is behind the driver but some loaders have the engine in front and an air compressor built in behind the driver. An *excavator* can slew 360° without its tracks. Loaders cannot slew more than about 250° with their rear-end equipment and very little if at all with their front-end equipment. When using the backhoe, the weight of the machine should not be carried on the wheels but on the two stabilizers (outriggers) and on the lowered loading bucket to give rock-steady stability. Tyres in good condition on average ground need about 65% more weight on them to produce the same pull or push as crawlers. See also **backhoe loader, FOPS, four-wheel drive, ROPS**.

load–extension curve [mech.] A curve showing the results of a *tensile test* on a metal test piece, and relating the load to the elongation of the piece throughout the test. For mild steel it generally shows the *yield point*. The curve is very like the *stress–strain curve*.

load factor, f. of safety (1) [stru.] One distinction between a load factor and a *factor of safety* is that load factors are often partial, so that the final safety factor can be a *weighted* combination of many load factors. Thus BS 5950, *Structural use of steel in buildings*, recommends 1.6 for *live load* alone, but 1.2 for live load + wind + crane load; 1.4 for wind alone; and 1.2 for forces due to temperature change. Crane loads are factored at 1.6 for vertical or horizontal loads alone but at 1.4 for the two together.
(2) [elec.] The ratio of the average to the maximum electrical load over a certain period for a certain consumer or power station.

load–haul–dump See **LHD**.

load-indicating bolt A development in 1963 of the *high-strength friction-grip bolt*, which gives a reliable, easily inspected, direct indication that the specified minimum tension has been achieved. The head of the bolt carries small projections on its underside, which compress as the bolt is tightened. The gap thus indicates the amount of the bolt tension, and this can be measured quite simply by feeler gauges pushed under the bolt head. See **load-indicating washer**.

load-indicating washer In the *high-strength friction-grip bolting* of structural steelwork, a special washer with dimples that crush when the bolt is tightened to the correct tension. This can be easily measured with a feeler gauge. These washers were preceded by the much more expensive *load-indicating bolts*.

loading boom Any overhanging structure from which material is loaded into wagons or lorries.

loading gauge [rly] The cross-sectional outline of a loaded wagon including its loads. It is smaller than the *structure gauge*.

loading shovel A *loader*.

loam [s.m.] A roughly equal mixture of sand, clay and *humus*.

local attraction [sur.] Deviation of the magnetic needle from the magnetic north at any point. It may be due to iron or steel in the ground or a penknife in the surveyor's pocket.

lock [hyd.] A *chamber* separated from the two reaches of a canal or river each side of it by gates through which barges or ships can pass upstream or downstream, usually for a small payment. See **slackwater navigation**.

lockage [hyd.] The water lost from the upper to the lower reach of a canal by passing a vessel through a lock.

lock bay [hyd.] A lock *chamber*.

lock cut See **cut**.

locked-coil rope [mech.] A steel-wire rope which is not stranded but built of concentric rings of specially shaped, closely packed wires in opposite *lays*. The outer ring is of S-shaped wires locked together so that if one breaks it cannot work loose. The surface of the new rope is very smooth but the ropes are stiff; they are therefore well suited for use as shaft guide ropes or as ropeway or *cableway* tracks.

Locke level [sur.] A *hand level*.

lock gate [hyd.] Two gates separating the water in the upper and lower *reaches* from that in the lock chamber. Sometimes a third lock gate divides the chamber into two compartments.

lock nut [mech.] A second nut screwed on to a bolt after the nut which carries the load, to prevent it from unscrewing. Measured along the bolt length, it is usually half the depth of an ordinary nut. A castle nut is a particular type of lock nut with notches cut in its outer face through which a split pin can enter a hole in the bolt. The nut in this case is doubly secured.

lock paddle [hyd.] A *sluice* for filling or for emptying a lock chamber.

lock sill, clap s., mitre s. [hyd.] That part of the floor of a lock chamber against which the gates bear when shut.

locomotive [rly] A railway engine which draws trains by its own steam engine, diesel engine, electric motor or gas turbine.

locomotive crane A diesel-powered crane which travels on *standard-gauge* track. It is commonly used for its simplicity and robustness, but its lifting range is limited. It weighs about five times the maximum load which it can lift. See **mobile crane**.

locomotive haulage [rly] Haulage by loco is feasible and economical where tracks are strong, straight and flat, preferably under 3% gradient, always less than 6%, above which a rope or *rack railway* or rubber-tyred haulage must be used.

loess [s.m.] Wind-borne *silt*, found in Europe, the USA and Asia. When dry it can stand vertically in cuts, having some cohesion. Its grain size is between 0.02 and 0.006 mm. See **classification of soils**.

log chute, logway [hyd.] A way through or beside a dam, for logs and driftwood.

logway A *log chute*.

London clay [s.m.] See **stiff-fissured clay**.

long column, strut [stru.] A column which fails when overloaded, by buckling rather than by crushing. In reinforced-concrete work this is assumed to happen with columns which are longer than 15 times their least dimension, but with *lightweight aggregate* concrete the ratio is 10. The permitted stress on a column is reduced as its length increases. See **reduction factor, short column, slenderness ratio**.

long dolly A *follower*.

longitudinal bead test, slow bend t. [mech.] A test for *weldability*. A steel plate on which a welding bead is deposited is bent double. If the plate or weld metal is not weldable it will crack.

longitudinal profile, l. section, vertical alignment A section vertically through the centre line of a road, railway, pipeline, etc. to show the original and final ground levels.

long-line method of prestressing See **pretensioning**.

long span, [stru.] **limiting s.** A span is long for a particular slab, beam, girder or *bridge* type if it is near the greatest economical length for that type. Some long spans are: solid concrete slab, 10 m; hollow-tile concrete slab, 11 m; arches, 300 m; *cantilever bridges*, 600 m; *suspension bridges*, the *Golden Gate Bridge* (1937), 1280 m between centres of support. See **Parramatta Bridge**.

long-strip flooring Laying an industrial *prestressed concrete* floor in lengths of 60 m (200 ft) as recommended by the Concrete Society. Even greater areas are possible by following the society's recommendations. See **high-bay warehousing, prestressed concrete pavement**.

long ton or **gross ton** The ton of 2240 lb equivalent to 1.016 metric tons (tonnes). Cf. **short ton**.

loop detector, inductive-loop vehicle detector A convenient, inexpensive vehicle detector which needs no maintenance, gives an output while a vehicle is present and does not detect pedestrians. First installed in automatic car parks it has become normal for actuating traffic lights. The inductance of the wire installed in a slot in the road is altered by the steel and iron in the vehicle over it – by 4% for a saloon car, by 0.02% for a bicycle. These loops can operate automatic barriers, control factory doors, detect queues, etc. The cost of installing the loop in its slot in the road is much more than its purchase price so it is advisable to give it a first-class installation. Loops must not be within 50 mm (2 in.) of steel reinforcement.

loop take-up See **conveyor loop**.

loose-boundary hydraulics [hyd.] The study of the flow of a fluid along uncertain boundaries involving airborne or waterborne transport of sand. In a stream with banks and bed of loose sand, with rising water level the sand is picked up, while with falling water level sand is dropped, so that the stream boundaries change with the water level. It includes the study of *antidunes, dunes, bed load, hydraulics, layered flow, saltation* and *turbulent flow*.

loose core Large aggregate which has become segregated from the concrete by mishandling between mixing and placing.

loose ground Granular soil with a relative compaction below 90%, which can be dug with a shovel. Unlike *compact material*, a pick is not needed to break it up.

loose-sleeper relaying [rly] The laying of *continuously welded rail* by first spacing *sleepers* at their correct positions, and bringing in long lengths of rail 183 m (600 ft) long. These are joined by *thermit* welding and *destressed* after welding.

lorry loader A light crane (e.g. 7 tons lift) fixed behind the cab of a lorry for loading or unloading it.

Lossier's cement See **expanding cement**.

loss of ground In excavating soft ground the volume of silt, clay or sand dug out is greater than the volume of the excavation because material flows into it even when no *boil* occurs. This ground is lost from outside the excavation and may cause settlement of neighbouring buildings. To avoid this, excavation must be done in sheet-piled *cofferdams*, *Gow caissons*, etc. See **groundwater lowering**.

loss of head, lost head [hyd.] *Hydraulic friction* in a pipe or channel expressed as a loss in potential energy (metres head).

loss of prestress [stru.] Losses of prestressing force after *transfer* arise mainly through elastic shortening, *shrinkage* and *creep* of the concrete and creep of the steel. The use of high-quality concrete can reduce these losses considerably. Correspondingly, if lightweight aggregate is used, the losses can be expected to be 50% more.

low-alkali cement A *hydraulic cement* with less than 0.6% Na_2O equivalent (*ACI*).

low-carbon steel [mech.] Steel with 0.04–0.25% C, generally called *dead mild* or *mild steel* according to its carbon content.

lower explosive limit Fuel gases in air have two explosive limits. For example, *methane* is explosive between 5% and 13%, 5% being its lower explosive limit. The lower limit is the most hazardous. The upper limit is rarely reached because the air is often unbreathable by then. Petrol (gasoline, benzene) vapour in air has its lower limit at 1.3% so road tunnel electrical equipment must be either *flameproof* or *intrinsically safe*, and smoking should be forbidden. Refrigerants have the following lower limits: 1% upwards for hydrocarbons; ethyl chloride, 3.5%; methyl chloride, 8%; ammonia, 16%. The last three are also poisonous.

Lowermoor See **Camelford**.

lower plate The *limb* of a *theodolite*.

low-heat Portland cement, ASTM P. c. type IV A cement that is made with different chemical composition from *ordinary Portland cement* so as to liberate less heat in dams or other massive concrete blocks. These may take many months or years to cool to the ambient temperature, and therefore could crack badly if built with other cements. Even though it is more finely ground, it is not so strong as ordinary Portland cement. It is only made to special order in the UK, but *sulphate-resisting cement* is sometimes a good substitute. See BS 1370.

low-water valve [mech.] See **fusible plug**.

luffing *Derricking*.

luffing cableway mast (see diagram) A *cableway* tower hinged at the foot and held up by guys which can be adjusted to allow some lateral movement either way at the head so as to move the *track cable*.

luffing jib crane, derricking j. c. A crane with a jib hinged at the foot to allow the hoisting rope to work at different radii. Cf. **level-luffing crane**.

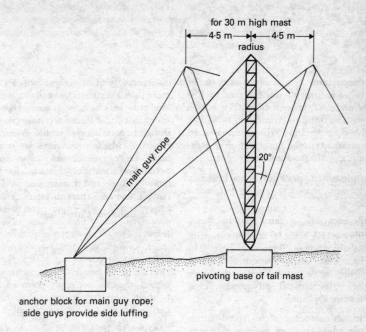

for 30 m high mast

main guy rope

20°

pivoting base of tail mast

anchor block for main guy rope;
side guys provide side luffing

Luffing cableway; tail mast. (F. C. Harris, *Construction Plant*, Granada, 1981.)

lump-sum contract A contract usually without a *bill of quantities* or *schedule of rates*, therefore from the *client*'s viewpoint suitable only for very small, very urgent work. Cf. **measurement contract**.

lurching [rly] Momentary transfer of part of the weight of a train from one rail to the other. The total load is unaltered. See also **impact factor**, **nosing**.

Lytag A *lightweight aggregate* made by sintering *fly-ash*.

M

macadam Uniformly sized stones rolled to form a road. The road may be *waterbound*, cement bound, or coated with tar or bitumen and may be laid by hand or machine. *Tarmacadam* is the commonest road construction in Britain. See **single-sized aggregate**.

macadam spreader A machine that spreads *macadam* received from a lorry.

machine [mech.] A tool or device for overcoming force by applying another, usually smaller, force elsewhere. The lever, the screw, screwdriver, wheel, lifting tackle and jack are all machines. See **mechanical advantage**, **contractors' plant**.

machine tools [mech.] Metal-cutting machines such as boring machines, drills, grinders, planers, key seaters, hobbing machines, tapping machines, shapers and slotters, lathes, etc.; in general, those machines which make other machines.

macrodefect-free cement Thin-walled products resembling strong plastics, but made of *high-alumina cement* mixed only with an organic plasticizer and water.

macro-engineering The American Society for Macro-engineering, based at the Polytechnic Institute of New York, investigates subjects such as the *Channel Tunnel* or the development of offshore oilfields by tunnelling to *artificial islands*.

made ground, made-up g. Ground which has been raised by *fill*.

maglev, magnetic levitation A transport system which costs only 10% more than a conventional railway with steel rails, with 60% of its energy costs and only 15% of its maintenance costs. But extra costs arise from the elevated structure. Maglev *guideways* are normally above ground. One exists between the Birmingham National Exhibition Centre and Birmingham Airport. There are others in Japan, Vancouver and West Germany, and one is conceived for Atlantic City, New Jersey. They are electrically driven, quiet, smooth, vibrationless and non-polluting. Speeds of up to 480 kph (300 mph) are possible. Being without wheels, they do not apply point loads to their supports; in movement they may be 10 cm (4 in.) above the rail.

magnesite [min.] $MgCO_3$, magnesium carbonate. When heated it is used for making a basic *refractory* lining to steel furnaces. It can be used at temperatures up to $1800°C$. As with lime the heating drives off CO_2, so that magnesite refractory is really magnesia (MgO).

magnetic bearing (1) [sur.] The *bearing* of a line measured from magnetic north.

(2) [mech.] Mechanical bearings under electronic controls, which provide magnetic suspension to counteract the weight of the rotating unit, have enabled machine tools to turn at 30,000 rpm (two radial bearings with one thrust bearing). Such bearings may be used also on compressors, centrifuges or pumps. A 40 kW centrifuge needs only 1.4 kW of magnetic energy to carry the weight magnetically. Small rotors at 100,000 rpm are thus now feasible.

magnetic compass [sur.] See **compass**.

magnetic declination [sur.] See **declination** (2).

magnetic north/south pole [sur.] Centres of magnetic attraction in the north of Canada and in Antarctica. They are continually moving and these movements account for the continual alteration of *declination* throughout the world.

magnetic variation [sur.] (1) *Declination* (2).
(2) Small daily variations in the declination which nowhere exceed 1°. Annual variations also exist and the dip also varies regularly.

magnetite Magnetic iron oxide (Fe_3O_4), which can be used as a dense aggregate for building a *biological shield*. It weighs about 5.1 tons/m³.

magneto [elec.] or **magneto generator** A small alternating current generator which carries its magnetic field in a permanent magnet.

magnetometer [sur.] An instrument for measuring the intensity of a magnetic field. It is particularly useful in *geophysical prospecting* in conjunction with an air survey of the ground. An aeroplane carrying a recording magnetometer flies low over the ground and its indications are later plotted as contours of magnetic force over the *mosaic*, showing in a striking way where magnetic rocks are most likely to be found.

MAG (metallic active-gas) welding *MIG welding* in which carbon dioxide or oxygen either alone or mixed with argon are used to shield the weld. It is so called because oxygen is never inert and carbon dioxide is rarely so.

Maihak strain gauge An *acoustic strain gauge* from Germany.

main A large supply pipe for water, gas, etc. For water it may be a *gravity main*. The word is also used to describe the 'service main' bringing the gas or water service along each street.

main beam [stru.] A beam which bears directly on to a column or wall, not on to another beam.

main canal [hyd.] An irrigation conduit taking water from the supply. It delivers water to the *laterals*.

main contractor, general c., prime c. A *contractor* who is responsible for the bulk of the work on a site, including that of the *subcontractors*.

main drain, trunk sewer A *sewer* which leads to a *sewage treatment works*, sewage pumping station or *outfall*.

mainframe computer A large, expensive, general-purpose computer with many times the speed and memory capacity of a *microcomputer*. For the *microcomputer* owner an advantage of using his or her micro as a *terminal* to a mainframe is the enormous amount of information available from the many databases accessible through mainframes. The essentials for the connection are a suitable *modem* and a *communications program*. The emphasis is on 'suitable' because not all UK modems can access databases in the USA, and not all communication programs work in the way one wants.

main holes [min.] *Relief holes*.

mains-borne signalling Use of the house wiring to carry switching signals which allow reading of meters, internal telephone calls, activation of a burglar alarm, etc. even in the householder's absence. In France, sound radio travels along the lighting mains to hotel rooms. One electronics maker sells a 3-pin plug which can use the mains as a local area network for computer communications – advanced telemetering in the home.

main sewer See **main drain**.

main steel, m. reinforcement Steel bars usually of at least 10 mm ($\frac{1}{2}$ in.) dia. in beams, slabs, columns or piles, as opposed to *links*, which though often essential are regarded as secondary steel. Slabs do not have links, but *one-way slabs* have *distribution steel* at right angles to the main steel.

maintenance period The *defects liability period*.

main tie [stru.] The tension members joining the feet of a roof *truss*, generally at wall-plate level.

malleability That property of a metal, resembling ductility, which allows it to yield without breaking when hammered.

malleable cast iron [mech.] Iron castings which have been heat treated by packing in an oxidizing agent and then holding at 800°C for about four days. Surface oxidation and change of internal chemical structure are both believed to cause the improvement in toughness and strength. The average tensile strength is more than double that of ordinary *cast iron*. It is not really malleable but has noticeable ductility in the tensile test.

mammoth pump [min.] The *air-lift pump* within the hollow boring rods of a *trepan*.

management by exception, exception reporting A computer technique which saves the time and reduces the worries of busy managers. A *microcomputer* scans large volumes of data automatically, but reports only results which are outside the limits it has been given. The manager thus needs only to think about matters on which urgent

action is needed. It has been used for controlling the maintenance of civil engineering plant and for *critical path scheduling*.

management contract, m. fee contract An agreement between a *client* and a trusted *main contractor* (paid by a fee) who works not at first as a builder but as an adviser to the *consultants* and later as a manager, programming and organizing construction. The main contractor can offer useful advice and therefore is appointed early, and later supervises construction as well. The system is best for urgent work with complex mechanical and electrical plant that is hard to price.

Procedures can be telescoped and time can be saved because of the intelligent cooperation from the start between consultants and contractor. Various parts of the job can be at quite different stages simultaneously, either at design or detailing or being built. This is impossible with ordinary *linear* contracts, in which the job must be completely billed and in part designed and detailed before the contractor can price it. The *contract documents* include, apart from the management contract, a works contract of the rates and prices for the work to be done by the works contractor who, unlike the management contractor, bears the risk. Management contracting is increasingly popular, so much so that a local council (Cambridge), in 1989, acted as its own management contractor to build a bridge for cyclists, hoping both to improve the bridge and to save time and money.

mandrel, mandril (arbor USA) An accurately turned rod or thick pipe which serves as a *pile core* for a pipe pile being driven.

manganese [min.] Chemical symbol Mn. A metal of which at least 0.8% is added to nearly all steels.

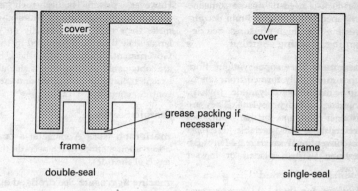

double-seal   single-seal

Manhole covers.

manganese steel [mech.] Steels containing more than 1% Mn, which are appreciably tougher than ordinary steels with only about 0.8%. Colliery *drawgear* made of 1.5% manganese steel has the advantage that it does not need annealing every six months, unlike wrought iron. Alloy steels with about 10% Mn are used for dredger buckets and similar hard use.

manhole (1) [mech.] An access hole to a tank or boiler drum, just large enough for someone to enter. It is normally covered with a cast-iron or steel plate called the manhole cover.
(2) An *inspection chamber*.

manhole cover (See diagram) A cast-iron plate fitting into a cast-iron frame bedded in the concrete slab over an *inspection chamber*. Manhole covers over foul drains are sealed by a tongue projecting down from the cover all the way round its edge into a grease-filled groove corresponding with it on the frame.
Ductile iron is now commonly used, but in the elegant pedestrian precincts of central Budapest the manhole covers are of bronze, while in 1988 in poverty-stricken Managua, the capital

of Nicaragua, cast-iron manhole covers were being stolen at night as the only source of metal. See **double-seal manhole cover**.

manhole spacings Manhole spacings vary with the diameter of the sewers they give access to. Sewers smaller than 1 m (39 in.) dia. cannot be entered and their manholes should be at every change of diameter or direction and not further apart than 100 m. Spacings may be larger if high-pressure jetting is available for cleaning the sewer. On sewers larger than 1 m, spacings can normally be 200 m; if ventilation is good, even more. But if workers must wear *respirators* in the sewer, 200 m should be the absolute maximum.

manifold [mech.] A thick pipe, generally curved, into which or from which several smaller pipes lead. See **header**.

manipulator A jig to which work can be clamped, for it to be easily raised, rotated, etc. during *welding*.

man-lock An *air lock* through which people, not materials, pass. It is an access to a tunnel, shaft or *caisson*, driven in *compressed air*. A separate man-lock is needed because someone may spend more than 60 min decompressing from high pressures. See **decanting**, **working chamber**.

manmade fibre rope Synthetic fibre ropes are usually formed from one of three materials: polyamide (nylon), polyester or polypropylene. They are stronger than, and for many purposes preferable to, the vegetable *fibre ropes* of conventional cordage. Many BS describe twisted, plaited or hawser types.

manometer [hyd.] A *pressure gauge* with a U-tube usually containing a liquid such as kerosene or mercury which does not mix with water. The two ends of the U-tube are joined to the two points in the pipe between which the pressure difference is required.

manual metal-arc (MMA) welding (shielded metal-arc (SMA), stick w. USA) A common but highly skilled manual welding method using *covered electrodes* coated with material that shields the arc and excludes the atmosphere with its harmful oxygen, nitrogen and water. No hoses or gas cylinder are needed, consequently welds can be made even on remote parts of a building site. The equipment is relatively cheap and portable. All steels except high-alloy steels can be welded, also cast iron, aluminium, copper and their alloys. The best welding voltages are between 20 and 45 V DC and welding currents can vary from 20 to 600 A.

manufactured sand *Sand* made by crushing rock, stones or slag (*ACI*).

map [sur.] A drawing of part of the earth's surface. Countries are usually mapped by Government organizations such as the Ordnance Survey of Great Britain or the French General Staff.

However, private bodies such as mining or air survey companies also make their own maps, usually on a larger scale than those provided by the Government. OS maps of all the UK are obtainable at 1:10,000 scale; all except mountains at 1:2500, and city centres generally at 1:1250.

margin See **verge**.

marigraph [sur.] A gauge at a tidal observation station which records the levels of the tides.

marine aggregate See **dredged aggregate**.

marine borers Molluscs and crustaceans which live usually in warm water and destroy wood by eating it. The gribble (Limnoria lignorum) and shipworm (Teredo navalis) are the best known, since they can penetrate any wood in favourable water. The best antidote usually is full-cell treatment with creosote, but concrete or Monel metal encasement have sometimes been found necessary. Light explosive charges set off in the water will kill the borers but will not ensure that they do not return.

marine surveying, m. hydrographic s. [sur.] The mapping of the sea bed and the estimation of the currents, *scour* and *silting* near the coast.

maritime plants [hyd.] Plants which grow on foreshores in salty conditions may prevent or reduce *scour*. Rice grass (cord grass or Spartina townsendii) grows best in foreshore muds 0.6–1.2 m below high water. Marram grass (Psamma arenaria) flourishes in the arid conditions of the sand dunes which it helps to stabilize. Shrubby seablite (sea gorse or Suaeda frutosa) will grow in shingle beaches. These plants may be a very cheap *revetment*.

marl [s.m.] Now any soil or rock containing a vague quantity of lime. The term originally meant any soil or rock suitable for liming (marling) arable land.

marram grass See **maritime plants**.

masonry cement A cement, usually a *Portland*, that hardens slowly and holds water well, sometimes containing a plasticizer and other *admixtures*.

mason's putty See (*B*).

mass centre *Centre of gravity*.

mass concrete Any volume of concrete thick enough to need measures to cope with the heat generated by *hydration* of the cement, and with the prevention of cracking (*ACI*). Cf. **plain concrete**.

mass curve, m. diagram [hyd.] A graph of accumulated monthly inflow volumes plotted as ordinates against the time in years or months of the period likely to be critical for a *reservoir*. The mass curve provides a first approximation to the size of reservoir that will be needed.

mass-haul curve A curve which shows the amount of excavation in a cutting which is available for fill. The abscissa shows the distance from the centre line, and the ordinate shows the amount of fill available, i.e. the amount of excavation up to that point. See **balance point**.

mass transit A light railway, monorail or other public transport system, sometimes driverless.

mast [stru.] (1) A slender tower held upright by guy ropes, for instance in a *cableway* or *guyed tower*.
(2) The upright member in a *derrick* from which the *jib* is supported.

master programme As soon as the

contract has been signed, the *contractor* draws up the master programme under which it is hoped the job will be built. It is the basis of the *critical path scheduling*.

masthead gear Any gear carried on a mast or jib; this has come to mean two drums carried on a crane jib to control two subsidiary ropes over pulleys near the head of the jib. These two ropes can exactly orient a chisel grab digging a trench for a *diaphragm wall* and can thus control the direction of the wall. Any trench during excavation is liable to be overloaded by the crane standing near its edge. This device enables the crane to stand well away, so that it does not *surcharge* the edge of the trench.

mastic See **sealant** and (*B*).

mastic asphalt A wearing course to a road or a waterproofing to a roof, consisting of liquid *asphalt* spread hot by hand floats. Some clean sand or chips may be mixed with it. It is laid in two or three coats, the latter being 18 mm thick in all.

mast platform See **rack-and-pinion mast platform**.

mat (1) US term for a *raft*, or a footing of steel or concrete under a post. A *blinding*.
(2) A heavy mesh of reinforcement in a concrete slab.

match lines [air sur.] The lines along which mosaics are cut so as to hide the joints between them. Edges of fields or roads, railways or streams are best.

materials handling, mechanical h. [mech.] Machinery for handling stone, coal and other granular material, including *conveyors* of every sort, cranes, *transporter cranes*, grabs, elevators, railways, *cableways* and ropeways. It merges into *earth moving*.

materials lock, muck l. An *air lock*

through which rubbish is sent out and building materials sent into a pneumatic caisson or shaft being driven in compressed air. See **working chamber**.

matrix The material of a solid, in which the larger grains are embedded. *Cement paste* is the matrix of concrete or mortar, tar or bitumen the matrix of blacktop. Quartz is the matrix of a *quartzite*, often also of a granite.

mat-supported jack-up platform An *offshore platform* used where the seabed is weak and bearing pressures must be kept low. The mat or base is connected to all the legs, unlike the *spudcans* of other *jack-up rigs*.

mattress (1) A concrete slab at ground level used as a base for a transformer or for other plant.

(2) A layer of *blinding* concrete.

(3) A large flexible sheet that prevents *scour*, formerly made of brushwood, now often of *geotextile* weighted down with rocks or other *armour*.

maturing See **curing**.

maturity of concrete (ASTM standard C 1074–87). A figure in degree-days, degree-hours or other unit which shows roughly the amount of *hydration* of a concrete and thus an approach to its strength. It may help in deciding when to prestress tendons or remove formwork or cold-weather protection. Microprocessor-controlled maturity meters are mentioned in ASTM C 1074–87.

maximum cement content To reduce *shrinkage*, *cracking*, *thermal cracking* and *creep*, most specifications insist on a maximum cement content of 550 kg (or less) of cement/m³ of concrete. Cf. **minimum cement content**.

maximum dry density [s.m.] The

dry density obtained by a stated amount of compaction of a soil at the *optimum moisture content*.

mean [stat.] An arithmetic mean is an *average* in which all signs are taken as positive. In an algebraic mean the signs of the quantities are considered and the mean may be either positive or negative.

mean depth [hyd.] The cross-sectional area of a stream divided by its width of free surface.

mean velocity [hyd.] See **current meter**.

measurement contract, admeasurement c. A *contract* based on the measurement of work on completion, the commonest and fairest type of contract. It includes either a *bill of quantities* or a *schedule of prices*, and for civil engineering work normally uses the *ICE Conditions of Contract*.

measuring chain [sur.] See **chain, Gunter's chain**.

measuring weir [hyd.] A *weir* built to measure the flow of a stream. Since the weir crest is horizontal, the water depth over it is more easily measured than from the rough, stony or muddy bed. It may extend across part of the width of the stream, in which case it has *end* (or side) *contractions*. If it covers the full width of the stream, the contractions are suppressed and it becomes a *suppressed weir*. Near a weir, the water level drops fairly steeply. The *head* at a weir must therefore be measured upstream of this part. This is best done by a float gauge that continuously records the depth of water. See **Venturi flume**.

mechanic (1) [mech.] A *fitter*, a man skilled in mechanical engineering.

(2) A *tradesman* (B).

mechanical advantage [mech.] The

ratio of the load raised by a *machine* such as *lifting tackle* to the applied force. The mechanical advantage divided by the *velocity ratio* is the *efficiency* of the machine.

mechanical analysis [s.m.] The determination of the proportions of the different particle sizes in a given soil sample to help in *classification* of the soil. Grains larger than 0.06 mm can be sieved. Smaller particles cannot be measured by sieving; their effective diameters are estimated from their *terminal velocities* in water by *wet analysis*. See **consistency limits, fineness modulus, grading, mesh**.

mechanical bond, m. key Adhesion obtained by surface roughness, e.g. from *deformed* reinforcement bars, from scratching plaster, the serrated shapes of Lewis bolts, or the rough surface of a slab intended to receive a *monolithic topping*. Cf. **specific adhesion**.

mechanical dredger A *bucket-ladder dredger* or a *grab dredger*.

mechanical efficiency [mech.] See **efficiency**.

mechanical equivalent of heat, Joule's equivalent The amount of mechanical energy which can be transformed into one heat unit. In imperial units it is 778 ft-lb per British thermal unit. In metric units, one kilowatt-hour is one kilowatt for 3600 seconds, so 1 kWh = 3.6 MJ. See also **joule**.

mechanical handling *Materials handling*.

mechanical key See **mechanical bond**.

mechanical loading See **excavator**.

mechanical properties Properties concerned with the strength of a material. For steels the main require-

ments are either the *yield point* or the *ultimate tensile strength*; the percentage *elongation* (an indication of *ductility*); and the *Charpy* V-notch value for the relevant metal thickness and working temperature. Minima for all these are laid down in BS and other standards for each grade and thickness of steel. The thicker the steel, the less hot working it has undergone, so the more likely it is to be unreliable. Similarly, the lower the working temperature, the more likely it is to suffer *brittle fracture*.

mechanical rammer A machine carrying a weight which it lifts and allows to drop on the soil. See **frog rammer, power rammer**.

mechanical sampler See **sampler**.

mechanical shovel, power s. An *excavator*.

median strip A *central reservation*.

mediation A less formal procedure than *arbitration*. Wholly voluntary, not enforceable, it was introduced by the American Arbitration Association for the early stages of a dispute. The mediator, a construction man or woman, meets the two sides first separately and then together, and tells each about the flaws and strengths of its case without disclosing the position of the opponent. If mediation is not acceptable, the parties may still go to law or to arbitration.

medical air lock, decompression chamber, compression c., hyperbaric An air chamber consisting of a steel cylinder of 2 m dia. and about 5.5 m long, closed at one end, with airtight doors (or one door) at the other end. It is connected to a compressed-air supply and fitted with a bed, telephone, pressure gauge and clock. See **caisson disease**.

medium-carbon steel [mech.] See **high-carbon steel**.

meeting post, mitre p. [hyd.] Vertical timbers at the outer end of each of a pair of lock gates. They are chamfered so that they fit each other to close the gate.

mega- A prefix meaning one million times.

Mekometer An *EDM* instrument developed at the UK National Physical Laboratory and commercially made by Kern of Switzerland. The latest (1988) model can measure lengths up to 5000 m with an accuracy of 1 in 1 million or better.

member [stru.] A structural member is a wall, *column*, *beam* or *tie*, or a combination of these. Any structure or frame consists of a number of members fixed together.

membrane [stru.] A thin film or skin, such as the skin of a soap bubble or a waterproof skin. See **sealant** (3), **slip membrane**.

membrane analogy [stru.] In the determination of the stress in a twisted bar, lines of equal shear stress are well represented by the contour lines of a soap bubble (membrane) of the same shape. There is therefore an *analogy* between these contour lines and the stress lines. This enables the lines of stress to be more easily understood than would otherwise be possible.

membrane curing of concrete Spreading tarry, resinous, waxy or plastic sheets or substances over wet concrete to prevent it drying and improve *curing*. See **sealant** (3).

membrane processes The four main *desalination processes* that use a filtering skin or membrane are *reverse osmosis*, *ultrafiltration*, *microfiltration* and *electro-dialysis* (ED). ED filters *ions* inwards, the other methods filter them out. The membrane is always artificial, often cellulose acetate, but many other synthetic organic membranes are used. All can purify water to better than 500 ppm of dissoved solids, and they are becoming increasingly popular.

membrane theory [stru.] A theory of design of thin *shells* which is based on the assumption that the shell cannot resist bending because it deflects like a balloon or bubble. Only shear stresses and direct tension or compression therefore can exist in any section. The conditions necessary for a membrane state of stress are uniform loading and edge support, and uniform or smoothly varying thickness and curvature.

meniscus See **surface tension**.

meridian, m. of longitude, m. plane, true m. [sur.] A plane passing through the earth's axis of rotation and the point under consideration. It coincides with the true north–south line through the point.

meridian passage [sur.] The *culmination* of a star.

Merrison committee See **box girder**.

mesh (1) A woven wire cloth used for screening sand or gravel or for the laboratory *mechanical analysis* of soils. See **fineness modulus**.
(2) Light woven or welded *wire-mesh reinforcement*.

metal Broken stone for making roads.

metal-arc welding Electric-*arc welding*, often with a consumable metal electrode, generally using low-voltage DC. See **manual metal-arc**, **automatic welding**.

metallic aggregate Steel or cast-iron punchings, filings or chips, sometimes used for making concrete twice as dense as usual – up to 4.8 tons/m.^{3}, e.g. for a *biological shield*.

metallic-electrode inert-gas welding. See *MIG welding*.

metalling A road surface.

metallurgical cement *Supersulphated cement*.

meteoric water Rain, snow, hail, dew, etc.

meter (1) An instrument which measures the flow or quantity of a fluid. See **current meter, flow meter, integrating meter**.
(2) US spelling of *metre*.

methane CH_4, known as **marsh gas** in marshes or **firedamp** in coal mines. Firedamp is released from coal seams and the rocks around them into the mine air. It occurs also in the natural gas from oil formations, and has been ignited during tunnelling under the Thames, in the Orange-Fish *tunnel*, for the north anchorage (in dolerite) of the *Forth Road Bridge*, and in 1984 at the Abbeystead waterworks in Lancashire, killing 16 visitors and injuring 28. Methane can explode in air at any percentage between 5% and 14%; cf. **petroleum vapour**. See **flameproof enclosure, intrinsic safety, landfill gas**.

method of slices [s.m.] A method of calculating the stability of an earth slope. The wedge of soil is divided into vertical slices of uniform width. The equilibrium of the whole wedge is calculated by summing the effects on all the slices. Cf. **circular-arc method**.

methyl methacrylate A binder for *polymer concrete*, which bonds well, does not stick to tools, and 'feathers' to zero thickness. In winter it can be used even at $-10°C$ ($14°F$), and in summer up to $38°C$ ($100°F$). It resists water and salt. Unfortunately the aggregates must be bone dry before mixing (eight hours' drying), the liquid is flammable and it stinks horribly.

metre (meter USA) The length unit that is now coming to be used throughout the world. It is equal to 39.37 in., 1000 mm or 100 cm. 1000 m are equal to 1 km. See also **meter**.

metric system A system of weights and measures devised in France in 1795 which is decimal and fully interrelated. Thus a cubic metre of water weighs 1 tonne (1000 kg), and a litre of water 1 kg. A pressure of 1 kg/cm^2 is equivalent to that of a column of 10 m of water (1 atm, 1 bar or 0.1 MP). The current version is the SI system.

micro- A prefix meaning one millionth. See **microcomputer**.

microcomputer, micro By far the commonest type of computer, with all the essentials of a large computer. Every micro has peripherals including a keyboard resembling that of a typewriter, as well as a monitor or television screen, a cassette recorder or disk drive for its *floppy disk* or minidisk memory, and usually also a printer.

microfiltration A *membrane process* for desalination of water to high purity, using filter cartridges which start off creating a pressure drop of only 0.1 bar and are thrown away or washed for reuse when the pressure drop rises to about 3 bar. Microfiltration is best for the total removal of *suspended solids*, but the cartridge filters cannot remove more than about 100 ppm of particles in the size range 0.04–100 microns.

microfloppy, minidisk, microdisk See **floppy disk**.

micrometer (1) A measuring device which works on the principle of the *micrometer gauge*.
(2) [sur.] A fitting on the eyepiece of a good-quality *dumpy level* which enables the levelling staff to be read to high precision.

(3) One *micron*, 0.001 mm, sometimes written 1 μm.

micrometer gauge, micrometer [mech.] A length-measuring instrument which measures to one thousandth of an inch or less. It consists of a G-shaped steel frame, one leg of which is a round bar accurately threaded with 100 divisions per inch. A nut running on it is marked outside with lines which represent 0.001 in. Precise micrometers are obtainable with metric graduation.

micron (1) **micrometer** One thousandth of a millimeter, 0.001 mm, now written 1 μm.

(2) A pressure equivalent to 0.001 mm of mercury, written 1 μmHg.

microorganisms, microbiota [sewage] The purification of sewage in *biological treatment* is largely done by these creatures too small to be seen by the naked eye. The smallest are viruses, then come bacteria. Larger yet are the first primitive plants, *algae*, visible only because they accumulate in millions on pond surfaces. Useful, in the same way as any plants they convert the carbon dioxide in the water into oxygen. Larger again are mosses, ferns and the first tiny animals – rotifers and crustaceans – feeding on each other. Skilful sewage engineers ensure the right balance of these populations by their treatment, encouraging one creature to harvest another. It is only under the microscope that these organisms can be seen as individuals rather than masses.

micropile, minipile, pin pile Small piles, usually *bored*, some as small as 60 mm (2½ in.) dia., have been used in *soil nailing* to pin through a *slip surface* (soil dowelling), thus strengthening the soil mass. They are useful also for *underpinning* a bridge or quay wall, sometimes by drilling a sloping hole. The preferred slope is up to 15° from the vertical. The length should not exceed 75 times the diameter. (*BRE Digest* 313, 1986.)

microptic theodolite [sur.] An extremely compact, lightweight *theodolite*, weighing only 4 kg, with an 11.5 cm long telescope. Both the horizontal and the vertical circles are of glass, and have optical reading to 1 second of arc.

microscope, optical m. An instrument which gives a magnification of several hundred times and is indispensable to the mineralogist. See **petrographic microscope**.

microsilica (MS), silica fume, condensed s. f. (CSF) (see diagram) A by-product from the smelting of silicon and ferro-alloys, amounting annually to about 200,000 tons from Norway and 500,000 tons from the USA and Canada. This extremely fine *artificial pozzolan* first aroused interest as an air pollutant, a *fume*. Its fineness explains why it is so much more reactive than *fly-ash*. Its specific surface is about 20,000–25,000 m²/kg or 50–100 times as much as a good Portland cement, and it is therefore that much more difficult to handle than fly-ash, which is relatively coarse, so it is often sold as a slurry in water, 50/50 by weight of dry solid; the slurry has a density of about 1.3. Its water demand is not high if used in proportions of less than 10% by dry weight of cement and it does not then need a *water reducer* or *superplasticizer*. These tiny particles, properly used, reduce the *segregation* of concrete and improve its strength, chemical resistance, imperviousness and durability. It is even claimed that microsilica provides a more hard-wearing surface than a *granolithic* topping. In Iceland, with its hard conditions for exposed concrete, cement is made with 7–8% of CSF to reduce *alkali–aggregate reaction*. Though microsilica is

outline of pore-blocking calcium silicate
hydrate (CSH) resulting from chemical combination
between the free lime in cement and the
spheres of microsilica

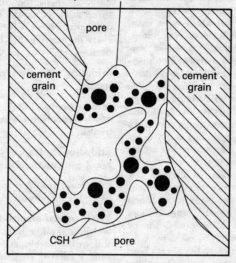

Presumed mechanism for the impermeability of concrete containing microsilica. (L. Hjorth, *Nordic Concrete Research*, No. 1, 1982 (from a photomicrograph).)

more expensive than cement, builders in areas like Quebec or Norway, where CSF is readily available from smelters, use it merely to save money, as at *Seattle*. For the best results, the coarse aggregate should be added to the mixer first, then the slurry and the water. After a short mixing the cement is added, then the sand. This takes slightly longer than standard mixing, but it improves the plasticity and final strength of the concrete.

microstrainer, microscreen [sewage] A drum of very fine stainless steel wire mesh rotated on a horizontal shaft and partly submerged in the effluent or water being screened. This common method of *tertiary treatment* removes very fine solids and is an alternative to sand filters. Often, when clean drinking water is drawn from an upland stream, this may be the only treatment apart from *chlorination*. But to maintain throughput during maintenance, extra microscreens are needed. Ultraviolet lamps at the screens prevent the build-up of *algae* and other *microorganisms*.

microtunnel, non-man-entry tunnel Any machine-made tunnel too small for a person to work in. The absolute minimum, even for someone small, is 90 cm (3 ft). A microtunnel must usually start from a *launch pit* and finish in a *reception pit*. *Pipe jacking* methods have been used to drive very small pipe without excavating. See **microtunnelling machine**.

microtunnelling machine, microtunneller A machine for driving a tunnel smaller than a person can work in (900 mm, 3 ft). Many types exist, and they may include *augering, jetting, pipe jacking, fullface tunnelling machines*, etc. In 1988 a small microtunneller of 135 mm (5⅜ in.) dia. could drive lengths up to 150 m (490 ft). The microtunneller with its rotating electrical cutting head has to drag an electric cable, a high-pressure water pipe and a pipe containing bentonite slurry to lubricate the walls of the tunnel. It cannot drill rock or loose gravel.

At this small diameter the cost was estimated to be twice the cost of trenching. But in Berlin in 1988 microtunnels at 150 mm (6 in.) dia. were cheaper than trenching. Microtunnels can be enlarged indefinitely by reamers.

Since deep trenching has always been more expensive than shallow trenching there has hitherto been no incentive to lay pipes or cables deeper than the minimum 1 m. This situation has now changed because a deep microtunnel is little more expensive than a shallow one. When a microtunneller meets an obstacle and cannot advance, it is usually withdrawn, and an attempt is made to drive more deeply. The important implication is that records of the locations of pipes and cables must now be more accurate than formerly to avoid conflicts underground. See **supersonic air knife**.

middle strip [stru.] That part of the *reinforcement* layout of a *plate floor* or *mushroom slab* which is half the span in width. It usually has less steel than the *column strip*.

middle third [stru.] That part of the thickness of a wall or arch which is one third of the total thickness and is cen-

tral in it. If all the forces, including the wall weight, combine to form a resultant which is everywhere within the middle third there can be no tensile stress in the wall. This is aimed at in the design of walls without reinforcement, whether brick, masonry or concrete. See **core**, **gravity dam**.

middling frame See **tucking frame**.

mid-ordinate [d.o.] An ordinate which occurs half-way between the extremes.

midpoint, midrange [stat.] The average of two extreme observations.

midspan [stru.] A point half-way between the supports of a beam.

MIG (metal-electrode inert-gas) welding *Welding* of *light alloys, mild* and carbon steels, and other metals with a *covered electrode* to suit the metal, and with inert-gas shielding. Cf. **MAG welding**.

mil [air sur.] 0.001 radian.

mild steel [mech.] (see diagram) Steel containing from 0.15 to 0.25%C, which because of its low carbon content cannot be hardened by quenching but is much more ductile than *high-carbon steel*. Steel made to BS 4360, 'Weldable structural steels', has a yield point of 230 N/mm^2, but higher yield points are obtainable and the standard also states the chemical analyses of the steels concerned. Mild steel is used for making pipes, joists, etc. See **dead-mild/high-tensile steel**, **round bar**, **yield point**.

mill (1) [mech.] A machine for grinding or crushing, such as a ball mill, hammer mill or pug mill.

(2) A building containing (particularly woodworking) machinery.

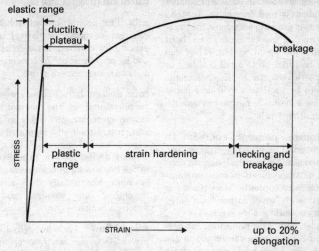

Idealized stress–strain curve for mild steel in tension. (D. Nethercot, *Limit States Design of Structural Steelwork*, Van Nostrand Reinhold, 1987.)

millimetre Abbreviation mm. One thousandth of a metre. 25.4 mm = 1 in.

millimicron One millionth of a millimetre, formerly written mμ or μμ. See **micron**.

milling [mech.] Removing metal shavings from a surface by pushing it on a moving table past a rotating toothed cutter.

milling machine [mech.] A *machine tool* for *milling* invented by E. Whitney in 1818. It can be an extremely complicated machine used for making drills and gears as well as for cutting channels and making plane surfaces.

mill scale Black iron oxide formed on steel sections during *hot rolling*. Because of this scale, bars which come untreated from the rolling mill are called *black* bars.

mill tail [hyd.] A *tail race*.

millwright A skilled man who could handle an axe, saw or plane, as well as forge, turn or bore iron. He understood machines, geometry, levelling and the layout of watercourses. See **early civil engineers**.

mineral admixtures *Pozzolans*, natural or artificial. True *admixtures* never amount to more than 5% of the weight of cement and pozzolans can amount to 30%, so pozzolans truly are *cement replacements*.

mineral separation, m. beneficiation Treatment of a mineral to improve its value and remove as much as possible of the useless rock.

miner's dip needle A *dip needle* which shows the dip of the earth's magnetic field and the presence of magnetic material in the ground.

minidisk, microfloppy See **floppy disk**.

mini-excavator An excavator smaller than about 5 tons in weight. In 1988 the fastest growing section of the excavator market was for those from about 700 to 1300 kg in weight. They can easily be lifted on to a tall building or other awkward site by another crane, there to be used as a *hydraulic power pack* for a *hydraulic hammer* in demolition or other work.

minimum cement content To ensure good *workability*, *compaction*, low *permeability* and long-lasting concrete, most specifications insist on minimum cement contents. More cement is needed with small rather than large aggregate, and for sites exposed to rain, seawater, etc. The minimum for *prestressed concrete* varies from 300 to 410 kg/m³, for reinforced concrete from 260 to 410 kg/m³, and for plain concrete from 220 to 350 kg/m³. These values can be reduced by 30–40 kg/m³ where the W/C ratio is under strict control (BS 5400). For precast piles to be driven hard, the minimum is 400 kg/m³ with a 28-day strength of 40 MPa (5800 psi); for normal driving the minimum is 300 kg/m³ with a 28-day strength of 25 MPa (3600 psi). *Trémie* concrete should have at least 350 kg/m³. *Fly-ash* or other *cement replacements* are not counted in the calculation. Cf. **maximum cement content**.

minipile See **micropile**.

minus sight [sur.] This term is recommended by the ASCE instead of the term *fore sight* in levelling.

misclosure [sur.] *Closing error*.

miser [s.m.] A large hand *auger* for exploring loose soils.

mistake [sur.] See **gross error**.

mitre post See **meeting post**.

mitre sill [hyd.] A *lock sill*.

mix The proportions of a batch of concrete (or mortar or plaster), usually varying for concrete between 200 and 400 kg of cement/m³ of concrete, with a stated water/cement ratio.

mix design [stru.] The choice and proportioning of the cement, sand, coarse aggregate and water in a concrete mix, usually the responsibility of a specialist civil engineer. This specialist may design for low heat evolution or for low cracking, not necessarily for strength, but usually does aim at a minimum *cube strength*, and may specify a *minimum cement content*.

mixed-flow turbine [hyd.] An inward-flow reaction *water turbine* in which the water acts on the runners both radially and axially. Various types exist.

mixed liquor In an *aeration tank*, the sewage *effluent* containing *flocs*.

mixer See **concrete/soil mixer**, **pug-mill** (*B*).

mix-in-place [s.m.] *Soil stabilization* in which the soil is mixed by a *travel mixer* without being removed from the site. Cf. **plant mix**.

MMA welding *Manual metal-arc welding*.

mnemonic [d.o.] A device to aid the memory. Mnemonics are often used in engineering formulae; the symbols being chosen so that they are easily remembered. One engineering mnemonic is the *bending formula*:

$$\frac{M}{I} = \frac{f}{y} = \frac{E}{R} \text{ remembered as } Mifyer$$

Others are:

Thirty days hath September,
April, June, and November;

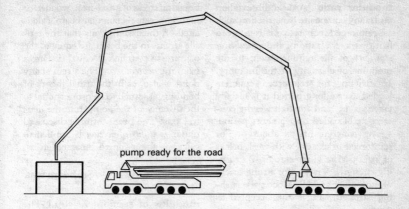

pump ready for the road

Mobile concrete pump.

All the rest have thirty-one
Excepting February alone.

Sir Humphry Davy
Abominated gravy.
He lived in the odium
Of having discovered sodium.

mobile concrete pump (see diagram) A truck-mounted *concrete pump* may place concrete with a long, folding placing boom having four hinges, and able to slew 340°. The capacity varies up to 150 m³/hour, the horizontal reach from the centre of slew is from 12 to 26 m (39–85 ft) and the vertical reach is from 17 to 30 m (56–99 ft). *Truck mixers* or *agitating trucks* supply it with ready-mix.

mobile crane, truck-mounted c. A crane which is driven by petrol or diesel engine and travels on either *crawler tracks* or rubber tyres. When lifting at one point it is usually stabilized by *outriggers*, often with jacks to lift it off its tyres. One large British mobile crane is powered by two diesel engines providing electric power to the separate electric motors of each motion.

Its biggest lifting capacity is 200 tons at 4 m radius. Not all mobile cranes can slew through 360°. Loaders although fully mobile are not called mobile cranes; nor are *locomotive cranes*. Mobile cranes generally weigh only about twice the maximum load which they can lift.

modal split In traffic analysis, the proportion of all journeys made by the different methods (modes) of travel available. A modal-split model is a way of forecasting these proportions.

model analysis, modelling Creating either mathematical (written) or visual, small-scale (often architectural) shapes of a proposed construction to help in understanding it. A third type is the *harbour model*. See also **deformeter, dimensional analysis, photo-elasticity**.

modem (modulator–demodulator) An electronic device connected between a computer and a telephone. It enables data to be sent between computers. See **terminal** (2).

modular ratio [stru.] (abbreviation *m*) In any composite construction such as reinforced concrete or reinforced brickwork, the ratio of the *modulus of elasticity* of the reinforcement to the modulus of elasticity of the masonry. In Britain, for reinforced concrete loaded over a long period it is usually taken as 15, and for lightweight concrete 30. Loaded over a short period, strong concrete has *m* about 6. For reinforced brickwork with weak bricks of 10–15 MPa (1450–2200 psi) crushing strength, *m* is 33. For strong bricks of more than 55 MPa (8000 psi) crushing strength, 12 is the accepted *m* value.

modulus of elasticity, stiffness, Young's modulus (*E*) [stru.] For any material the ratio of the stress (force per unit area) to the strain (deformation per unit length). It is expressed in units of stress, and is usually constant up to the *yield point*. (See **Hooke's law**.) The values for some common materials in GN/m² are about as follows: steel 200; *light alloys* of aluminium, 69, of magnesium, 45; greenheart, 23; Douglas fir, 11; English oak, 10; most softwoods about 10; English ash and beech, 13; resin-bonded chipboard, or *lightweight concretes* (of density around 960 kg/m³) about 3. See **elastic constants, grades of steel and concrete**.

modulus of incompressibility [s.m.] The ratio of the pressure in a soil mass to the volume change caused by the pressure.

modulus of resilience [mech.] See **resilience**.

modulus of rigidity, shear m. [stru.] The ratio of the *shear stress* to the *shear strain* in a material. See **Poisson's ratio**.

modulus of rupture [stru.] The breaking stress of a cast-iron, wooden or mass concrete rectangular beam, calculated on the assumptions that the tensile strains in the beam are equal to the compressive strains at equal distances from the *neutral axis*. The strain at any point is also, as in the usual theory of bending, assumed to be proportional to the distance of the point from the neutral axis. The beam formula discussed under *modulus of section* is applied to calculate the assumed breaking stress.

Breaking stress $f = \dfrac{M}{Z}$. See **beam test**.

modulus of section, Z [stru.] The second moment of area of a beam section (also called its *I* or *moment of inertia*) divided by the distance *y* from the extreme fibre to the neutral axis, used in the formula $f = \dfrac{M}{Z}$. For a symmetrical beam there is one section modulus:

$$Z = \frac{I}{y = \text{half beam depth}}$$

but for an unsymmetrical beam there are two, one corresponding to each distance from the neutral axis. (*M* = bending moment, *f* = stress.) See **mnemonic, modulus of rupture, plastic modulus**.

modulus of subgrade reaction, *k*
A design method for foundations of roads, etc. based on the assumption that the ground sinks in proportion to the load on it after the first 1.3 mm settlement of a 30 in. (76 cm) loading plate (*ACI*). It is used in Westergaard's method of designing road foundations. Soil pressure = *k* × settlement.

modulus of volume change, coefficient of v. decrease [s.m.] The *coeffi-*

cient of compressibility divided by (1 + initial *voids ratio*).

Mohr's circle of stress [stru.] A graphical construction which enables the stresses in a cross-section oriented in any direction to be easily determined if the *principal stresses* are known. It is commonly used for determining stresses in two directions, but also, with slight further complication, can be used for determining three-dimensional stresses. See **ellipse**.

moil See **gad**.

moisture barrier See **damp course**, **vapour barrier** (*B*).

moisture content [s.m.] The weight of water in a soil mass divided by the weight of dry solids and multiplied by 100. See **optimum moisture content**; also (*B*).

moisture index [s.m.] See **index of liquidity**.

moisture meters Electrical meters exist for determining the moisture content of concrete, timber, plaster, etc., but they generally measure only the moisture of the surface material.

moisture movement A property which causes a material to increase in length when its *moisture content* increases. Most of these materials also shrink when their moisture content falls (all soils, cement products, timbers, wood products, bricks). The reversible moisture movement of concrete or concrete bricks varies from 0.01 to 0.055%, the first and non-reversible drying *shrinkage* being from 0.02 to 0.08%. The moisture movement of clay building bricks is usually below 0.01% and that of sand-lime bricks may vary from 0.001 to 0.05%. Clinker concrete may have up to 0.2% and for this reason clinker blocks must be kept dry before they are built in,

otherwise severe cracking of the plaster over the blocks must be expected when the blocks dry out as the house is warmed by its occupants. Redwood along the grain has 0.17%, but across the grain the much larger value of 0.7%. This is the expansion which causes panelled softwood doors to be too tight in winter if they fit the frame well in summer. Soils, particularly clays, expand with increasing moisture content. *Clays* shrink with decreasing moisture content until the *shrinkage limit* is reached. See **bulking, equilibrium moisture content**.

mold US spelling of *mould*.

mole (1) A *breakwater*.
 (2) Any *microtunnelling* or even large *tunnelling machine*, but usually a *pneumatic impact mole*.

mole drain A drainage slot cut in a stiff clay by drawing a *mole plough* through it. This method is cheaper and quicker than laying *field drains* and lasts for many years.

mole plough A vertical knife blade carrying a horizontal bullet shape at its lower end. The bullet is adjustable in diameter from 50 to 175 mm (2–7 in.), and the knife can be shortened or lengthened to suit the *gradient* of the *mole drain*. Mole ploughs may be tractor-mounted or wheeled and towed. A caterpillar tractor making an 8 cm dia. mole drain at 0.6 m depth uphill (and this is the correct direction) needs to be of at least 50 hp. Mole ploughs have also been used for laying copper or polythene water-pipe by towing it on a rope behind the mole plough. The pipe is uncoiled from a drum set up at the exit from the drain to the ditch. Speed of laying can reach 36 m/min, and diameters up to 35 cm can be laid to a depth of 2 m.

moling *Microtunnelling*, *tunnelling*, *pipe renewal* or using a *mole plough*.

moment distribution [stru.] A method of solving the *bending moments* in continuous beams and redundant frames by successive approximations. Every span is first assumed *end fixed* at each support. Each end fixing is then relaxed in turn and the released support moments are distributed to the different supports in proportion to the *stiffnesses* of the spans, the stiffness being the $\dfrac{I}{L}$ (inertia/length). See **Hardy Cross**

moment of a force [stru.] The turning effect of a force about a given point. It is equal to the force multiplied by the shortest distance between the force and the point. See **bending moment**.

moment of inertia, I, second m. of area [stru.] About an axis in the plane of a section, it is equal to the sum of the products of all the elementary areas times their distances squared from the line. When no axis is specified, the least moment of inertia is usually intended, that is, the I about an axis through the centroid. This is the I of the *bending formula*. Its unit is length to the 4th power. See **polar moment of inertia**, **static moment**.

moment of resistance [stru.], **resistance moment** The couple produced by the internal forces in a bent beam when it is bent to the highest allowable stress. It is the highest bending moment which the beam may carry.

momentum [mech.] The product of the mass of a body and its velocity.

monitor [min.] A *giant*.

monitoring well A well sunk only to watch the water quality, not to draw water for consumption. Cf. **disposal well**.

monkey (1) A *drop hammer*.
(2) Scots term for the trip gear which detaches the drop hammer from the rope which raises it.

monocable An *aerial ropeway* with one rope which fulfils the double function of haulage and track rope. A large monocable installed in 1930 from Tilmanstone colliery to Dover harbour delivered 120 tons of coal per hour to dockside bunkers. This unusually long monocable passed through a tunnel, and was laid out in two sections of 6.5 and 5.2 km.

monolith A hollow masonry mass, usually circular or rectangular and of concrete, sunk as an *open caisson* and excavated by grabbing crane. It has several compartments or wells which are sealed to the bedrock or firm ground when this is reached, often by filling with concrete. A monolith may also be sunk as a *box caisson* or be built up from the foundation within a cofferdam or *pneumatic caisson*. A cylinder is a circular monolith.

monolithic Forming a single block without joints, a description often applied to the whole of a reinforced-concrete structure between successive *movement joints*.

monolithic topping, m. screed A *screed* placed on a concrete slab within an hour or so of casting, to ensure perfect *bond* between the two. But on old concrete, if a bonding layer is laid on the roughened surface, equally good monolithic action is possible.

monorail Many types of monorail exist, the simplest being one for placing concrete, a deep rail laid on sleepers on the ground. But most monorails are overhead – used as people-movers – or in tunnels for transport or at the face of a tunnel drivage carrying a *backhoe* to dig out the face. See **telfer**.

monotower crane A *tower crane*.

montmorillonite Sodium montmorillonite is a *clay* mineral found in *bentonite*. It swells when mixed with water, making a *thixotropic* suspension. But it can be the cause of *swelling soils*.

monument, beacon [sur.] A large stone or mass of concrete set by a surveyor to mark a corner or line of a site boundary or *triangulation* point.

mooring forces See **berthing impact**.

mortar A paste of cement, sand and water laid between bricks, blocks or stones, and usually now made with *masonry cement*, formerly with cement and lime putty. *Cement paste* can be regarded as the mortar of concrete. See also (*B*).

mortar lining See **pipe lining**.

mosaic [air sur.] A map made by fitting together a large number of *vertical photographs* which have been enlarged to the same scale. It is not called a map because its inaccuracy though slight may be more than that of a map of the same scale, or at least of a different sort. A mosaic is much more quickly made than a map.

motion study The study of the movements of workers with the intention of reducing their labour and increasing their output by laying out their work carefully or providing them with work-saving tools.

motor-generator set, genset An internal combustion engine, usually a piston engine, driving an electrical generator. The smallest size is about 1 kW, and such a unit can be easily carried by one person. For a construction site away from mains electricity, gensets are essential, providing the power for electrically driven tools, welding sets, etc.

motor grader A *grader*.

motor starter [elec.] See **starter**.

motorway (interstate USA) A road with *dual carriageways*, limited access even for abutting landowners, and hard shoulders. No vehicle may travel at less than 50 mph (80 kph).

mould A temporary, usually wooden structure built to hold concrete or plaster while it is setting. The use of this word is usually restricted to factory work. For site concrete the word *formwork* is more usual.

mouldboard In a farmer's plough, the steel blade which turns over the slice of earth. Hence the flat blade of a *dozer* which pushes the earth in front of it. Mouldboards may be straight or angled. They can usually tilt as well as roll, now that they are hydraulically controlled.

mould oil Oil, soft soap, worthless paint or similar liquids which are not absorbed by *formwork*, laid over it to prevent it sticking to concrete. See **release agent**.

mound breakwater A *rubble-mound breakwater*.

mountain railway [rly] A railway so steep that trains are pulled by ropes or a rack locomotive. See **locomotive haulage**, **rack railway**.

movable bridge [stru.] Many types of *drawbridge, swing bridge, traversing bridge* and others that can be moved to allow vessels to pass. They often have a metal plate deck of steel or *light alloy* stiffened by joists below it.

movable dam [hyd.] A dam of which part or all may be removed during floods to increase the flow past it. See **Fabridam**, **stop log**.

movement joints in concrete Joints have many tasks but any joint may fulfil more than one task. *Contraction joints* allow *shrinkage*, though in some of them the steel continues through the joint. The *expansion joint* is a complete break in both concrete and steel. *Sliding joints*, *hinges* and *settlement joints* also exist.

moving forms A term which has been used in several different senses – *climbing forms*, *travelling forms*, *slipforms*. According to *ACI* it means travelling forms.

moving load [stru.] A *live load* which is moving and so, for *structural design*, must be increased by an *impact factor*.

MTBF; MTBR; MTB(U)R; MTTF; MTTR Mean time between failures; mean time between repairs; mean time between (unscheduled) removal (of a replaceable part); mean time to failure; mean time to repair.

muck Waste rock, rubbish or excavated earth.

mucking, m. out [min.] Removal of *waste* rock in tunnelling.

muck-shifting plant See **earth-moving plant**.

mud flow See **flow slide**.

mud flush [min.] *Drilling fluid*.

mud jacking Boring a hole through a concrete road slab which has sunk and connecting it to a mud jack by flexible pipe. The mud jack is a lorry on which is a mixer for water–soil–cement slurry and a pump which forces the mix under the slab and raises it (USA).

mudsill A *sill*.

mud slab, m. mat *ACI*'s terms for the UK's *blinding concrete*.

Mulberry harbour A harbour made on the north-west coast of France in June 1944 by the invading Allies, who towed precast concrete quays from England and sank them in position offshore.

multi-bucket excavator, scraper e. A machine like a *bucket-ladder dredger* designed for digging sand or ballast in long cuttings for road, railway or canal excavation. One large machine digs 80 m^3 per hour on a slope 8 m high. The bucket chain is sloped at the angle of repose or slightly flatter. The driving engine is at the upper end, which is the delivery end of the bucket ladder, and is carried on a truck moving on rails along the top of the excavation. Cf. **rotary excavator** (2).

multi-face tunnelling machine Two or more *fullface tunnelling machines* can dig overlapping tunnels, with one machine ahead of the other. This is claimed to save 13% of the excavation needed for one large circular tunnel. *New Civil Engineer*, 10 Nov. 1988.

multi-layer filter, multi-media f. A *sand filter* for drinking water in which usually the top layer is anthracite. After backwashing the anthracite remains at the top because it has only half the density of sand.

multi-phase flow [hyd.] Underground this refers to the movement of oil, gas and water, sometimes also to fresh water, brackish water and salt water.

multiple-arch dam [hyd.] A *dam* built of repeating arches whose axes slope at about 45° to the horizontal. The arches are carried on parallel buttress walls. This is a lightweight dam suitable for a weak foundation, as for example the dam of La Girotte at 1750 m altitude in the French Alps. See **flat-slab deck dam**, **round-headed buttress dam**.

multiple-dome dam [hyd.] A development of the *multiple-arch dam*.

multiple-expansion engine [mech.] An engine driven by steam or compressed air which expands in two or more stages. A *compound engine* has two, a triple-expansion engine three, a quadruple-expansion engine four stages. See **compounding**.

multiple wedge [min.] The *plug and feathers*.

multiplying constant [sur.] In *stadia work*, a constant by which the staff *intercept* is multiplied to give the distance between the staff and the *tacheometer*. It is usually 100.

multipurpose bucket See **four-in-one bucket**.

multi-wheel roller A *pneumatic-tyred roller*.

municipal engineering The design and maintenance of roads, streets, sewers, water supply, public transport, lighting, airfields and other public services for a town, with concern for the environment.

mushroom construction, m. slab, flat slab [stru.] Reinforced-concrete solid slabs carried by columns which may be *flared* at the top but are not joined by beams. The slabs may be thickened round the columns with *drops*. The *plate floor* and other floors with a flat soffit are now preferred, both by architects and by contractors.

mushroom slab See **mushroom construction**.

mushroom valve, poppit v. A mushroom-shaped valve seated on an opening in the floor of a water or *effluent* tank, and fitted in tanks based below ground level. It prevents outward leakage while the tank is in use but allows inward leakage when it is empty, preventing flotation or damage to the tank.

N

nailable concrete *Aerated concrete* or concrete made with *lightweight aggregate*, sometimes containing sawdust.

nailer *ACI*'s term for a *nailing strip* (*B*) made of wood or other nailable material that may be cast into concrete or bolted to steel as a base for fixings.

nail gun An electrical or compressed-air machine which drives nails from a strip or magazine with no need for someone to hold the nail.

nappe The sheet of water flowing over a *weir* or *dam* crest.

narrow gauge [rly] A railway gauge narrower than 1.435 m, the *standard gauge*.

National Measurement Accreditation Service (NAMAS) Operated by the National Physical Laboratory, NAMAS is a merger of NATLAS, the National Testing Laboratory Accreditation Scheme, and BCS, the British Calibration Service, which together have accredited some 400 very varied laboratories, including those concerned with concrete technology.

National Rivers Authority (NRA) NRA in September 1989 took over from the *water authorities* (which then became *water service plcs*) their duties for control of *pollutants* and of *river basin management*, including flood prevention. It can prosecute and fine water service plcs for pollution or non-fulfilment of their duties. The NRA therefore now controls the Thames Barrage, a flood-prevention structure of Thames Water.

NATM (New Austrian Tunnelling Method) Two-stage *tunnel lining*, starting with *shotcrete* applied immediately after blasting and followed often by a thicker concrete lining cast in *formwork*. It was probably first used by Anton Brunner, an Austrian mining engineer, who patented it in 1958, but has been successfully used throughout the world.

natural asphalt *Asphalt* as it is excavated. See **natural rock asphalt**.

natural frequency of a foundation The frequency of *free vibration* of a foundation-soil system. It is important that this frequency should be appreciably different from that of any machines which the foundation carries, to avoid *resonance*.

natural gas [min.] Gas which flows from the ground and can be used as a fuel. It is mainly methane, with other paraffins and olefins.

natural harbour A harbour protected by the surrounding coast line. Cf. **artificial harbour**.

natural rock asphalt Rock such as sandstone or limestone containing *asphalt* or bitumen in its voids.

natural scale [d.o.] A drawing made to equal vertical and horizontal scales is said to be to natural scale.

Nautical Almanac [sur.] An astronomical calendar published annually several years in advance by the Astronomer Royal, for astronomers, surveyors and navigators, like the *American Ephemeris*.

Navier's hypothesis An assumption made use of by engineers in the design of beams. The stress (or strain) at any point due to bending is assumed to be proportional to its distance from the *neutral axis*. This hypothesis is nearly true and together with Bernoulli's assumption and Hooke's law it greatly simplifies beam calculations. Galileo made the first recorded attempt at formulating this idea early in the 17th century.

navigation [hyd.] A river canalized for shipping. See **slack-water navigation**.

navvy (1) A labourer who specializes in digging foundations, trenches, sewers, drains, railways and roads with pick, shovel and *graft*, and who may also be able to do concreting, timbering and tunnelling. The word is believed to be derived from the *navigation* labourers who dug the English canals in the 19th century.
(2) Any crawler-mounted crane with *excavator* attachments.

NDT *Non-destructive testing*.

neat lines, net l. The lines defining the sides of an excavation to be paid for in tunnelling. Any material removed beyond the neat line is *overbreak*.

necking [mech.] The *contraction in area*, accompanied by *elongation*, which occurs on a *ductile* test piece failing in tension. This *plastic deformation* occurs with good mild steel.

needle gun, Jason hammer [rly] A compressed-air tool with many needles that clean *thermit welds* before inspection.

needle traverse [sur.] A *compass traverse*.

needle valve [hyd.] A cone-shaped valve which ends in a point, used for regulating the flow to large turbines.

needle weir [hyd.] A fixed *frame weir* which carries heavy vertical timbers (needles) in contact. The timbers can be withdrawn as required to vary the water level.

needling To insert a *needle* (*B*) into a wall.

negative-buoyancy coating A concrete coating with steel mesh or polypropylene fibre reinforcement cast around a submerged pipeline to keep it safely sunk when it contains anything lighter than water – air, gas, light oil.

negative moment [stru.] *Hogging moment*.

negative skin friction [s.m.] *Fill*, unless well compacted, tends to settle. Therefore a pile or caisson driven through recent fill may be pulled further down by the settling material. This negative friction is added to the other vertical loads in the design of a deep foundation under fill.

negotiated contract A *contract* based on an agreement discussed between a *client* or *consultant* and a *contractor*.

Netcem A substitute for asbestos cement, made of *polypropylene fibrillated fibre* and cement, and backed by Shell Chemicals.

net duty, farm d. [hyd.] The water supplied to a farm expressed as the depth of water over the irrigated area.

net lines See **neat lines**.

net loading intensity [stru.] At the base of a foundation, the additional pressure caused by the weight of a structure, including backfill. Usually it is the difference between the load in kg/m² before digging starts (vertical earth load only) and the load per m² after the structure is complete and fully loaded. Cf. **gross loading intensity**.

net ton The US *short ton* of 2000 lb. (909 kg).

network [sur.] A *control network*.

neutral axis, n. plane, n. surface [stru.] In a beam bent downwards, the line or surface of zero stress, below which all fibres are stressed in tension and above which they are compressed. In homogeneous material the neutral axis passes through the centre of area (*centroid* of the section).

neutral pressure, n. stress [s.m.] The *hydrostatic pressure* in the pore water of the soil. See **effective pressure**, **hydrostatic excess pressure**.

neutron-absorbing aggregate, high-density a. Iron chips, steel punchings, lumps of barytes or iron ore can help to produce a concrete density of 4.8 tons m³, about twice as dense as common *dense concrete*.

Newlyn datum [sur.] The levelling datum now used by the *Ordnance Survey* of Great Britain, determined by several years of observation of mean sea level at Newlyn, Cornwall. It differs by more than 0.3 m from levels at different points (e.g. in London) based on *Liverpool datum*.

newton (N) The unit of force in the SI system, equal to about 0.225 lb. A pressure or stress of 1 N/mm² or a MN/m² is 145 psi, the same as a megapascal (MPa) or 10 bar (10 atm).

Nicol prism [min.] A crystal of Iceland spar which has been cut and cemented with Canada balsam so as to exclude the ordinary ray of light and to allow only the extraordinary, plane-polarized ray to pass through it. It is also called a polarizer or an analyser from its function in the microscopic analysis of rock sections. See **petrographic microscope**, **polariscope**.

night soil See **non-sewered sanitation**.

nip A wound caused to a diver by a fold of the diving suit at great depth crushing his skin. Nips are prevented by wearing several layers of very thick underclothing.

nitrates [sewage] A European directive limiting nitrate discharges by sewage works necessitates heavy investment by the privatized water companies. See **nitrification–denitrification**.

nitriding, nitrarding, nitrogen hardening [mech.] *Case-hardening* of alloy steels containing aluminium by holding at about 500°C for $2\frac{1}{2}$ days in ammonia gas to introduce nitrogen into the surface of the metal. No subsequent heat treatment is needed. It distorts the steel less than *carburizing* or *cyanide hardening*.

nitrification–denitrification [sewage] Nitrogen in sewage originates mainly as ammonia (NH_3), but oxidizes to nitrites ($-NO_2$ compounds) and *nitrates* ($-NO_3$ compounds). This oxidation to $-NO_2$ and $-NO_3$ (nitrification) can take place in *trickling filters* and *aeration tanks*. Nitrates and nitrites pollute drinking water and must be removed by denitrification, sometimes by conversion to nitrogen gas. Disposal on land in warm weather allows small quantities of nitrates and other *pollutants* to be removed by absorption into vegetation. But nitrates in drinking water wells from East Anglia to a line joining Gloucester to Burton-on-Trent affect four million people. The nitrates come from fields draining excess nitrate fertilizer or possibly from crops of clover or beans.

nitrocellulose, cellulose nitrate, guncotton [min.] A solid *high explosive* used with nitroglycerin in gelatine explosives.

nitrogen hardening See **nitriding**.

nitrogen method of freezing ground [s.m.] See **liquid nitrogen**.

nitroglycerin [min.] A liquid *high explosive* which evolves carbon dioxide, nitrogen, water and oxygen when detonated. It is the main part of *dynamite*.

nitrous fumes [min.] Reddish fumes of NO_2 and N_2O_3 which are produced when *nitroglycerin* explosives burn instead of detonating. They are poisonous and may be fatal to breathe.

node [stru.] A *panel point* in a framed structure, particularly a truss, where two or more members meet.

no-fines concrete See (*B*).

nominated subcontractor A *subcontractor* chosen by a *client* or *consultant*.

nomogram, alignment chart [d.o.] A diagram used like a graph for eliminating or shortening calculations. In its simplest form it consists of three straight lines each graduated for one of the variables in a relationship. By joining any two of them with a straight edge the third can be read off. One line showing the velocity of flow in a water pipe can be related to a second showing the internal diameter and to a third showing either the quantity flowing or the friction loss, or both (one value each side of the line). Nomograms are very much simpler to read than graphs showing the same amount of information, because they have no confusing background of squared lines. However, they give no picture of variation, as a curve does, and this is a disadvantage.

non-biodegradable That which does not rot, e.g. plastic bags, bottles or the pesticide DDT, which remains as a poison in animals.

non-cohesive soil [s.m.] A *frictional soil* (sand, gravel, etc.).

non-destructive testing (NDT) A branch of physics that includes any way of checking the quality of materials that does not damage them, e.g. the weighing of coins in a bank, the tapping of railway wheels to see if they 'ring', the filling of a length of drain with water to find leaks, etc. There are at least seven main groups of methods: electrical (conductivity, eddy current, electro-chemical, dielectric); magnetic; penetrant dye; *radiography* and *radiometry*; thermal; ultrasonic and sonic; and visual. See also **cover meter, destructive tests, hardness, rebound hammer, semi-destructive testing, strain gauge, ultrasonic testing**.

non-intrusive techniques See **trenchless pipelaying**.

non-linearity See **linearity**.

non-metallic minerals [min.] Abrasives, asbestos, asphalt, building stone, clay, lime, coals and carbon minerals, petroleum, gems, fossil gums, natural gas, pigments.

non-return valve [mech.] A *check valve*.

non-rusting reinforcement Various fibres of plastic, glass or stainless steel have been used in concrete, either random in lengths of less than 50 mm (2 in.) or in long lengths oriented along the unit. The difficulty with all plastics is that they stretch too much to be practical. But drawn polypropylene bar can achieve a strength of 500 MPa (72,500 psi) and both aramid and carbon fibre are stronger than steel, though these two would have to be embedded in resin to protect them from the alkalinity of concrete. In seawater bare galvanized steel rusts after two years and has to be replaced after five. See **Caricrete, epoxy-coated steel, fibrillated film, glassfibre, Netcem, Polystal, pultruding**.

non-sewered sanitation The cost and scarcity of water will probably prevent sewerage from reaching even half the people of the developing contries by 2050. Other methods are used where water is scarce. The contents (night soil) of earth, chemical or ash closets are removed at night by the local authority for burial. In flat water-logged areas of England, *vacuum sewerage* is less costly than gravity sewerage.

non-slip floor, non-skid floor. A concrete floor surface treated with iron filings or carborundum powder or indented while it is wet to roughen it.

non-tilting mixer A drum-shaped *concrete mixer* with two openings which rotates about a horizontal axis. The mixed concrete is extracted from it by inserting a chute which catches the concrete as it drops from the side baffles. There are several standard sizes in Britain, up to about 4 m³.

nonwoven Synthetic film used in soil engineering, sometimes formed with punched, rectangular or rounded openings. Not being woven, these *geosynthetics* can hardly be called *geotextiles*.

normal curve See **Gaussian curve**.

normalizing [mech.] Heating of steel to above the range of *critical points* followed by cooling in air. It softens the steel and makes it less brittle. Other metals also are normalized. Cf. **stress relieving**. See **heat treatment**.

normal law of error [stat.] The equation of the *Gaussian* error curve.

normal stress [stru.] *Direct stress*.

North Sea pollution Although other coastal states agreed to stop dumping sewage *sludge* into the North Sea in 1984, the UK continues to do so as well as releasing 1.4 million m³ of sewage into it daily. Sea birds are killed by oil slicks from oilfields and tanker ships. The Humber and other estuaries are poisoned by heavy metals and chemical wastes. Herrings have been scarce for years, mackerel are becoming

scarce, and up to 40% of the flounder, plaice and dab caught off Holland are cancerous. Not civil engineers but society as a whole is responsible for these disasters. See **river pollution**.

nosing [rly] A horizontal force applied by a locomotive perpendicularly to curved track, usually taken as 100 kN (10 tons) at rail level. Cf. **lurching**. See also (*B*).

no-slump concrete Very stiff concrete, which sinks less than 6 mm (1/4 in.) in the *slump test*.

notch, n. plate [hyd.] A small *measuring weir* with its upper edge above water level.

notched bar test [mech.] An *impact test*.

notched weir [hyd.] A *measuring weir*.

notch effect [stru.] The locally increased stress at a point in a member which changes in section at a sharp angle. Close to a right angle notch the stress can be three times as high as the average across the reduced section. Notches are therefore avoided in highly stressed members. See **impact test**, **stress concentration**.

notcher A machine in a steel-fabrication shop, which strips the flanges from the end of a rolled-steel joist.

notching (1) Excavating a cutting in a series of horizontal steps advancing in sequence.

(2) [mech.] Cutting a steel joist in *fabrication* with a *notcher*.

NO$_x$ Oxides of nitrogen, gaseous *pollutants* produced by furnaces, cars and other internal combustion engines.

nozzle Any opening in the shell of a *pressure vessel*. So as not to weaken the shell, the nozzle is often strengthened by a tube welded on the outside.

NPSH (net positive suction head)
The suction lift of a pump, the greatest
distance below it, from which it can
raise water.

N-truss [stru.] A *Pratt truss*.

nuclear decommissioning At least
two levels of nuclear decommissioning
exist. At the Shippingport reactor, 40
km (25 miles) from Pittsburgh, USA,
decommissioning began in 1983 with
the removal of reactor fuel and should
be completed in 1990, when the site
will have become 'green grass', with all
radioactivity removed. At Gentilly in
Canada, however, which is the first
commercial reactor to be decommis-
sioned, the reactor will be left in 'static
state', in other words still radioactive,
for 50 or 100 years until the next stage
of decommissioning. The job was completed in 1986 for about $25 million, and
took only two years. Three US reactors
apart from Shippingport have been de-
commissioned: Elk River in 1984, Los
Alamos in 1980 and Canoga Park, California, in 1982. These were all small
plants. Uranium mine tailings dumps,
that emit radon, have been capped by
the US Department of Energy with
seals claimed to last 1000 years.

For the large electrical generating reactors of the UK, decommissioning might
cost between £300 and £700 million.

nuclear density gauge A quick
gauge for concrete density, used on
site. It must be calibrated for each job,
but is used in *roller-compacted concrete*
where the *slump test* is useless and a
quick test is essential.

nuclear waste disposal Three types
of nuclear (radioactive) waste exist,
high-level, intermediate and low-level,
high-level being the most dangerous
and the smallest in bulk. The volume
of storage required for the disposal of
50 years of nuclear low-level and
intermediate-level waste in the UK
would require an excavation as large as
the Channel Tunnel. Intermediate
waste in the UK alone is accumulat-
ing at 6000 tons a year and low-level
waste at ten times this rate. From 1955
to 1982 disposal took place at sea in
depths below 2000 m in 100 litre metal
canisters enclosed in 10 cm (4 in.) of
concrete, until the National Union of
Seamen banned this method. Disposal
can be based on land or at sea, from
an *artificial island* or offshore platform. One proposal is for a number
of offshore boreholes of 15 m (49 ft)
dia. Whichever method is used, the
waste repository must be watertight
or radioactive waste may be leached
out. (*Society of Underwater Technology*,
Sept, 1988.)

O

oblique photograph, or aerial p. [air sur.] An air photograph taken with the camera axis inclined away from the vertical. A high oblique is one which shows the horizon, a low oblique does not include the horizon. Cf. **vertical photograph.**

oblique projection [d.o.] A pictorial view of an object showing its elevation, plan or section to scale with parallel lines projected from the corners (at 45° or any other angle) to indicate the other sides. See **projection.**

OBM *Ordnance Bench Mark.*

OD *Ordnance Datum.*

oedometer [s.m.] A *consolidation press.*

offset [sur.] A horizontal distance measured at right angles to a survey line to locate a point off the line.

offset scale [d.o.] A short scale used for plotting details on a map, which have been fixed by measured *offsets* in the field.

offshore construction, o. platform (see diagram) Drilling at sea is done either from drill ships or *semi-submersibles*, from platforms built on land and floated out, or by starting from an *artificial island.* In 1965 the deepest offshore drilling was in 193 m (632 ft) of water. By 1988 the greatest depth had reached 6 km (4 miles) in exploration drilling, but this was too deep for oil or gas production. Generally offshore platforms are specialized for exploration in the early stages and later when the presence of oil or gas is proven they become (or are superseded

by) production platforms which extract the gas or oil. See also **gravity platform, guyed tower, jacket, jack-up p., mat-supported p., subsidence, tension-leg p.**

off-tamp finish See **tamped finish.**

OFWAT The 'watchdog' of the privatized water industry, headed by the *director-general of water services*, with 79 staff to supervise 39 companies, compared with OFGAS which has more than 100 to supervise one company.

Ohio cofferdam A *double-wall cofferdam* built of two lines of vertical timbers, held by tie-rods across the gap, anchored to horizontal *walings* along the outside of each wall. It is built on land or near the shore, floated into the river and fixed by filling it with material dropped in by crane or pumped in by a suction dredger.

oil debris analysis A *condition-monitoring* technique in the maintenance of heavy machines, such as *excavators* or mining machinery, that has helped British Coal to reduce its time-based maintenance routines. In 1988 British Coal saved £84 for every £1 spent on such methods. The oil from sumps is drained, filtered and its debris examined to find both the rate of wear and the source of any metal fragments, using spectroscopic, metallurgical and magnetic methods – information which allows a prediction of when a part will need to be changed. (A. W. Tuke, *Mining Engineer*, Sept. 1988.)

oil-well cement A *hydraulic cement* (*B*) which sets much more slowly than

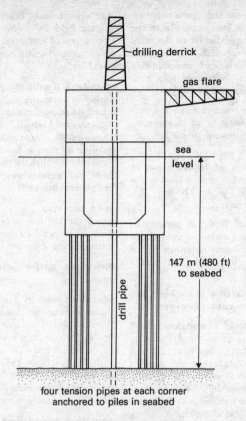

- drilling derrick

gas flare

sea level

147 m (480 ft) to seabed

drill pipe

four tension pipes at each corner anchored to piles in seabed

Offshore platform in the North Sea, Hutton Field, moored by tension legs (tethers). (D. A. Fee and J. O'Dea, *Technology for Developing Marginal Offshore Oilfields*, Elsevier, 1986.)

ordinary Portland cement at the high temperatures in oil-wells. A *casing* is fixed (cemented) into an oil well by lowering it to its full depth, then pumping the calculated volume of cement slurry followed by *drilling fluid* in through the drilling rods and up outside the casing until all the drilling fluid outside the casing is displaced by cement slurry. This is when the cement slurry begins to appear at the surface. Half an hour's pumping may well be needed before the slurry can be allowed

to set, therefore a slow initial set at high temperature is essential. See **slug**.

oil-well derrick See **derrick** (5.)

on center [d.o.] A US term for *centres*.

one-ply roofing *Single-ply membrane*.

one-way slab [stru.] A *reinforced-concrete* slab in which one pair of sides is much shorter than the other pair. The *main steel* therefore goes the short way to save steel – the slab spans one way. The *distribution steel* is parallel to the long sides, perpendicular to the main steel. A two-way slab, on the other hand, is one that is nearly or completely square, and its main steel goes both ways – the slab spans two ways.

opal A highly reactive form of amorphous *silica*, liable to *alkali–aggregate reaction*, therefore undesirable in a concrete aggregate.

open caisson A *caisson* which may be a *cylinder*, *monolith* or *drop shaft* and is open at both top and bottom. Cf. **pneumatic caisson**.

open channel [hyd.] A conduit in which the upper surface of the water is not in contact with the crown of the pipe but with the air. It cannot be a pressure pipe.

open cut Excavation in the open, not in a tunnel. See **cut and cover**.

open-face shield The original type of *shield*, with no steel *bulkhead* protecting the tunnel from the face. In soft ground, if the face is unstable, water or mud may rush in unless the tunnel is full of *compressed air*. *Earth-pressure-balance* or *slurry shields* are closed-face shields, with a steel bulkhead closing off the face. They have made compressed air unnecessary for keeping water out of a tunnel.

open-frame girder [stru.] A *Vierendeel girder*.

open-graded aggregate An *aggregate* containing large voids after compaction (*ACI*).

open hole, uncased h. A *borehole* without *casing*.

open jetty (see diagram) See **jetty**.

open sheeting (1) Vertical poling boards not touching each other, held up by walings and struts.

(2) Horizontal boards at open spacing held by *soldiers* and struts. Both types of timbering can be used only in ground which neither flows nor crumbles.

open-tank treatment Immersion of timber in warm zinc chloride solution, a method of preserving timbers which requires only small capital outlay. *Pressure-tank* treatment with zinc chloride solution is quicker and often more effective than the open-tank treatment but requires twice the capital. See **preservatives**.

open traverse [sur.] A *traverse* in which the last line is not joined to a point of known coordinates or to the starting point of the traverse.

open-web girder [stru.] A *lattice girder*.

operating system For a computer, the set of *programs* that make it work. Sometimes they are loaded in by disk or cassette tape, but vendors of small *microcomputers* usually provide them with the machine. Different systems are used for different tasks but any program bought must be compatible with the machine's operating system.

operation waste [hyd.] Water lost from an irrigation system through spillways or otherwise, after being diverted into it.

optical coincidence bubble [sur.] A *prismatic coincidence bubble*.

optical compensation [sur.] [1] *Autocollimation* of a level.
[2] The use of an *automatic compensator* in a *level* or the vertical circle of a *theodolite*.

optical distance measurement [sur.] *Stadia work*, or distance measurement with a *subtense bar* suitable for detail mapping. Electro-optical dis-

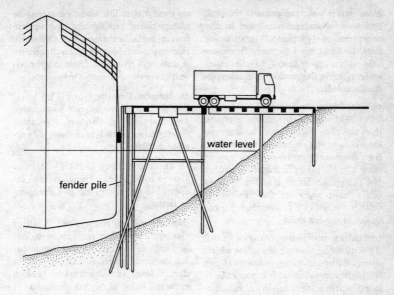

Open jetty, built on driven piles.

tance measurement, on the other hand, is a type of *EDM*.

optical microscope See **microscope**.

optical plummet, autoplumb [sur.] (1) An auxiliary telescope, of small magnification, now provided on most *optical-reading theodolites*, which has a line of sight, via prisms, through the centre of the hollow vertical axis of the theodolite. A centring device, giving some 25 mm of movement, enables the instrument, previously levelled, to be centred while the point on the ground (or in the roof) is being sighted through the small telescope.
(2) Other means of optical plumbing include the fitting of a pentagonal prism to the object end of the telescope, turning the horizontal line of sight

through 90°, thus eliminating the auxiliary telescope.

optical probe A device for looking into inaccessible places such as the inside of a water pipe or *sewer* or the cavity of a cavity wall which lets water through. See **endoscope, fibrescope**.

optical-reading theodolite [sur.] Modern theodolites that have largely superseded vernier or micrometer theodolites since they are smaller, lighter, more accurate, quicker and more convenient to read. Also, being totally enclosed, they need less maintenance than older instruments. The horizontal and vertical circles are of glass, which is more precisely graduated than the earlier circles of metal, and the physical centre of a glass circle is also more easily found. An optical image of the

optical square

glass circle is transmitted through prisms into an eyepiece next to the main telescope eyepiece. The surveyor thus does not need to move from one side of the instrument to the other when reading the circles. See **microptic theodolite**.

optical square [sur.] A compact, hand-held instrument, more accurate than the *cross staff* but less so than the *theodolite* or the *sextant*, which enables the observer by mirrors or prisms to view a point straight ahead as well as one perpendicular to it, and thus to set out a right angle on the ground. It usually measures no more than 7.6 cm in dia. and 2 cm thick.

optimum moisture content [s.m.] That moisture content of a soil at which a precise amount of *compaction* produces the highest *dry density*. It is particularly important to achieve this in *soil stabilization* before the road is completed. See **dry-density/moisture-content relationship**, **Proctor compaction test**.

orange-peel bucket A *grab* shaped like a half orange cut into segments, used for digging sand, gravel or clay inside small *cylinders* and elsewhere.

ordinary lay [min.] A description of the *lay* of a steel wire *rope* in which each strand twists in the opposite direction from the wires of which it is made. It is therefore much less likely to untwist than a *Lang lay rope* and can be used without guides to hoist a bucket.

ordinary Portland cement, type I ASTM P. c. A *cement* made by heating to clinker in a kiln a slurry of clay and crushed limestone, the most generally available and the cheapest *Portland cement* used in building.

ordinates See **graph**.

Ordnance Bench Mark [sur.] A levelling *bench mark* in Britain established by the *Ordnance Survey* at a level shown on their maps. The latest value may be obtained by buying the appropriate bench mark list from the Ordnance Survey, available for each kilometre square on OS maps. Ordnance bench marks are often brought up to date.

Ordnance Datum [sur.] The levelling datum for the agency responsible for mapping in Britain (*Ordnance Survey*). See **Newlyn datum**, **Liverpool datum**.

Ordnance Survey [sur.] The former Government agency (now civilian) responsible for surveying and mapping in Britain.

organic silt, o. clay [s.m.] A silt or clay containing plant remains, occasionally animal remains. It is recognizable by its stink when disturbed and its dark colour. It has very low bearing capacity, being highly compressible.

oriented-core barrel [min.] A *borehole surveying* instrument which takes and marks a core to show its orientation, and at the same time records the bearing and slope of the hole.

orifice meter, o. plate [hyd.] A plate with a hole in it, placed across a pipe of flowing fluid. The pressure difference, measured between the two sides of the plate, indicates the flow rate. It corresponds to the *measuring weir* of an open channel. The head loss of an orifice plate may reach 50% – a disadvantage. Cf. **Venturi meter**.

orifice rheometer An instrument to show how concrete flows at a site using a *water reducer*. It shows whether there is too much or too little of the *admixture*. If the wet concrete does not flow or flows too slowly, it has not enough *workability*. If it flows well but leaves coarse *aggregate* behind (*segregation*), there is either too much *admixture* or too much water. Intermediate conditions between these two extremes may imply satisfactory concrete; they can be evaluated by the time taken for the concrete to flow.

origami panel Origami is the Japanese game of folding paper to pretty shapes. But thin panels of hardened concrete – *ferrocement* – or of *glassfibre-reinforced concrete* also have been folded along joint lines to make successful canoes, cladding panels, etc. by Prof. R. Wheen of Sydney.

origin [d.o.] A point of intersection between two axes of a *graph*, the zero point of a graph where both *y* (abscissa) and *x* (ordinate) equal o. See *Cartesian coordinates*.

orthogonal [d.o.] At right angles, a term sometimes used for orthographic.

orthographic projection [d.o.] The usual method of mechanical drawing by making *projections* such as plans, elevations and sections on to a horizontal or vertical plane. The projection lines are at right angles to the plane on which the drawing is made. See **plan**.

orthophoto [air sur.] A differentially rectified near-vertical aerial photograph. In contrast to a rectified photograph, which removes distortions due to non-verticality only, the orthophoto removes distortions due to height variations of the ground inherent in the camera's central perspective geometry, enabling measurements to be scaled off the orthophoto as if it were a map. Orthophotos have superseded conventional maps for many purposes where they are cheaper to produce and where the photograph is more useful than a drawn map. See **rectification**.

orthophotoscope, orthoscan, orthophotomat, orthoprojector, etc. [air sur.] Trade names for *plotting instruments* for *rectification*.

orthotropic [stru.] Abbreviation of 'or-thogonally anisotropic', implying that there are noticeable differences in strength, stiffness or both in two directions at right angles, as in a beam and slab bridge. Cf. **isotropic**.

orthotropic plate floor, o. p. deck [stru.] A floor or bridge deck which is markedly stiffer in one direction (the direction of the span) than in the direction perpendicular to the span, e.g. a hollow-tile floor or a steel-plate deck with joists welded under it. *Plate floors* are an example of an *isotropic* floor.

OS *Ordnance Survey*.

Osaka Bay See **Kansai International Airport**.

osmosis In chemistry, the diffusion of a solvent or of a dilute liquid through a skin (permeable in only one direction) into the more concentrated solution. It is analogous to the movement of water in soil during electro-osmosis. See below and **thermo-osmosis**.

osmotic pressure The pressure in *osmosis* exerted by a solvent when its entry through the skin into the more concentrated solution is prevented.

outburst bank [hyd.] The middle part of the slope of a sea embankment, between the footing and the *swash* bank.

outcrop [min.] An exposure of a stratum or orebody at the earth's surface. A buried outcrop is one which would be seen if the recent loose deposits over the *bedrock* were removed.

outfall [hyd.] The point at which a sewer or land drainage channel discharges to the sea or to a river. It may be controlled by a gate or flap valve to prevent tidal water backing up.

Because most effluents are warmer than the sea or lake they flow into, the effluent is usually seen on the surface as a slick, and drifts with the current.

But in summer the effluent may be cooler than the top layer of sea or lake and so may remain invisible below it. An effluent outfall to a small stream must be carefully designed not to scour the bead or opposite bank. Hard *revetment* of brick, stone, concrete or timber is needed. Outfall pipes may be made of steel, concrete, high-density *polythene, polyvinyl chloride, glassfibre-reinforced plastic*, etc. and must have a *negative-buoyancy coating* so that they do not float.

output transducer See **actuator**.

outriggers Usually four stiff beams extending some metres out from the base of a *mobile crane* or *access platform*, bearing on the ground and thus providing stability. They are often operated hydraulically from the driver's cab.

ovality Pipes compressed too much become oval. This often happens to sewers. A crude test of ovality is to clean the pipe and then roll a ball down it. If the ball stops, the pipe is probably oval.

oven-dry soil [s.m.] Soil dried in an oven at 105°C.

overbreak, overbreakage [min.] The amount of rock excavated beyond the *neat lines* of a cutting or tunnel. This additional excavation is not paid for and its cost must therefore be included in the price tendered by the contractor for the rock within the neat lines.

overburden [min.] (**capping** USA) (see diagram) The worthless rock or soil over the valuable material in an open cut or mine.

overburden pressure The weight per unit horizontal area of the rocks at a certain level. In mines it is often taken as about 2.5 ton/m² for each metre depth.

over-consolidated clay [s.m.] A clay that in previous geological times was loaded more heavily than it is now and consequently has a tendency to expand if it gets wet.

overfall A *nappe* or the structure over which it flows.

overflow [s.m.] See **classifier**

overflow stand [hyd.] A stand pipe in which water rises and overflows at the *hydraulic gradient*.

overhaul (1) In excavation, a distance of haul in excess of the *free haul*, for which, therefore, extra payment must be made.

(2) [mech.] **overrun** The condition in a haulage engine when the load runs towards the engine faster than the rope and thus slackens and tangles the rope on the drum.

overhead ropeway An *aerial ropeway*.

overheads, overhead expenses, establishment charges, oncosts The costs of electric light, roads, supervision, accounting, directors' fees, etc. which cannot be fairly charged to one item, and must be spread over all items and perhaps over all contracts.

overhead travelling crane Lifting plant which is usually power-operated at least in its hoisting motion. It is carried on a horizontal girder spanning between rails above window level at each side of a workshop and consists of a hoisting *crab* which can itself travel from end to end of the girder. The whole area of the workshop between the rails can thus be traversed by the crab. See **gantry, traveller**.

overland flow [sewage] A method of *land treatment* of sewage *effluent*, suitable for clay soils, which uses no furrows or other laborious digging but usually sprays the water at the head of

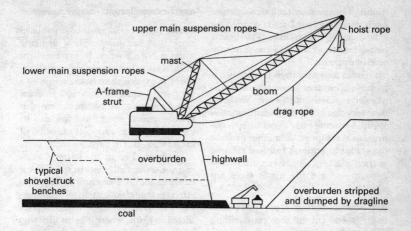

upper main suspension ropes

hoist rope

mast

lower main suspension ropes

A-frame
strut

boom

drag rope

overburden — highwall

typical
shovel-truck
benches

overburden stripped
and dumped by dragline

coal

Large dragline stripping overburden, showing its advantage over a face shovel and wagons doing the same work. (*Planning and Operation of Strip Mines* (conference proceedings), South African Institution of Mining and Metallurgy, 1986.)

a gentle slope, preferably one at 2–8%. Grass filters out the solids from the effluent, some of which infiltrates harmlessly to the groundwater. The remainder is collected in ditches or *field drains* for recirculation, other treatment or discharge.

overload In any engineering construction, a *load* greater than the device or structure is designed for.

overload trip [elec.] *Protective equipment* on a motor starter or on a circuit breaker, which ensures that the power is cut when the current exceeds a certain value. It is usually operated by a *solenoid*.

overplanted See **underplanting**.

overrun [mech.] See **overhaul**.

overrun brake, overriding b. [mech.] A brake fitted to a towed vehicle such as a rubber-tyred concrete mixer or a trailer. It operates as soon as the towing lorry slows down and the

towed vehicle tends to push into it. This relative movement of the towed vehicle towards the towing vehicle applies the overrun brake. High-speed towing is thus safe.

oversize, underflow [s.m.] In classifying mineral, aggregate, etc. into two sizes, the larger size; the smaller size is the undersize or overflow. For screened material the terms undersize and oversize only are used. See **classifier**.

overturning [stru.] Tipping over. Every structure, in its stability calculations, must be checked to ensure that it cannot turn over.

owner Usually the *client*.

oxidation ditch, ring d., Pasveer d. A Dutch development of the *activated sludge process* in which a rapidly rotating steel brush on a horizontal shaft (Kessener brush) aerates the *effluent* from *primary treatment* and sweeps it round. The *aeration tank* or oxidation ditch itself is a broad ditch divided by a wall down the middle. The liquor circulates round the wall continuously or intermittently.

If continuous operation is essential, a higher investment must be put in to pay for a *sedimentation tank*. *Sludge* is then recycled to the ditch continuously. With intermittent operation the brushes are stopped sometimes to allow the sludge to settle. A batch of effluent is then pumped out, fresh effluent is brought in, and the brush starts up again.

oxy-acetylene flame, oxy-gas welding A metal-cutting or metal-welding flame obtained from compressed oxygen and acetylene or other fuel gas in separate steel cylinders. Oxy-coal gas and oxy-hydrogen flames are also used. These gas flames are used for *brazing* (*B*) or welding copper, but for structural steel, *arc welding* methods are usual.

oxygen injection into water/effluent Pure oxygen, though much more expensive than air, is five times more concentrated and has been injected into the start of a long pumped sewer to prevent it going septic. Since the 1970s it has been injected into *activated sludge*. The March 1988 *Civil Engineering* (ASCE) described oxygen injection into the reservoir of the 60 m (195 ft) high Russell Dam, Georgia. Two horizontal 366 m (1200 ft) long PVC pipes of 10 cm (4 in.) dia. fitted with *diffusers* were fixed 1.5 m above the bed, upstream of the dam. The dissolved oxygen was raised to 6 mg/l.

oxythermic lance *Thermic boring.*

ozone (O₃) A poisonous blue gas used since 1907 for *disinfection* of drinking water at Nice, France. It is an excellent disinfectant but completely decomposes to oxygen in an hour or less, so a very small chlorine injection is usual also. Ozone is an air *pollutant* formed in photochemical smog, and can also be formed by the ultra-violet radiation of *arc welding*. But in Paris for many years it has been used in the *tertiary treatment* of sewage effluent. In the USA it has been found to be cheaper and more effective than chlorination for disinfecting *raw water*, e.g. at Los Angeles filtration plant, where it helps filtration by formation of a strong, sticky *floc*.

Some botanists think ozone kills pine trees. This has been attributed also to *sulphur dioxide* and to NO_x.

But ozone is most helpful to humanity in one way. The ozone layer in the stratosphere, 10–35 km (7–22 miles) up, absorbs ultra-violet rays and thus protects people from the skin cancer that they cause. The CFCs (chlorofluorocarbons) mentioned under *greenhouse effect* contain chlorine. One molecule of chlorine can destroy 100,000 ozone molecules. But it is estimated that CFCs released at the earth's surface take 50–150 years to reach the stratosphere. Consequently we do not yet know the effect of today's releases of CFCs. This is why CFCs should be banned.

ozonizing *Disinfection* of water by injecting it with ozone; often used in France.

P

pachometer A *cover meter*.

pack, packing A steel plate inserted between two others to fill a gap and fit them tightly together.

package In computing, a set of *programs* with their documentation.

package deal See **design and build**.

packer [s.m.] An inflatable rubber ring, like a small inner tube, that is fixed on to a perforated grouting tube to go into a borehole, and inflated when the tube has been lowered to its correct level. The inflated packer prevents grout passing beyond it. The *tube à manchette* uses packers.

packing (1) [mech.] Hemp or similar material inserted in the stuffing box of a pump to make it watertight.
(2) A sand cushion between the helmet and the head of a driven *pile*. See **pack**.
(3) [rly] Tamping the ballast below rail sleepers to achieve the correct line and level for the rails.

pack-set cement, sticky c. Cement which flows badly from a bunker, possibly because of mechanical compaction, particle interlock or electrostatic attraction of the particles (*ACI*).

paddle [hyd.] A wooden panel for closing the water passage in a lock or sluice or culvert.

paddle hole [hyd.] A hole which allows water to pass into or out of a lock.

pad foundation An isolated foundation for a separate column.

page A small wooden wedge used in timbering trenches.

pallet A lifting tray used for stacking material with the *fork-lift truck*.

pallet fork Equipment on a *loader* for lifting and moving factory *pallets*.

PAM (pressiomètre autoforeur marin) The self-boring *pressuremeter* developed in France by the Laboratoire des Ponts et Chaussées and the Institut Français du Pétrole. It is worked from a ship by winches. See **stressprobe**.

pan Steel **telescopic centering** (*B*).

Pandrol clip [rly] A high-tensile steel spring fastening driven into a steel base plate to hold a rail. It can be fixed by two people with an automatic machine at the rate of 750 sleepers per hour so is superseding other rail clips.

panel [stru.] In a *lattice* girder, the amount of the girder enclosed between adjoining vertical members. See **panel point**.

panel point, node [stru.] A junction on a *truss* chord, particularly one where a vertical meets a chord.

panel relaying, prefabricated track [rly] Relaying rail track by replacing both rails with sleepers attached in lengths (panels) 18.3 m (60 ft) long. The panel is built in the workshop and lifted into place by crane.

pan head A head shaped like a cut-off cone, to a rivet or screw.

pannier A basket used like a *gabion*.

pantograph (1) [d.o.] Rods connected like a parallelogram, used for copying a drawing to the same or any other scale.
(2) [rly] An expanding, hinged,

diamond-shaped structure over an electric loco for collecting power from an overhead trolley wire. It is of the same shape as (1).

parabola [d.o.] The shape made by cutting a cone. It is also the curve of the *bending-moment diagram* for a uniformly distributed load on a simply supported beam. For this reason a uniformly loaded arch is often made parabolic, to avoid tension in it.

Parafil rope Aramid fibre (made by ICI) enclosed in a tough polymer sheath. It has been proposed for the cables in very long-span suspension bridges. Aramid fibre strength is about 2800 MPa (406,000 psi) compared with 3000–4000 MPa for carbon fibre but is less dense, 1.45 against 1.79 g/cm^3 for carbon fibre.

parallel-flanged beam A *universal beam*.

parallel-motion equipment Equipment fitted to a *drawing board* that is a considerable improvement on the tee-square and less expensive than the *draughting machine*. It consists of a thick, transparent, Perspex straight edge, some 10 cm wide, lying flat on the board and some 15 cm longer than it. It runs absolutely parallel to itself. This is arranged by two tensioned endless wires fixed to it, one each side, which run over pulleys at the top and the bottom of the board. There are thus four pulleys. The two at the top are locked to one shaft so that their motion and that of the wires is exactly the same. Tensioning is either by a spring or by a weight on the bottom part of the endless wires.

parapet See (*B*).

park-and-ride In *traffic engineering*, a stratagem to encourage private motorists to abandon their cars at car parks in the suburbs close to bus, rail or underground stations from which public transport to the city centre is easy and fast.

parkway (1) In the USA a *freeway* which passes through a park and is administered by a local authority, often a park authority. It is not open to commercial traffic. Cf. **street**.
(2) [rly] A railway station with parking spaces which serves a country district rather than a town.

Parramatta Bridge, Gladesville B. (1964) An elegant arch, and the longest concrete span in the world, 305 m clear span, over the Parramatta river at Gladesville, Sydney, New South Wales, a short distance from the *Sydney Harbour Bridge* over the same river. This arch bridge, built of 50-ton precast concrete blocks, is a revival of the ancient technique of the *voussoir arch*, which had been moribund for 50 years. The 6-lane carriageway is 22 m wide and has 1.8 m footways. Maunsell and Partners of London designed it.

Parshall measuring flume [hyd.] The improved *Venturi flume* of the US Department of Agriculture for measuring the flow in *open channels*. It has a contracting length and an expanding length separated by a throat at which there is a sill.

partial load factor See **load factor**.

partially fixed [stru.] An end support to a beam or column which cannot develop the full *fixing moment* to the beam or column is called partially fixed.

partially separate system A drainage system in which the rainwater from house roofs and backyards flows away with the house sewage, and the rest of the rainwater flows off in a different *sewer*. See **combined/separate system**.

partial prestressing [stru.] Prestressing to a stress level such that under design loads, tensile stresses exist in the precompressed tensile zone (*ACI*).

particle-size analysis, grading, p.-s. distribution [s.m.] The proportions by weight of the different particle sizes in a soil or sand determined by *mechanical analysis* so as to build up a *grading curve*.

parting agent, p. compound A *release agent*.

Pascal's law [hyd.] A law of hydrostatics (1646) which states that in a perfect fluid the pressure exerted on it anywhere is transmitted undiminished in all directions.

passing place (1) A local widening of a narrow road where vehicles can pass each other.
(2) A railway siding.

passivation In connection with the corrosion of *reinforcement*, passivation implies that the pH in the concrete is above 10, preferably about 13, which prevents corrosion of the steel. Carbonation or other processes may lower the pH to about 9 and corrosion then starts.

passive earth pressure, p. resistance [s.m.] The resistance of a vertical earth face to deformation by a horizontal force (usually due to *active earth pressure*). The passive earth force in sand is equivalent to that of a fluid weighing three to four times as much as the sand. It is the reciprocal of the coefficient of active earth pressure in *Rankine's theory*,

i.e. $\dfrac{1 + \sin\phi}{1 - \sin\phi}$. *See* **earth pressure**.

passive resistance *Passive earth pressure*.

Pasveer ditch An *oxidation ditch*.

patenting *Heat treatment* to soften steel before and during *cold drawing*.

pavement (1) In the UK the lay term for a *footway*.
(2) The whole construction of a road or airstrip including *stabilized soil* and the surface whether of asphalt, concrete, wooden or stone blocks, etc. Rigid pavements are of *reinforced concrete*. A flexible pavement may be of any other material including *lean concrete*. The first part of a pavement is the subgrade (soil) it is built on, then comes the sub-base if any, then the *base* and finally the *wearing course*. See **prestressed concrete pavement**.
(3) Any hard floor of tiles, bricks, concrete, wooden blocks, etc.

paver See **concrete paver, slip-form paver**.

paving A surfacing over the ground of wood blocks, stone setts, bricks or precast slabs, or a layer of concrete, asphalt or coated macadam.

paving brick A hard brick of *engineering brick* (*B*) quality for paving.

paving flag or **flagstone** A thin flat stone for surfacing a footway. Cast-stone flags are cheaper than natural stone and they are made without reinforcement 50 or 63 mm thick, 0.3 or 0.6 m wide and 0.3–1 m long, and other sizes.

paving train See **fixed-form paving train, slipform paving train**. Cf. **roller-compacted concrete**.

pay line The *neat line* of an excavation.

PBAC *Polystyrene-bead-aggregate concrete*.

peat blasting, bog b., swamp shooting [s.m.] A method of road building over peat deposits. Hard fill is first dumped on the road site over the peat to a height equal to the depth of the peat. Holes are then drilled or jetted (see **jetting**) through the fill into the bottom of the peat at the centre and edges of the road. Charges of explosive are fired at the edges, followed one second later by the detonation of the charges at the centre. The peat is displaced outwards and the fill sinks into the space which was occupied by the peat. If exploratory drilling shows that the fill has not settled sufficiently, the process can be repeated until the settlement is complete. This was done in the Middle West of the USA in the 1920s and in Germany for the autobahn roads of about 1935 in peat to some 20 m depth.

pebbledash See (B).

pedestrian-controlled dumper, buggy, power barrow A small *dumper* of about 250 litres capacity controlled by someone walking beside it. By moving a full mixer-load of concrete it greatly increases the mixer output above what is possible even with 3–6 barrowmen, each wheeling a barrow.

peer review A North American activity which seems much quicker and easier for *consultants* to organize than *quality assurance*, as well as demanding less paperwork; it is simple, effective and approved by insurers.

Representatives of a competing firm or other independent organization examine either one project (PPR – project peer review) or the whole of a firm's activities. A firm requesting a peer review in 1986 paid $400 plus expenses each day for each reviewer, making the total cost of a typical review between $1000 and $3000. A management consultant's report would cost 10–50 times as much and be less valuable because the consultant would not understand the work.

Solid benefit from a peer review comes from the fact that some insurance companies will not provide insurance unless the firm concerned has had a peer review in the last four years. Improvements almost always occur in the firm reviewed, sometimes even before the review takes place. Two of the institutions that use peer review are the American Consulting Engineers' Council, since 1984, and the Association of Soil and Foundation Engineers (ASFE). ASFE was in fact founded in 1960 partly because its members found themselves to be uninsurable, but it was not until 1978 that it adopted peer review from the American Institute of Certified Public Accountants, and with complete success. The American Institute of Architects also uses it.

The results of the peer review, usually in the form of a report, are completely confidential though they may be made available to the insurance company concerned. The reviewers never take over the original designers' work but tell them of possible improvements. (*Civil Engineering* (ASCE), Oct. 1986 and Nov. 1988; also ASCE's *Quality in the Constructed Project, Guidelines for Owners and Constructors, Preliminary Edition for Trial Use and Comment*, ASCE, 1988.)

peg [sur.] A short pointed wooden rod driven into the ground to mark a line or a level. A nail driven into the top of the peg usually shows the position of the point.

peg-top paving Paving with very small *setts*.

pellicular water, adhesive w. [hyd.] A film of water over soil particles, so thin that it is under strong forces of molecular adhesion and cannot be removed in a centrifuge even with a power of 70,000 g. It can, however, be removed by evaporation and the 'pellicular zone' is the depth to which evaporation effects penetrate.

Pelton wheel [hyd.] The commonest *impulse turbine*, a wheel carrying buckets at its perimeter which are struck by a fast-flowing water jet. When the wheel is to be stopped the jet must first be deflected away from the wheel and then shut off slowly enough to avoid *water hammer* in the pipeline. The Pelton wheel is used for medium to very high heads, of 250 m to over 1000 m. See **water turbine**.

penetration (1) Of sheet piling, its *cut-off depth*.
(2) Of a *monolith* or *caisson*, its depth below ground surface.

penetration needle See **Proctor plasticity needle**.

penetration-resistance testing, Windsor probe [stru.] Finding the strength of a concrete by shooting a bolt into it and measuring the depth of penetration, a method recognized by the *ASTM*. As in all *semi-destructive tests*, the damage caused by the bolt must be repaired.

penetration tests [s.m.] Tests of the soil in place which give a surer indication of its load-bearing capacity than tests in the *soil mechanics* laboratory. Broadly they can be divided into *static* and *dynamic penetration tests*. Evidence from penetration tests should always be interpreted with the help of information from boreholes.

penetrometer [s.m.] The wash-point penetrometer is a cone-shaped instrument which is *jetted* into the ground to the required level and then forced in at a measured pressure. It is thus a *static penetration test*.

penning See **pitching**.

penning gate [hyd.] Specifically UK term for a rectangular sluice gate which

opens by lifting upwards. See **penstock**.

penstock [hyd.] (1) A pressure pipe which supplies water to a *water turbine*.
(2) A *penning gate*. In the USA only the first sense is used, and this first sense is coming to be accepted in Britain.

peptizing agents Materials which can lower the viscosity of a liquid when they are added in small amounts. This may occur chemically by depolymerization or by reducing *flocculation*.

perched water table [min.] *Groundwater* maintained temporarily or permanently above the *standing-water level* in the ground below it, usually by an impervious stratum between them.

percolating filter [sewage] A *trickling filter*.

percolation [s.m.] The movement of gas or water through the pore spaces of the ground. See **Darcy's law**.

percussion drill See **cable drill**.

percussion tools Tools which work by striking rapid blows. Most of them are driven by compressed air like the *hammer drill*; but whether they are driven by electricity or compressed air or hydraulically, they drill rock, chip slag or excess metal from a weld, caulk joints between steel plates, close rivets, and *bush-hammer* stone or in other ways tool the surface. See **hydraulic hammer**.

percussive-rotary drilling [min.] *Rotary drilling* (2) combined with a vibratory or percussion motion on the bit; a fast modern method of rock drilling.

percussive welding [mech.] See **resistance percussive welding**.

perfect frame [stru.] A *frame* which is stable under loading from any direc-

peripheral equipment

tion and would become unstable if one of its members were removed or if one of its *fixed ends* became hinged. Cf. **redundant frame**.

peripheral equipment, p. device Any device or accessory which can be electrically connected to a computer, such as a cassette recorder, disk drive, keyboard, *light pen*, printer, etc.

peristaltic pump, squeeze pump A pump named after the peristaltic contractions which force food along the animal intestine. It can pump *sludge*, concrete and other semi-solid mixtures, and can be used for dosing measured quantities of e.g. *admixtures*. A pair of rollers at one end squeezes its sturdy rubber tube to force fluid along, while the pair at the other end releases its pressure.

perlite See (*B*).

permafrost A quarter of the world's land area, the perennially frozen ground of Alaska, Antarctica, northern Canada, Greenland, Iceland, Siberia, etc. *Foundations* must be built below the *active layer*, which thaws in summer. Good insulation must be placed below the ground floor to prevent thawing of the foundations. See also **Alaska pipeline, frost heave, refrigerator foundations**.

permanent adjustment [sur.] An *adjustment* to a surveying instrument which is made only occasionally and not at each *set-up*. Cf. **temporary adjustment**.

permanent formwork *Formwork* that encases concrete throughout its life, either for appearance, construction speed or warmth (woodwool, insulating board). Steel permanent formwork in *composite floors* is quick and convenient, looks good, and can provide *composite action*. See **Seattle**.

permanent set *Plastic deformation*. It does not include *elastic strain*.

permanent shuttering A lining to *formwork* which encases concrete throughout the life of the structure. Woodwool or insulating board have been used as an insulating permanent shuttering.

permanent way [rly] The *rails, sleepers, ballast, switches, points* and *crossings* laid for a railway, not the temporary way for building it.

permeability [s.m.] In cm/s or other unit of speed, the rate of diffusion of a fluid under pressure through soil, concrete etc. It is measured in the field by pumping at a constant rate from a borehole until the level of the water in the borehole is constant. The amount of flow related to the slope of the water table provides the data from which the *coefficient of permeability* in *Darcy's law* can be obtained. Several other boreholes must be drilled to determine the level of the water table during pumping. For sands or silts, the permeability generally varies inversely with the *specific surface*, but is always highly variable. See also **Hazen's law, permeameter**.

permeability-reducing admixture See **water-repellent cement, pore filler**.

permeameter [s.m.] A laboratory instrument for measuring the coefficient of *permeability* of a soil sample. The constant head permeameter is used for permeable materials like sand or gravel, the falling head permeameter is used for impermeable materials like clay or fine silt.

permeation grouting Highly skilled *grouting* at low pressures which fills the soil voids with the minimum of disturbance. Detailed knowledge of the soil is needed to design the treatment, and excellent supervision. Cf. **consolidation grouting**.

permeator See **reverse osmosis**.

permissible stress, working s. *Elastic design* is based on permissible stress, sometimes regarded as the stress at failure divided by a *factor of safety* of 3. Design in this way is much simpler than *plastic design* but sometimes less economical in material and is not accepted by *Eurocodes*.

personal equation [sur.] In accurate instrument work (not only surveying) it is found that each observer tends to measure every value with a consistent difference from the average of all observers. An observer's personal equation states the correction which must be applied to all his or her readings to bring them up or down to the average. (The average is the surveyor's closest approximation to the truth.)

pervibration A term sometimes used for internal *vibration* of concrete.

pervious macadam A road surface of high quality with much reduced splashing in rain and consequently fewer accidents. The macadam has a large volume of interconnected voids through which the rainwater drains. But the voids slowly block up so the life of the spray-reduction is about six years on main roads or three years on motorways.

PETN [min.] Pentaerythrite tetranitrate, used in *detonating fuse*.

petrographic microscope [min.] A *microscope* used for studying *thin sections* of rocks, concretes, etc. to determine their minerals, particle size, etc. It is fitted with two *Nicol prisms*, a polarizer and an analyser, in addition to lenses which enlarge the image.

petroleum vapours More dangerous than *methane*, because they can explode at only 1% in air, these vapours occur in drains, sewers, tunnels etc. A drainage pumping station was destroyed by one such explosion in 1989.

PFA (pulverized fuel ash) See **fly-ash**.

phenolic resins, phenol formaldehyde r. The first *thermosets*, brown or black plastics, were produced commercially in 1916 as bakelite. They have many uses in building as water-resisting glues for wood, etc. and resist fire and chemicals better than many other plastics.

photo-elasticity [stru.] A technique for examining by *model analysis* the distribution of stresses in unusual shapes under load. *Polarized light* is passed through a transparent model which shows *isochromatic* and *isoclinic* lines. This technique gives the directions of the axes of *principal stress* at any point and the magnitude of the difference of principal stresses. As in the *petrographic* microscope a polarizer and an analyser are needed. The method has been used for determining the stresses in models of soil structures, dams and other exceedingly complicated structures.

photo-electric cell, photocell, photodiode An electronic device in which the electric current varies with the amount of light falling on it. A door in a corridor of a British Rail Intercity train opens when the passenger's leg interrupts the light beam falling on the controlling photocell.

photo-electric effect When light falls on certain electrical conductors a current flows in them; similarly when light ceases to shine the current is interrupted. This interruption of a circuit is used in the *photo-electric cell*.

photogrammetry [air sur.] The making of maps or the measurement of objects from photographs. Certain points of known coordinates on the ground or the object must be recognizable on the photos. With the help of a *plotting instrument*, a contoured map or dimensioned drawing can be produced from the photos. Photogrammetry from a land camera was first used by Col. Laussedat in France in 1849 only ten years after Daguerre had invented his photographic process. Laussedat also took photos from the air using kites and balloons but abandoned them in favour of ground photos. Aeroplanes were first used for photography about 1913. See **ground control**, **mosaic**, **oblique/vertical photo**, **orthophoto**, **radar**, **timing**.

photomicrograph A photograph of an object as seen under, and magnified by, the microscope.

photosynthesis The growth of green plants and *algae* in sunlight with the release of oxygen from the carbon dioxide they absorb. Photosynthesis in water plants oxygenates the water, reducing the organic pollutants, making the water more habitable by fish and more drinkable by humans. Much photosynthesis takes place in the sea but only at depths reached by sunlight.

photovoltaics See **PV**.

phreatic surface The *water table* (mainly USA).

phreatic water [s.m.] *Groundwater*.

phreatic zone, z. of saturation The ground below the *water table*. It is saturated with water, but a small amount above it, the lower part of the *capillary fringe*, is also saturated.

pH scale The chemical way of measuring acidity and alkalinity, on a scale from 0 to 14. Neutrality is at pH 7, neither acid nor alkaline. Numbers from 7 to 14 indicate increasing alkalinity, 14 being the most alkaline. Numbers below 7 indicate acidity, with 0 implying the strongest acid. It is impor-

tant to detect acidity in waters because acidity usually means *corrosion* of metals. pH meters exist.

picket [sur.] A *range pole*.

pickling of metal The dipping of steelwork in hot sulphuric acid, then hot water, then hot phosphoric acid to remove scale as a preparation for painting or galvanizing. The steel is then dried and primed while still warm. Cf. **phosphating** (*B*).

pier (1) A wide column or short wall of masonry or plain or reinforced concrete for carrying heavy loads, such as a support for a *bridge*.
(2) A *breakwater* used as a quay.

pier cap, p. template The top part of a bridge *pier* which distributes uniformly over the pier the concentrated loads from the bridge. Pier caps used to be of large thick stones but generally now they are of reinforced or mass concrete. Granite pier caps are allowed to be loaded at about the same stress in compression as lightly reinforced concrete, at 5 N/mm². Other types of rock are allowed much less than this.

piercing See **probing**.

pier template *Pier cap*.

piezocone A development of the *cone penetration test* (*CPT*) in which the *pore-water pressure* or its dissipation is measured as the probe advances into the ground. First used in Sweden in 1977, it provides more accurate estimates than the CPT.

piezo-electric sensor A suitably cut piezo-electric crystal (Rochelle salt or quartz) under mechanical strain develops an electromotive force (piezo-electric force) between two of its faces. Such a crystal can be used as a pressure *sensor*.

piezometer tube [hyd.] An open-topped tube (*stand pipe*) for measuring moderate pressures. It contains water or kerosene for low, or mercury for higher pressures.

piezometric surface [hyd.] An imaginary surface above or within the ground, at which the water level would settle in a *piezometer tube* whose lower end passes below the *water table*. It indicates the level to which the water from an *artesian well* could rise. It corresponds to the line of the *hydraulic gradient* in a pipe, except that the water in the pipe flows. A piezometric surface is more responsive to pumping than a water table. It can be lowered in a few minutes by pumping 2 km away, though the water table may take months to settle.

pig (1) An iron or lead block cast at the smelting furnace for remelting at the foundry, etc. The word *ingot* is used for precious metals and steel.

(2) (**pipeline internal gauge**), **go-devil** (see diagram) A device forced through a pipeline under pressure from fluid or compressed air. After concreting, a ball or plug of sacking or cement bags is forced through the pipeline of a *concrete pump* by water or air pressure to clean out the pipes. Those at the discharge end must be warned before this is done. In the oil industry, go-devils are for cleaning, pigs are for other purposes. A batching pig separates two types of oil flowing through a pipeline. The diagram shows a pig suitable for a welded steel pipeline. For a water main or sewer a foam swab is commonly used.

An inspection pig (intelligent pig) continuously records the position and type of faults in the pipe wall and its coating. A calliper pig or gauging pig measures and records changes in the pipeline bore.

Before the oil at 70°C (158°F) could flow into the *Alaska pipeline* the pipe had to be purged of oxygen to eliminate explosion hazard. This was done by injecting hot nitrogen to warm the pipe, separated from the air in front by a pig and from the oil behind by another pig. But when the first oil arrived at Valdez after travelling 1280 km (800 miles) it was only a few degrees above freezing point, though later the pipeline warmed up considerably.

In the cleaning up of a Surrey water main heavily encrusted with calcium carbonate from water out of the chalk formations, 25 pigs of gradually increasing diameter had to be forced through, resulting eventually in complete recovery of the original 0.4 m diameter.

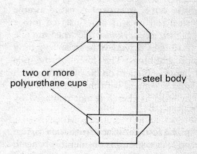

two or more polyurethane cups — steel body

Pig (pipeline internal gauge) for cleaning a pipeline, suitable only for small bores, e.g. welded steel gas or oil pipe. (C. Kershaw, *Measurement and Control*, November 1987.)

pig-trap A connection to a pipeline enabling a *pig* to be put in or taken out; the installation includes pig signallers, pressure gauges, controls, valves, etc.

pilaster See (*B*).

pile A timber, steel, *reinforced* or *prestressed concrete* post *driven*, *jacked*, *jetted* or cast into the ground as a *bored pile*. See also **batter/displacement/ sand/screw pile** and **cylinder, pile-placing methods**, **precast concrete**, **pretesting**, **reamer**.

pile bridge A bridge carried on *piles* or piled *bents*.

pile-cage spacer A precast concrete or plastic disc threaded on to a *main steel* bar of a *bored pile's cage of reinforcement* to maintain the concrete cover.

pile cap (1) A *footing*, usually a *reinforced-concrete* beam, resting on *piles* and distributing loads down to them from the structure above.

(2) A steel plate fixed on top of a steel cylinder to distribute the load on to the concrete filling within it.

pile cluster See *pile sleeve*.

pile core, mandrel A withdrawable steel rod or stiff pipe inserted into a hollow steel *cylinder* pile when the pile is sunk by driving rather than by jetting or grabbing. The force of the pile hammer is spent on the pile core, which makes contact with the pile shoe. The cylinder is therefore not damaged by driving.

piled foundation A *foundation* carried on *piles* to ground considerably beneath the surface.

pile-drawer A *pile extractor*.

pile driver, p. frame A hoist and *leaders* which can handle the weight of a pile and drive it into the ground. See **dolly**, and below.

pile extractor, p.-drawer Any *pile-driver* which strikes a pile upwards and loosens the grip of the pile on the ground. The actual pull of withdrawal of the pile is done by the crane from which the pile extractor hangs. It is therefore essential to have a crane which is strong enough for the job. If the pile and extractor weigh 3 tons and the total lift of the crane is 4 tons, only 1 ton of upward force will reach the pile. This is the minimum required –

more than 1 ton lift should be available. Piles can be extracted by other methods including upward jacking from the ground.

pile foundation A *foundation* carried on *piles*.

pile frame A *pile-driver*. It must be appreciably taller than the longest pile to be driven.

pile group Several driven or bored piles placed close together to take a heavier load than a single pile could carry. A pile group whether of steel, concrete or timber piles is generally capped by a reinforced-concrete *pile cap*.

pile hammer Generally either a *drop hammer* or a *double-acting* hammer. A steam hammer may be fully automatic or semi-automatic. Semi-automatic hammers are single-acting, have a steam admission controlled by hand line, give about 40 blows per minute and drive timber or concrete piles. They are extravagant with steam but very simple, trouble-free and easy to use. Steam hammers can also work on compressed air. A Diesel hammer is a drop hammer in which the ram is raised by a *Diesel-engine* piston. About 60 blows are delivered per minute. The blow of the drop hammer is slightly cushioned by compression of the combustible charge in the cylinder.

pile head The top of a pile. The heads of precast-concrete piles are protected by *packing* under a *pile helmet* during driving and sometimes also by a timber *dolly*. The heads of wooden piles are encircled by a *driving band*.

pile helmet A cast-steel cap which covers and protects the head of a precast-concrete pile during driving

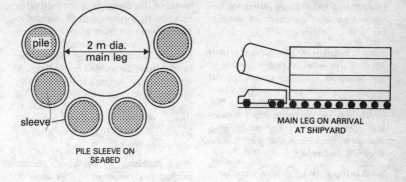

PILE SLEEVE ON
SEABED

MAIN LEG ON ARRIVAL
AT SHIPYARD

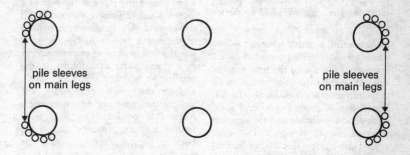

Pile sleeve of Eider platform: plan at seabed.

and holds the *packing* in place between it and the pile head. It is recessed on top for a plastics or hardwood *dolly* to cushion the blow from the drop hammer.

pile hoop A *driving band*.

pile-placing methods Driving by *drop hammer* or stream hammer, *jetting*, jacking (*pretesting*), boring, pulling down, vibrating, washing out, blowing out, coring, drilling, grabbing out with a grabbing crane or explosives. See **bored/driven/screw pile, silent pile-drivers**.

pile ring A *driving band*.

pile shoe A high-grade cast-iron point on the foot of a wooden or concrete *driven pile* to help it penetrate the soil. Its sides slope at about 1 horizontal in 6 vertical. For driving through uniform clay, silt or sand, it may be unnecessary.

pile sleeve, p. cluster (see diagram) In an *offshore platform*, a nest of tubes several metres long, through which piles are driven and then grouted, to hold the platform.

pillar [min.] A mass of coal or ore left unworked to support the overlying ground. At great depths, pillars (except *shaft pillars*) are little used because they crush, sometimes dangerously.

pilot circuit [elec.] A control circuit such as that used in *remote control* or for *contactors*. It carries a small (in coal mines *intrinsically safe*) current which operates a *solenoid* or other device that breaks a large current in e.g. a contactor. The contactor may produce an *incendive* spark, but the pilot circuit does not. See **relay**.

pilot lamp [elec.] A small electric lamp which shows whether the power is turned on to a circuit, often used to indicate the attitude of a *circuit breaker*.

pilot tunnel, p. shaft A tunnel, *microtunnel* or *shaft* driven sometimes at a fraction of the size of the final excavation. It enables the final tunnel to be driven from either end at will, with full knowledge of groundwater and rock, with good ventilation and the simplest method of soil removal. The pilot tunnel also ensures that the final tunnel is in the right place. Some microtunnels are reamed several times before the final diameter is achieved. But in the *Channel Tunnel* the pilot is driven at its final diameter. It is nevertheless a pilot because it precedes the running tunnels.

pinchers Two *poling boards* strutted apart on opposite sides of a trench.

pin joint [stru.] A *hinge* in a structure.

pinned, pin jointed [stru.] A description of a beam or a column which has a *hinge* at the end.

pipe bursting, p. expansion *Pipe renewal* that begins with a *pneumatic impact mole* or a hydraulic pipe burster, sometimes two in succession, the second one expanding to a larger diameter. A disadvantage of the method is that nearby services may be damaged. It is possible to burst stoneware pipes unless they are set in concrete blocks, but not cast iron or steel pipes. The burster may be guided by pulling. A 200 mm pipe can be expanded to 280 mm and a 300 mm pipe to 380 mm. If the new lining, usually of medium-density *polythene*, is inserted in 200 m lengths, access shafts have to be available every 200 m.

pipe jacking, p. pushing, thrust boring (see diagram) In soft ground, building an underground pipeline by assembling rigid pipes at the foot of a *launch pit*, and jacking them through the ground to a *reception pit*, instead of digging a trench to lay them. It was the first method of *microtunnelling*. A 0.9 m (3 ft) dia. tunnel is the smallest in which a miner can work, and for any smaller pipe either pipe jacking or another microtunnelling method is needed. The jacks react against a strong concrete wall at the back of the launch pit. Usually the soil inside the pipe is excavated by a microtunnelling machine, but thin steel rods or pipes can sometimes be pushed without prior excavation. Where the thrust becomes too much for the jacks available an *interjack station* must be inserted to recommence pushing. *Polythene* pipe, the most popular replacement pipe, cannot be jacked, being too flexible, though it can be pulled. Jacked pipelines are usually straight or very slightly curved. See also **buried shield, pipe ramming, steel sleeve jacking, vibrating pile-driver**.

pipe liners Polythene (polyethylene) pipe liners of up to 500 mm dia. have been inserted into old gas mains and sewers, effectively making them tight, prolonging their useful life, and making them smooth inside.

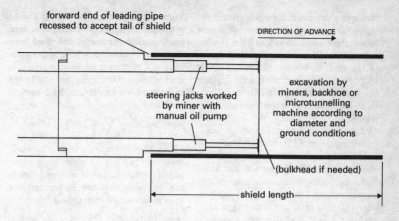

Pipe jacking with a shield and steering jacks. (Pipe Jacking Association.)

pipeline transport, slurry t. Coal, clay, chalk and other minerals have long been moved considerable distances by pipeline. Cement-makers have known this for years. In 1957 the first long coal pipeline carried coal successfully from the mine in Cadiz, Ohio, to the consumer in Cleveland. Mammals pump solid blood corpuscles round in the liquid serum of their blood. Most pipeline transport systems use water as the fluid and it is both corrosive and abrasive. Pumped slurries usually contain 20–30% solids, rarely more. Slurry transport saves labour in moving excavated soil from a tunnel face but has sometimes been found uneconomic for distances of more than 3 km (2 miles). Belt conveyor or locomotive transport of muck by wagon can then be used.

pipe lining Cast-iron sewers or water pipes need linings inside to protect them from abrasion or corrosion or both. A common lining is carefully formulated cement mortar applied by spray-machine or centrifugally spun. In France *high-alumina cement* has been used. The outside of pipes is protected from corrosion by a zinc coating sprayed on in the works or in more corrosive water by a polythene sleeve. The worst conditions are in *pipeline transport* – corrosion combining with abrasion.

In water pipes, mortar linings have been used since the mid-1930s in pipes of from 75 mm (3 in.) to 2 m (6½ ft) dia. and they should last for 50 years. A trowelling machine can produce a smooth lining that improves the flow of water. Mortar acts as a corrosion inhibitor over the whole surface and can cover cracks 2 mm wide. But consumers must be cut off for 24–30 hours and there may be problems with alkalinity of the water for a short time afterwards.

For small pipes of below 150 mm (6 in.) dia. the minimum cement mortar lining is 5 ± 1 mm; for pipes from 150 to 450 mm it is 7 ± 1 mm; for pipes larger than 450 mm (18 in.) the thickness should be 10 ± 1 mm. Both ends of the pipe should be carefully sealed to prevent air circulating and drying the mortar. Curing is usually complete in 18 hours with a correctly formulated mortar.

Protection against the sort of acid attack experienced in *Kuwait* was provided in the Istanbul sewer built in 1988 to 2.4 m (8 ft) internal dia. It is lined first with concrete segments, as usual grouted outside, but a further 100–150 mm (4–6 in.) concrete is cast in place inside this to carry a ribbed polyvinyl chloride sheet whose ribs are held in the concrete. The pvc covers only the top 270° of the sewer, which might suffer acid attack.

Epoxy resin pipe linings are not yet fully accepted for water pipes but they may be used in certain conditions in pipes from 75 mm (3 in.) to 375 mm (15 in.) dia. There are no pH problems but the lining is only 1 mm thick, as against the 5 mm or more of the cement mortar. Also it is easy to remove a bad mortar lining but hard to remove a bad epoxy lining. To ensure there are no gaps in the lining, a closed-circuit television camera can be pulled through the pipe after spraying. Cameras as small as 44 mm ($1\frac{3}{4}$ in.) exist.

If the groundwater has more than $0.2\%\,SO_3$, concrete pipe must be made with *sulphate-resisting cement*.

In 1988, in the Los Angeles metro tunnels, *polythene* (HDPE) lining all round was used to exclude such gases as methane and hydrogen sulphide which are present in the ground in California. The metro is to start up in 1992. See **pipe relining**, **spray lining**.

pipe pile (*ACI*) A steel cylinder of from 25 to 60 cm (10–24 in.) dia., generally driven open-ended to a firm bearing, then excavated and filled with concrete. It may have several sections from 1.5 to 8 m long, joined by cast-steel sleeves. One type has its lower end closed before driving, by a conical steel shoe.

pipe pushing A steel rod of about 38 mm ($1\frac{1}{2}$ in.) dia. can be pushed into silt or clay by *pipe-jacking*, *pipe ramming*, etc. but not for more than about 30 m (100 ft) because of aiming difficulties. Suitable soils can be reamed out to as much as 200 mm (8 in.).

pipe ramming Use of a *pneumatic impact mole* for *pipe jacking*. This may be the cheapest way of pipe jacking since no expensive concrete abutment wall is needed for jacking, though a *launch pit* and *reception pit* are needed as usual. Steering is more difficult than with jacks.

pipe relining, p. renovation Removal of the deposits inside a pipe followed by the spraying inside it of a *pipe lining*. See **U-liner**, **sliplining**.

pipe renewal Because water pipes, sewers and gas pipes began to be laid in quantity around 1850, even cast iron pipes are in urgent need of renewal in the 1990s. Various methods of renewal include *pipe bursting*, *rolldown*, *sliplining*, *sock lining*, *spiral winding*, *spray lining* and *swaging*. If no relining method can be made to work, a *microtunnelling* or *pipe jacking* method of installing a new pipe will cause less surface disturbance than trenching for it. British Gas calculates that pipes can be renewed at half the cost of excavation and replacement of the whole pipe, using swaging or rolldown of up to 200 m of polythene pipe at a time. British Gas Northwestern was investigating in 1988 how to reline pipe as large as 1067 mm (3 ft 6 in.).

The limits of HDPE (*polythene*) pipe manufacture were continually being extended at the end of the 1980s. It was then considered an achievement that 180 mm (7 in.) pipe could be delivered in a 500 kg coil 120 m long and not in short lengths. See **sewer pipe materials**.

pipe wrapping In very corrosive environments, pipes or other metal objects may be wrapped with tapes impregnated with substances that prevent corrosion. These last better than paint but do not look beautiful.

piping, or subsurface erosion [s.m.] The movement through a dam or cofferdam of a stream of water and sand. Piping is a subsurface *boil*, and may, like boiling, become disastrous. See **critical hydraulic gradient, dispersive clays, toe filter**.

PIR foam *Polyisocyanurate foam.*

Pista grit trap An advanced design of *detritus tank.*

pit (1) A surface excavation to obtain sand, clay, gravel, etc.; a *borrow pit* or a *trial pit* for exploration of the ground.

(2) A rectangular hole dug for *underpinning* or for building foundations, *retaining walls*, etc.

pit boards *Well curbing.*

pitch See (*B*).

pitched work A stone *revetment* for the slopes of a reservoir or river bank, also called *pitching*.

pitcher A granite *sett.*

pitcher paving *Paving* with granite *setts.*

pitching (1) Lifting a *runner* or a *pile* and placing it in position for driving. In a small steel sheet pile *cofferdam*, all the piles are pitched before any are driven, ensuring that they fit.

(2) **penning, soling** Large stones 18–45 cm deep placed on edge and wedged by small stones called spalls (or rolled) to form a road foundation or a *revetment* to protect an earth slope from scour. Road pitching is usually placed directly on the ground. Embankment or river bank pitching should be placed on about 23 cm of quarry rubble, chips or pebbles. For heavy waves the stones should be at least 50 cm (20 in.) deep and 15 cm (6 in.) wide. See **beaching**.

pitching ferrules One or two short lengths of galvanized steel pipe cast into a reinforced-concrete pile and used as holes for lifting it. The exact position is calculated so as to give the least possible *bending moment*, since this is the biggest bending stress which occurs in the life of the pile.

Pitot tube [hyd.] (see diagram, p. 318) A device used in measuring the pressures and the velocities of flowing air or water. Of the tubes in the stream, one orifice faces the current, the other is at right angles to it. The latter measures the static pressure, the former the total (velocity + static heads). The difference between the two readings is the velocity head and from it the velocity can be calculated. Because the Pitot tube is so small, it can be used without disturbing the flow in small pipes or air ducts and can therefore be very accurate.

pivot bridge A *swing bridge.*

placing boom The hinged delivery pipes on a *mobile concrete pump.*

placing plant Plant for placing wet concrete in position. This may consist of *cableways, concrete placers, concrete pumps, conveyors, dumpers*, lorries, *pedestrian-controlled dumpers, placing booms, platform hoists, telfer* cranes, *trémies* or *vibrators*, or merely of one

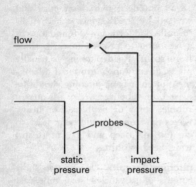

flow →

probes

static pressure

impact pressure

Pitot tube probes. (E. A. Parr, *Industrial Control Handbook*, vol. 1, *Transducers*, Collins, 1986.)

barrow and a road made of scaffold boards.

plain concrete Concrete with no reinforcement to carry weight or bending forces but with light steel to reduce shrinkage and temperature cracking, generally about 0.6% of the volume of the concrete. Mass concrete has less steel than this, sometimes none.

plan [d.o.] A view from above of an object or an area in *orthographic projection*.

plane frame [stru.] A frame in which the centre lines of all the members are in the same vertical plane. Most building frames are of this type for the reason that calculation and fabrication are simple. Three-dimensional frames (*space frames*) usually cost more in labour of design and construction, though they may also provide a roof over an area which could not be covered in any other way. A single plane frame is not stable against wind but needs support from walls, from bracing to neighbouring frames or from a shear wall. See **core wall**.

plane of rupture In the *wedge theory* the plane along which during the design of a retaining wall the retained earth is imagined to fail. The angle of elevation of the plane determines the weight of earth retained.

plane of saturation [s.m.] The *water table*.

planer (1) A machine fitted with milling cutters for smoothing a road surface which is usually first heated to soften it.
(2) [rly] A machine used on the track and in the works for planing rails and thus restoring their usefulness.

plane surveying [sur.] The measurement of areas on the assumption that the earth is flat and has no curvature. This assumption leads to no noticeable error for small areas. Cf. **geodetic surveying**.

plane table [sur.] A drawing board set on a tripod for use in the field. The fixing to the tripod includes a swivel head which enables the board to be rotated and oriented as required to the surrounding country. An *alidade* enables the bearings of objects sighted to be ruled in pencil on the drawing board. The positions of the objects along the ruled line can be fixed by taping or by *stadia work* from a tacheometer beside the plane table, or by a second plane table *set-up*. See **intersection, progression, radiation, resection**.

plane-tabling [sur.] Mapping with a *plane table* and *alidade*. It is a very quick way of mapping small areas of open country and is one of the earliest methods of surveying, being several hundred years old.

planimeter [d.o.] An *instrument*

which measures the area of a plan whose perimeter has been traced out by its moving arm.

planish [mech.] To smooth or polish by light hammering with a smooth hammer.

planning engineer A civil engineering assistant or *chartered engineer* who plans for a contractor's requirements of cranes, mixers, and other plant and materials in a contract.

planometric projection [d.o.] A pictorial view of an object showing it in *plan* with oblique parallel lines from the corners showing the front, side and thickness. See **projection**.

plant See **contractors' plant**.

plant mix *Soil stabilization* by carrying the soil to a stationary mixer, returning it to the site and respreading it. See **soil mixer**.

plant-mixed concrete *Ready-mixed concrete*.

plasma Ionized gas, consisting of positive ions and electrons in equal number, therefore electrically neutral but conducting. For welding, cutting, metal-spraying, etc., a common gas is 85% nitrogen, 15% hydrogen.

plasma torch A type of *TIG welding* torch in which the arc *plasma* is forced through a water-cooled opening to create an intensely hot flame for welding, cutting or spraying of metal or ceramic.

plastic cracking, p. shrinkage c. Cracks in freshly compacted (*green*) *concrete*, caused either by settlement of the slab or by its drying *shrinkage*. Although they may pass right through the slab, they do not seriously weaken it and may be healed at least partly by good curing. Polythene sheet put down before the surface water has dried out can prevent cracking. Any retarding *admixture* should be removed from the mix to reduce the time available for cracking. After cracking has occurred the cracks can be closed by cement grout applied immediately, or by *revibration* if the cracks were noticed soon enough. See **autogenous healing**.

plastic deformation, p. flow, p. yield [stru.] The flow without fracture of a material during loading. Mild steel and other *ductile* metals show plastic yield during tensile testing above the *yield point*. Clays also are plastic. Any metal which yields plastically is usually preferred in structural design to one which breaks suddenly as in *cleavage fracture*. See **plastic fracture**.

plastic design, collapse d., limit d. [stru.] The design of a structure on the assumption that *plastic hinges* form at full load. In a *continuous beam* or other continuous structure the hinges redistribute the bending moments by comparison with the elastic (pre-plastic) state. An economically designed frame has many simultaneous plastic hinges. Cf. **elastic design**; see **limit states**, **load factor**, **plastic modulus**.

plastic flow See **plastic deformation**.

plastic fracture Breakage of a metal in tension by drawing out (*necking*). The necking produces a work-hardening effect on steel and the eventual failure is more gradual than the abrupt and therefore dangerous cleavage fracture. See **cup-and-cone fracture**.

plastic hinge [stru.] A point of maximum bending moment, which is assumed in *plastic design* to be stressed up to the yield point of the steel. If the hinge is in fact stressed to the yield point it does yield slightly and throws some bending moment on to other parts of the structure, forming eventually more plastic hinges. However, the greatest load applied to the structure (design load) is never more than two thirds of that which causes a plastic hinge and usually considerably less.

plasticity [s.m.] A soil is plastic if, like clay, when squeezed in the hand it does not break up. A concrete is plastic if it never suffers *segregation*. One highly plastic mortar, made with *microsilica*, was plastered on a vertical face, 10 cm (4 in.) thick, without difficulty. Cf. **elasticity**.

plasticity index [s.m.] See **index of plasticity**.

plasticizer, densifier Either an expensive and highly effective *superplasticizer* or a cheaper, less effective *water reducer*.

plastic limit [s.m.] The water content at the lower limit of the plastic state of a clay. It is the minimum water content at which a soil can be rolled into a thread of 3 mm dia. without crumbling. See **consistency limits**.

plastic modulus [stru.] A value used in the *plastic design* of steel structures, which is a constant for each particular shape of section and corresponds to the *modulus of section* used in *elastic design*. The plastic modulus of a rectangular beam is 50% larger than its section modulus. For I-section beams the plastic modulus is generally within 1% of $1.15 \times$ the modulus of section.

plastics, polymers Of the two main classes of plastics, *thermoplastics* and *thermosets*, thermoplastics soften on heating and harden again on cooling without chemical change. But thermosets are chemically changed during moulding to become inert, non-softening materials like melamine formaldehyde table tops, *epoxides* or *polyesters*. All common plastic pipes are thermoplastic and melt in a fire.

Probably all plastics burn but not all burn easily. Some, like PVC, *glassfibre-reinforced plastic* (GRP) laminates, phenolic laminates, nylon or polycarbonate, burn with difficulty or are self-extinguishing. Cast phenolic resins are non-flammable, phenolic foams do not give off fumes in fire and perhaps do not burn. Most plastics have high *creep* and a high *coefficient of thermal expansion*. Because of their low stiffness (*modulus of elasticity*) compared even with timber, most structural designs in GRP are aimed merely at limiting the deflections.

By international agreement, plastic pipes are described by their outside diameters. See also **sewer pipe materials** and (*B*).

plastic settlement In concrete, sinking of the solids, forcing the water to move upwards (*bleeding*) after vibration and before the *initial set*. If there is too much plastic settlement, cracks may appear above the steel bars in a slab or beam. Cf. **plastic cracking**.

plastic welding [mech.] Welding with steel or iron in the plastic state, such as *forge welding* and, usually, *pressure welding*. (Not to be confused with heat welding or fusing of *polymers*.)

plastic yield [stru.] The usual term for *plastic deformation*.

plat [sur.] Term used in the USA for a plan showing land ownership, boundaries and subdivisions with their descriptions but nothing else.

plate (1) [mech.] Copper thicker than 10 mm and wider than 0.3 m, or steel thicker than about 3 mm (11 SWG, but see **steel sheet**). Cf. **sheet** (*B*).

(2) [sur.] The upper plate of a *theodolite* is the vernier plate carrying the telescope and the vernier within the

lower, outer plate, which is graduated from 0° to 360°. The upper plate is often called the plate, the lower plate is the *limb*. See also (*B*).

plate bearing test [s.m.] An old method of estimating the bearing capacity of a soil by digging a pit down to the proposed foundation level, placing a stiff steel plate about 0.3 m square on the foundation, and loading it until it fails by sinking rapidly. The method has been somewhat discredited since it has been proved that on compressible soils the results apply to a depth of only about 1½ times the plate width. If the soil below this depth is softer, the building as a whole will probably settle faster than the plate at the same load. However, if the soil is homogeneous to a depth, below the building, equal to twice the building width, or if the underlying soil is harder than that at the plate, the method may be useful. See **ultimate bearing pressure**, **side jacking**.

plate bonding Strengthening a concrete beam or slab by gluing a steel plate on to it with epoxy resin, after scrupulous preparation of the concrete surface.

plated beam See **compound girder**.

plate floor, beamless f., flat slab [stru.] A *reinforced-concrete* floor with a flat undersurface, commonly used in office buildings because it allows room shapes to be altered easily. Careful structural design is needed for the slab with its hidden beams containing much steel, but some contractors find it the cheapest floor to build. There is a saving in *formwork* because of the smooth undersurface. There is also no laborious placing of the blocks of a *hollow-tile floor*. A saving in weight equivalent to that obtained from hollow blocks may come from using *light-*

weight aggregate. See **one-way slab**, **two-way slab**.

plate girder [stru.] Formerly a girder with angles riveted to plates, now usually a unit with two flange plates welded to a web plate. See **Britannia Bridge**.

platen [mech.] In a compression-testing machine such as those used for concrete cubes, the smooth steel plates which are in contact with the concrete faces during the test. See also (*B*).

plate screws [sur.] *Foot screws*.

plate vibrator [s.m.] A self-propelled mechanical *vibrator* with a flat base, used for compacting fill which is to be built on.

platform, North Sea p. See **offshore platform**.

platform gantry A *gantry* built to carry a *portal crane* or for other purposes.

platform hoist, goods h. Many types of hoist are used in building. Those with a tower fixed to the building being erected can be used for wheeling a loaded truck on at one level and off at another and can carry 1 ton or more. Whether self-propelled or towed, *access platforms* can reduce the delay and expense of scaffolding, e.g. for fixing the *cladding* over a building surface. *Scissors lifts* rise only vertically, other types have considerable outreach.

plot [d.o.] (1) To draw a map of the ground from field notes made by a surveyor.
(2) To draw a *graph*.
(3) An area of land for building, called a 'lot' in the USA.

plotting instrument, plotter, stereoplotter [d.o.] A large automatic drawing instrument which can accurately measure land or objects on air photographs, as well as providing *stereoscopic* vision. Optical and mechanical versions enable scale drawings to be made from photos by reproducing the geometry of the camera in two-dimensional space. Computer versions use programs to calculate three-dimensional coordinates from the two-dimensional photos, and, combined with the variables of the camera or imaging system, generate either a drawing or a computer file of coordinates. See also **computer graphics**, **orthophoto**.

Plougastel Bridge A bridge first built in 1930 over the Elorn river in northwest France, having three arch spans at 187 m centres. It is of reinforced concrete with a very small amount of reinforcement (less than 0.3% total) and is also known as one of the earliest sites where *Freyssinet* used *flat jacks*. The *formwork* for one arch was built on the shore and floated out on barges into position under the first arch. When this arch was built, the formwork was lowered and floated away to the next arch. The bridge was destroyed in 1944 but rebuilt to the original design.

plough Farm equipment for loosening *topsoil*. See also **mole/snow plough**.

plough steel [mech.] A name used for the steel from which wire is drawn for making steel ropes, or for *prestressed concrete*.

plug and feathers, multiple wedge [min.] Feathers are two steel bars which fit together. They are flat on the inner face, curved on the outer face, and inserted into a hole drilled in rock. The plug is a wedge which is inserted between them and driven in. A line of holes is drilled and then wedged simultaneously to split the rock along the line. It is used for cutting out rock in foundations where the rock must not

be shattered, like *broaching*, or for cutting dimension stone. The plug can be driven by hand-hammer or with a *jack-hammer*.

plum, displacer A large stone of any shape dropped into a mass concrete structure such as a dam to reduce the volume (and cost) of the concrete.

plumbago Graphite used as a *refractory* for making crucibles.

plumb bob, plummet [sur.] A weight hanging on a string called a *plumb line* to show the direction of the vertical.

plumb box [sur.] A flat glass mirror with cross-hairs. It is accurately levelled and placed on the ground below an opening in the floor above it. When its cross-hairs are sighted from above, the line of sight is vertical. This little used, simple device deserves to be better known.

plumbing [sur.] Transferring a point at one level to a point vertically below or above it, usually with a *plumb bob*, *optical plummet*, etc. See also (*B*).

plumb line [sur.] A string on which a weight is hung to stretch it in a vertical direction. The string should be braided like fishing line to avoid spinning of the bob. Wire is used for *shaft plumbing*.

plummet See **plumb bob**.

plunger See **ram** (3).

plus sight [sur.] In levelling, the term recommended by ASCE for what is called in Britain a *back sight*.

ply-web beam (see diagram) A built-up timber beam with a web of strong plywood, and flanges of timber glued and/or nailed to it. They can be built up on the site or in the factory for spans of 12–15 m (40–50 ft). Box beams, double-web beams or multi-web beams exist.

 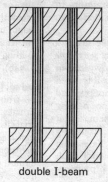 

| I-beam | double I-beam | box beam |

Ply-web beams, nailed and/or glued.

plywood Widely used for lining *formwork*. See (*B*).

pneumatic Acting by the pressure of air or gas.

pneumatic breakwater See **bubble barrier**.

pneumatic caisson A *caisson* in which the *working chamber* is kept full of *compressed air* at a pressure nearly equal to the water pressure outside it. Its advantage over the *open caisson* is that men can work in it in the dry and without wearing diving suits, but they are under an air pressure considerably above atmospheric. Its limiting depth is about 35 m of water, 3.5 bars of pressure. Open caissons have been sunk to much greater depths – as much as 80 m below water level. Pneumatic caissons are much more expensive so are used only when circumstances compel for one or more of the following reasons: (i) when the *formation* below water level has to be inspected and other means of dewatering are not possible; (ii) when the ground contains rock or boulders, making a sheet pile *cofferdam* impracticable; (iii) when the caisson forms the starting point for a tunnel to be driven under compressed air; (iv) when the adjoining ground must be properly supported and not allowed to flow in, as may happen with grabbing from an open caisson. See also **decanting, hydraulic ejector, weighting, sealing pneumatic shafts to rock**.

pneumatic conveyor [mech.] A tube through which powder or granular material is transported by an air blast. It is used for cement, wheat, pulverized coal, etc.

pneumatic drill [mech.] A compressed-air drill. See **rock drill**.

pneumatic impact mole, p.i. pipe burster, pipe expander A compressed-air-driven *microtunnelling* or *pipe renewal* device. It is fast, powerful and effective but may damage nearby services, and at large sizes can freeze up underground when working at full power because of the large flow of compressed air. A 152 mm (6 in.) pipe burster can be lifted only by crane. They exist in sizes from 40 to 250 mm (1.5–10 in.) and can advance at about 10 m/hour. A 130 mm (5 in.) dia. machine weighs 105 kg in all and its 45 kg hammer delivers 350 blows/min. In *pipe ramming* it can drive a *steel sleeve* of 300 mm (1 ft) dia.

A 95 mm (4 in.) dia. mole needs a *launch pit* 1.9 m long, but if *polythene* pipe is to go in, a shallow lead-in trench is also needed, possibly even longer. These moles cannot expand cast-iron pipe or clay pipe passing through a concrete encasement, and it is not easy to guide them accurately over long distances. They have been used by British Telecom and British Gas. See **hydraulic mole**, **pipe ramming**.

pneumatic mortar *Gunite* or *shotcrete*.

pneumatic pick A light *concrete breaker* (9–14 kg).

pneumatic sewer ejector An *air-lift pump*. Cf. **hydraulic ejector**.

pneumatic shaft sinking The use of a *pneumatic caisson*.

pneumatic structure See **inflatable**.

pneumatic tool [mech.] Any tool worked by compressed air, usually a hand tool.

pneumatic-tyred roller, rubber tyred r., multi-wheel r. A towed *compaction* unit with many independently sprung wheels. It can be loaded with water tanks or other ballast and may weigh up to 200 tons. It compacts earth dams and similar fills in 10–15 cm thick layers by from 8 to 10 passes of the roller. See **wobble-wheel roller**.

podger, construction spanner A single-ended, open-jaw spanner with a pointed handle, 25 cm long for 10 mm bolts, up to 1 m long for 35 mm bolts. The pointed end is used for aligning two or more drilled steel plates which are to be bolted, such as fishplates to

rails, or stanchions. The point is inserted through the holes and moved about until they are brought into line. It is used for railway work and building steel frames. See diagram (p. 349).

podzol [s.m.] A relatively acid, surface soil from the *A-horizon* of temperate climates, from which much soluble material (iron and aluminium oxides) has been leached into underlying soils. It has more silica than iron and aluminium oxides together, and may have up to seven times as much.

point-bearing pile An *end-bearing pile*.

point gauge [hyd.] A sharp point fixed to an attachment which slides on a graduated rod for measuring the water level. The point is lowered until it barely touches the water surface. See **hook gauge**.

point load [stru.] A *concentrated load*.

point load strength test The *splitting test*.

point of inflexion A point of *contraflexure*.

points, switch tongue. [rly] Hinged, tapered *rails* which can be arranged to direct a train on to one or another of the tracks at a junction. The hinge is at the heel of the rail, the toe of the movable rail being locked against the *stock rail*. According to the direction of approach of a train, points are called facing or trailing points. If the train first meets the heel they are trailing points, if it first meets the toe they are facing points. See **catch points**, **spring points**, **turnout**.

Poisson's ratio, p [stru.] For elastic materials strained by a force in one direction, there will be a corresponding strain in all directions perpendicular to this, equal to p times the strain in the

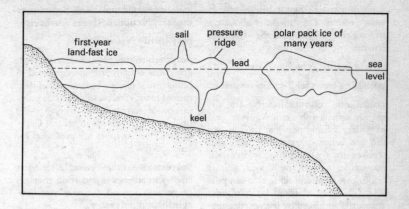

first-year land-fast ice · sail · pressure ridge · polar pack ice of many years · lead · sea level · keel

Sea ice terms. (J. Gaythwaite, *The Marine Environment and Structural Design*, Van Nostrand Reinhold, 1981.)

direction of the force. Poisson's ratio for steel and aluminium is 0.30. The relationship between the *modulus of elasticity E*, the *shear modulus G*, and Poisson's ratio is as follows:
$$E = 2G (1 + p).$$

poker vibrator The usual name for an *internal vibrator*.

polar ice (see diagram) Of the world's fresh water, 98% is in polar ice caps.

polariscope The optical apparatus used in the *petrographic microscope* and in the *photo-elastic* analysis of stressed models. The object to be examined under polarized light is placed between two *Nicol prisms*, a polarizer and an analyser.

polarized light Light in which the vibrations are in one plane only is said to be polarized. A *Nicol prism* filters out all light except that vibrating in one plane, and thus conveniently polarizes the light in a *petrographic microscope*.

polarizer One of the *Nicol prisms* used in a *polariscope*.

polar moment of inertia [stru.] The polar moment of inertia of a plane section is its second moment of area about an axis perpendicular to its plane. If the axis passes through the *centroid*, the polar moment is equal to the sum of the other two moments of inertia about axes passing through the centroid and in the plane but perpendicular to each other.

polder Low-lying land reclaimed from the sea by enclosure with *dykes* (Holland), followed by pumping.

pole (1) [elec.] A terminal of an electric supply, an *electrode*.
(2) See **derrick**.

poling back Excavating behind timbering which has already been placed.

poling board (1) A steel or wooden board 1–1.5 m long driven outside the leading timber frame of a *heading* in weak or running ground, to hold up the roof or sides when *forepoling*. See **jet grouting**.
(2) Vertical boards which support the

sides of a trench or pit being sunk in loose ground. They measure about 1.2 m long and 3–5 cm thick. See **tucking frame**.

poling frame A *tucking frame*.

polishing [sewage] *Tertiary treatment*.

pollutants, contaminants In air, mainly smells (poisonous gases), and in water, soluble poisons, though *fly-ash* and other dusts, solid *fumes* and smokes are air pollutants, and carbon monoxide, a most poisonous gas, is odourless. Heat and noise also can pollute. They are checked by 'emission standards' enforced by the *environmental health officers* of local authorities, or by the *Pollution Inspectorate* of the Department of the Environment or the *National Rivers Authority*. In Scotland river purification boards control river pollution, regional authorities control water and sewage treatment, and district councils control the disposal of solid waste. In the USA the main control is through the very powerful EPA, the Environmental Protection Agency.

All pollution except that from volcanoes is manmade. Until about 1650 the world population increased by doubling every 15 centuries. It now doubles every 38 years, thus increasing pollution. Another worry is that 80% of the world's arable land is already under cultivation. See **Camelford, flue gas desulphurization**, NO_x.

Pollution Inspectorate, HMIP (Her Majesty's Inspectorate of Pollution) A body controlling all types of pollution, which came into existence as part of the Department of the Environment on 1 April, 1987. It combines the functions of various bodies, now defunct, called the Alkali, Radiochemical, and Hazardous Wastes Inspectorates, and is interested in the quality of discharges by *water service plcs* and others. See **cancer cluster**, **National Rivers Authority**.

polyamide *Nylon (B)*.

polycarbonates Plastics stronger than glass, and equally transparent, used in bullet-proof glazing in banks. But they cannot resist alkaline solutions. A polycarbonate semi-tube protects a cycle bridge over rail tracks in central Cambridge. Similar tubes are proposed for large bridges.

polyesters *Synthetic resins (B)* used as *binders* in adhesives and resin mortars or *polymer concretes*. See **glassfibre-reinforced polyester**.

polyethylene The chemical name for 'polythene'.

polyisocyanurate foam, PIR f. A thermally insulating foam (80 kg/m^3) made from a modified polyurethane, used inside hollow *glassfibre-reinforced concrete* panels, or blown as a 25 mm (1 in.) thick strengthening layer to hold the underside of a tiled or slated roof with rotted nails.

polymer concrete Monomer or resin, mixed with aggregate, and polymerized (cured) after placing. Polymer concretes though expensive are used in many different ways in civil engineering, especially for fast repairs to roads and bridges, or for precasting wall panels or pipes for water or sewage. Neat polymer concrete hardens in 30–90 min. More than 100 bridge decks in North America have been given overlays of polymer concrete some 13 mm ($\frac{1}{2}$ in.) thick, mostly with polyester, about 20 with epoxy, and more recently with *methyl methacrylate* or styrene, sometimes with *isobutyl silane* added. The aggregates must be dry and their grading must be workable. Rock flour or even cement may be added to give workability. Sometimes the polymer wearing surface is attached to the steel bridge plates in the fabricating shop. Electrically conducting polymer concrete is in use with *cathodic protection*. Steel-fibre-reinforced polymer concrete has been found to be strong, with a bending strength of 85 MPa (12,000 psi). This is achieved by using a high proportion of steel fibres, from 7 to 18%, introduced not by mixing but by grouting with the polymer mortar.

Many polymer concretes need aggregate containing less than 0.5% water, achieved by laborious drying at 110°–112°C for 4–8 hours. The mix contains from 6 to 10% resin. Such concretes feather down to thicknesses of 4 mm or even less, but for thin layers two or three coats are usual. They bond well to concrete or masonry, are tough, impervious, resist chemicals and frost, and have good water retention while hardening. But their compressive strength may be less than Portland cement concrete and their *curing* needs are opposite. An anti-foaming agent may be needed to reduce *air entrainment*. Steel tools need cleaning with solvents even while work is in progress when epoxy or polyester are used. Hardening is usually within an hour, but during that hour *workability* is good because of the entrained air.

polymer grid reinforcement [s.m.] *Reinforced earth* using *nonwoven* plastic grids.

polymeric linings to pipes Plastic *pipe linings* can be of many types: GRP (glass-reinforced plastic), or a *sock lining*, or polyester resin concrete, or polypropylene or *epoxy*, but not all, are approved for drinking-water.

polymer-impregnated concrete Ordinary hardened, dry concrete which has been soaked with a liquid plastic (monomer) such as *methyl methacrylate* (MMA) or styrene. The liquid which has soaked in is polymerized (the monomer becomes a polymer) by gamma radiation or other means to form either polymethyl methacrylate (PMMA) or polystyrene. The procedure is expensive but the concrete is strengthened in tension, compression and *E*-value, has less *creep* and drying *shrinkage*, high resistance to freeze-thaw, to chemicals and abrasion, but low fire resistance and a high coefficient of expansion. To reduce the cost, partial impregnation can be applied to highly stressed areas.

polymer injection See **flow enhancer, grout, polymer stabilization** and above.

polymer-modified glassfibre-reinforced cement GRC made with a polymer dispersion mixed in. Several have been tried; some of them allow the cement slurry to contain much less water than usual, making *vacuum* dewatering unnecessary and improving the long-term weather resistance. Antifoaming agents must be used. (BRE Information Paper 10/87.)

polymer Portland cement concrete (PPCC) *Polymer concrete.*

polymer stabilization [s.m.] Ground improvement by injection of a *grout* or grouts containing an organic compound which may strengthen the soil or reduce its *permeability*, or both.

polyolefins *Thermoplastics* that include *polythene, polypropylene* and polybutylene. Polyolefin pipes can be joined by butt-fusion welding, induction-welding, heat-shrink sleeves, *solvent cementing* or mechanical joints such as Viking–Johnson connectors.

polypropylene One of the most interesting and least expensive plastics used in construction. A polypropylene cord will last years, and it has high *tensile strength* – 34 MPa (5000 psi), rising for the fibres, to 550 MPa (80,000 psi). The *modulus of elasticity* is about 15

Polystal

| Density, kg/m³ | Tensile strength, MPa | | Bending modulus, N/mm² | |
	short-term	long-term	short-term	long-term
LDPE 915–926	–	–	–	–
MDPE 926–940	19	8.5	500	100
HDPE 940–965	25	6.5	800	180

GPa (2,200,000 psi). Nets made of it can reinforce a road wearing surface. See **Caricrete, Netcem**.

Polystal An advanced *composite*, a prestressing *tendon* made of glass-reinforced plastic and used for *post-tensioning* concrete. The first bridge using Polystal is in Ulenbergerstrasse, Düsseldorf, built in 1986, with continuous spans of 21 and 25 m (70 and 84 ft). The 59 Polystal tendons each have 19 glass-reinforced plastic rods of 7.5 mm dia. anchored in a special block. Each tendon exerts a force of 600 kN (60 tons). Since there is no risk of corrosion, such tendons need no concrete cover, but they are not cheap. See **fibre-reinforced plastics**.

polystyrene-bead-aggregate concrete (PBAC), Styropor c. Lightweight polystyrene beads set in cement mortar, sometimes sold as an insulating board. It weighs only 400–600 kg/m³ and may be used in hollow insulating wall panels, e.g. of *glassfibre-reinforced concrete*, with which it bonds well. The hollow GRC panels should have weepholes at the base to allow surplus water to escape.

polysulphide sealant Many firms make polysulphide *sealants* of high resilience that will accept movement up to half the thickness of the joint.

polythene (polyethylene), low-density p. (LDPE), medium-density p. (MDPE), high-density polythene (HDPE) A common material in its various grades for making pipes or even spherical silage tanks or *septic tanks* (*B*) as large as 80 m³ The properties are briefly given above. Pipe-jointing methods are mentioned under *polyolefins*.

polyurethane Polyurethane sealants in British mines were banned after a fire in the Michael Colliery, Fife, killed nine miners because of poison gas released from burning polyurethane. In the Kinross Mine, South Africa, in 1986, 170 miners died from a similar cause.

polyurethane foam grout A weak (0.6 MPa (90 psi)) grout successfully used for filling cavities behind an old 2 m (6½ ft) wide water tunnel repaired in California in 1986. In spite of its low strength it was preferred to cement grout because fewer injection holes were needed and injection was fast. It weighed only 80 kg/m³ (5 lb/ft³).

polyvinyl acetate (PVA) A chemical with many uses in construction, e.g. for gluing wood, making emulsion paints or mixing with cement mortar in a *dashbond coat*.

polyvinyl chloride (PVC) One of the cheapest plastics, versatile and obtainable in different hardnesses depending on the amount of plasticizer and other components added to make it suitable for wall sheet, flooring, *water stops*, corrugated roofing sheet, window frames, pipes, etc. See **post-chlorination, unplasticized PVC**.

polyvinylidene fluoride (PVdF) A long-life finish for cladding steel, aluminium, wood or reinforced plastic, which has been found to be colour-stable for more than 15 years.

pond (1) **reach, pound** [hyd.] The stretch of water between two locks on a canal or a river.

(2) A German unit of force equivalent to one gram, thus one kilopond (kp) is equal to one kilogram force.

Pont de Ré (see diagram, p. 330) A bridge more than 3 km long (from La Rochelle on the west coast of France to a holiday resort, the Île de Ré), built in only 19 months by the contractor Bouygues in 1987–8. It has 29 concrete box-girder spans of 110 m each. Its extraordinarily fast construction speed was probably in large measure due to Bouygues's *launching gantry*, designed for the job and weighing only 500 tons, carried on 48 wheels, each one motorized. Only six hours were needed to advance the gantry 110 m from one span to the next.

pontoon A vessel, generally flat-bottomed, for carrying plant or materials or for carrying a part of a floating bridge (*pontoon bridge*).

pontoon bridge A temporary or permanent bridge which floats on pontoons moored to the river bed. Permanent bridges are built in this way when the foundation material is very poor. In this case the pontoons may be of reinforced concrete (Lake Washington Bridge, near Seattle).

popcorn concrete *ACI*'s term for the UK's *no-fines concrete* (*B*).

pop-out Breakaway of small fragments from a concrete surface, leaving conical pits (*ACI*).

pop rivet, blind r. A tubular metal rivet with a notched shank, used for fastening ordinary or chromium-plated steel or nonferrous metal sheets. It is inserted into a predrilled hole and pulled from the far side with a special tool; the extension then pops off, leaving a smooth finish. It can rivet ventilation ducting or other sheet metal which is not easily accessible on the far side. Cf. **huckbolt**.

population [stat.] A number of units from one source which have one variable characteristic (at least) which is the subject of *statistical* examination, e.g. a consignment of bricks or a number of concrete cubes (crushing strengths).

pore filler, permeability-reducing admixture A mineral powder added to a lean concrete mix to reduce leakiness. *Bentonite, fly-ash, diatomite* (*B*), *hydrated lime* (*B*) or *microsilica* have been used. See **fillerized cement**.

pore refinement Conversion of large pores in concrete to small ones, usually by adding *pore filler* or an *air-entraining agent*. Pore refinement improves the *workability* and imperviousness of the concrete.

pores Voids, usually small.

pore-water pressure [s.m.] The pressure of water in a saturated soil, sometimes measured by inserting open-topped tubes into the soil, or tubes leading to a *Bourdon pressure gauge*. Pore-water pressure is often measured in earth dams during and after their construction because the amount of the pressure gives an indication of the *consolidation* process. If the pressure is zero, consolidation is complete. See **hydrostatic excess pressure**.

pore-water-pressure cells [s.m.] Sensitive instruments for measuring pore-water pressures due to load

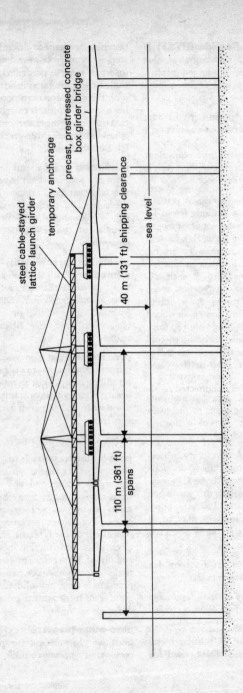

Travelling launch gantry building the Pont de Ré from La Rochelle to the Île de Ré. (*New Civil Engineer*, 22 September 1988.)

steel cable-stayed lattice launch girder

temporary anchorage

precast, prestressed concrete box girder bridge

40 m (131 ft) shipping clearance

sea level

110 m (361 ft) spans

changes such as the rise and fall of the tide.

porosity [s.m.] The ratio of the volume of voids to the total volume of a soil sample. In sands it is from 30% to 50% but in clays may be above 90%. If the *voids ratio* is e, the porosity is

$$\frac{e}{1+e}.$$

The pore space is the volume available for storing fluids – oil, gas or water. The saturation is the fraction of the pore space taken by storage.

Porosity in concrete is caused usually by excess water, especially with high W/C *ratios*. The worst porosity is usually at the edges, and may cause the steel to rust.

portable crane A crane which is not self-propelling but can be moved about on wheels. It has a power-driven hoist and sometimes also power slewing and *derricking*.

portal (1) **portal frame** [stru.] (see **Titan crane**) A *frame* consisting of two uprights rigidly connected at the top by a third member which may be horizontal, sloping or curved. It is a *redundant* frame.

(2) [min.] The entrance to a tunnel.

portal crane, p. jib c., gantry c. [mech.] A *jib crane* carried on a four-legged *portal*. The portal is built to run on rails set parallel to the quayside in the floor of a quay. Wagons and lorries can pass under the portal or the portal can pass over them.

Portland blast-furnace cement A *cement* with at least 35% of *ordinary Portland* and up to 65% of blast-furnace slag crushed with it. This is good for making the concrete in dams, which does not need high early strength. It resists leaching well and has a lower heat of hydration than Portland cement. See **Trief process**.

Portland cement Many different cements now in use are Portland cements or at least contain some; the varieties include: *ordinary, rapid-hardening, ultra-high-early-strength, Portland blast-furnace, sulphate-resisting* and *water-repellent cements*, apart from coloured cements. Portland cement is made by heating to clinker, in a kiln, a slurry of crushed chalk or limestone and clay. The clinker is finely ground and some gypsum $(CaSO_4)$ is added.

Portland cement concrete The commonest *concrete* in use today.

Portland pozzolan cement *Portland cement* of which up to 30% has been replaced by natural or *artificial pozzolan*. The pozzolan combines with the free lime in the cement and may improve the concrete by reducing its *permeability*. It hardens more slowly than Portland cement concrete but eventually reaches or exceeds the strength of Portland cement concrete. In non-standard mixes the percentage of pozzolan may rise well above 30%. See also **microsilica** and BS 6610.

position head [hyd.] The *elevation head* of a fluid.

post-chlorination A chemical process which applied to *PVC* enables it to withstand 120°C and to be used for hot-water pipes.

post-cooling See **precooling of concrete**.

post-hole auger A hand tool for making post holes or for *site exploration*. It is used down to about 4 m for 15 cm (6 in.) dia. holes, but holes are drilled faster with a *truck-mounted drill* or *shell-and-auger boring*.

post-stressing *Post-tensioning*.

post-tensioning A method of *prestressing concrete* in which the cables are pulled or the concrete is jacked up after it has hardened. This method is usual for bridges and heavy structures which are poured in place. The losses of prestress are slightly smaller than with pre-tensioning. See **anchorage** (2), **pre-load tank**.

pot A *hollow tile*.

potable water Drinking-water.

potential energy [mech.] *Energy* due to position such as the *elevation head* of water or the elastic energy of a spring or structure caused by its deformation.

potentiometer, pot, voltage divider A precise, variable electrical resistance, usually a moving contact arm (the slider or wiper) which travels over a coil of fine resistance wire wound on a hard spool. It is used in many *sensors* to measure voltages.

pot floor A *hollow-tile floor*.

pound (abbr. lb) A unit of weight in some English-speaking countries equal to 454 grams (0.454 kilogram). 2240 lb are equivalent to 1 long ton. Also see **pond**, and conversion factors, p. xiv.

pound-foot [stru.] A unit of *bending moment*, the effect of 1 lb force at a distance of 1 ft. Cf. **foot-pound**.

powder spreader A *bulk spreader*.

power [mech.] Mechanical power can be provided by portable *internal-combustion engines* or by stationary plants such as electric power stations or air compressors with power lines running out from them to small motors. The sources of power are generally only fuel and falling water, though occasionally the sun, winds, and the warmth of rivers are used. In electricity, power is the product of current squared times resistance, I^2R, which is the same as voltage times resistance, EI, and is expressed in watts or *kilowatts*. Mechanical power is expressed in *horse-power* or kilowatts. See conversion factors, p. xiv.

power barrow, p.. buggy A *pedestrian-controlled dumper*.

power borer, pipe reamer A rotating pipe-cleaning tool with sprung cutters, pushed for distances up to 150 m through pipes as large as 300 mm (12 in.). It can clean very heavily encrusted water mains or sewers, being pushed, not pulled, and is claimed not to harm the pipe.

power earth auger [s.m.] A *truck-mounted drill*.

power float, trowelling machine Orbiting steel trowels driven by a petrol engine can give a very smooth, sometimes polished finish to a concrete slab, suitable for *high-bay warehousing*.

power pack, p. supply, p. source Any power supply, whether from electrical mains or from a *prime mover*. Hydraulic *excavators* or *loaders* have a power pack consisting of a high-pressure pump and relief valve transferring their input power into high-pressure moving *hydraulic fluid*. Computers and word processors need low-voltage direct current, so they have a power pack which converts the mains alternating current to their low-voltage d.c.

power rammer A hand-operated compacting machine which weighs about 90 kg and is raised by its own internal-combustion engine. It rams the earth by dropping on it. See **frog rammer**.

power shovel A loose term for any *excavator*.

power take-off [mech.] An external splined shaft on a *tractor* used generally while the tractor is stationary for driving plant such as winches, pumps, threshing machines, etc.

power wrench See **impact spanner**.

pozzolan, pozzolana, cement replacement A volcanic dust found at Pozzuoli, Italy, and used since Roman times as a *hydraulic cement* when mixed with lime. All pozzolans contain silica. (Sio_2) and siliceous or aluminous minerals. Artificial pozzolans are industrial *fumes* like *fly-ash* or *microsilica*. Both artificial and natural pozzolans harden more slowly than *Portland cement*, but where this does not matter money may be saved or concrete quality improved, or both, by using these cement replacements in suitable proportions. See **Portland pozzolan cement**.

pozzolanic activity, pozzolanicity The ability of the silica in a natural or *artificial pozzolan* to combine with the free lime in a cement to form calcium silicate hydrate, the strong part of hard concrete.

pozzuolana See **pozzolan**.

Pratt truss, Linville t., N-truss, Whipple-Murphy t. [stru.] A bridge-truss or roof-truss with vertical struts separating the panels. Many variations of the panel shape exist, one extreme being the *queen-post truss* (*B*) with only two verticals.

preamble In *bills of quantities*, the introduction to each *trade* (*B*), describing its workmanship and materials.

pre-boring for piles *Spudding* or boring holes for timber piles through ground which is too hard for them to be *driven* in without damage.

pre-camber, upward deflection Beams may be built with an upward deflection, making them slightly concave beneath, so as to avoid any appearance of excessive downward deflection under load. For relatively short spans up to 10 m (33 ft), the upward deflection in mm should equal roughly the number of metres in the span. For spans longer than 10 m it is about double this.

precast concrete Concrete beams, columns, lintels, *piles*, and parts of walls and floors which are cast and partly matured on the site or in a factory before being lifted into their position in a structure. Where many of the same unit are required, precasting may be more economical than casting in place, may give a better surface finish, reduce shrinkage of the concrete on the site, and make stronger concrete. See **pre-tensioning**. Cf. **cast-in-situ**.

precast floors See **pretensioned floors**.

pre-chlorination Clean *raw water* usually does not need chlorination. But if it is not clean the usual practice is to add a fairly large dose of chlorine (2–5 mg/litre). Added upstream of the settling basins this should oxidize the pollutants, precipitating manganese and iron, killing *algae* and bacteria, reducing colour and slime, and helping settlement. But pre-chlorination is not advisable for a water containing much silt. The chlorine is lost by absorption into the silt.

precipitation All the water which falls as rain, hail, snow or dew, expressed as daily, monthly or annual millimetres of rainfall.

precise levelling [sur.] With the best instruments and procedures an accuracy of about 0.5 mm/km is possible. In ordinary levelling the allowable difference between two successive measurements of the level difference between two points is about $12\sqrt{L}$ mm where L is the distance between the points in km.

precision [sur.] The precision of a measurement is the fineness with which it has been read. For instance a precise tape measurement may be taken to 0.001 m but may nevertheless be grossly inaccurate because of *tape corrections* which have not been allowed for. Precision is therefore different from *accuracy*.

pre-coated chippings, grit *Coated chippings* or *grit*.

pre-consolidation load [s.m.] By drawing the curve of compression of a clay which is compressed in the *consolidation press*, an estimate can be made of the highest load to which it has been subjected in the past. This pressure can be translated into a depth of soil and compared with the existing depth of soil. The difference is the erosion caused by glaciation and other processes.

pre-cooling of concrete in hot-weather concreting, the coarse aggregate and water may be cooled. Ice chips instead of mixing water can cool the concrete by $5.6°C(10°F)$. In hot climates concrete is commonly placed at temperatures below $10°C$ ($50°F$). Post-cooling is much more elaborate and expensive and implies building cooling-water pipes in the concrete mass, common in large dams in addition to pre-cooling. In the rebuilding of Maentwrog Dam, North Wales, in the hot summer of 1989, Kier used 3000 litres of liquid nitrogen at $-200°C$ in a pour of 600 m^3, adding £2500 to the cost. It was injected into the mixed concrete at the batcher in 40 s bursts but cooled it by only $6°C$.

prefabrication Apart from *precast concrete* units, manufacture in the factory also of bathrooms or toilet units, etc., completely finished and wrapped in plastic film. Such prefabrication can enormously speed up site work. Two men with air skates can place a bathroom pod without difficulty. Only the connections for electricity, hot and cold water, and drains have to be made on the site.

preflexed beam, prebeam [stru.] A type of *composite construction* in which a steel beam is loaded to bend down (sagging) with a concrete encasement around its lower flange until this hardens. The load is then released and the beam straightens, compressing the concrete round the bottom flange. When the beam is built in and the slab is cast, the top flange and web are concreted in the usual way. In Japan some 250 simply supported bridges have been built thus since 1967 and the technique has begun to be applied to *continuous-beam* bridges.

pre-formed rope [min.] A steel-wire rope made of strands which are bent to their *lay* before they are laid together. The rope therefore does not spin or kink.

preliminary treatment (1) For *raw water* many pre-treatments are possible. Some (but not all on the same water) of the following treatments may be used: *screening*, storage in a *reservoir*, *pre-chlorination*, *aeration*, reduction of *algae*, preliminary *sedimentation*, *coagulation*, mixing or *flocculation*.

(2) [sewage] The first treatment of sewage includes grit removal by a *detritus tank*, removal by *screens* of large solids, plastic film, leaves and twigs, and the use of a *scum weir* to float off grease, expanded polystyrene, etc.

pre-load tank A circular concrete tank prestressed by winding round the walls a single *post-tensioned* high-tensile wire in a continuous spiral. The tension is applied by a machine hung from wheels travelling round the top of the wall, with the wire gradually climbing to the top. The wire is tensioned on this machine by passing (for instance) a 5 mm dia. wire through a *cold drawing* die of 4.8 mm bore during winding. A 15 cm (6 in.) thick coating of *shotcrete* or *gunite* protects the prestressing wire.

prepacked concrete *Grouted aggregate concrete*.

prescribed mix A concrete in which the proportions and types of aggregates, cement, *admixture* and water are specified. Strength is usually not the main aim. Cf. **designed mix**.

preservatives for timber Civil engineering timbers for harbour work or others that do not need much handling are best treated against fungus and termites by *pressure creosoting*, but such timbers are unpleasant to handle. Copper–chrome–arsenic solutions are said to be as effective as creosote but cannot be used for harbour work, being soluble. General guidance is given by many BS. Railways now use boron pellets for extending the life of wooden sleepers in the track. New softwood sleepers are pressure creosoted.

press In steel *fabrication*, presses are used for punching holes, notching, shearing, joggling and so on. See **hydrostatic press**.

pressure (1) *Force* per unit area. In this sense it is similar to stress, but stresses are usually estimated on solids, pressures on fluids. See conversion factors, p. xiv.

(2) [min.] The air pressure for compressed-air *rock drills* can be from 5 to 7 bars, the most economical pressure being 6.2–6.6 bars, since although breakages are high at this pressure, drilling speeds are also high.

(3) [min.] *Drilling fluid* pressures require to be balanced against the gas pressure in an oil-well. Gas pressures vary between 90 and 120 mbars per metre depth.

(4) [s.m.] Ground pressure in shafts may reach, in loose ground above the water table, 5 KN/m^2 of shaft exposed per metre of depth. Below the *water table*, 10 KN/m^2 per metre depth is the usual maximum but this figure rises to 15 in fluid silt or clay. See **effective pressure, hydrostatic excess pressure**.

pressure creosoting The most effective way of preserving timber by creosoting, under pressure in tanks. See **preservatives for timber**.

pressure filter A *rapid sand filter*.

pressure gauge [mech.] An instrument for measuring fluid pressure. The *Bourdon pressure gauge* is a simple and popular device, but needs calibrating for precise work. See **manometer**.

pressure head [hyd.] The head of water at a certain point in a pipeline due to the pressure in it.

pressuremeter, push-in p. A soil-testing device that produces a *stress–strain curve* for a soil. It began with the work of Louis Ménard in France in the 1950s. He inserted a cylindrical 'balloon' into a pre-drilled hole and for ordinary soils inflated it to a maximum pressure between 2.5 and 10 MPa (360–1450 psi). For very hard soils or weak rocks the pressure can reach 20 MPa (2900 psi). The balloon is connected to a 'black box' which produces the data for the stress–strain curve. A more recent development, the 'push-in pressuremeter', makes its own hole in the seabed in water as deep as 1000 m. The 16 ton seabed frame measures 6 × 7 m (20 × 23 ft) in plan and carries a measuring probe 6 m high and of 160 mm ($6\frac{1}{2}$ in.) dia. See also **PAM, stress-probe**.

pressure tank (1) A tank in which timber is inserted for impregnating with creosote, zinc chloride or other *preservative*. Cf. **open-tank treatment**.

(2) A closed tank for heating tar or bitumen and spraying it through jets on to a road. See **tank sprayer**.

pressure vessel A usually metal tank to contain fluid under pressure. Steam boilers were the first, followed by *prestressed concrete* pressure vessels of 25 m (82 ft) internal dia. and 3 m (10 ft) wall thickness. The *Boiler & Pressure Vessel Code* of the ASME is the international standard. The *ACI* also has a useful guide, No. 361R–86, mentioning composite steel and concrete pressure vessels with a gastight steel lining buried in the concrete. See **nozzle**.

pressure welding, solid-phase w. [mech.] Welding by pressing the joint parts together while the weld metal is plastic. Many methods of electrical *resistance welding* use pressure. 'Cold welding' is the term used for softer metals which can be welded by pressure without heat. See **forge welding**.

prestressed concrete [stru.] Concrete in which cracking and tensile forces are eliminated or greatly reduced by compressing it by stretched cables within it, or by pressure from *abutments*. The two main methods both use bars or wires in the concrete: *post-tensioning* and *pre-tensioning*. Prestressed concrete is economical for spans which are large or where the beam depth must be reduced to a minimum.

A possible development is the prestressing of concrete by the use of *expanding cement*. The *cold drawn* wires used in prestressing are of carbon steel containing 0.7–0.8%C with a breaking strength of 1540 N/mm² (222,000 psi) at 7 mm dia. increased by further cold drawing to 2300 N/mm² (333,000 psi) at 2 mm dia. The 0.2% *proof stress* is often used to define the strength of the wires. The working stress allowed is generally 0.65 of the breaking strength or 0.80 of the 0.2% proof stress, whichever is the lesser. The behaviour of these wires on stressing varies considerably with the period which has elapsed since cold drawing. See **steam curing**, **tendon**.

prestressed concrete cylinder pipe, cylinder prestressed concrete pipe A welded steel pipe, used in pipelines, at whose ends a steel socket and a spigot are welded. Inside, it is lined with spun concrete. Outside, it is radially prestressed by high-tensile wire wrapping covered by cement mortar.

prestressed concrete pavement The success of the prestressed concrete pavement at Schipol Airport, Netherlands, since the 1960s has brought favour to this construction which costs more than reinforced concrete and needs more thought but may last better in the long run. Such slabs are centrally prestressed, usually in both directions, and laid on a *slip membrane* of two polythene sheets. Only one joint is needed every 140 m instead of ten with reinforced concrete. A prestressed-concrete slab 20 cm (8 in.) thick suffices where a reinforced-concrete one would need a 35 cm (14 in.) thickness.

A warehouse in Wakefield, Yorkshire, of 37,000 m² area was built in 1988 with such a prestressed-concrete slab 225 mm (9 in.) thick, using concrete of 39 MPa (5600 psi) 28 day strength. The minimum prestress in the concrete in the middle of the slab was 1 MPa (145 psi), applied by straight unbonded strands of 15.7 mm dia. in both directions. The slab was cast by the *long-strip* method, in widths of 5 m, but 6 m was tried and found satisfactory. No joints were visible in the slab after the prestressing. The only indication was a change in colour.

The slab was prestressed in three stages. The first stressing was longitudinal, to one third of the final stress, after about 40 hours when the concrete had reached 10 MPa (1450 psi) crushing

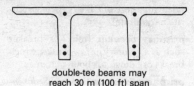

double-tee beams may
reach 30 m (100 ft) span

hollow-core beams are
often 1·2 m (4 ft) wide

topping provided on all
pretensioned floors

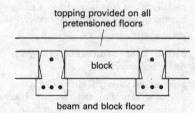

block

beam and block floor

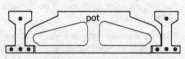

pot

beam and pot floor; pots
may be concrete or clay

Pretensioned floor beams. All have prestressing wire in the bottom, and a mortar or grano topping. Some are also reinforced elsewhere.

strength. When the infill bays between the first slabs had been cast they were stressed up to half the final stress. At about 72 hours the full prestress was applied, including transverse prestressing which closed up the construction joints. The dead-smooth finish was suitable for the *high-bay warehousing*, as intended.

prestressing [stru.] Applying forces to a structure to deform it in such a way that it will withstand its working loads more effectively or with less total deflection. When concrete beams are prestressed they deflect upwards slightly by an amount about equal to their total downward deflection under design load. Downward deflection is thus less than half that of a reinforced-concrete beam of the same shape. The struts (or braces in the USA) to deep excavations in bad ground are prestressed to prevent settlement of the surface and damage to neighbouring structures. See **pretesting, tendon** and above.

pretensioned floor beams (see diagram) In 1990 there were three common types of economical, pretensioned, precast, concrete floor and roof beams, the longest being double-tee beams for spans up to 30 m (99 ft), usually on roofs; then hollow-core beams, the most popular for fast construction, often 1.2 m (4 ft) wide; and third, for short spans, simple I-beams with unprestressed hollow pots or blocks between them. All these beams are likely to break if placed the wrong way up. They should be laid with the main wires in the bottom.

pretensioning, Hoyer method of prestressing [stru.] Concrete members are precast, in a works, with the tensioned wires embedded in them. The wires are anchored either against the moulds or against permanent abutments in the ground. After hardening, the concrete is released from the mould and the wires are cut at the anchorage. This method may give a larger loss of prestress than with post-tensioning but is usually economical for small members and may produce better concrete since it is always factory controlled. In long-line prestressing, used for the precasting of *pretensioned floor* slabs or beams, the casting bed may be as much as 180 m long, enabling units to be cut to any desired length with a *diamond saw*. The width of the units may be 1.2 m, and their thickness 15, 20 or 25 cm. They usually have tubular voids running down the length and occupying about 30% of the cross-section.

pretesting A term used by Lazarus White and Edmund Prentis to describe their patented method of *underpinning* tall buildings in New York. Steel-cylinder piles are sunk by jacking or grabbing (or both) to the rock, which may be 12 m (40 ft) down. The piles are filled with concrete. When the concrete has hardened they are jacked against the structure one at a time and pinned against it while the load is on the jack. In this way the elastic shortening of the pile, which in 12 m is appreciable, has no effect and the building is underpinned to new foundations without movement. Pretesting is the jacking of the piles, so called because it tests them to a load about 50% more than their design load. The word is also used for the prestressing of struts (braces in the USA) in timbered excavations.

pretreatment See **preliminary treatment**.

priced bill A *bill of quantities* is sent by a *consultant* to a contractor for him or her to *tender*. The contractor enters the price (rate) opposite each *item*, extends the prices and sums the totals. This priced bill, with the other *contract documents*, is the contractor's offer, *tender* or bid (USA) to do the work for the sum stated.

Primacord fuse [min.] US *detonating*

fuse with a core of PETN (pentaerythrite tetranitrate).

primary breaker [min.] A breaker such as a *gyratory* or *jaw breaker* which breaks ore down to about 5 cm max.

primary treatment [sewage] *Sedimentation*, the settling out of solids from sewage and their removal as a *sludge*, usually at least 96% water. The efficiency of primary treatment is improved by *preliminary treatment*. Sedimentation removes three quarters of the *suspended solids* and nearly halves the *biochemical oxygen demand* of the *effluent*. Sedimentation is also a common treatment for drinking-water, especially when it is taken from a river, but the quantity of sludge is smaller.

prime contractor A *main contractor*.

prime cost contract A *cost-plus-fixed-fee contract*.

prime cost sum (PC sum) The actual cost of an article of stated quality. It is a price entered into a *bill of quantities* by a *consultant*, to pay for the cost to a contractor of a specific article such as a set of gold-plated bath taps, or for work to be done by a *nominated subcontractor*, plus the *contractor*'s percentage defined in each *contract*.

prime mover [mech.] An *internal-combustion/steam engine*, *water wheel*, *steam/water turbine* or windmill which converts fuel or other natural energy into mechanical power.

primer (1) [min.] A cartridge which sets off the other cartridges in a hole. It is one in which a *detonator* is inserted and is usually either the first or the last of the charge.

(2) [s.m.] In *soil stabilization* a bituminous spray with which soil is covered after compaction so as to waterproof it. See also (*B*).

priming (1) Filling a pump or *siphon* so that it flows.

(2) [hyd.] The first filling of a canal or reservoir with water. It may occur annually or only once in the lifetime of a reservoir.

principal point [air sur.] The point on an air photograph where the optical axis of the camera intersects the film.

principal stress [stru.] If a piece is loaded by several forces in different directions, these *stresses* may be resolved into three simple *direct stresses* which are in planes at right angles to each other. Of these three planes, two are called principal planes and the stresses across them (which are greater than that across the third plane) are called principal stresses. See **Mohr's circle of stress**, **photo-elasticity**.

principle of Archimedes [hyd.] When a body is immersed in a fluid it loses weight by an amount equal to the weight of the fluid which it displaces, called its *buoyancy*. This principle applies to floating or submerged bodies as well as to those which neither float nor are submerged. See **displacement**.

principle of superposition [stru.] See **superposition**.

prismatic coincidence bubble, split b. [sur.] A refinement fitted to most modern, precise levelling instruments. An arrangement of prisms enables the surveyor to sight half of each end of the bubble at the same time, with the images of the two halves conveniently alongside each other. Exact levelling is thus made easier.

prismatic compass [sur.] A pocket compass by which bearings as close as $1°$ of arc can be read by holding the instrument to the eye and looking through the prism at the compass card while sighting the object.

prismatic telescope [sur.] A *theodolite* telescope with an eyepiece fitted

with a prism reflecting at $90°$. By this means steep sights can be taken easily.

prismoidal formula A formula for obtaining the volume of earth from the length of the excavation L, the two end areas A_1, A_2, and the area Am at the midpoint.

$$\text{Volume} = \frac{L}{6}(A_1 + A_2 + 4 \times Am).$$

Cf. **Simpson's rule**.

prism square [sur.] An *optical square* containing a prism.

privatization Sale of a nationalized industry to individuals who may run it at a profit. The first private electrical generating station was preparing to start up in Wales in 1988. Another, of 1000 MW, planned at Barkingside on the Thames, was to burn natural gas.

privatization of the water industry On 1 September 1989 the *water authorities* were converted to *water service plcs* with the loss of their *river basin management* duties to the *National Rivers Authority*. The sale of the plcs on the Stock Market was expected to follow shortly after. Water charges were expected to rise by 30–50% to meet EEC requirements for drinking-water and sewage discharges. See **Camelford**, **director-general of water services**, **prosecution of a water authority**.

probability [stat.] In *statistics*, the estimation of the number of chances of an event happening compared with the total number of chances. It is used in engineering for estimating probable strengths from occasional test results, and in *surveying* for estimating *probable error*.

probable error [stat.] A *deviation* equal to 0.6745 times the *standard deviation*. For a *Gaussian* distribution of values, ordinates placed at amounts equal to the probable error each side of the average value include half of the observations. It is therefore the most probable error. The idea is used in *surveying*, in construction and in sampling.

probe See **fibrescope, sonde**.

probing, piercing [min.] Pushing or driving a pointed steel rod up to 6 m long into the ground for determining the position of bedrock or hard lumps. In Burma, bamboo rods are used in probing for gem gravels.

process-control engineer A person who must be familiar with pipework, mechanical engineering, electricity, electronics, computing, hydraulics, pneumatics, chemistry, physics, etc. Similar qualifications are needed by sewage-works managers.

processing [s.m.] The various operations of *soil stabilization*, including *pulverizing*, moisture control, addition of *stabilizer*, mixing, rolling and covering with a *primer*. See **plant mix**.

Proctor compaction test [s.m.] A way of compacting soils in the laboratory, standardized to give results which can be compared. By weighing different compacted samples of the same soil the *optimum moisture content* can be determined. The test is used for soils in earth structures such as *earthen dams*, *soil-stabilized* roads and so on. See **compaction**.

Proctor plasticity needle, penetration n. [s.m.] A rough but convenient instrument for measuring the resistance of a soil to penetration at a standard rate of 13 mm/s. Needles of area varying from 645 to 32 mm^2 are used, and a spring balance shows the force needed to push the needle in. Granular soil must be screened to remove coarse material which gives erratic results. Cf. **California bearing ratio**.

production platform An *offshore platform* from which oil or gas, or both, are sent away by ship or pipeline.

profile (1) See **longitudinal profile**.
(2) [sur.] Two upright posts driven into the ground, joined by a horizontal board (*sight rail*) nailed to them. With other profiles it shows the amount of earth that must be dug out along a trench, using a *traveller* (2) of the right length.

profile paper [d.o.] Squared paper on which profiles of ground levels can be drawn.

profilometer A towed instrument which records the surface irregularities of a road.

program, software The instructions to a computer which enable it to fulfil a task. Programs are of very many types but nearly all have to be bought. An operating system, the first program, is usually provided with a *microcomputer* when this is bought. All other programs are written to conform to it. With it there could be a GEM program (graphics environment manager) which 'surrounds' the operating system to make it 'friendlier'. GEM is used by many inexpensive micros. A WIMP (windows, icons, mice, pull-down menu) program, also very friendly, is based on GEM. All programs require memory; often the better and friendlier the program is, the more memory it needs. Before buying a program it is advisable to check that your micro has enough memory for the program in question and whether it is possible to buy and connect more if needed.

programme See **master programme**.

progression, traversing [sur.] Doing a *traverse* with a *plane table* set up at each station in turn, using the *alidade* to align the backsight and foresight and taping (or determining by *stadia*) the distance between stations. The traverse lines are drawn on the plane table as they are made.

projection [d.o.] Drawing on a plane surface (usually paper or cloth) such objects as machines, buildings or the earth's surface, which are not plane. To make the drawing practicable it is assumed that parallel lines are projected from each point on the object towards the paper. See **axonometric/ isometric / oblique / orthographic / planometric projection**.

projection welding [mech.] An electrical *resistance welding* process like *resistance spot welding*, except that projections are formed at the places to be welded before they are put in contact between the *electrodes*.

project manager The *contractor*'s most responsible representative on a site; usually several *agents* on the site work under him or her.

project network analysis *Critical path scheduling*.

promoter, employer, owner The initiator of a *civil engineering project*.

proof stress [mech.] A means of comparing the strengths of metals which have no definite *yield point*. The 0.2% proof stress, for example, is that stress which causes a *permanent* set of 0.2% in the material; it is often used as a basis for the safe working stresses of light alloys, both in compression and in tension. See **high-strength friction-grip bolts**, **prestressed concrete**, **secant modulus of elasticity**; also BS 4461 and BS 4486 on cold-worked steels.

propeller fan, axial-flow f. [min.] A modern fan evolved from the airscrew. It is easy to reverse and adjust, efficient but noisy. Several airscrews can be placed in series and the power of the

fan thus increased without difficulty and at reasonable extra cost.

proportionality [stru.] See **limit of proportionality**.

proportioning The measurement by weighing or volume-batching of the constituents of a concrete, mortar or plaster before they are mixed. See **batching plant**.

prosecution of a water authority Officially allowed levels of pollution for effluent discharged into a river are 30 mg/litre of *biochemical oxygen demand* and 45 mg of *suspended solids* per litre. If these levels were exceeded, private prosecutions, e.g. by local authorities or riparian owners, could be brought but only within six months of the date of the offence. The main evidence usable against a guilty *water authority* came from tests made by the water authority and reported by it. Not all water authorities were conscientious about making these reports. But many were found guilty on the basis of such evidence, though punishments were few. See also **National Rivers Authority**, **privatization**.

protective equipment [elec.] Electrical circuits and switches, such as relays, *circuit breakers*, *overload trips* or *earth-leakage protection*, which protect a machine or its operator from faults or overloads.

proving ring [stru.] A steel ring accurately turned, *heat-treated*, ground and polished. It is precisely *calibrated* in a testing machine by measuring its diametral deformation for different loads. It is used with a *dial gauge* for measuring a load applied to a structure during testing, as well as for small-scale laboratory work.

provisional sum A sum of money set aside in the *bill of quantities* by a *consultant* for unforeseen work such as *subcon-*

tracting or pumping in the event of a flood.

proximity switch A *limit switch* operated electronically, usually by capacitance or inductance. It may be preferable in a dirty environment which could harm a mechanical limit switch or jam it.

Prüfingenieur, checking engineer In Germany some of the *responsibility* for *construction* is accepted by 700 or more independent consultants (Prüfingenieure) retained by local councils and paid one third of the design *consultant*'s fees to check the design, drawings and construction. Prüfingenieure have existed since 1934; they take half the risk. The Prüfingenieur's first task is to help the owner get a building permit from the local council. His or her report confirms that the design documents are complete and agree with the architect's design and the soil tests. When the council accepts the report the contractor can start digging.

public works *Civil engineering* construction.

pudding stone See **conglomerate**.

puddle, pug To pack with *clay puddle* and thus to make watertight.

puddle clay, pug *Clay puddle*.

pug Bricklayer's mortar.

pull [min.] The depth of ground which can be shattered at each cut, i.e. 'pulled' out of the face of a tunnel or shaft during driving. When the *cut* is well designed it is generally not more than 15 cm shorter than the longest hole. With the popular *wedge cut*, pulls of 2 m can be obtained in almost any ground.

pull-lift A chain-operated or rope-operated pulling device, giving a pull of up to 5 tons, and light enough to be carried by one person.

pull-out testing [stru.] A *semi-destructive test* for concrete, popular in the USA and Denmark as the Lok-test. A 25 mm dia. disc, 8 mm thick, with a threaded rod passing through its centre, is cast 25 mm deep under the surface of concrete. The rod and disc are pulled out by a jack bearing on a 55 mm dia. ring. A disadvantage is that the method must be planned before the formwork is complete. In the CAPO test (cut and pull), a hole is drilled into the concrete and a bolt fixed in and pulled out without preplanning.

pultruding, pultrusion [stru.] A combination of 'pull' and '*extrusion*', to describe structural sections made since the 1950s of fibre-reinforced plastics that have been pulled through a hot die. They contain 45–75% by weight of continuous longitudinal fibre reinforcement, usually in a matrix of polymer resin but sometimes of vinyl esters which resist fire and corrosion better. Large pultruded sections have been successfully used in a power-station *flue-gas desulphurization* plant, a highly corrosive environment, but a pultruded ladder failed after only six weeks in a Cairo sewer manhole. Pultruded sections can be made to measure.

They weigh only a quarter as much as concrete and are slowly getting cheaper. Many types of reinforcement and *binder* resin have been used, including polyester, vinyl ester and *methyl methacrylate* resins. *Glassfibre-reinforced concrete* units have been made by pultrusion. Pultruded units can be sawn, drilled, threaded, bolted, or joined by glue or *solvent welding* (*B*). One maker is Fibreforce Composites, Runcorn.

pulverized coal Finely ground coal of which 99% is smaller than 0.25 mm dia. At power stations most of the *fly-*

ash is caught in *electrostatic precipitators*.

pulverizing mixer A soil mixer used in *soil stabilization* for pulverizing the ground over which it passes with its revolving tines, and mixing it with the *stabilizer* spread over the ground. See **spotting**.

pump [mech.] A machine used for raising liquids in a pipe. The main types are *air lift*, *centrifugal pump*. *diaphragm pump*, *displacement pump*, *reciprocating pump*, *rotary pump*, *hydraulic ejector*.

pumpability See **thickening agent**, **workability**.

pumped storage A 'peak-lopping' and energy-saving activity of electrical generating authorities. It involves pumping water up to a mountain reservoir at times of low energy demand at nights and weekends. During peak power demands, the water is released to drive a turbine turning an alternator generating electricity. The 1800 MW Dinorwig pumped storage plant in North Wales, completed in 1983, is the largest UK example. Cf. **compressed-air energy storage**.

pumping of a slab on the ground Ejection of water or mud through joints or cracks in a slab when a vehicle passes over it, caused by water being forced up by the slab moving down (*ACI*).

pun To ram wet concrete or earth with a rod or *punner* so as to consolidate it by driving the air out.

punch, puncheon (1) An upright from a waling to the ground in a timbered excavation.

(2) A *follower*. See also (*B*).

punching [mech.] Forming rivet holes in metal with a press. The method is quicker than drilling but weakens the metal round the hole. The strength can be increased to that of a drilled hole if the hole is reamed out after punching, removing the weakened metal.

punching shear [stru.] When a heavily loaded column punches a hole through a base, the base is said to fail by punching shear. Punching shear is prevented by thickening the base or enlarging the foot of the column so that the shear stress (assumed uniform) round the perimeter of the column does not exceed twice the allowable shear stress in concrete.

punner (1) **hand rammer** A wood or metal block at the bottom of a handle. It is raised and dropped to compact earth or to bed paving slabs.

(2) A steel bar plunged up and down in wet concrete to compact it.

purlin See (*B*).

pusher tractor A crawler tractor used for pushing, particularly during the filling of a large *bowl scraper* digging difficult soil. Since one bowl scraper does not provide enough work for one pusher tractor, it usually pushes two or more scrapers in turn.

push shovel A *face shovel*.

puzzolane See **pozzolan**.

PV (photovoltaics) The use of the sun's energy. Probably 1000 photovoltaic pumps were working in sunny countries in 1988.

PVA *Polyvinyl acetate*.

PVC *Polyvinyl chloride*.

PVdF *Polyvinylidene fluoride*.

pycnometer [s.m.] An instrument used for determining the density of soils, the simplest type of *relative density* bottle. It may be merely a jam jar. It is weighed three times: empty, then full of soil, then full of soil and water. If the density of the soil particles is known, these three weighings will also enable the moisture content of the soil to be calculated, otherwise a fourth weighing is needed, after drying.

pyramid cut [min.] A method of blasting several rings of holes in tunnelling or shaft sinking. The holes of the central ring (*cut holes*) are shaped like a pyramid, with their toes close together.

pyrometer A *sensor for furnace temperatures* that does not need to touch the furnace because it responds to radiant heat. Its sensor is a slice of a lead zirconate titanate.

Q

qanat See **khanat**.

quadrant (1) A quarter circle; a 90° segment.

(2) A granite sett 46 × 23 cm shaped like a quadrant.

quadrantal bearing, reduced b. [sur.] A *bearing* less than 90° measured from east, west, north or south.

quadrilateral [d.o.] A four-sided figure of any shape. Its area is equal to the product of the diagonals times half the sine of the angle between them.

quality assurance The systematic action needed to give confidence of satisfactory service. It includes *quality control*. See **classification society**, **peer review**.

quality control See BS 5750 (many parts 1981, 1987); also **type testing**, **quality assurance**.

quantities The materials and labour put into a building, translated into the units of the *bill of quantities* in mm, m, m², m³, kg, tonnes, etc. Whether quantities are provided in the *contract* depends on the amount of information available at the time it was written. Different types of contract are used when quantities, approximate quantities or none are included in the *contract documents*.

quantity surveying Writing bills of quantities, measuring work, settling the payments between *client* and *contractor*, and *arbitrating* disputed points.

quantity surveyor A person trained in construction costs and building payment procedures who advises the *client* and *consultants* on costs and *contract documents*, especially the *bill of quantities*. During construction he or she controls costs, provides financial statements, and at the end helps to settle the final account. Even at the early stage of the *feasibility study* the quantity surveyor can give a rough estimate of cost.

quarry An open pit from which building stone, sand, gravel, mineral or fill is taken. See also (*B*).

quarter The *flank* of a road.

quartering, coning and q. A way of reducing the bulk of a sample of gravel, sand, etc. in such a way that it remains *representative*. The material is shovelled into a conical heap and divided into four equal quarters. Diagonally opposite quarters are rejected. The other diagonally opposite pair is kept and requartered until it is small enough.

quartering way See **rift**.

quarter peg Pegs placed at the quarter width of a road which, in conjunction with centre pegs, define the road surface.

quartz SiO_2. Crystalline *silica*. The main part of sand, gravel or sandstone; the transparent part of *granite*; the commonest known mineral and one of the hardest. Silica exists in many crystalline or non-crystalline forms, some of which are gems.

quartzite [min.] A strong *sandstone* cemented by quartz, therefore about 98% silica.

quay A *wharf* or *jetty*.

Quebec Bridge One of the largest *cantilever bridges* in the world, built in 1917. The clear opening of its central span is 549 m, which is larger than that of the *Forth Bridge*, but the Forth Bridge has two main spans and is therefore considerably longer.

quenching [mech.] Cooling steel rapidly from above the *critical point* to harden it. This is done for most carbon-steel cutting tools such as miners' picks, drill steels, carpenters' chisels, and so on. See **tempering**.

quicksand [s.m.] A sand through which water moves upwards so fast that the sand is held in suspension by the water. It therefore can carry no weight. Quicksands are made stable by reducing the flow of water. Fine sands are more dangerous than coarse sands, and uniform sands are more dangerous than well-graded sands. The worst sands so far investigated have a *coefficient of uniformity* of less than 5 and an effective size of less than 0.1 mm. See also **boil**, **critical hydraulic gradient**, **uplift**, **wellpoint**.

quick test, undrained shear t. [s.m.] A *box shear test* or a *triaxial compression test* of a cohesive soil carried out without allowing the sample to drain. See **consolidated quick test**, **drained shear test**.

quoin post [hyd.] A *heel post*.

R

race (1) [hyd.] A channel to or from a *water wheel* (headrace or tailrace).

(2) [mech.] A groove in which a machine part moves (ball race).

rack (1) [mech.] A toothed bar which in a *rack railway* is engaged by a cog wheel.

(2) [hyd.] A *trash rack*.

rack-and-pinion-mast platform A type of *access platform* which can rise higher than a *scissors lift*. Raised by cogs acting on a rack built into a steel lattice mast, it may be stabilized up to 20 m height by *outriggers*, and above this level by tying into the building. They are used for hoisting men to work on a ten-storey building or, with two masts and perhaps two decks, for placing large panels of cladding. Their great advantage is that the difficult, regular inspection of a hoist rope is avoided, but a flexible power cable has to lead to the platform.

racked timbering *Timbering* which is diagonally braced to prevent deformation.

racking [stru.] See **wracking forces**.

racking course A layer of graded stone spread over stone *pitching* to fill cavities and form the road shape before surfacing it.

rack railway A mountain railway which can safely be used on gradients from 8% to 15%, at which gradient it becomes uneconomical and a rope haulage is preferable. A toothed *rack* bar beside or between the rails provides extra adhesion by connecting with a driving pinion on the locomotive called the climber. See **locomotive haulage**.

radar [sur.] RAdio Detection And Ranging – the measurement of distance by reflection of radio waves.

For civil engineers the most important use of radar is in *EDM*; all EDM instruments use it, some with visible light, others with VHF radio or infrared radiation. The position and height of an aircraft at the moment of exposure of a *vertical photograph* are fixed by radar, reducing the need for *ground control* points. Highway engineers and water engineers can be helped by weather radar. The unmanned radar station on the southern shore of Lough Neagh, Northern Ireland, can detect rainfall as far away as 210 km and can estimate its intensity within a range of 85 km all round.

Ground radar can find the faults in a concrete ground floor slab. Frankley Reservoir, near Birmingham, built in 1904 to contain 900,000 m³ of water from the Elan Valley in mid-Wales, 130 km (90 miles) away, was losing 540 litres/s when it was examined successfully by ground radar in 1987.

radial gate, segmental sluice g., Tainter g. [hyd.] A dam gate with a curved water face and a horizontal pivot axis which is usually also the centre of curvature of the water face.

radial-sett paving Paving of small *setts* laid in concentric arcs to form fan shapes.

radiation [sur.] Plotting the surrounding points on a *plane table* set up by radiating lines drawn with the *alidade*, marking off on each line the distance of the point to scale.

radioactive testing of concrete

radioactive waste

Several radioactive methods of testing are possible for concrete as well as for welding. All are expensive, and need highly skilled operators and strict safety precautions. See **radiography**, **radiometry**.

radioactive waste See **nuclear waste disposal**.

radio control A type of *remote control* used where cables would be awkward or impracticable, e.g. for an underwater excavator, shotcreting in large tunnels, or work inside a furnace. A crane may be radio-controlled by its banksman if the crane driver cannot see the crane hook.

radiography *Non-destructive testing* methods involving the detection of radiation by photography rather than by the instruments of *radiometry*. *Gamma radiography* is used for testing welds, for detecting variability of *compaction* in concrete up to 600 mm thick as well as for measuring concrete density, depth of corrosion or the quality of grouting around prestressing *tendons*.

radiometry *Non-destructive testing* methods involving detection of radiation by instruments such as Geiger counters or scintillation counters, combined with radiation measurement by electronics. Gamma radiation is used for testing concrete because its sources, cobalt 60 or caesium 137, are much more portable than X-ray equipment.

radius [mech.] The horizontal distance from the centre of the crane hook to the centre of the slewing pivot, a distance needed for calculating the allowable load on a crane hook to prevent the crane tipping over.

radius-and-safe-load indicator See **safe-load indicator**.

radius of gyration, r_x or r_y [stru.] A value used in calculating the *slenderness ratio* of a strut. If A is the cross-sectional area and I the moment of inertia of the strut, the radius of gyration equals

$$\sqrt{\frac{I}{A}}\Psi$$

radius point The centre of a circular curve, used in setting out e.g. a *tangent point*.

radon (Rn) (called **thoron** in the thorium series) A daughter product of radium, this radioactive, colourless, odourless gas is believed to cause from 5000 to 20,000 deaths a year in the USA from lung cancer and 1500 in the UK. It is generally confined to rocky areas and may be dissolved in drinking-water, but can be removed by a *GAC* filter, which should be outside the house.

raft foundation A continuous slab of concrete, generally reinforced, laid over the ground as a foundation for a structure. It is as large as, or slightly larger than, the area of the building which it carries. A *buoyant raft* is a particular and very expensive raft foundation.

rag bolt A *Lewis bolt*.

rail [rly] One of two parallel steel bars in a railway, laid on *sleepers* to form a track for wagons or trams with flanged wheels. For surface work on passenger track, rails weighing up to 55 kg/m are used in Britain, usually *flat-bottomed*, and up to 77 kg/m elsewhere. In mines on steep gradients in the USA, wooden rails of 5 × 8 cm section are used, where their high friction is helpful. Also, when a locked wheel does skid, no flat is formed on the wheel. Early railways used wooden rails and, later, cast-iron ones. See **continuously welded track**.

rail bender [rly] (see diagram) A *jim crow* can bend *bullhead rails* but *flat-*

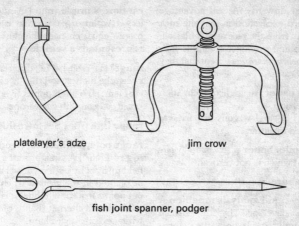

platelayer's adze   jim crow

fish joint spanner, podger

Railway tools.

bottom rails are more difficult to bend.

rail chair [rly] See **chair, rail fastening**.

rail fastening [rly] For the British *bullhead rail* the cast-iron rail chair with steel or wooden keys to wedge it to the rails was essential. It was held down by spikes or screws through three holes. *Flat-bottomed rails*, now standard in Britain, are fastened down by various proprietary fixings, usually with a resilient pad between rail and sleeper, made of rubber sheet. On timber sleepers there is also a metal baseplate, next to the wood.

rail gauge [rly] On construction sites, gauges vary from 0.3 m up to the *standard gauge* (1.435 m). See **broad/narrow gauge**.

rail key [rly] A wedge which fixes a bullhead rail into a *rail chair*. It may be a steel spring or a hardwood wedge. The key is driven in on the outer side of the rail so as to maintain the correct gauge.

rail test, falling weight t. A test for

the brittleness of rails in which, at 10°C, a 1 tonne weight is dropped from increasing heights on to a 1.3 m length of rail on supports 0.9 m apart.

rail tie [rly] US term for a *sleeper*.

railway curves [d.o.] A set of arc or spiral shapes cut out in wood or transparent plastic, of different radii, for drawing plans of rail tracks to scale.

railway transit [sur.] A *theodolite* with no vertical circle (USA).

rainfall, intensity of r. Rain, snow, hail, etc. measured in mm/hour for a *unit hydrograph*, or in other units for other purposes.

rain gauge An instrument which collects the rain falling on it and can thus indicate the *rainfall*. It usually consists of a funnel from which the rain drips into a cylinder graduated in millimetres of rain falling on the funnel area.

rainwash [s.m.] The movement of surface soil and rock down a slope with the help of rain.

raise [min.] A vertical or sloping shaft

excavated upwards. A real advantage in upward excavation is that the rock falls away from the face as it is blasted. Raise-drilling machines resemble *full-face tunnelling machines* except that they work upwards.

rake (1) **batter** An angle of inclination to the vertical.
(2) [rly] Several wagons or carriages connected together.

raking pile, batter p., r. prop A *pile* (or prop) which is not vertical.

raking shore See (*B*).

RAM (1) **rapid analysis machine** A machine developed by the British cement industry. It finds the cement content of fresh concrete in ten minutes.
(2) The immediate, fast, read-write memory inside a computer, not disk or tape.

ram (1) The cast-iron weight of a *drop hammer*.
(2) To compact loose material with a punner, to tamp.
(3) [hyd.] **plunger** The moving cylindrical block in the working chamber of a hydraulic press or of a ram pump.
(4) [hyd.] See **hydraulic ram**.

rammer (1) A *punner*.
(2) See **power rammer**.

ramp (1) A steeply sloping track or floor.
(2) A short length of drain laid much more steeply than the usual gradient.

ram pump [mech.] A *single-acting* reciprocating pump which has no piston but a ram. Unlike pistons the ram is of constant diameter and does not fit closely in the cylinder. The ram passes through *packing* in the end of the cylinder without a connecting rod. Rams pump by *displacement* only. Pistons pump by end pressure and displacement.

random sample [min.] A *sample* selected without bias so that each part has an equal chance of inclusion. See **representative sample**.

range (1) [stat.] The difference between the highest and the lowest value.
(2) [sur.] To align points by eye with *range poles* or with a telescope.

range line [sur.] See **township**.

range pole, banderolle, r. rod, ranging rod [sur.] A straight staff 2 m long held upright by a surveyor's chainman, or planted in the ground when setting out points in a straight line. It is usually painted with alternate bands of red and white, exactly 0.5 m long.

ranging a curve [sur.] Setting out a curve for a road. Railway usage is 'pegging' or 'stringing' a curve.

ranging rod A *range pole*.

Rankine's theory [s.m.] or state of stress theory of granular earth pressures developed by Rankine about 1860. His value for the pressure on a vertical wall retaining earth with a horizontal surface is $\dfrac{1 - \sin\phi}{1 + \sin\phi} \times$ soil density for each metre depth of earth retained, where ϕ is the angle of friction of the soil. This value $\dfrac{1 - \sin\phi}{1 + \sin\phi}$ is called the coefficient of *active earth pressure*. Cf. **passive earth pressure**, **wedge theory**.

rapid-hardening cement, high-early-strength c., ASTM Portland c. type III Portland cement which hardens more quickly than *ordinary Portland cement* and is more costly because it is more finely ground, having a *specific surface* of at least 325 m^2/kg. Vibrated cubes should have a 3-day strength of at least 21 MPa (3000 psi) compared with only 15 MPa (2200 psi) demanded of ordinary Portland cement.

rapid infiltration, r. filtration, land f. [sewage] A method of *land treatment* which is best used over sandy subsoils. The *effluent* is led through parallel furrows and ditches directly through the sand to *field drains* buried at a suitable depth. Some is absorbed by plant roots but *pollutants* will be partly removed even if there are no plants. The field drains are laid parallel to the furrows and may remove the effluent for further treatment or recirculation. Crops may grow on the ridges between the furrows. This method directly *recharges* the groundwater.

rapid sand filter, pressure f. Equipment developed from the *slow sand filter* but much less bulky and contained in a steel, covered tank. Also the sand is slightly coarser, more deeply submerged, and cleaned by *backwashing*. One disadvantage is that a polluted backwash has to be disposed of. It is used for filtering drinking-water or swimming-pool water and in the *tertiary treatment* of sewage *effluent*. Some rapid sand filters are automatic; partial blocking by filtered dirt raises the water pressure and starts the backwash which may include *air scouring*. The slow sand filter produces a dirty sand which can be smelly, but it does not pollute water.

rapid-transit system US term for (i) a suburban railway, (ii) express trams that travel in special traffic lanes, or (iii) other fast public transport.

ratchet-and-pawl mechanism [mech.] A cogwheel (ratchet) with which a single tooth (pawl) engages to prevent it turning backwards and allow it to turn forwards. The pawl is held down on to the ratchet by a spring.

rate of spread See **spread**.

rating (1) [hyd.] The relationship between the water level and the discharge of a stream, well or *aquifer*, or the work of taking the observations and making the calculations needed to establish this relationship.

(2) [hyd.] The greatest amount of water per unit time that can safely be drawn from a stream, well or aquifer without damage to it. If the water source is overrated, the pipeline and pumps will be underworked and consequently will have cost too much.

(3) [elec.] The power output of an electric motor over a certain period as certified by the maker. A continuous rating of so many kilowatts means that the motor is designed to work permanently without overheating at the power stated. A half-hour rating is the possible power output for half an hour only. Since the rating is based on the rate of heating and cooling of the electrical part, the half-hour rating of a motor is appreciably higher than its continuous rating.

rating curve, stage-discharge c. [hyd.] A graph of the water level of a stream against its flow rate. See also **duration curve**.

rating flume [hyd.] (1) A *control* flume.

(2) A flume containing still water in which *current meters* or *Pitot tubes* are drawn at known velocity so as to calibrate them.

rational formula The *Lloyd-Davies formula*.

rat-tail file A round *file* tapering almost to a point.

ravelling See **fretting**. Cf. **scabbing**.

Rawlbolt The Rawlbolt is a heavy-duty fixing device, manufactured by the Rawlplug Company Limited, for attaching structural and non-structural items to masonry and other construction materials. It comprises a bolt (which may be captive or removable) enclosed within a segmented expansion shield. The bolt is either fabricated with a conical end-section or carries a separate cone-shaped nut. The action of tightening the bolt draws the end-section or nut into the shield, which expands to set the anchor. Rawlbolts may be used in suitable concrete, brickwork or stone and are available in a range of sizes. Versions can be supplied with hook or eye bolts.

raw water Water which, after treatment at a *waterworks*, becomes a *water supply*.

Raymond standard test A *dynamic penetration test* used by Terzaghi to compare the bearing capacities of soils. A hole is first bored to the proposed foundation level, and a *soil sampler* of 5 cm outside dia. is then lowered into the hole and pressed 15 cm into the soil. Its position is then measured and it is given a number of blows from a 64 kg hammer which drops 76 cm. The number of blows required to drive it 0.3 m into the soil is recorded and gives a measure of the *relative density* of a sand. The sample tube is withdrawn with a 0.3 m long soil specimen inside it.

RBC *Rotating biological contactor*.

reach, pond [hyd.] One stretch of water between two locks.

reaction [stru.] The upward resistance of a support such as a wall or column against the downward pressure of a loaded member such as a beam.

reaction turbine [mech.] A steam or *water turbine* in which the jets or nozzles are on the moving wheel, as opposed to an *impulse turbine*, which has only fixed jets.

ready-mix (r.m.) concrete Concrete either mixed at a *batching plant* and carried to the site in an *agitator truck*, or batched, but not mixed, and carried to the site in a *truck mixer*. For either method, BS 5328:1981 specifies a maximum of two hours between putting the cement into the wet aggregate and placing the concrete. ASTM C94–83, however, specifies a maximum 300 revolutions of the mixer for mixing between 7 and 16 rpm or the concrete must be placed within 90 min. See also **shrink-mixed concrete**.

realignment An alteration to the line of a road, railway, etc., which may affect only its slope (vertical alignment) but more usually alters its layout in plan (horizontal alignment).

real-time computing system If a computer can respond with a sensible answer to a query within about five seconds it is working in real time, which is needed for activities like *CAD*. *Batch systems* are not designed for quick response but to process a large output cheaply. The same *mainframe computer* may work in real time by day and in batch mode by night.

ream [mech.] To enlarge or smooth a *borehole* or a hole in metal with a *reamer*.

reamer (1) [mech.] A hand or machine tool for finishing a drilled hole. It has a cylindrical or conical shaft with cutting flutes or teeth.

  (2) **belling tool** An under-reamer for *bored piles*. Large ones can enlarge the base of a 1.5 m dia. pile to 5 m diameter.

rebars *Reinforcement* – steel bars.

rebound hammer, Schmidt impact h. A quick testing method that, in conjunction with careful cube or *ultrasonic* pulse velocity testing of the same concrete, enables its strength to be determined within ±3 MPa

(435 psi). There may be differences in rebound from faces that are trowelled and those that are cast against form-work. Other differences come from variations of moisture content, curing, concrete composition, surface texture and depth of carbonation from the surface. The hammer should not be used on concrete weaker than about 20 MPa (2900 psi) because it will damage the surface. The hammer is a good way of comparing concrete units provided that charts of all readings are kept, showing their relations to the concrete cube strengths.

recalescent point [mech.] See **critical point**.

receiver [min.] An air container interposed near the rock drills in a compressed-air pipeline to store air for the short but frequent periods when the air demand exceeds the supply. It is also convenient to use it as a sump to drain *condensate* from the pipeline. A drain cock is provided for this purpose.

receiving water Any stream, estuary, lake or sea that receives *effluent*.

reception pit, receiving p. A pit dug at the opposite end from the *launch pit* of a microtunnel or *pipe-jacking* job.

recharge [hyd.] Refilling of an *aquifer* either by *artificial recharge* or by the natural recharge of rainfall and *infiltration*.

reciprocal levelling [sur.] Eliminating instrumental error in levelling between two points by taking levels on them from two set-ups, one near each point. The difference in level is averaged.

reciprocating [mech.] A description of anything with a to-and-fro

motion. See **percussion tools**, **single-acting**.

reciprocating engine [mech.] A *steam engine* or an *internal-combustion engine* with cylinder and piston, not a turbine or a jet engine.

reciprocating pump/compressor [mech.] A pump or compressor worked by pistons or rams (see **ram pump**), but not a centrifugal pump or compressor, nor an *air-lift pump*.

recording gauge, recorder [hyd.] A gauge which records automatically the level of the water in a stream or tank, sometimes also the velocity and pressure in a pipe. It often works by a float or by a submerged air tank with a rubber diaphragm which moves inward as the water level rises, causing a pressure increase in the air, which is recorded by the pen. See **stilling well**.

recovery peg, reference p. [sur.] A peg placed at a known relationship in level, direction and distance to another peg to enable this one to be re-established, if disturbed.

rectangular coordinates [d.o.] *Cartesian coordinates*.

rectangular hollow sections [stru.] See **tubular sections**.

rectangular weir [hyd.] A *measuring weir* with a rectangular notch, normally a *contracted weir*.

rectification, differential r. [air sur.] Elimination of the image displacements in a *vertical photo*. Rectification corrects displacements caused by *tilt*, whereas differential rectification corrects those caused by hills and valleys, See **orthophoto**.

recycling, pavement r. The immediate reuse of scarified road surfacing or base course by rotavating it to a depth of some 20 cm (8 in.) and mixing in hot or cold bitumen binder, sometimes as well as cement. Recycling uses only 1–3% binder, against 6% for quarried stone, and the total cost can be half that of conventional methods, though the quality of the recycled road may be extremely high.

Concrete from demolished buildings can also be recycled, saving on the high cost of tipping. Some demolition contractors have bought crushers and screens for this reason.

red shortness [mech.] *Hot shortness* of steel or other metal.

reduced bearing See **quadrantal bearing.**

reduced level [sur.] A level stated in relation to a known *bench mark* or other datum.

reduction [min.] The extraction of a metal from its ores, a chemical term meaning the removal of oxygen from an oxide to produce a metal or other element.

reduction factor [stru.] A factor by which for a given *slenderness ratio* the permissible stress on a long column is reduced below that allowed on a short column, so as to prevent *buckling*. Building regulations and BS state the reduction factors for any given slenderness ratio in any material; steel, timber, masonry, reinforced concrete.

reduction in area [mech.] The *contraction in area* of a *tensile test*-piece.

reduction of levels [sur.] The calculation of the differences in level between various points from the staff readings in a level book.

redundant frame, statically indeterminate f. [sur.] A frame which has more members or continuity than is needed for it to be a *perfect frame*. It is therefore necessary to remove a member or members or some fixity to make it perfect. See **superposition.**

reeving thimble, r. link [mech.] An end loop to a crane *sling*, through which the remainder of the sling can be passed. It is usually pear-shaped. Round *thimbles* are non-reeving.

reference mark, r. object [sur.] A distant point chosen so that the *bearings* to other points can be measured from it at a station.

reference peg [sur.] A *recovery peg.*

reflection cracking Where a road is repaired with new material laid over it, cracks from the old road may continue through into the new material, but of course there is no 'reflection'.

reflux valve [hyd.] A check valve.

refractive index of air [sur.] A measure of the amount by which light or other electromagnetic waves bend as they pass from one type of air to another. It is a factor calculated from observations of temperature and barometric pressure and is used to correct *EDM* results affected by weather.

refractory Description of any material that resists heat.

refractory clay A clay used for making a *refractory lining.*

refractory concrete Common refractory concretes can withstand temperatures from 300° to 1300°C. They fail usually because they begin to shrink at some 80°C below the softening point of the aggregate.

refractory linings, refractories Bricks or rocks that are hard to melt and therefore used for lining furnaces. Service temperature limits are controlled more often by the aggregate than by the cement. *High-alumina cement* can withstand 1200°C (but concrete made from it with silica gravel and sand should not be used at temperatures above 300°C), limestone 500°C, and blast-furnace slag (dense or foamed), brick or calcined *diatomite* (*B*), 800°C; some igneous rocks, including basalt, dolerite and pumice, as well as expanded clay aggregate, will resist at least 1000°C, dead-burned magnesite 1400°C; bauxite 1500°C, and chromite 1600°C. See also **basic refractory, bauxite, calcium aluminate, Carborundum, silica brick, silicon carbide, zirconia.**

refrigerator foundations Since a large refrigerator can, even if fairly well insulated, freeze the ground beneath it and thus cause *frost heave*, which may raise the building dangerously, it is advisable to ventilate foundations or otherwise to ensure that they are not frozen. With boiler houses on clay a similar problem occurs owing to the shrinkage of the clay by drying out when heated. In this case the building sinks unevenly, more below the boiler than elsewhere. A similar problem occurs with the foundations of every heated building in a *permafrost* region. One solution is to have shallow foundations and to ventilate over them, keeping the ground floor of the building above ground. Another, rather more expensive, is to carry the building on piles. Both foundations should rest on permanently frozen soil, well below the *active layer*.

refuge (1) An island in the traffic, separated from it by a *kerb*, sited so as to divide the traffic streams and to provide a safe area for pedestrians. See **bollard** (2).

(2) [rly, min.] **manhole** An excavation in the side of a tunnel where someone can shelter from passing traffic.

refusal A pile which sinks only about 13 mm in five blows is said to be driven to refusal.

regime [hyd.] (also **regimen** USA) A stream or canal is said to be in regime if its rate of flow is such that it neither picks up material from its bed nor deposits it. This is unusual in the whole length of a stream and only short lengths of it can be expected to be in regime at any one time. However, in the course of the years after construction, canals generally tend towards a regime condition.

reglette [sur.] The short, usually 12 in. scale, divided into hundredths of a foot, formerly used for accurate measurements of length with a steel band graduated in feet only. See **graduation of tapes.**

regulated-set cement *Jet cement.*

regulating course A layer of stone put over an old road to shape it before it is resurfaced.

rehabilitation See **pipe renewal.**

reheater, interheater [mech.] An accessory to steam or compressed-air engines, which greatly reduces their consumption of air or steam. A reheater superheats the steam or reheats the air between expansion stages, increasing its pressure and reducing the likelihood of freezing.

reinforced brickwork Brickwork with expanded metal, steel-wire mesh, hoop iron, or thin rods in the bed joints. Rods also can be arranged to pass vertically through a properly bonded wall at the intersection lines of vertical joints. Reinforced brickwork has been used in India (*Quetta bond* (*B*)) and in California to ensure against the collapse of walls in earthquakes.

reinforced concrete

Hoop iron was occasionally used in the bed joints of English brickwork for hundreds of years. *Wall ties* (*B*) are not considered to be reinforcement but they do strengthen walls which they tie together. See **reinforced grouted-brick masonry**.

reinforced concrete Concrete containing more than 0.6% by volume of reinforcement consisting of steel rods or mesh. The steel takes all the tensile stresses (theoretically) so that the cracks which always occur in reinforced concrete do not appreciably weaken it. In good design the reinforcement is distributed well enough for the cracks not to be conspicuous. See **bond, creep, plain concrete, prestressed concrete, shrinkage**, and below.

reinforced earth (see diagram) In the USA road or railway cuttings often have sides of reinforced earth. The cut may be faced with precast slabs with *tiebacks*. The earth is reinforced by the wall and holds it at the same time. Bars used near the sea should be non-rusting.

The Romans built reed-reinforced earthen *levees* along the river Tiber, and 1000 years earlier the people of Iraq built earthen ziggurats. We now use *geosynthetics* of the most varied types, laid horizontally, vertically or sloping. They may allow drainage without *piping* or merely strengthen the soil mass. Probably the originator of reinforced earth was Henri Vidal, in France, who in 1963 with his children saw the effect of building sand castles reinforced with pine needles and then in 1966 built a reinforced-earth structure and patented his methods.

reinforced grass Grass can be grown through gaps between interlocking precast concrete slabs to reinforce an earthen surface over which vehicles rarely pass. A less expensive method on a dam or levee only occasionally flooded is to use one of various plastic grids or mats. Mats are 10 cm (4 in.) thick or more – interlocking masses of plastic threads through which grass grows easily. Such plastic grids or mats must be pinned to the ground at intervals.

reinforced grouted-brick masonry [stru.] In the USA brickwork built like the *cavity wall* (*B*) except that the cavity, known as a collar joint, is filled with fluid mortar as the wall rises and reinforced with mesh or rod reinforcement. The bricklayer can work on only one side of the wall at a time; nevertheless this construction is much used in the earthquake region of California. The cross joints have been found to be competely filled by grout. See **lateral-force design, reinforced brickwork**.

reinforced plastic See **glassfibre-reinforced polyester**.

reinforcement (1) **rebars** Rods or mesh embedded in concrete to strengthen it, usually of steel. *Glassfibre-reinforced concrete* exists but bamboo may be used where it grows well. Reed or *hessian* (*B*) have been used in *fibrous plaster* (*B*). Unlike *tendons*, reinforcement puts no stress on the concrete. See also **cold working, coupling, deformed bars, fibre-reinforced concrete, lap, round bar, wire-mesh reinforcement**.

(2) In a *weld*, the amount by which the weld metal rises above the parent metal.

reiteration, repetition [sur.] A method of increasing the *precision* of a measurement of an angle with a given instrument by repeatedly measuring it. (The precision is proportional to the square root of the number of readings.) With a *theodolite* the technique is as

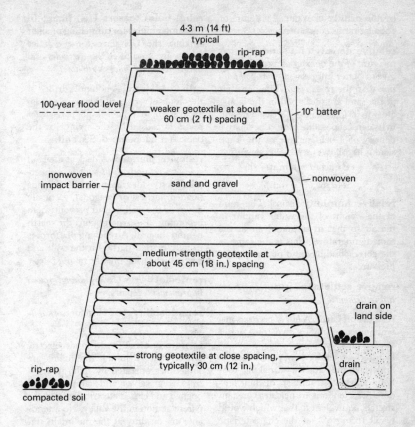

Ohio river flood wall. (*Civil Engineering of ASCE*, August 1988.)

follows: after the first measurement the lower plate is unclamped (the vernier being kept clamped) and the telescope turned back to the first point. The lower plate is then clamped, the upper plate released, and the angle is measured again. The total angle should now be double the first angle. The procedure is repeated until the desired precision is obtained.

relative compaction [s.m.] The *dry density* of the soil in situ divided by the maximum dry density of the soil as determined by the *Proctor compaction test* or other standard test. It is generally expressed as a percentage. Cf. **degree of compaction**.

relative density (formerly **specific gravity**) The weight of a substance divided by the weight of the same volume of water at 4°C. It is therefore a number without units – a ratio. To determine the density of a substance its relative density must be multiplied by the density of water − 1 g/cm³ or 1000 kg/m³ as convenient.

relative density of a sand [s.m.] A measure of the density of a sand which gives a better impression of its *compaction* than the *voids ratio*. The densities of the sand are measured in the laboratory in its loosest possible dry state and in its densest possible state. The relative density of a field sample of the same sand is as follows (d = field density):

$$\frac{d\max (d - d\min)}{d\,(d\max - d\min)}$$

relative humidity [min.] The ratio of the weight of the water vapour in the air to that in saturated air at the same temperature. This is the same as the corresponding ratio of the vapour pressures.

relative settlement [stru.] *Differential settlement*.

relaxation [stru]. A *loss of prestress* in a tendon, caused by its own *creep* under stress. Since creep is uncertain, manufacturers of tendons may be asked to undertake the expense of making 100-hour constant-strain tests of their tendons. The tendons are held at a constant stretch, equivalent to that which would result from 70% of the characteristic load specified by e.g. BS 4486, 'Cold-worked high-tensile alloy steel tendons'.

relay [elec.] A device which responds to a specific current or power and opens or closes a *pilot circuit*. The pilot circuit operates a second circuit often of a *solenoid* of a *circuit breaker* or *contactor* or other *protective equipment*.

release agent, parting a., p. compound, bond breaker A general term that includes any greases, *mould oils* or *sealants* (2) laid over forms or form linings either to ensure a good finish to the concrete or to improve the durability of the *form*, or for both reasons.

relief holes (easers UK) [min.] In breaking ground for tunnelling or shaft sinking, the US term for the holes which are fired after the *cut holes* and before the *lifter holes* or *rib holes*.

relief well [s.m.] A borehole drilled at the foot of an earth dam or of an excavation to relieve high *pore-water pressures* caused by the weight of the dam. It is not pumped. See **boil**.

relieving platform A deck at the land side of a *retaining wall* such as a sheet-piled jetty to transmit heavy loads (loaded lorries or wagons) vertically down to the wall and to prevent them becoming a *surcharge* on the earth behind the wall. A relieving platform is usually carried partly on the wall and partly on bearing piles or raking piles.

remedial works *Underpinning, ground improvement*, repairs, etc.

remoldability *ACI* term for *workability*.

remote control [hyd.] Hydroelectric power stations in the mountains are sometimes operated by radio (*telecontrol*), the valves and circuits being opened and closed electrically from the parent station in the valley. No attendants are on duty at the mountain station, which is visited only at weekly or monthly intervals by a maintenance gang for greasing and replacing electrical contact-points.

Remote control of *hydraulic hammers, rock drills*, etc. enables the operator to be at a point where he or she can supervise the work safely without obstruction from the machine. The control is usually through an electric cable but may be by *radio*.

remote reading of meters See **mains-borne signalling, telemetry**.

remote sensing [air sur.] Using *sensors* to detect from the air what is on the ground, rather than its precise dimensions and mapping. Besides aerial photography, *infra-red thermography* and airborne *magnetometers*, an increasing number of sensors and cameras are carried on satellites. Geology, hydrology, vegetation, mineral resources, environmental destruction or even the weather can be investigated. Air photos can show up a geological *fault* that could cause a disaster if a dam were to be built over it. See **ground truth**.

remoulding [s.m.] The disturbance of the internal structure of a clay or silt. When remoulded a clay loses shearing strength and gains compressibility. For this reason pile-driving is inadvisable in some clays. See **remoulding index**.

remoulding index [s.m.] The ratio of the load per millimetre of compression of undisturbed clay to the load per millimetre of compression of the remoulded clay. See **sensitivity ratio**.

rendering, cement r. See (*B*).

renewable energy Tidal or hydroelectric sources, or wind or landfill gas can provide power that is naturally renewed. In 1989 four landfill gas sites were each selling up to 2 MW of electricity generated from refuse to the national grid. UK renewable energy supplies may reach 450 MW by the year 2000, including 200 MW from wind parks, 200 MW from landfill gas, and 50 MW of hydroelectricity.

repetition [sur.] *Reiteration*.

replacements, cement r. *Pozzolans*, natural or artificial.

repose [s.m.] See *angle of repose*.

representative sample [stat.] A *sample* which can be selected only by planned action to ensure that a fair proportion is drawn from the various parts of the whole. The sampling within the parts may be *random*.

requisitioning a sewer Where no sewer exists, an owner or developer of property or a district council in England or Wales (not Scotland) may requisition the local *water services plc* to provide a sewer, subject to various conditions, one of which is sharing the cost.

resection [sur.] Locating a point, either on the *plane table* or with a *theodolite*, by angular observations from the required point to points of known position. A minimum of three known points is needed to provide a solution and a fourth known point provides a check. If only two known points are available, then a *base line* must be measured to provide a solution.

reservoir A tank or artificial lake where water is stored. *Artificial recharge* of an *aquifer* may be cheaper and more effective sometimes than building an *impounding reservoir*. See **service reservoir**.

reservoir roofs Reservoirs for drinking-water are usually roofed to keep out leaves and other dirt. Lightweight roofs are cheap and quick to build because they need no foundations. A plastics sheet of 1500 gauge black polythene was installed on Horsley reservoir by the Newcastle and Gateshead Water Co. in 1973. The 30 × 45 m sheet floats on the water. Rainwater on top of it helps to filter out ultraviolet light and so protects the polythene. Another type of lightweight roof, more complicated but with the advantage of allowing access to the water, is the *air house* (*B*), which has often been used over reservoirs in earthquake areas.

resident engineer A civil engineer who watches the interests of the *client* at a site, working under the *consulting engineer*.

residual chlorine After *dechlorination* of drinking-water, the small amount of chlorine remaining which maintains the *disinfection* is its residual. *Ultraviolet radiation* and *ozone* leave no residual.

residual errors [sur.] Errors which cannot be eliminated from a measurement despite the most careful work.

residual stress [stru.] (see diagram) Stress that remains in a material from its manufacturing process. In steel joists the outer parts of flanges and the centre of a deep thin web cool faster than the junctions between web and flanges. Cold straightening often relieves the stress. Heating in a stress-relieving furnace is usually not practicable so stress relief is usually achieved by heavy proof loading or by the onset of the first working loads. *Buckling*

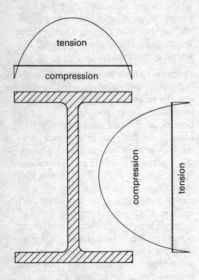

Typical residual stresses of a rolled-steel section. (D. Nethercot, *Limit States Design of Structural Steelwork*, Van Nostrand Reinhold, 1987.)

strength is reduced by residual stresses, particularly those caused by welding, but the residual stresses of *cold expansion* are helpful.

resilience [stru.] The *strain energy* stored in an elastic material per unit of volume. The modulus of resilience is the greatest quantity of energy per unit volume which can be stored in a material without *permanent set*. Steel can store 0.027 kg-m/cm^3, rubber about 0.54 kg-m/cm^3.

resin anchor, capsule a. A fixing into a hole drilled in masonry or rock. Its plastic capsule has separate compartments containing fast-setting epoxy resin and its hardener. The bar to be fixed is pushed into the hole containing the capsule by an electric drill. The rotation of the bar breaks the capsule and mixes the epoxy with its hardener. In cold weather, the bar should be heated to 15°C. The more expensive *roof bolts* are of this type.

resin coatings See **epoxy-coated steel**.

resin injection Low-viscosity resin may be injected into concrete or masonry to fill and seal cracks. The accepted narrowest crack that can be sealed with resin is 0.2 mm. Other resins are used in *artificial cementing*.

resistance (1) **r. losses** [hyd.] The resistance of a pipe or channel to flow is usually expressed in metres head of water.
(2) [rly] See **rolling resistance**, **tractive resistance**.
(3) [elec.] The opposition to the flow of electric current that heats an electrical conductor.

resistance-flash welding *Flash welding*.

resistance losses [hyd.] See **resistance**.

resistance moment [stru.] See **moment of resistance**.

resistance percussive welding [mech.] *Resistance welding* in which the heavy welding current is passed at the same time as a mechanical pressure forces the parts together.

resistance projection welding [mech.] See **projection welding**.

resistance seam welding [mech.] *Resistance welding* in which the welding electrodes consist of two rollers moving along the seam to be welded and that are pressed together at the moment when the welding current passes. It is often *automatic*.

resistance spot welding *Resistance welding* with *electrodes* of small end area which melt the metal where they touch.

resistance strain gauge [stru.] See **electrical-resistance strain gauge**.

resistance temperature detector, r. thermometer A length of wire wound on a bobbin and housed in a protective sleeve. Platinum wire is used for accurate work because of its stability and *linearity* over a wide range of temperatures. Nickel and copper are also used. All three have positive *temperature coefficients* that enable the *sensor* to work over a range of nearly 1000°C, from −250° to +730°C. But they are not very sensitive and may be disturbed by shock or vibration.

resistance welding The general term for the welding methods mentioned above, as well as *stud welding* (*B*).

resistivity logging, electrical l. A *well logging* method in which electric current is applied to the drilling fluid in an uncased hole in order to determine the resistivity of the rocks penetrated. Wet clays have low resistivity,

granites very high, often one million times as large. Other electrical methods include induction logging and the investigation of electric currents that exist naturally in the well.

resoiling, soiling Levelling the ground after building or dredging is finished and replacing topsoil so as to make the ground fit for vegetation.

resonance [stru.] When the *vibration* of a machine corresponds closely with the *natural frequency* of its *foundation* it is very likely that the foundation will vibrate excessively. It is therefore important to design foundations so that their natural frequency is different from that of the machines which they carry. See **soft-suspension theory**.

respirator, self-contained breathing apparatus Steel cylinders containing compressed oxygen, carried on the back, used by rescue workers in mining to avoid breathing poisonous air. The water industry also uses respirators; consequently any *manhole* for entry to a reservoir or sewer should be at least 0.6 × 0.9 m (2 × 3 ft) throughout for wearers of respirators to pass.

responsibility for construction In France and Germany building law is different from that in English-speaking countries. In 1804 Napoleon in his *Code Civil* placed full liability on the *contractor*. Divided responsibility was eliminated, as was the need for professional liability insurance by *consultants*. The code was updated in France in 1978 by the '*Assurance Construction*' law which requires compulsory insurance as follows: ten years for the structure and anything that affects its stability; two years for finishings (*second oeuvre*) and 30 years against deliberately concealed defects. Consultants' calculations are checked twice more, by the contractor before tendering and again by the Bureau de Contrôle (*classi-*

fication society), such as the *Bureau Veritas* or *CEP* (*Centre d'Etudes de Prévention*). Failure to insure may result in six months' jail.

In Germany such checks are made by many local offices (Prüfamt für Baustatik) under the local province (*Land*) and by the local TÜV (Technischer Überwachungsverein). Like a state in the USA, each *Land* has responsibility for enforcing its building laws. See **Prüfingenieur**.

retaining wall (see diagram) A wall built to hold back earth or other solid material (a dam holds back liquid). Important differences in design exist between free and *fixed retaining walls*. Fixed retaining walls are supported both at top and bottom and they cannot tilt, so that the earth pressure does not fall to the reduced value of the *active earth pressure*. *Free retaining walls*, however, can be *gravity retaining walls*. They need not be designed to bend, but must be able to tilt or slide enough to bring on the reduced value of the active earth pressure. They may be of *sheet-pile* construction (timber, steel or reinforced concrete). Bridge abutments are usually free retaining walls except for those which are fixed arches and portal frames. Cantilever walls are free and may be counterforted or buttressed, sometimes with a keel in the base to prevent sliding, or a *relieving platform* at the top to reduce earth pressure due to *surcharge*. See **diaphragm wall**, **flexible wall**, **Rankine's theory**, **wedge theory**.

retarder (1) **r. of set** An *admixture* which slows up the setting rate of concrete, sometimes applied to *formwork* so that when it is stripped the *cement paste* which has been in contact with it can be removed by light brushing. The rough texture thus formed may be needed for its visual appeal or to make a good bond for plaster. Some retarders accelerate the initial set. This can be counteracted by adding the retarder as late as possible. An admixture used in the USA in 1988, Delvo, made by Master Builders of Cleveland, Ohio, allows concrete to be used even after several days. It must be added before the concrete is three hours old, but can be added to concrete at up to 32°C. Ready-mix concrete makers use it. See also (*B*), **jet cement**.

(2) [rly] **wagon retarder** Remotely controlled hydraulic pistons can brake wagon wheels by pressure on their flanges. Heavy motor vehicles also may have retarders for use on long downhill slopes.

retention sum A sum held by the *client* for a certain period to offset costs resulting from any failure by the *contractor* to fulfil the *contract*.

reticulation A distribution network for water, gas, sewer or other pipes or cables.

reticule [sur.] A set of fine intersecting lines on the *diaphragm* at the optical focus of a telescope. The intersection defines the line of sight (*collimation line*) of the telescope. In modern telescopes the reticule is provided by lines etched on glass, no longer by spider webs.

retreading roads A method of repairing roads whose surface is breaking up or which are misshapen. The surface is first scarified 7 cm deep, and the road is then shaped with a grader, rolled and surfaced.

return drum See **tension drum**.

return sheave [min.] A pulley fixed at a distance from a haulage drum. The tail rope passing round it enables the drum to pull away from the engine.

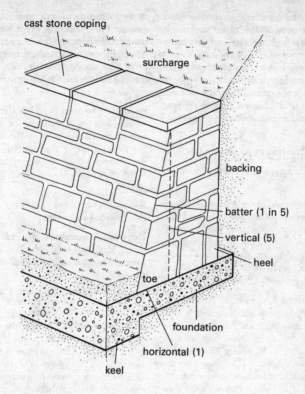

Retaining wall.

reverse curve, reversed c. [sur.] A *curve* of S-shape.

reversed filter See **loaded filter**.

reverse face, reversing f. [sur.] See **change face**.

reverse osmosis (RO) Probably the most popular and economical *membrane process* of *de-salination*. It can reduce the total dissolved solids usually by 95%, sometimes by 99%. Bacteria, viruses and pyrogens are 100% rejected. Seawater of up to 50,000 ppm (5%) can be upgraded to tap water at 200 ppm. The main costs are the pump to pressurize the water at 10–30 bar pressure and the pre-treatment processes. The first commercial RO plant was built by Havens Industries of San Diego, California, in 1968. Since the essence of the process is reversing an *osmosis*, the membrane is all-important. Called a 'permeator' it is a cartridge filter of cellulose ester, nylon or polyphenylene oxide, etc.

reverse-rotary drilling *Rotary drilling* in which the direction of flow of the drilling fluid is reversed, in other words the fluid is pumped down outside the drilling pipe and rises up inside it.

Because the pipe bore is smaller than the area outside the pipe, the rising mud has a high velocity and can carry large pieces easily. The method is suitable for work in rocks that break easily into coarse lumps or for deserts where water is precious and air must be used instead. It is claimed that the rising fluid is less adulterated with chips from the walls.

reversible tramway See **jig back**.

revet To protect a surface with *revetment*.

revetment A covering to a soil or rock surface to protect it from *scour*. It does not withstand earth thrust. Common revetments include: asphalt, beaching, concrete slabs precast or cast in place, *mattresses*, *maritime plants*, *pitching*, *reinforced grass* and *rip-rap*.

revibration According to A. M. Neville revibration of concrete 1–2 h after placing may increase the compressive strength by 15%, especially in concretes liable to bleeding, by expulsion of entrapped water. Despite these advantages it is not widely used, probably because if too late it damages the concrete. A concrete column can settle up to 8 mm ($\frac{1}{3}$ in.) in a normal storey height during placing, vibration and setting, mainly because of drying *shrinkage*, roughening the top 30 cm (1 ft) of the column. Revibration is useful when casting against rough boards. It then dramatically improves the surface finish, but the moment to revibrate must be carefully chosen by an experienced concretor. It can close some types of *plastic cracking* if this is noticed soon enough.

revolver crane A crane of unusually large lifting capacity. Two at Graythorp on the Tees, near Hartlepool, in use in 1973 for building North Sea oil production jackets, could lift 800 tons at 21 m radius or 112 tons at 61 m radius. They had jibs 73 m long and travelled on concrete tracks 23 m apart. The cranes were erected on steel towers 49 m high above the dock floor, which travelled under load, at 5 m/min by jacking.

revolving screen [hyd.] A *trash rack* which may be a cylinder or a belt, turned mechanically or by the force of the water passing through it. As the screen appears above the water it is freed of rubbish by a scraper, a water jet, etc.

Reynolds critical velocity [hyd.] See **critical velocity**.

Reynolds number, R. criterion [hyd.] Reynolds established the principle used in hydraulics and aerodynamics that

$$\frac{\text{fluid velocity} \times \text{pipe radius}}{\text{viscosity}}$$

is the same for all fluids at the *critical velocity*. This number, which differs with the velocity, is the Reynolds number for the given flow condition. It is much used in *model analysis*, since the Reynolds number in model and structure should be the same. See **Froude number**.

rheology The study of *deformation* and flow, including *creep* in solids, fluids and semi-solids like *mastics* (*B*) and fresh concrete.

rheometer See **orifice rheometer**.

rib (1) **r. centre** The curved beam which carries the centering of an arch.
  (2) A small beam projecting below a reinforced-concrete slab, whether flat or arched. It stiffens the slab.

ribbed slab A *hollow-tile floor*.

rib holes (trimmers UK) US term for holes in tunnelling or shaft sinking which are drilled at the sides of the tunnel or shaft and therefore are fired last, after the *cut holes* and *relief holes* (and *lifter holes* if any).

rice grass See **maritime plants**.

rider sewer A shallow *sewer* receiving house connections. Connected to a parallel deep main sewer at intervals, it reduces the amount of deep excavation and the cost of connections.

riffle sampler [min.] A *sampler* designed to reduce a batch of ore to half its original size. It consists of a rack of vertical parallel plates a uniform distance apart, each of which diverts the ore falling on to it. Alternate pairs of plates divert the ore to alternate sides. The volume reduction is rapid for dry ores crushed to suitable fineness. To reduce the sample to one quarter or one eighth its original size the sampler can be built of several riffles in series. Riffle sampling is much quicker and less laborious than coning and quartering.

rift, grain, quartering way Quarrymen's words for the roughly perpendicular joints and cleavage planes in most rocks. Rift is the easiest plane of splitting; grain may be a secondary cleavage at right angles to it. Some writers mention a third plane, head. See also **tough way**.

rigger (1) A semi-skilled or skilled worker who erects and maintains *lifting tackle*, hand-operated *derricks*, *ropes*, *slings*, *cradles* (B) and other scaffolding.

(2) A less widely skilled man who erects tubular or other metal scaffolding.

rigid arch [stru.] An arch without *hinges*, *fully fixed* throughout.

rigid design [stru.] A design of a steel structure in which the connections are assumed able to develop the stiffness and to carry the bending moments required by full *continuity* (BS 5950).

rigid frame [stru.] A structural frame in which the columns and beams are rigidly connected together without *hinges*.

rigidity [stru.] The resistance of a material to shearing or twisting.

rigid pavement Road or airstrip construction of concrete slabs. Cf. **flexible pavement**.

rigid pipe Unglazed or glazed pipes made of vitrified clay, concrete, asbestos cement or grey cast iron. Flexible or rigid joints may be used, as with *flexible pipe*.

rimmed steel, rimming s. Partly deoxidized, low-carbon (0.15–0.3%C) steel which has a core containing impurities and a rim of relatively pure steel. The steel round the central gas pores is not oxidized and the pores close up eventually during rolling. Cf. **killed steel**.

ring In the lining of tunnels and shafts with cast iron, pressed steel or precast concrete segments of circles, a complete circle formed by these segments is called a ring. The cast-iron rings of the London tube railways are 508 mm (20 in.) wide, 25 mm (1 in.) thick and of 3.65 m (12 ft) bore.

ring ditch An *oxidation ditch*.

ring main A ring layout of *mains* in populous districts, which makes supplies of water, gas, electricity, etc. secure, supplying each consumer from two directions. If the supply is cut from one side, there is still a supply from the other. The ring main for water under London, begun in 1987, has the further advantage that being large (3 m (10 ft)) bore it will reduce the water pressures in the mains. When the 76 km main is completed in 1994 it should deliver 1300 Ml/day to half the population of London from several waterworks. One section in north London, begun with pipe jacking (later changed to conventional tunnelling), used concrete pipes 5 m long and 340 mm (14 in.) thick; 25% of the cement in the concrete was replaced by *fly-ash* to resist the sulphates in the soil.

ring tension, hoop t. [stru.] The tension which occurs in the wall of a circular bunker or tank containing solid or liquid. The force of the material pressing outwards on the walls, at the top of an open tank, is carried wholly by ring tension, at the bottom wholly by tension in the floor, and intermediately in proportion.

riparian [hyd.] Concerning the banks of a stream, lake or canal. A riparian owner owns at least part of one bank.

ripper (1) A single massive steel point (rarely two) up to 1 m (3 ft) long and 80 mm (3 in.) thick, towed by a bulldozer to break rock. Since ripping is much cheaper and quicker than drilling and blasting, it is helpful to discover whether ground is rippable and this can be done by *seismic prospecting*. See **impact ripper**. Cf. **scarifier**.

(2) A *concrete breaker*.

ripple [hyd.] A small wave on a water surface or a repetitive wave on a sand bed, with a flat upstream face and steep downstream face, caused by slow flow slightly above the least flow needed to move the sand.

rip-rap [hyd.] Stones for *revetment* from 7 to 70 kg in weight. They protect the bed of a river or its banks from *scour*.

rise The vertical distance from the crown of a road to its lowest point.

rise and fall [sur.] A method of reducing staff readings by working out the rise or fall from each point to the one

following it. The rises or falls are entered in the level book in special columns parallel to those for the staff readings. Cf. **collimation method**.

rising groundwater under cities Under central London the standing water level is rising at 1 m (3 ft) yearly. Similar conditions in Liverpool have caused difficulties in the underground railway. Structural troubles arising from this could include reductions in friction on piles and their *bearing capacities*, *swelling* of clays resulting in the lifting of buildings, and the trapping and pressurization of hazardous gases from old waste tips beneath cellars or basements. Tube railways might be unaffected in clay but all must pass through gravel overlying the clay and the gravel is increasingly waterlogged. The water level is rising because of the heavy licence charges for private wells, imposed since 1974 under the 1973 Water Act. The water level might drop if licence fees were to be reduced.

river basin management (see diagram) Economic development of a river basin, often involving electrical generation from water power, as well as building dams, *levees*, bridges or aqueducts, controlling river flows, and digging canals for irrigation or navigation.

river pollution The UK does not have a good record for river pollution but its sea pollution is probably even worse. In 1987 there were 23,253 cases of river pollution; 1402 of these were regarded as serious but only 288 prosecutions took place. Many of the guilty parties were water authorities' sewage works. The punishments meted out by the courts were generally not large enough to prevent repetitions. A fine of a few hundred pounds every few years is well worth paying, from the

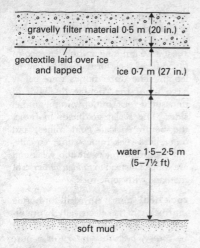

gravelly filter material 0·5 m (20 in.)

geotextile laid over ice
and lapped

ice 0·7 m (27 in.)

water 1·5–2·5 m
(5–7½ ft)

soft mud

Use of geotextile to divert the river Vantaa, Finland. A layer of gravelly filter material, laid on lapped geotextile sheet over the ice every winter, sank uniformly in spring with 200–250 mm (8–10 in.) of unmelted ice. After several years the river had a new alignment.

viewpoint of a large polluting works, if it saves hundreds of thousands of pounds of investment in waste treatment plant. See **North Sea**.

river purification boards (RPB) Bodies in Scotland that control the purity of rivers, like the *National Rivers Authority* in England and Wales.

rivet [mech.] A round bar of *dead-mild* steel with a cup-shaped, conical or pan-shaped head, driven while red-hot into a hole through two pieces of steel which are to be joined. The head is held in position by a holder-up with a dolly while a second person strikes (sets up or *upsets*) the other end, forming a head on it with a hand hammer or pneumatic or hydraulic riveter. Rivets of smaller diameter than 18 mm can be driven cold. Aluminium, copper and other materials are also used for making

rivets. *Light alloy* is riveted with light-alloy rivets to avoid corrosion at the rivet hole, but steel rivets can be used with protective painting. Riveting, except *pop riveting* is rarely if ever used now on building sites; it has been superseded largely by *high-strength friction-grip bolts*.

RL loading A railway *bridge loading* like that of London Transport, lighter than *RU loading*, used on track carrying only passenger traffic. For a 100 m bridge or shorter, the main load is a distributed load of 50 kN/m (5 tons/m). If the bridge is longer than 100 m the extra length is loaded at 25 KN/m. In addition to the distributed load a *point load* of 200 kN (20 tons) is placed at the most unfavourable position on the bridge (BS 5400).

road (1) A concrete, *stabilized soil*, earth, *hoggin*, tarred or other surface for vehicles or animals to travel on.
(2) [rly] A rail track.

road base [s.m.] A *base course*.

road bed, trackbed [rly] The *ballast* which carries the *sleepers* and *rails* of a *permanent way*.

road breaker See **concrete breaker**.

road forms, side f. Wooden or steel planks on edge which form the side of a concrete slab and are set with their upper surfaces at the correct levels for the road so as to guide the men screeding.

roadheader A *boomheader*.

road heater A machine which heats a road surface by blowing flame or hot air on to it.

road junction A meeting of roads at the same level. Cf. **grade-separated junction**.

road-making plant Specialized equipment, including *concrete pavers*, *con-*

crete spreaders, gritters, planers, pressure tanks, pulverizing mixers, road heaters, rollers, scarifiers and *smoothing irons*.

road panel An area of concrete laid in one operation and bounded by free edges or joints on all sides, some of which may be *dummy joints*.

road pricing, ERP (electronic road pricing) A concept for restricting the use of congested roads by motorists, with an electronic device installed in every car notifying its presence to roadside computers. These are connected to a central computer which periodically invoices the motorist for his or her use of a congested road. After serious attempts to install road pricing, Hong Kong failed to overcome public objections to 'invasion of privacy'. But Hong Kong succeeded by other means in reducing car registrations from 211,000 in 1982 to 170,000 in 1984. This was achieved by raising the tax on motor fuel and tripling the car tax. Singapore uses other methods of *traffic restraint*.

road roller A power-driven *roller* weighing from ½ to 12 tons. See **compaction**, **roller**, **tandem**.

road surface Also called the carpet, topping or *wearing course*.

rock A mass of different grains cemented together by a *matrix*. Sandstone is a rock consisting of quartz grains cemented by an iron oxide (red sandstone) or by calcite (pale sandstones). Quartzite is a sandstone cemented by quartz.

rock anchor, r. bolt See **roof bolt**.

rock breaker A *hydraulic hammer*. See also **breaker**.

rock burst [min., r.m.] Below about 1000 m (3000 ft) in rocks as strong as granite, an explosion of rock that may involve from 50 kg to many tons.

Severe rock bursts were suffered in the driving of the Mont Blanc tunnel between France and Italy, opened in 1965, 11.6 km long and under 2000 m (6500 ft) of rock for more than half its length. The worst bursts were below 2450 m of the Aiguille du Midi in unfissured rock, but only in the tunnel sides. The Kolar goldfields in India, the South African gold mines, and other hard-rock mines suffer equally, but coal mine rocks are too weak and are subject instead to *sudden outbursts* of methane with some rock.

rock drill [min.]. Any drill for boring into rock, whether electrically, hydraulically or compressed-air driven. It may be *rotary*, *percussive-rotary* or a pure percussion tool. Compressed-air drills blow air down the centre of the drill steel to clean the hole and stop the bit jamming.

rocker bearing A bridge or truss support which is free to rotate but not to move horizontally unless it is also carried on rollers.

rocker pipe To adjust the settlement of a building to the smaller settlement of a buried drain outside it, a short drain pipe is placed at the entry to the building. It should be 0.5–0.75 m long. Where much settlement is expected, several rocker pipes may be needed and the pipe slope may have to be increased to prevent any backfall.

rocker shovel [min.] A high speed mechanical shovel used in tunnelling.

rockfill dam [hyd.] (see diagram) An *earthen dam* built of any available stone or broken rock. If built of random stone it is likely to be faced with a flexible watertight skin of steel plate or timber on the water side. If the watertight skin is brittle, like reinforced concrete, or if the dam has a watertight concrete core wall instead of a water-

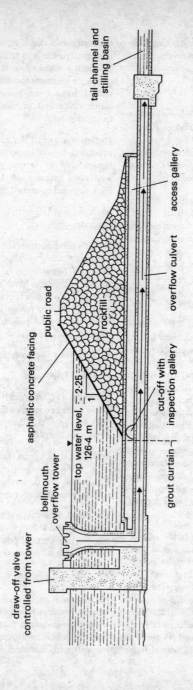

Rockfill dam, Roadford, Devon. (*Construction News*, 20 October 1988; Alfred McAlpine's 3½ year contract, worth £16 million, started in February 1987.)

draw-off valve
controlled from tower

bellmouth
overflow tower

top water level,
126·4 m

asphaltic concrete facing

public road

rockfill

2·25
1

cut-off with
inspection gallery

grout curtain

access gallery

overflow culvert

tail channel and
stilling basin

tight skin, the stone must be carefully compacted so that there is no chance of movement.

rock flour Crushed rock of *silt* size.

rock fragmentation Breaking rock in mining, tunnelling or quarrying. *Principles of Rock Fragmentation*, by G. B. Clark (Wiley 1987), discusses wave equations, percussive or rotary drilling, water-jet cutting and explosives.

rock mass classification (rmc), rock quality designation (rqd) Many rqds exist to help tunnel design. They classify rock masses according to rock strength, cracks or other discontinuities, groundwater, geological folding or faults, and are described in *ASTM Special Technical Publication 984*, a report of a 1987 symposium.

rock mechanics [r.m.] The science of the rocks in the ground, their fabric, strengths, stresses, strains, friction, elasticity, equilibrium, creep, ductility and flow under pressure, and the use of these properties in tunnelling and mining generally, often helped by laboratory testing. It includes the study of blasting, *rock bursts*, *rock noise*, *bumps*, rock breaking and the stability of rock slopes. Rock mechanics makes use of many testing methods and instruments that are used also in *soil mechanics*, as well as the evidence of boreholes and *geophysical* surveys, *well logging*, etc.

rock mill, hydrofraise In the excavation of a *diaphragm wall*, *secant wall*, etc. before grabbing begins, a pair of contra-rotating toothed drums (the rock mill) can be lowered into the trench to break up the stiff soil or soft rock that would delay the grab. Sizewell B power station in Suffolk has a diaphragm wall 30 m (100 ft) deep, in places much deeper and claimed to be

the longest in the UK, where the diaphragm trench was sunk in this way. Without a diaphragm wall at least six series of *groundwater-lowering* pumps would have been needed, involving much delay and unnecessary expense in excavation.

rock noise [r.m.] *Rock bursts* are often preceded by differences in the minor noises that are given out by rocks. Listening instruments have therefore been devised to record them.

rock pocket US term for minor *honeycombing*.

Rockwell hardness test [mech.] A *hardness* testing method in which the depth of penetration of a conical diamond point (for hard metal) or of a steel ball (for softer metal) is measured. Like the *Vickers hardness test* it is suitable for *case-hardened metal*. See **Brinell hardness test**.

rock wheel See **trencher**.

rod [sur.] See **staff**.

rodding (1) Unblocking drains with *drain rods*. See **inspection chamber**.
(2) In the USA the *punning* of concrete. See also *(B)*.

rodding eye A pipe connecting a bend in a *drain pipe* to the ground level. *Drain rods* can be pushed through it to unblock the drain if need be.

rod float, velocity r. [hyd.] A wooden rod weighted at its lower end, designed to float in a vertical position with most of its length submerged so as to average the velocity in the stream throughout its submerged depth. See **float**.

rolldown A method of *pipe renewal* in which a standard medium density *polythene* pipe is passed through a series of rolls to reduce its diameter and thicken its wall with the minimum of elongation. Lengths as much as 180 m can be rolled down and stored on the site until needed since they last for weeks without re-expansion. A 104 mm (4 in.) pipe can be rolled down to fit into a 95 mm bore iron pipe, gaining 4% wall thickness and 2% in length. Re-expansion is not difficult using water pressure at 1.4 MPa (200 psi) but problems may be caused by the pipe shrinking back to its original length. Consequently low-frequency pulses of water are used, encouraging slippage between the old and the new pipes. One long rolldown job drew 1135 m of pipe into a 268 mm (10½ in.) gas main at Wokingham, in 180 m lengths welded together on site.

rolled asphalt A wearing course or base consisting of hard *asphalt* and *coarse aggregate* laid hot and rolled until it is nearly free of voids.

rolled concrete See **roller-compacted concrete**.

rolled-steel joist (r.s.j.) An I-beam made of one piece of steel passed through a *hot-rolling* mill. Cf. **cold rolling**.

rolled-steel section Any hot-rolled-steel section, including *rolled-steel joists*, *channels*, *angle sections*, bulb angles, zeds, rails and so on. This is usually the cheapest steel, since the rolling process is continuous. Cf. **tubular sections**.

roller The main equipment for *compaction* of soils or roads. Rollers weigh from 500 kg upwards. *Pneumatic-tyred rollers* improve the density in depth as well as on the surface. *Sheepsfoot rollers* can compact cohesive soils but the surface remains loose. See also **high-speed / road / vibrating / wobble-wheel roller**.

roller bearings [mech.] Hard steel cylinders in bearings which have low friction. Compared with the best plain bearings their *rolling resistance* is at most one half and usually about one quarter. The advantages of roller bearings are outstanding where locomotives have replaced rope haulages. Small changes in resistance make a very large difference to the number of trams which a loco can pull, but even large changes in resistance do not alter the number which a rope can pull.

roller bit [min.] In *rotary drilling* for oil-wells and in other deep holes (down to 5 km), a drilling bit with several toothed rollers which are rotated by the turning of the drill rods. They are used for hard rock, since in spite of their complicated shape they last longer. This fact may be decisive even if they drill more slowly because the withdrawal of the drill pipe, changing of the bit, and sending the bit back into the hole may stop drilling for 5–10 hours. Roller bits are rented from the makers in the USA, since most oil companies are not equipped for refacing them with *hard facing*. See **turbodrill**.

roller-compacted concrete, dry lean c. Concrete roller-compacted dams built since the early 1970s have cost one third less than conventional concrete dams. No side forms are used. Since then, roller-compacted concrete roads have been built in the USA, using continuous mixers of at least 250 tons/h capacity. The concrete is extremely stiff, so must be mixed in a twin-shaft pugmill as for asphalt road material. It is placed with a *concrete spreader* or modified asphalt paver and elaborate *paving trains* are not needed. Dump trucks are covered to protect the concrete against sun, wind, rain or frost. The mixer must be less than 15 minutes away from the placing point and the concrete should be placed within 45 minutes, but fresh concrete placed against concrete laid down less than 90 minutes earlier will not make a *cold joint*. Four passes of a 10-ton two-drum vibrating roller are usually satisfactory.

The roller drum overhangs the concrete edge by 20–50 mm (1–2 in.). The maximum thickness of a lift is 250 mm (10 in.). If two lifts are used, the joints should be aligned. A light water spray starts the first 24 hours' curing followed by a week's *membrane curing*. No traffic is allowed for 14 days. Careful attention is paid to the aggregate grading, which is measured several times daily. With such stiff concrete the *slump test* cannot be used so a *nuclear density gauge* is used instead. US experience up to 1987 was that the road surface is less smooth than with conventional concrete, limiting its use to car parks and secondary roads. In roller-compacted dams fly-ash contents have reached 60–80% of the cementitious material.

roller gate [hyd.] A hollow cylindrical *crest gate* for dam spillways. It is carried on large toothed wheels at each side which mesh with steeply sloping racks up which the gate travels as it is being opened. See **sector gate**.

rolling lift bridge [stru.] A bridge of which the lifting part has at its shore end a segment bearing on a flat (rolling) surface.

rolling load [stru.] A *moving load*.

rolling resistance [min.] Friction between rails and wheels. On level track it is the bulk of the resistance. For old colliery tubs it may reach 50 kg/ton but for modern mine cars on good track it may drop to 2 kg/ton.

rolling-up curtain weir A *frame weir* in which the frame remains upright. The barrier consists of horizontal planks which increase in thickness with their depth below water. They are connected by chains and drawn up to open the weir.

rollway [hyd.] The overflow portion of a dam; an overflow spillway.

Roman cement See (*B*).

roof bolt, rock b., rock anchor A device for holding rock in position that originated in mines but is now used also for holding up shaky cliffs. Roof bolts may be end-anchored or anchored for their full length. Full-length anchored bolts are usually held by a resin which hardens in a few minutes, providing immediate safety to the people fixing the bolts. End-anchored bolts may be held by an expansion cone at the point or by resin. The usual pattern for rock bolting with 2 m long bolts is to place them 1.2 m (4 ft) apart both ways. A recent development is a tubular bolt, used also as a drill steel to drill the hole – of 30 mm dia. with a 52 mm dia. drill bit forged on the end. During drilling the tube is used for water injection; afterwards for cement grout injection. Having a threaded end, this type can be extended indefinitely. See **resin anchor**, **split-set stabilizer**.

roof control, strata c., ground c. [min.] *Rock mechanics* used for solving mining problems, such as forecasting when and how the roof will cave or squeeze (converge) towards the floor, with a view to improving both safety and mineral production.

root The part of a dam which merges into the ground where the dam joins the hillside.

rooter A towed *scarifier*.

Rootes blower [mech.] A *rotary blower* consisting of a pair of hourglass-shaped members which interlock and rotate together. It gives a higher pressure than the average fan of (0.07–0.2 bar) 0.7–2 m water gauge and delivers fairly large volumes, so it is suitable for forcing ventilating air through long pipelines.

rope [mech.] Steel-wire rope is generally used for most hoisting or haulage purposes, *fibre rope* being more flexible but less strong. For the steel of which ropes are made, see **prestressed concrete**. Ropes are described in the following way. A 25 mm 6/19 rope is of 25 mm dia. with six *strands* each of 19 wires. See **lay**.

rope diameter [mech.] The diameter of a steel-wire rope is the greatest diameter obtainable across the strands. This is increasingly so now also for *fibre ropes*.

rope fastenings [mech.] The best fastening between a steel-wire rope and its socket is the white metal capping used for man-winding shafts in Britain. Where this method of capping rope is inconvenient, ropes are usually doubled back on themselves round a steel *thimble* and fixed with *bulldog grips*.

ropeway See **aerial ropeway**.

ROPS Roll-over protective structure for a *loader*, *bulldozer*, etc. Cf. **FOPS**.

Rossi–Forel scale [stru.] This scale for the intensities of earthquakes, evolved by Rossi and Forel, grades earthquakes from 1 (very slight) to 10 (catastrophic). In a grade 8 earthquake, large cracks appear even in ordinary houses and they may be damaged beyond repair.

rotary blower [mech.] An air compressor suitable for pressures up to 0.7 bar above atmospheric, often a *Rootes blower*.

rotary drilling [min.] (1) A method of drilling deep holes (including shafts)

for oil, gas or mining, from 0.15 m (6 in.) to several metres diameter. Lengths of heavy hollow drill pipe screwed together pass down the hole and carry a cutting bit at their tip. (See **fishtail/roller bit**.) *Drilling fluid* is pumped down the pipe, passes round the cutting edge of the bit, cools it, and brings up the broken rock. The first crude rotary drill was invented and used by the Baker brothers, drilling contractors in Dakota, in 1882. It required much development and was first effectively used at Spindletop, Texas, in 1901. Rotary drills are always used for holes deeper than 1800 m, often for shallower holes, and they have drilled down to 4.5 km. (See **cable drill**.) Cores can be obtained by fitting *corers* instead of ordinary drilling bits. See **Betws**, **kelly**, **reverse-rotary drilling**, **turbo-drill**.

(2) Drilling a hole for blasting by a rotating (usually electrically driven) anger-shaped drill steel. The detachable bit usually has a *hard facing* on the cutting edge. In such hard rocks as granite, rotary drills cannot compete with *percussion tools*.

rotary excavator (1) See **tunnelling machine**.

(2) **bucket-wheel excavator, wheel e.** A machine with a vertical digging wheel, on a horizontal axle, which carries large buckets on its rim. These machines, used for quarrying, particularly in German brown-coal mines, are perhaps the world's biggest excavators, weigh over 5000 tons, and are able to move over 5000 tons of earth or coal per hour. They are carried on crawler tracks.

rotary meter A *current meter*.

rotary pump See **rotodynamic pump**.

rotary screen, trommel [mech.] The commonest screen for sizing gravel or broken stone. It is a rectangular plate bent round to a cylindrical shape, with holes punched in it of suitable size. If the holes are of two different sizes, the larger holes are at the lower end, and the screen thus classifies into three sizes: oversize, intermediate and undersize.

rotary snowplough A snowplough with a rotating blade to shoot the snow off the road.

rotating biological contactor (RBC), r. b. disc [sewage] An aerobic *biological treatment* in which a slowly rotating horizontal shaft just above the surface of *effluent* being treated, carrying closely spaced discs of up to 3 m dia., alternately submerges them and exposes them to air.

rotational slide, cylindrical s. [s.m.] A failure of a clay slope which involves slipping of the earth on a curved surface. Much research was done on this sort of failure in Sweden after several hundred metres of Gothenburg harbour slid into the sea in 1916. The Swedish *circular arc* method was then evolved by Petterson and developed by Fellenius (1927).

rotation recorder An instrument which measures the very slight rotation of the support of a bridge during loading. See **spread recorder**.

rotative drilling [min.] See **rotary drilling** (2).

rotodynamic pump, rotary p. There are three types of rotary pump: *centrifugal*, axial-flow (like ships' propellers) and mixed-flow pumps – a mixture of the two designs. All three are cheap, small and lightweight, can be set horizontally or vertically, and can adjust their flow rate and head by variation of the runner diameter or speed or both.

rotor The rotating part of an electric motor, as opposed to its stationary part,

the stator. For turbines or *rotodynamic pumps* the rotating part is the *runner*.

roughcast See (*B*).

roughness [hyd.] A value which enters into every formula for flow in pipes or channels and expresses the friction or resistance to flow due to the surface texture of the channel or pipe.

round [min.] In tunnelling or shaft raising or sinking, a set of holes drilled, loaded and fired together, generally involving breaking the rock for a *pull* of 1.2–2 m. It includes *cut/lifter/relief/rib* holes.

round bar The commonest and cheapest steel *reinforcement* for concrete – *mild steel* with a *yield point* of 230 MPa (33,400 psi) and a *tensile strength* of 288 MPa (42,000 psi). Most other types of reinforcement can be much more highly stressed.

round-headed buttress dam A mass-concrete dam built of parallel buttresses which are thickened at the water end until they touch each other. The appearance is like the *multiple-arch dam*. The spillway may be a curved slab which passes over, joins and stiffens the downstream ends of the buttresses. This dam may be of mass concrete except for the spillway slab and is considerably cheaper than the multiple-arch dam but slightly heavier.

routing, flood r., storm r. Calculating the changes of flow along the line of a channel (river, stream, sewer, etc.) with the help of a *hydrograph* or other means.

roving A spool of glassfibre, e.g. for making *glassfibre-reinforced cement*.

Royal Commission recommendation [sewage] The Royal Commission on Sewage Disposal (1898–1915) recommended that sewage *effluents* should contain no more than 30 mg/l of *suspended solids*, nor have a *biochemical oxygen demand* (BOD) above 20 mg/l – the so-called '30:20 standard'. In addition the effluent must, on discharge, be diluted by at least eight times its own volume of *receiving water* having BOD of less than 2 mg/l. If this volume and purity of receiving water are not available, the effluent must be made purer.

A 20:20 recommendation is widely accepted in Europe but the effluent at Dubai in the Arabian Gulf is much better – 10:10 so as to irrigate the city's parks without pollution or smell. This is achieved by *activated sludge* treatment followed by *trickling filters* and finally *sand filters*.

r.s.j. A *rolled-steel joist*.

RTD *Resistance temperature detector*.

rubbed finish A smooth finish on a concrete surface. See (*B*).

rubber linings Steel or cast-iron pipe which transports *hydraulic fill* may last only three weeks. With a rubber lining it can last seven years. Another use for rubber linings is on the inside of steel *tanks* for *mixed liquor* or *water treatment*. Rubber linings 1 mm thick have also been put inside concrete tanks to prevent leakages. One type of rubber – polysulphide – has been reinforced with carbon fibre to bridge gaps in leaky tanks (Thiokol).

rubber-tyred rollers *Pneumatic-tyred rollers*.

rubber tyres The tyres of *loaders*, *bowl scrapers*, etc. can be responsible for a third of their running costs. In bad conditions tyres need replacing every 1500 hours of work; in good conditions they last 5000 hours. On rocks they last longer if fitted with chains. It is an offence under the UK Motor Vehicle Regulations to use a vehicle with wrongly inflated tyres, and

20% under-inflation takes off a third of the tyre life, wearing out walls and increasing fuel consumption. Over-inflation may cause aquaplaning on a wet road. Very large tyres are retreaded (remoulded) several times.

rubble concrete Masonry consisting of large stones set in joints of 15 cm of concrete and faced with ashlar. Used in massive work such as dams.

rubble drain, blind/spall/stone d. A trench filled with stones selected so that they allow water to flow through it. See **drain**.

rubble-mound breakwater, mound b. [hyd.] A *breakwater* built of heavy heaped stones, weighing up to 10 tons. Between tide marks the slope is nearly flat (1 in 12) but at more than 5 m below low water it is much steeper, about 1.25 horizontal to 1 vertical. Below water the rubble may be strengthened by injection at 200°C of fluid bitumen mixed with sand. Rubble may be a good foundation for *blockwork*. See **armour**.

ruling gradient [rly] The *limiting gradient*.

RU loading [rly] The *bridge loading* for European mainline railways. It includes four 250 kN (25 ton) point loads at 1.6 m (5 ft) intervals, preceded and followed by a distributed load at 80 kN/m (8 tons/m) beginning at 0.8 m (3 ft) from the extreme point loads thus:

(*distributed*)

0.8 m	1.6 m	1.6 m
8 tons/m	25 tons	25 tons
1.6 m	1.6 m	0.8 m
25 tons	25 tons	8 tons/m

rumble strip, serrated s., jiggle bars On a highway, a slightly raised strip of asphalt, plastic, etc. across the lane of traffic that is approaching a hazard. The strip is 10–15 cm wide and 1–3 cm high. Some 10–20 are placed together at a spacing sufficient to warn the driver of the approaching hazard.

rummel Scots term for a *soakaway*.

runner, rotor [mech.] The rotating part of a *water wheel, turbine, centrifugal pump, fan,* etc.

runners Vertical timber sheet piles 5 cm or more thick driven in by hand ahead of the digging at the edge of an excavation. They may be pointed with an iron shoe at one side of the timber so that as each pile is driven in it is forced tightly against its neighbour.

running ground [min.] Very wet or very dry sand or silt which flows like a liquid. It is usually dealt with by *forepoling*, close timbering or other special methods of support; by *artificial cementing* of the ground; or by *wellpoints* or similar drainage methods which strengthen the soil.

run-off The amount of water from rain, snow, etc. which flows from a catchment area past a given point over a certain period. It is the rainfall less *infiltration* and *evaporation*. In a stream it can be increased by springs of *groundwater* or reduced by loss to the ground. See **influent stream**.

run-off coefficient The *impermeability factor*.

S
-----

sack rub, bagging, bag wash Smoothing a concrete wall surface by filling in all the pits. The surface is first damped and mortar is rubbed all over it. Before it dries, a final mix of dry cement and sand is rubbed over with a sacking wad or a sponge-rubber float to fill the pits and remove surplus mortar (*ACI*).

sacrificial anodes The anodes used in *cathodic protection*.

sacrificial protection The property of zinc, cadmium, aluminium or tin coatings to protect a steel surface. The protecting metal is dissolved first (sacrificially) by the water, like an anode in *cathodic protection*.

saddle A steel block over the tower of a *suspension bridge* or aerial ropeway which acts as a bearing surface for the rope passing over it. See also (*B*).

safe-load indicator A painted steel sheet, rigidly fixed to its crane jib and easily seen by the crane driver, on which are shown the safe hook loads at various angles (radii) of the jib. A heavy hinged pendulum or needle changes its slope as the jib changes slope, thus pointing to the maximum safe load at each jib angle. If the crane driver observes the safe loads shown on the indicator, the crane should not overturn. If the jib is changed the new jib is installed with its own safe-load indicator. With *outriggers* working, the safe load normally increases and the indicator cannot be relied on. Cranes can also overturn because of soft ground. See **crane safe working load** (diagrams).

safety belt [min.] A harness or belt worn by someone to prevent him or her dropping more than 0.6 m while working. A remote anchorage belt is sometimes used by quarrymen. Since this belt allows a drop of 1.8 m it is provided with a shock absorber.

safety factor See **factor of safety**.

safety fuse, blasting f. [min.] A train of black powder enclosed in a waterproof braided textile, used for firing *detonators* in small-scale work in quarries or metal mines. It burns at 0.6 m/min ± 10%. For underwater work, special fuse must be used.

safety rail See **check rail**.

safety valve [mech.] A valve fitted by law to all boilers. It opens at a pressure slightly above that at which the boiler works. It thus releases the excess of steam and reduces the pressure. The noise of escaping steam also draws attention to the fact that the boiler pressure is too high.

sag bar [stru.] An *anti-sag* bar.

sag correction, catenary c. [sur.] A *tape correction* to the apparent length of a level base line due to the sag in the tape. The measured length is longer than the straight length between supports by $\dfrac{w^2 l}{24 t^2}$ metres where w is the is the weight of the hanging length of the tape in kg, l is the apparent length between supports in metres, and t the tension of the tape in kilograms.

sagging moment [stru.] A bending moment which causes a beam to sink in the middle. Usually described as a positive moment. See **hogging moment**.

sag pipe, inverted siphon [hyd] A *sewer* or irrigation channel carried in a pipe under a road or other obstacle. It needs very careful design to avoid blockage, using a flow which at its slowest will still be *self-cleansing*. Even so, it should be provided with a draw-off valve at the foot to clean it out if necessary.

St-Nazaire Bridge [stru.] See **cable-stayed bridge**.

saline wedge, salt water w. In estuaries where the freshwater of a river flows out to sea, the seawater being saline and denser sinks below the river water. In the Mississippi this saline wedge may extend 160 km (100 miles) upstream when the river flow is low.

saltation The bouncing movement of sand grains carried by wind or water, usually hitting the bed at a flat angle, between 10° and 16° to the horizontal. The result of many impacts in the same direction is to cause the sand bed to creep forward. A fast grain will shift one on the bed which is six times its diameter – 200 times as heavy.

sample [min.] A small amount taken from a rock exposure or mass of broken mineral or other material which is analysed for its content of valuable mineral. In civil engineering work, samples of sand, coarse aggregate and other materials are regularly taken for analysis. See **random sample**, **representative sample**.

sampler (1) **mechanical s., sample splitter** [min.] A device for reducing the volume of a crushed sample to a *representative sample* which can be handled in the laboratory, e.g. a *riffle sampler*. It may also be a device for taking samples at fixed time intervals from a stream of solid or liquid.

(2) [s.m.] See **soil sampler**.

(3) [hyd.] An instrument for examining the bed or water on it in hydrographic surveying. It may be a *bottom sampler*, a *sand catcher*, etc.

sampling spoon [s.m.] See **soil sampler**.

sand [s.m.] Granular material (composed mainly of quartz) of 2–0.06 mm in size. It has no cohesion when dry or saturated but has 'apparent cohesion' when damp. See **classification of soils**, **graded sand**, **uniform sand**; also (*B*).

sand blast [mech.] A compressed-air jet which throws sand or flint on to a surface to be smoothed, cleaned or etched for decoration (glass). Shot or steel grit are often used because they are less dangerous to the lungs than *silica* dust. See **shot blasting**.

sand box See (*B*).

sand catcher, s.-grain meter [hyd.] A hydrographic instrument through which the water flows and deposits sand. The instrument is then brought up to the surface and the sand quantity measured.

sand drain [s.m.] See **vertical sand drain**.

sand fill, s. filling [min.] See **hydraulic fill**, etc.

sand filter, slow s. f. Many types of sand filter exist, but for water purification only the *rapid sand*, the *multilayer*, and the slow sand filter are mentioned in this book. The slow sand filter has existed since 1829 and is valued for drinking-water since it removes both unpleasant smell and taste as well as half the organic *pollutants*.

It is a large open tank, about 3 m deep. From the bottom upwards, above the burnt clay or precast concrete underdrains are first a layer of coarse gravel, then fine gravel covered by uniform sand 0.7 m deep of about 0.3 mm particle size, drowned under 1.2 m of water which slowly passes down through it. Every month or so the filter is drained and the top few centimetres of sand are skimmed off with their dirt and rejected. This is repeated until the sand bed becomes too thin and has to be rebuilt. The slow sand filter can filter dirty water such as sewage *effluent* at about 1 m/day. With clean drinking-water its rate is about 2 m/day.

sand-grain meter See **sand catcher**.

Sandö Bridge [stru.] Until the completion of the *Parramatta Bridge*, this was the largest reinforced-concrete arch bridge built, having a clear span of 264 m. The concrete in the 200 mm test cubes had an average crushing strength of 38 N/mm^2 at only seven days. It spans the Angerman river, north Sweden.

sandpaper surface A road surface from which sharp pieces of aggregate protrude not more than 5 mm.

sand piles A means of deep *compaction* of a silty soil by dropping into it a heavy weight such as a pile-driver ram. The ram makes a hole in the ground into which sand is poured. The ram then drives the sand farther in and the process is repeated as required. Using damp concrete the process makes a *driven cast-in-place pile*. See **dynamic consolidation**, **Vibroflot**.

sand pump, bailer, shell p., sludger A long tube, open at the top, fitted with a *check valve* at the bottom. It is lowered into a borehole drilled with a *cable drill* or similar rig to extract mud, cuttings and water.

sand pump dredger A *suction dredger*.

sandstone [s.m.] A sedimentary rock formed mainly of *quartz* grains of *sand* size cemented with calcite or iron compounds, usually both, since pure calcite and pure sand are both white or colourless and white sandstone is rare. See **matrix**, **quartzite**.

sand streak A *blemish* – a streak of sand visible in a concrete wall surface, a result of *bleeding* (*ACI*).

sand trap An enlargement in a channel to slow the current. It may be a *catch pit*.

sandwich construction See (*B*).

sandwick [s.m.] A type of *vertical sand drain*.

satellite–Doppler system [sur.] (see diagram, p. 380) Locating points on the earth by special radio receivers which detect and record the changes in frequency of radio signals transmitted by artificial satellites as they approach, pass over and recede from the receivers. The artificial satellites at about 1100 km altitude orbit the earth in a period of 105 min, and cross both poles. Like the *inertial surveying system*, this also was designed with a military purpose, to guide Polaris submarines, and became commercially available in 1967. Using *translocation* procedures, a point may be fixed relative to a known point with an accuracy of ±0.3 m by observing 30–40 passes of satellites over a period of 2–8 days. The *global positioning system* is superseding this method.

saturated air Air containing the greatest amount of water vapour possible at the given temperature. It is therefore at a *relative humidity* of 100%.

saturation line [s.m.] The *water table*.

saucer [hyd.] A flat *camel* used for floating a ship past a shallow place.

sawdust cement See (*B*).

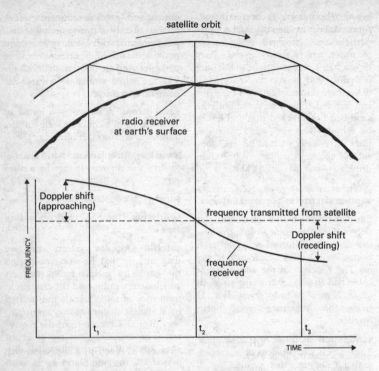

Location of a point on the earth's surface by satellite–Doppler system. (R. Brinker and P. Watt, *Elementary Surveying*, Harper and Row, 1984.)

saw files [mech.] Files with an extremely fine *cut*, and therefore a large number of teeth per centimetre.

scabbing The loss of patches of surface dressing from a road. Cf. **fretting**.

scabbler, hand s., pole s., pram s. A hand-held or self-supporting machine for removing varying thicknesses of hard concrete without damaging the structure. It is usually driven by compressed air and has one or more very fast-acting hammers. See also **floor scabbler**.

scabbling, scappling Stone flakes laid in a 23 cm thick bed under stone *pitching* in a *revetment*.

SCADA (Supervisory Control And Data Acquisition) Computer systems used in *telemetry* and elsewhere.

scaffolding A temporary platform for any construction purpose, including *falsework*. *Tubular scaffolding* with screwed couplers is very adaptable, but prefabricated scaffolding is quicker to erect even by unskilled people. Patented prefabricated types of tubular scaffolding have sockets at set heights to accept horizontal loadbearing tubes, but no screwing is needed. Many types exist.

scale (1) [d.o., air sur.] The ratio between the dimensions of a plan or a map and those of the object represented. The scale of an air photo is the focal length of the camera divided by the height of the aircraft at the instant the photo was taken. See **scale drawing**.

(2) [d.o.] A piece of boxwood, metal, ivory or plastic 15–30 cm long, graduated in accordance with precise ratios, proportionally to metres or other unit. See **fully-divided scale**.

(3) Mill scale.

scale drawing [d.o.] A mechanical drawing which shows an object with all its parts in proportion to their true size. Each dimension on the drawing is the same fraction of the true dimension of the object. This fraction is the *scale*.

scaling (1) [d.o.] Measuring dimensions on a drawing with a scale.

(2) [min.] Immediately after blasting, as soon as the air is clear but before the miners return to the face, the charge hand or foreman goes in alone and with a bar about 2 m long knocks down loose rock from the roof and sides so that the miners can safely work there.

(3) Local peeling of *cement paste* from a concrete surface.

scalping screen A *grizzly*.

scappling *Scabbling*.

scarf See (*B*).

scarifier An attachment on a *grader*, used for loosening soil. It has much shorter teeth than those on a *ripper*, and many more of them.

schedule A list or enumeration.

schedule of prices, s. of rates In the absence of a *bill of quantities* this document is provided by the *consultant* for *contractors* who are *tendering*. They fill in their prices against the *items* listed. A contractor who has no idea about the bulk of the work will have to set high prices. It is therefore best for the consultant to inform the contractors, even if only by a rough sketch, of the volume of the work to enable them to quote low prices.

Schmidt hammer A *rebound hammer*.

scissors crossover, s. junction [rly] A junction between two parallel tracks, shaped like a pair of scissors. It is possible to go from one track to the other when travelling in either direction, and in this it differs from the simple crossover, in which the passage from one track to the other can be made only along one junction track.

scissors lift A self-propelled or towed unit with a working platform that can be raised by bringing the ends of its legs close together like a pair of scissors. Street lamps or other high objects can be reached without the delay, expense and obstruction of scaffolding but access is available only vertically above the base, unlike some other *access platforms*.

scissors pipe [min.] (see diagram, p. 382) A flexible concreting pipe with 'concertina extension hinges' enabling it to concrete a 15 m (49 ft) length of tunnel lining without the delay of connecting additional pipe lengths (Putzmeister).

scotch block [rly] A wedge or block temporarily fixed to a running rail to scotch a wheel of rolling stock, i.e. to stop movement.

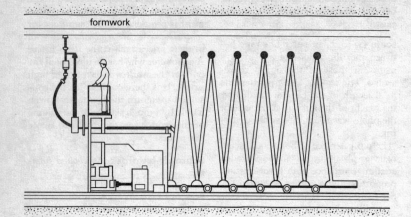

formwork

Rail-mounted scissors pipe for concreting a tunnel lining. (Putzmeister, *Tunnels and Tunnelling*, January 1989.)

Scotch derrick [mech.] See **derrick**.

scour, erosion Removal of the seabed or of a riverbed or river banks by waves or flowing water.

scouring sluice, scour pipe, wash-out valve [hyd.] An opening in the lower part of a dam through which accumulated sand, gravel or other rubbish can be occasionally expelled. Unfortunately, scour pipes are usually ineffective, and therefore some reservoirs silt up quickly. Boulder Dam reservoir (USA) is expected to have a useful life of only 100 years before it fills with silt.

scour protection Protection of earth or other submerged material by steel-sheet piling, *revetments*, *rip-rap*, a *mattress* or a combination of such methods.

scow A *dumb barge*.

scraper A *bowl scraper* or *scraper excavator*, or *scraper loader*.

scraper bucket The excavating part of a scraper or scraper loader.

scraper excavator A *multi-bucket excavator*.

scraper loader (slusher USA), scraper [min.]. A loader for underground or surface consisting of a *double-drum haulage* engine, a hoe-shaped or box-shaped scraper bucket dragged by the ropes, a *return sheave* and a ramp up which the rope pulls the bucket with its load of material. The ramp containing the engine may be mounted on wheels over the mine track. Mine cars run in under it and are quickly loaded.

screed, screeder, tamper (strike-off template USA) A wooden or metal templet raised, dropped and pulled along by two concretors, one at each end, to finish the surface of a slab being cast. Each end rests on *screed boards*. It is slightly longer than the width of the slab being finished. For a road slab the screed may have a cambered under-surface but it is usually straight. Mechanically vibrated screeds exist. A screed may also be a mortar *topping*.

screed board, s. rail A wooden or metal board at the edge of a concrete floor, set with its upper edge at the level of the finished concrete. Usually they are set upright to form the edge of a slab being cast, and removed after the concrete has hardened.

screed tester The *BRE*'s screed tester tests a *topping* that is at least 14 days old by a standardized 4 kg weight dropping 1 m (40 in.) four times. An indentation of 3 mm or less indicates a sound topping, 5 mm or more a failed one. Curled screeds may give a smaller depth of indentation than is justified because of the energy absorbed by the bending of the topping. If curling is suspected because of vibration of the topping or dust puffing up from a joint during the test, such facts are noted on the report. In 1988 the method was not recommended for the testing of *floating screeds (B)*.

screen (1) A sheet of wire mesh or punched steel plate on which granular material is sieved to separate it into various sizes (to classify it).

(2) In the treatment of sewage or *raw water*, a device that strains out some of the solids from a mixture of solids and water. It can be made of parallel, wedge-shaped wires or bars, or round wire, wire mesh or steel plate with holes punched through it as in the

trommel. At its most refined it is a *microstrainer*. In a sewage works the use of a *comminutor* eliminates the need for handling the solids. At reservoir intakes, screens are usually racks of massive bars. They should be installed parallel to flow lines, not perpendicular to them. Although this is expensive in screen area, it is less likely to result in blockage or over-topping of the screens or their collapse. At sewage works, screens are part of *preliminary treatment*. Coarse screens may be used to remove large solids from the crude sewage, so as to protect from damage the pumps at the works inlet. They are followed by fine screens to strain out smaller solids before the grit removal tanks. They are often cleaned by a rake mechanism. Screens are designed for a water speed through them of 1.2–1.4 m/s at maximum flow.

screen analysis, sieve a. [s.m.] The *mechanical analysis* of soils or ores by screens without *wet analysis*. It therefore lacks detailed information on the *clay* size particles.

screened material [s.m.] Material which has passed over or through a screen. See **oversize**.

screening The use of a *screen*.

screenings A reject from screening, either *oversize* or undersize.

screw conveyor [mech.] A *helical conveyor*.

screw pile, Braithwaite p. A spiral blade fixed on a shaft screwed into the ground by winch or capstan. They are suitable for use in soft silts or clays.

screws *Caisson disease*.

screw, set s. [mech.] A metal rod enlarged at one end to a head, with a *screw thread* (B) cut or forged on the full length of the shank, up to the head.

On a *bolt* the screw does not run the full length of the shank.

screw shackle, tension sleeve, turn buckle [stru.] A long cylindrical nut threaded internally with a right-hand thread at one end and a left-hand thread at the other. It connects two rods in a diagonal brace and can be used for lightly prestressing them. When the nut is tightened it draws the ends of the brace together.

screw spike [rly] A *coach screw* (*B*) which in the USA holds the foot of a flanged rail directly to the sleeper. In the UK it fixes only the baseplate or chair.

scrubber, wet s. A *flue-gas desulphurization* plant.

SCUBA (Self-Contained Underwater Breathing Apparatus), aqualung A SCUBA diver carries air cylinders on his or her back and is much more mobile than someone *diving* in standard diving dress, but can stay under water only for so long as the cylinders allow. Since the diver's body suffers the full pressure of the water, the depth he or she can dive to is also severely limited. Nevertheless a good SCUBA diver can make several dives a day of 10–20 min each to 40 m, and can work for 2 hours at 13 m. Below 5 m depth SCUBA divers should not work alone but in teams of two or three. The US Geological Survey uses teams of three, one of whom carries the writing tablet, the others the *clinometer* or other instruments. Rubber 'wet suits' should be worn in cold water to keep the divers warm.

scumboard [hyd.] A board which dips below the surface of a fluid to prevent scum flowing out.

scum weir, skimmings w. [sewage] A weir which can be lowered to remove scum or any floating objects, usually in *preliminary treatment*.

seal Any device that prevents leakage. In manhole covers the joint between cover and frame, which should be airtight. See *manhole cover*; also (*B*).

sealant, sealing compound (1) A fluid of plastic consistency laid over a joint surface or the outside of a *joint filler* to exclude water. Hot bitumen, rubber strip, plastic strip, hessian caulking and synthetic resins are used, but a more recent and sophisticated sort are the building *mastics* (*B*). Sealants may need to be renewed as some of them are short-lived.

(2) A durable coating of plastics such as epoxy resin or polyurethane, painted on the face of *form lining* or timber *formwork* to enable it to be re-used many times.

(3) **liquid-membrane curing compound** A coating (e.g. for roads, *bituminous emulsion*) over a damp, recently cast concrete surface, which prevents loss of water, and thus ensures proper *curing* of the concrete.

(4) A treatment for a set concrete floor, which strengthens the concrete surface or binds the aggregate, ensuring that it does not *dust*. Sodium silicate solution has been successfully used for many years. See also (*B*).

sea-level correction [sur.] A *tape correction*. It is a deduction from the measured length of a base line above sea level to bring it to its value at sea level. If R is the earth's radius at sea level, h the height above sea level and l the base length, the deduction is

$$\frac{h}{R} \times l. \; R = 6378 \text{ km at the equator,}$$

6357 km at the poles. This is about 1 ppm per 6 m of elevation above sea level.

sealing coat Bitumen, road tar or an emulsion of either, applied as a thin film to a road surface.

sealing drop shafts to rock [min.] When the cutting edge of a *drop shaft* has reached hard ground, a watertight joint must be made between it and the rock, since this is often the point of maximum water pressure. If the bedrock is a stiff clay, the cutting edge can seal itself automatically by pushing itself into the clay by its own weight, but this would be unusually lucky. Generally the bottom of the shaft must be filled with concrete. This plug of concrete must then be drilled through and injected with cement grout to seal the fissures in the rock. When all the fissures are sealed the concrete plug must be excavated by drilling and blasting.

sealing pneumatic shafts to rock [min.] Since workers have access to bedrock in the working chamber of the *pneumatic* shaft or *caisson*, a special technique is used which cannot be used with the open *drop shaft*. The shaft is first excavated 0.9 m into the bedrock and 0.3 m outside the shaft all round. During the excavation the shaft lining is held up on posts 0.9 m long. The rock outside the shaft is then lined with a smooth concrete wall just larger than the shaft. When the wall is completed, a ring of 12 cm of oakum is placed round the perimeter of the shaft. The workers leave the chamber after a light explosive charge is placed at each post to drive it into the centre of the shaft. All the shots are fired simultaneously and the lining drops on to the oakum making a watertight joint with the rock. The rock can then be grouted and the fissures sealed before the shaft is deepened.

seam welding [mech.] *Resistance seam welding*.

seating A surface which carries a large load.

seat of settlement [s.m.] That soil thickness below a loaded foundation within which 75% of the *settlement* occurs.

Seattle The 18 concrete main columns of the 62 storey building, 230 m (759 ft) high, at 2 Union Square, Seattle, are unique in several ways. The concrete costs three times as much as ordinary concrete but saved money by its exceptionally high strength of 131 MPa (19,000 psi) at 56 days though it was not vibrated. No *rebars* were used in the four central columns, each of 3 m (10 ft) dia. at base; instead they are enclosed in steel shells 16 mm (0.6 in.) thick with shear studs at 300 mm (12 in.) centres, and covered with the customary fire protection. The *water/cementitious ratio* was 0.22 but a *super-plasticizer* improved the slump to 250 mm (10 in.); 10% of the cement was replaced by *microsilica*, giving a 25% increase in strength. The aggregate pebbles were extremely strong and rounded, of 10 mm (0.4 in.) dia. Last, but probably not least, the concrete was pumped through holes in the bottom of the shells to exclude air. Two lifts of 3.7 m (12 ft) were cast at a time, making column pours 7.3 m (24 ft) high.

Progress was two floors every $3\frac{1}{2}$ days. Probably one reason for the construction speed was the absence of delays caused by column stripping and placing reinforcement. The floors are of I-beams carrying metal permanent formwork for the concrete slabs.

The demand for high strength came from the architect's need for low sway. In such tall buildings people have suffered from seasickness. A *Young's modulus* of 50,000 GPa (7,200,000 psi) was achieved to reduce deflection. Certainty before construction began was had by test casting, in the way proposed for the site, two lengths of column, of 3 m dia. and 3 m high, without vibration.

They were broken up and found to be adequate, with little *entrapped air*. One worry was that with their low W/C ratio they might 'dry out', shrink and separate from their shells. Other microsilica grouts have in fact lost strength after a year.

The structural engineer Skilling Ward Magnuson Barkshire has put up other buildings of the same structural type. It is believed that higher concrete strengths, possibly of 200 MPa (30,000 psi), are possible but only with special aggregate and, as in this case, with *microsilica, superplasticizer* and the best cement.

The frame cost $30 million but is claimed to be 30% cheaper than any alternative. (*Civil Engineering*, ASCE, Oct. 1987.)

secant modulus of elasticity [stru.] For materials like concrete or prestressing wire which have a variable *modulus of elasticity* (E), the value of E used must be either the slope of the tangent to the stress–strain curve (as for elastic materials with a straight-line curve like *mild steel* up to the *yield point*) or that of the secant. The secant is the line joining the origin of the curve to (for example) the 0.1% *proof stress* point on the curve. For materials within their elastic range the secant and tangent coincide.

secant piling, s. wall A deep, thick, watertight wall made with *bored piles* of at least 0.5 m (20 in.) dia., alternating with similar holes filled with concrete round a heavy rolled-steel stanchion.

The site for the British Library, Euston Road, London, with five successive sub-basements, had many hundred such piles forming a wall to protect the excavation for the foundation in 1983–4. The stanchions in the alternate piles measured 914 × 305 mm (3 × 1 ft) and the holes were bored to 1.18 m dia., some of them 30 m deep.

Where they are not needed in the upper part of a structure, secant piles can save concrete compared with a *diaphragm wall*. A diaphragm wall must be concreted for its full depth, but secant piling need be concreted only at the bottom, as happened in the British Library site for some piles. Secant walls are so called because the second series of holes cut (secant) the first concreted holes.

Holes are first bored for two stanchions at the correct spacing, leaving room for a concrete bored pile between them. The stanchions are lowered and concreted into the holes. The next day if possible, while the concrete is soft, the hole for the concrete bored pile between them is bored out, with the removal simultaneously of about 20 cm (8 in.) from the concrete around the stanchions each side. The cage of reinforcement is then dropped in and the pile concreted to make a watertight wall. The process continues with the drilling and concreting for the steel piles first.

secondary beam [stru.] A beam carried by other beams, not carried by columns or walls. Cf. **main beam.**

secondary reinforcement See **fibre-reinforced concrete**, **filament reinforcement**, **link.**

secondary sedimentation [sewage] A *sedimentation* at the end of *secondary treatment*. (All secondary treatment is preceded by *primary treatment* – the first sedimentation.) Secondary sedimentation is of two sorts. Sedimentation tanks called *humus tanks* settle the solids (*humus*) in the *effluent* from *trickling filters*. Other sedimentation tanks settle the effluent from *aeration tanks*.

secondary treatment [sewage] *Trickling filters* (biological filters) in small plants and the *activated-sludge process* in cities are the main means of treating the *effluent* that has flowed over the weirs of the *sedimentation tanks of primary treatment*. Both processes bring air (oxygen) into the water, reducing the effluent's *biochemical oxygen demand* and its content of *pollutants*.

second cut [mech.] A grade of *file* tooth. See **cut**.

second-foot [hyd.] A unit of flow, one *cusec*.

second moment of area The technically correct term for *moment of inertia* of a section.

second-order, secondary triangulation, trilateration [sur.] A *triangulation* or *trilateration* with sides about 10–20 km long. See **first order**, **third order**.

section (1) [d.o.] A mechanical drawing of an object as if cut at a position chosen to show certain details. It may be a *cross-section* or a longitudinal section.

(2) [stru.] The shape of a *rolled-steel section*, extruded light alloy member, etc.

sectional elevation [d.o.] A section of an object which shows, apart from the material cut in the section, various parts in *elevation*, which can be seen by looking beyond the sectioned part.

sectional tank A water tank built of pressed-steel, cast-iron or plastic pieces, usually 1.2 m square with external flanges, drilled for bolting outside the tank. The greatest depth usual is 5 m, at which depth frequent tie bars within the tank are needed. These tanks are very quickly erected and need little skilled labour.

section leader, checker, leading draughtsman [stru.] A *structural designer* in charge of a small group of draughtsmen or *designer* draughtsmen.

A section leader is usually also the checker but the details of organization are very variable.

section modulus [stru.] See **modulus of section**.

section properties [stru.] The area of a structural section, its moment of inertia and its modulus of section; in fact all those geometrical properties which affect the strength of a member in bending.

sector gate, s. regulator [hyd.] A *roller gate* in which the roller is a sector of a circle instead of a cylinder. Some types fit into a pit below sill level when open, but the usual arrangement is for the gate to be raised when open.

sediment (1) Any material which settles in a liquid, hence specially the material which is carried and dropped by a river, often called *silt*. The Yellow River in China, one of the world's worst for *siltation*, carries 54 kg of silt per m³ in the August high-water season but only 9 kg/m³ in February at low water, though even this is considerable. One Chinese solution has been to divert part of the river water to temporary settling ponds in low-lying ground, which is thereby gradually raised and made fertile. An explanation for the heavy silt load in Chinese rivers is that much Chinese soil is extremely fine wind-blown silt (loess), which is easily carried off in even a slow stream.

(2) [s.m.] A soil or rock that has been formed by laying down in water, often also described as *stratified*.

sedimentation (1) [s.m.] Settlement, the sinking of soil or mineral grains to the bottom of the water which contains them. Large particles settle much faster than small particles of the same shape and density. The principle, scientifically stated in *Stokes's law*, is made use

sedimentation tank

of in *classifiers*, *elutriators*, *cyclones*, coal washers, *wet analysis* and so on. See **flocculation**.

(2) Sedimentation is used in the treatment of *raw water* but more so in *sewage treatment*, for which every stage includes a sedimentation separating a more polluting *sludge* from a less polluting *effluent* (water). Primary sedimentation removes half the solid *pollutants*.

sedimentation tank, settling basin (clarifier USA) A tank through which sewage *effluent* or *raw water* passes slowly to allow solids to sink to the bottom, forming a *sludge* which drains out or is pumped away for further treatment and disposal. Many types exist, even some with upward flow.

seepage [hyd., s.m.] Groundwater flow, *infiltration*, leakage, etc.

seepage force [s.m.] See **capillary pressure**.

seepage line [hyd.] A *flow line*.

seepage loss A loss of water through the bank of a canal which is expressed as millimetres loss in depth per 24 hours, or as cubic metres lost per square metre of bank and bed (wetted perimeter). For the best clay the loss is 0.25 m^3 and increases with the permeability of the soil up to 5 m^3/m^2 per 24 h for gravels. Cf. **absorption loss**.

segmental bridge A bridge built usually of precast concrete segments held together by *prestressing tendons*, placed by *incremental launching* or by a launch gantry, or cast on a form-carrying gantry. The method is popular because it avoids the expense and trouble of supports in the spans, and can speed construction. Precasting of spans can begin while foundations are being dug. The *Pont de Ré* is a fine curved bridge of this type. The St-Jean Bridge at Bordeaux is 475 m (1560 ft) long but has no expansion joints except at each end. The intermediate supports have neoprene bearings, which allow temperature movement. A bridge at Cardiff has segments joined by glue.

segmental ring A *ring* for lining a tunnel or shaft, built up from segments of precast concrete, pressed steel or (originally) cast iron bolted to each other. Some concrete segments are not bolted but wedged. A 1 m dia. tunnel usually needs only three segments, but for a 3.65 m (12 ft) internal dia. tunnel, about 15 segments could be used and either one or two short key segments at the crown.

Concrete segmental rings have been adopted for much of the *Channel Tunnel* because of their fast erection speed (5 m/h), to keep up with the *tunnel-boring machines* (TBM). With segment-erection machines forming part of the TBM, they are almost as reliable as cast iron, and much cheaper.

segmental sluice gate [hyd.] A *radial gate*.

segregation During transport or placing, concrete may separate into areas rich in coarse aggregate and others rich in sand, resulting in *honeycombing*, because of either over-vibration, excess water or dropping through reinforcement, etc. Segregation can be stopped by correct *grading* of aggregates and careful placing, including *trémie* tubes for deep forms or the use of a *thickening* agent to achieve *plasticity*.

seiche [hyd.] The *wind shear* noticed in Lake Geneva.

Seikan Tunnel, Japan The longest public service tunnel in the world, nearly 54 km (33 miles) long, 23 km (14 miles) of it under the sea. Terrible geological difficulties, including flooding about 50 times, caused the tunnel driving to last 24 years. By then air

388

transport had become cheap and the 'bullet train' company no longer wished to use the tunnel. It was opened in March 1988 to other rail services.

seismic design [stru.] See **lateral force.**

seismic prospecting *Geophysical* methods of ground investigation that usually do not rely on cores from boreholes, although a short borehole is drilled sometimes for placing a small explosive charge in the ground. They are especially useful for investigating layered rocks like clays or coal seams. *Geophones*, placed at measured distances from the source of shocks or vibrations (shot hole), record the time of arrival of waves or their reflections and enable velocities to be calculated. Sound velocities vary from 150 m/s in earth to 2200 m/s in chalk or 8000 m/s in uniform, hard olivine rock. Where the information is required from soils that are less than about 15 m deep, a sledge hammer may be used instead of explosives, reducing disturbance to people near by. Rocks through which seismic waves pass at a higher speed than 2000 m/s are generally too strong to be broken by a *ripper*.

seismograph [min.] An instrument at the ground surface which records the electrical effects transmitted to it by a *seismometer* and thus shows the times and amplitudes of earth shocks.

seismology The study of *earthquakes.*

seismometer Either a *geophone*, used in *seismic prospecting*, or a device for detecting earthquake shocks. An early seismometer made in Japan about AD 136 consisted of balls dropping from a dragon's mouth into a frog's to show the direction of the shock.

selective tendering Competitive *tendering* for a chosen few *contractors*.

selenium Though needed as a trace element by humans, selenium is poisonous. In the arid districts of California and the Dakotas, farm animals have suffered selenium poisoning. The EPA raised the allowable level in drinking-water from 10 to 35 parts per thousand million in about 1987.

self-anchored suspension bridge [stru.] A suspension bridge with no anchorages because the cables are attached to the ends of the *stiffening girders* beyond the towers. Though this is a true bridge structure, it was used in Mantua, Italy, in 1964, for roofing a paper factory 250 × 30 m in plan, with the roof 19 m above ground. Like many other structures by its Roman designer, Pier Luigi Nervi, the building, overshadowed by its 50 m high towers, is spectacular, and probably expensive, but it is also practical, having few interior columns and being infinitely extensible sideways.

self-cleansing gradient [hyd.] A *gradient* at which the flow in a pipe of a certain diameter can carry out any solids which ordinarily come into it. The gradient should be neither too steep nor too sluggish. Designers now ordinarily aim at a flow velocity of 76 cm/s once daily in a circular sewer running full. In a 150 mm dia. sewer this is achieved at a slope of 1 in 174. As the sewer diameter increases, the necessary gradient diminishes, so that a 300 mm dia. sewer needs only 1 in 410 and a 600 mm dia. sewer only 1 in 965. The steepest allowable gradient for a 150 mm dia. sewer is 1 in 10, preferably 1 in 27.

self-docking dock [hyd.] A *floating dock* built in sections each of which can be docked (lifted up) on the others and repaired as need be. See **dock.**

self-reading staff [sur.] A *levelling staff* so graduated that the observer looking through the telescope of the level can read the elevation at which his or her line of sight cuts the staff. Nearly all staffs in building work are of this type. See **target rod**.

self-stressing concrete/mortar/ grout, chemically prestressed c./m./g. Concrete, mortar or grout made with *expanding cement (ACI)*.

semi-automatic welding *Welding* in which the machine controls the feed of wire but the welder controls the other variables, including the advance of the welded joint.

semiconductors Substances indispensable to electronics, like *silicon* and *gallium arsenide*, used in *transistors*, *integrated circuits* and photo-electric cells. They are less electrically conducting than metals, but more than insulators, and can be made to conduct electricity or to insulate as convenient. This 'electronic switching' is done by transistors in a tiny fraction of a second. Unlike metals, their conductivity improves as they warm up so they must not be allowed to overheat or they will conduct even more and burn out in a flash in 'thermal runaway'.

semiconductor sensor An electronic *sensor* for temperature, based on a diode or *transistor*. More recently developed than other *temperature sensors*, and based on an *integrated circuit*, they are small but their top temperature is limited as for all semiconductor materials.

semi-destructive testing of concrete [stru.] *Internal fracture testing*, *penetration-resistance testing* or *pull-out testing*. See **rebound hammer**.

semi-killed steel *Balanced steel*.

semi-permeable membrane In *osmosis*, a skin that is permeable in only one direction.

semi-skilled man A workman such as a *rigger* or builder's labourer, who has learnt his work by helping others and has a degree of skill rather less than that of a *skilled man* or a *tradesman (B)*. He has risen from labourer in his trade and may become a skilled man.

semi-submersible An *offshore platform* that may either rest on the seabed or float but always has a large underwater structure in the form of buoyancy tanks or a tower. Semi-submersibles can work in water 1000 m deep. They may be moored by anchors or by *dynamic positioning*, i.e. thrust from propellers or water jets on all sides. See **floating crane**.

semi-trailer See **articulated vehicle**.

sensible horizon, apparent/visible h. [sur.] The seen *horizon*.

sensitive clay [s.m.] A clay which loses its strength when disturbed. The most sensitive clays are subject to *liquefaction* in *flow slides*. See **dispersive clays**, **sensitivity ratio**.

sensitiveness, sensitivity [sur.] The responsiveness of an instrument to slight alterations in the quantity which it measures. The sensitiveness of a level or theodolite *bubble* is expressed in seconds of arc per millimetre of bubble movement.

sensitivity ratio [s.m.] A measure of how much *remoulding* may affect a clay. It is the ratio of the unconfined compressive strength in the undisturbed state to that in the remoulded state. Since clays are plastic, an exact failure point is not always found in the strength tests, and for greater accuracy the strengths at equal strains may be measured. For British alluvial clays the sensitivity ratio is from 2 to 5. For sensitive clays it is much higher.

sensor, sensing element, primary detector Any device for measuring or merely detecting a physical quantity such as temperature, light, line, level, etc. Sensors provide the signals needed for control, often through a *microcomputer* such as that in a *slipform paving train*. Every *transducer* includes a sensor. In tunnels, sensors detect carbon monoxide or diesel smoke. Traffic signals may be controlled by inductive-*loop detectors* which sense the arrival of vehicles. Sensors include accelerometers, Geiger counters, *liquid-level indicators, load cells, proving rings, strain gauges, thermistors, thermocouples*, thermometers and *ultrasonic* devices. See **remote sensing**.

separate system [sewage] A drainage system in which rainwater and sewage are carried in separate *sewers*. See **combined system**, **partially separate system**.

separator A *distance piece*.

septic sewage *Sewage* kept for a day or two in hot weather (longer in cold weather) loses all its oxygen, begins to blacken, stink, and produce hydrogen sulphide (H_2S) and methane, the first highly poisonous, the second dangerously explosive from only 5% in air. In part-full pipes, H_2S is converted to sulphuric acid (H_2SO_4), with serious damage to the concrete at the crown of the pipe. The gas is released most at *backdrops* (B) or at other points where the flow is turbulent. Maintaining a *self-cleansing gradient* helps to prevent septicity. In long pumped *sewers* pure oxygen is sometimes injected at the base of the sewer to prevent or delay septicity. See **flushing manhole**.

sequence of trades The order in which originally the various building trades worked. Different sequences have been laid down by different authorities and it is important to know which one has been chosen for the *contract documents* because the sequence fixes their layout. For civil engineering, the ICE's *CESMM* (*Civil Engineering Standard Method of Measurement*) is usual.

sequential starting A way of reducing the maximum electrical demand and ensuring safe working of electrical equipment, e.g. a series of conveyors feeding each other. British Coal use it under computer control. The final delivery conveyor starts first. It is also used in road tunnels for the switching of fans, lighting, pumps, etc.

serial contract A 'follow-on' *contract* for additional similar work, giving more favourable terms to the *client*.

serrated strip A *rumble strip*.

service pipe A gas or other supply pipe to a consumer from the main.

service reservoir, distribution r., clear-water r. A *reservoir* that is designed to be large enough to store at night the excess of day-time demand over night-time demand and to deliver it to consumers by day. It may receive its water by pumping or by *gravity main*, as convenient.

service road A small road parallel to a main road. It is used by traffic stopping at houses and shops so as to avoid obstruction to through traffic.

set (1) [hyd.] The direction of flow of water.

(2) [mech.] A bend in a piece of metal. See **cold sett, permanent set**.

(3) The penetration of a driven *pile* for each blow of the *drop hammer*.

(4) See **initial setting time**, **final setting time**.

set-control addition Calcium sulphate in any hydrated form between $CaSO_4$ and $CaSO_4 \cdot 2H_2O$. It is interground with *Portland cement* clinker during manufacture to modify the setting time of the cement.

set screw [mech.] See **screw**.

set square [d.o.] (**triangle** USA) A triangular piece of wood or transparent plastic material used for mechanical drawing. It is always made with one right angle but the other angles may be 60° and 30°, or both 45°, or adjustable.

sett (1) A block of dressed hard stone such as granite or diorite, sometimes used for *radial-sett paving*. It is generally much smaller than a *kerb* or *pitcher*, and may be 10–15 cm square on face but deeper than the face dimension.

(2) A follower used in pile driving.

(3) A blacksmith's hammer-shaped cutting chisel, either a cold sett or a hot sett.

setting The boards (*runners*, *sheeting* or *poling boards*) held in place by a pair of timber *frames* supporting an excavation. One setting consists of all the boards held by a pair of frames. See also **set** (4) above and (*B*).

setting out Putting pegs in the ground to mark out an excavation, marking concrete to locate walls, or preparing dimensioned rods for carpentry or joinery. Large-scale setting out is done by a civil engineer, small-scale work by a foreman of a trade.

setting up See **upsetting**, **set-up**.

settled sewage The *effluent* from *primary treatment*.

settlement See **subsidence**.

settlement crater Uniformly loaded *foundations* over clay, silt or other compressible soils settle in such a way that an originally flat surface becomes bowl-shaped. This crater of settlement is the main cause of the *differential settlement* which causes so much trouble in buildings. See **cone of depression**.

settlement joint A joint that allows part of a building, for example, to settle relatively to the neighbouring part in the event of mining or other subsidence.

settling basin [hyd.] One sort of *sand trap*.

settling velocity See **terminal velocity**.

set-up [sur.] A location at which a surveying instrument, particularly a *theodolite*, is stationed. Since several minutes are needed for centring an instrument over the station point, levelling it and adjusting it, surveyors take great care that their set-ups are not disturbed. See **temporary adjustment**, **three-tripod traversing**.

Severn barrage An 18 km long *barrage* proposed from Weston super Mare across the Bristol Channel to Cardiff, estimated in 1987 to cost £5500 million or 10% more than the *Channel Tunnel*. Its installed power of 7200 MW would be equivalent only to 1100 MW from a conventional power station according to the electricity authorities, because tides are not as uniform a power supply as steam. It could, however, last 100 years against the maximum 30 years of a steam power station, and would probably also be a river crossing less vulnerable to weather than a bridge. The main investigator is the Severn Tidal Power Group, including Sir Robert McAlpine, Balfour Beatty, Taylor Woodrow and Wimpey.

Severn Bridge A road bridge over the river Severn in the west of England, a suspension bridge of 988 m span, designed by Freeman, Fox & Partners for the consulting engineers Mott, Hay & Anderson, and completed in 1966. Fatigue damage to the deck was caused by 38-ton lorries for which the bridge was not designed. In 1987-9 this was repaired in contracts worth some £60 million, which added 2% to the steel in the towers, deck and hangers, but nothing to the main cables with their good reserve of strength.

In 1988 a second Severn bridge was under consideration by four consulting engineers who had been short-listed by the Department of Transport. One possible site was at English Stones, 5 km downstream from the existing Severn bridge and near the railway tunnel.

sewage (**waste water** USA) Waterborne waste – the water of a community after it has been used. In the UK it flows down the property owner's *house drain* into the *sewer* of the *water service plc*. Ordinarily it is equal in volume to the water consumption of the community, but in old sewers this *dry weather flow* is increased by the amount of the inward leakage – infiltration. In dry countries it may be reduced by outward leakage – exfiltration.

sewage disposal The disposal of (i) *effluent* and (ii) *sludge*. Effluent which forms over 99% of sewage can, by *sewage treatment*, reach a condition approaching Royal Commission *recommendation* in a modern works, and then passes to a river or other receiving water. Sludge is more difficult to dispose of because it starts at about 95% water, de-watering is difficult, and disposal on farms may be dangerous if there are poisonous metals from industry in the sludge. Much sludge from the UK is tipped at sea. In the USA this is forbidden and there has been much pressure from Continental countries for the UK to stop sea-dumping of sludge.

sewage farm A farm where crops are watered with sewage *effluent*, sometimes also with sewage *sludge*. But it may not grow crops eaten raw by

people. Sewage farms work well in the USA and Australia but are obsolescent in the UK because of the cold climate and the shortage of land.

sewage gas, sludge g. The gas from *digestion* tanks. Two thirds methane (CH_4) and one third carbon dioxide (CO_2) with small amounts of other gases, it has been effectively used on sewage treatment plants both for power generation in gas engines and for heating the digesters from which it originates. If there is much hydrogen sulphide gas (H_2S) this must be removed before the gas is used because it is extremely corrosive.

sewage treatment Separation of the 0.1% pollutants in sewage from the 99.9% water (*effluent*). It can include four treatments, *preliminary*, *primary*, *secondary* and *tertiary*. All methods result in a separation into two parts, the more polluting *sludge* and the less polluted effluent that flows over the weirs of *sedimentation* tanks.

sewer A pipe or underground *open channel* for carrying water or *sewage*. A sewer is usually the property of a *water authority* in the UK, a *drain* of an individual. Every sewer is fed by drains or by other sewers. See **combined/partially separate/separate system**.

sewerage The *water service plcs'* network of *sewers*; the means by which *sewage* is removed. Although up to 1815 discharge to sewers had been forbidden and wealthy people had cesspits under their houses, by 1850 Londoners building new houses were forced by law to fit WCs and to connect them to the sewers.

sewer lining See **pipe renewal, pipe linings**.

sewer pill A skeleton-framed wooden ball of nearly the diameter of the sewer which it floats down. As it passes it cleans the sewer walls. See **go-devil**.

sewer pipe materials *Ductile iron* of from 80 mm (3 in.) up to 1.6 m (63 in.) dia., *glassfibre-reinforced plastic* from 25 mm (1 in.) up to 4 m (13 ft), and high-density *polythene* (HDPE) from 16 mm to 500 mm (20 in.). The HDPE pipes may be fusion welded or mechanically joined. Unplasticized PVC pipes exist of from 110 mm (4 in.) up to 630 mm (25 in.) dia. and may be joined by *solvent welding* or by mechanical joints. Mechanical joints for plastic pipes resemble the *compression joints* (*B*) of copper tube. Concrete pipes have also been used but they resist corrosion less than plastics and with more joints they suffer more *infiltration*. See also **centrifugally cast GRP**.

sewer reamer A compressed-air driven tool with a tungsten-carbide-tipped cutting head hauled by rope (one in front, one behind) through a sewer to remove accumulations of grease, calcite, tree roots and any other obstruction before *pipe lining* or *pipe renewal*.

sexagesimal measure, sexagesimal graduation [sur.] Division of a circle into 360 degrees, each degree having 60 minutes, and each minute having 60 seconds, formerly universal in Britain. Cf. **centesimal**.

sextant, nautical s. [sur.] A hand-held instrument for measuring angles up to about 120° in any plane, and used at sea for measuring the altitude of a heavenly body such as the sun. Its main parts are two mirrors, i.e. one fixed *horizon glass* and one *index glass*, which is silvered all over and connected to an arm with a vernier moving along a scale of degrees. The objects between which the angle is required are brought to coincide in the index mirror and horizon glass, and the sextant reading

is then noted. The nautical sextant is accurate to about 30 seconds of arc, much more precise than the smaller *box sextant*.

shackle (1) A *wrought-iron* or *manganese-steel* chain or pinned coupling connected between wagons forming a train. The strength of a shackle in tons may be taken as 0.5 d^2 where d is the pin diameter in centimetres (BS 3032, 3551).

(2) A lifting ring on a crane hook or over a *kibble*.

(3) See **screw shackle**.

shaft Generally any slender or tapering object; a ventilating pipe; the part of a chimney which projects above the roof; the handle of a tool; and specifically:

(1) [min.] A vertical or sloping passage from the surface to the workings, used for ventilation, travelling or hoisting, or all three.

(2) [mech.] A cylindrical metal rod which rotates on its centre line and transmits power by its rotation.

shaft pillar [min.] An area of coal or other mineral left untouched around a shaft to protect it and the surface buildings around it.

shaft plumbing [min.] The operation of transferring one or two points at the surface of a vertical shaft to positions below them at the foot of the shaft. This can be done by a specially fitted theodolite, but the operation is called plumbing because, for shallow shafts, it is easily done with wire and plumb bob. A theodolite which is set up at the foot of the shaft is used for sighting the lower ends of the wires. A series of readings is taken on the wires as they sway backwards and forwards. Usually each bob is submerged in a bucket of water or thick oil to damp its movements. Accurate work is essential, since the orientation of the underground workings depends on these two points only 3 m or so apart. See **instrumental shaft plumbing, Weisbach triangle**.

shaft-plumbing wire [min.] Iron, steel, copper, brass or phosphor–bronze wire, not usually thinner than 0.8 mm. Copper or iron wire are ductile enough to unkink themselves without breaking and are suitable for shallow shafts. Steel wire as used for *prestressed concrete* is much stronger, but rusts more easily and with phosphor–bronze is more used for deep shafts.

shaft sinking [min.] Excavating a shaft downwards. This is the usual method of excavating a shaft – raises are rarely practicable for important shafts.

shaking test [s.m.] A rapid way of determining whether a sample of fine-grained soil is a *clay* or a *silt*. Coarse particles larger than 0.5 mm must first be screened out. The wet *remoulded* sample is held in the palm of the hand and tested in the following way. The surface of the soil is smoothed with a knife and the soil is shaken by tapping the hand. The soil then begins to glisten with the water which is pushed out of it as the grains slip past each other into a denser position due to the shaking. The soil pat is then squeezed, causing it to expand (in a direction perpendicular to the squeeze) like a dense sand during shearing (see **critical voids ratio of sands**). The result of the squeezing is that the water disappears from the surface, re-enters the soil and the sheen vanishes. Clays show none of these changes. Very fine, clean sands react most quickly. This property of a silt is called *dilatancy*.

shale [s.m.] A laminated clay or silt which has been compressed by the weight of the rocks over it. Unlike slate it splits along its bedding planes. Shales vary in hardness from slaty rock to hard clay.

shallow foundation [stru.] A *pad foundation, strip footing, short bored piles,* etc. Cf. **deep foundation**.

shallow manhole An *inspection chamber* of the same cross-section all the way up.

shallow well A shaft sunk to pump surface water only. See **deep well**.

shaping Forming an earth or road surface to the correct contour, usually with *graders* or *dozers*.

sharp sand Concreting sand – grit coarser than *soft sand*; see (*B*).

shear, s. force [stru.] The load acting across a beam near its support. For a uniformly distributed load or for any other symmetrical load, the maximum shear is equal to half the total load on a simply supported beam, or to the total load on a cantilever beam.

shear centre, c. of stiffness [stru.] A point such that, when the plane of applied horizontal load passes through it, the structure will deflect in the direction of the applied load without twisting. It may be inside or outside the plan area of the structure.

shear connector, s. stud A piece of steel welded to a *rolled-steel joist* or *stanchion*, or to steel *permanent formwork*, to ensure *composite action* between steel and concrete. It may be a loop welded on at both ends or a mushroom shape welded on only at the root. See **Seattle**.

shearing [stru.] Failure of materials under *shear*. It can be seen in the action of a pair of shears or scissors.

shear key A concrete *construction joint* shaped to transfer *shear* forces across the joint. A *joggle* joint is one type.

shearlegs, shears, shear legs A pair of poles lashed together at the top with a pulley hung from the lashing, used for lifting heavy loads. See **derrick** (2).

shear modulus [stru.] The *modulus of rigidity* (*G*). It is equal to the shear stress divided by the shear strain. See **Poisson's ratio**.

shear reinforcement *Links* are the commonest shear *reinforcement* but *bent-up bars* are used also in beams.

shears (1) *Shearlegs*.
(2) [mech.] A machine for cutting steel plates or sections.

shear slide [s.m.] A *landslip* in which a mass of earth slides as a block away from the material below it.

shear strain [stru.] The angular displacement of a member due to a force across it (shear force). It is measured in radians.

shear strength The stress at which a material fails in shear. It is the same in all directions for steel, very different in different directions for wood, and is usually measured on the glued face for glues. See also **cohesive soil**, **shear tests**.

shear stress [stru.] The shear force per unit of cross-sectional area, expressed in kN/m^2 or $lb/in.^2$ like other stresses. See **cube test**.

shear tests [s.m.] The shear strength of soil samples is often measured in the laboratory by the *box shear test* or by the *triaxial compression test*. However, since the behaviour of soils in place often differs greatly from their laboratory behaviour, the *vane test* has been developed for testing the shear strength of a soil at the foot of a borehole.

shear wall A *core wall* (2).

sheathing (1) Wooden boarding of any sort.

(2) A sheet-metal covering over under-water timber to protect it against *marine borers*.

sheave [mech.] A grooved pulley wheel, particularly the pulleys over which steel-wire *ropes* pass in mine hoists or haulages.

she bolt A combined wall tie and spreader for wall forms that leaves only a small hole in the concrete (*ACI*).

SHED (Shephard–Hill energy dissipator) A hollow concrete cube used as *armour*. It is 1.3 m (4 ft) across with a large spherical hole in the middle, making it 40% concrete and 60% void. All six sides have holes through them. A layer of SHEDs is much pleasanter to the eye than large rocks and is claimed to be more efficient in breaking up wave energy. The unit is concreted around a heavy reusable balloon. The SHEDs at the harbour of Bangor, Northern Ireland, are claimed to be the first reinforced armour. Each cube is reinforced by six stainless steel hoops held to each other by copper clips.

sheepsfoot roller. A towed roller for *compaction* of clay soils, with one or more drums which can be water ballasted. The drums have a number of steel bars about 10 cm long, welded radially to their surfaces. A typical sheepsfoot roller has a 1.1 m dia. drum, a loaded weight from 9000 to 22,000 kg per metre width of drum, and footprint pressures from 0.5 to 2 N/mm^2. Much heavier rollers have been built for use after the typical sheepsfoot roller; they have footprint pressures up to 7 N/mm^2. See **tamping roller**.

sheeters Light steel vertical poling boards for protecting trench sides, driven down by a pneumatic tool before the trench is excavated.

sheeting, sheathing, lagging Rough boards held against the earth in trenches by *struts*, *walings*, etc. *ACI* uses 'sheeting' or 'sheathing' also for the boards in *formwork* that touch the concrete.

sheet pavement Road surfacing like asphalt or concrete, which is very much smoother for traffic than stone *block pavement*.

sheet piles Closely set piles of timber, *reinforced* or *prestressed concrete*, or steel driven vertically into the ground to keep earth or water out of an excavation. *Bored piles* are often successfully incorporated with the concrete of a basement retaining wall. See **barge bed, bulkhead, clutch, diaphragm wall, secant piling, steel sheet piling**.

sheet-pile wall, sheet piling A wall of *sheet piles*, which may be a *cantilever wall* or anchored back at one or two levels to a *dead man* as a *tied retaining wall*.

sheet steel See **steel sheet**.

shelf angle [stru.] A mild-steel *angle section* riveted or welded to the web of an I-beam or channel section to support the formwork or the hollow tiles of a concrete slab.

shelf retaining wall A reinforced-concrete retaining wall with a *relieving platform* built on to its upper part.

shell, thin s. (1) [stru.] A thin, curved plate-like structure which can be an extremely elegant roof. Shells are usually designed by specialists, since the mathematics is advanced. Dischinger in Germany built the earliest shells, the first large shell being the Düsseldorf Planetarium 1926. Most shells are of concrete, in which 76 mm thickness for 30 m span is regarded as normal, but many timber shells have been built. Some steel shells exist, however, to which concrete is applied as an insula-

tion (foamed slag or vermiculite concrete or plaster). See **Leipzig market halls, membrane theory**.

(2) A *sand pump*.

(3) An 'empty' *expert system* sold to a buyer for the buyer to fill. This program when filled can manipulate a *knowledge base* according to rules set by the *knowledge engineer*. The knowledge base is added by the customer. Shells can be applied to very varied problems and when filled become expert systems.

shell-and-auger boring An old percussion drilling method, still used for drilling holes in *site investigation*. Hand rigs can bore to 24 m (80 ft) depth at 20 cm (8 in.) dia., as well as to 0.6 m dia., but less deeply. The *auger* is used for boring in clay, the *sand pump* or shell for sands below the water table. A rope can be used for the shell but sectional boring rods must be used for turning the auger. Small boulders are broken up by a chisel bit on the rods. A *shear legs* or light three-pole *derrick* carries the weight of tools and rods. Holes more than 100 m deep have been drilled by three people with a three-pole derrick but a *truck-mounted drill* is now more commonly used for deep holes if they can be reached by a lorry. See **cable drill**.

Shell-perm process [s.m.] Injection of bitumen emulsion into a permeable soil to reduce the flow of water into an excavation. The emulsion contains a coagulator to make it solidify in the ground. This effectively closes the pores but does not strengthen the ground.

shell pump A *sand pump* for bailing out boreholes.

shell roof [stru.] See **shell**.

shield A steel protective tube used in soft ground, inside which miners drive a tunnel with pick and shovel. Instead of miners now there is often a *tunnel boring machine* (TBM) or, in a small tunnel, a *microtunnelling machine*. The shield has jacks around the edge which push on the tunnel lining of *segmental rings* and advance the shield with the TBM. The protection of the shield eliminates timbering but other dangers, such as inrushes of water or mud, cannot be completely forgotten though they are greatly reduced by *earth-pressure-balance* or slurry shields. The original *open-face shields*, with miners loading the muck into wagons, had protective overhead half-moons and are still common with hand-mining. See also **blind shield**, **bladed shield**, **boomheader**, **silt displacement**.

shielded-arc welding *Arc welding* in which the weld is protected from the atmosphere by argon, helium or other gas liberated from the coating on the *electrode* or from a gas cylinder through a hose.

shielded metal-arc (SMA) welding, stick w. US terms for *manual metal-arc welding*.

shielding concrete See **biological shield**.

shift [sur.] The radial displacement from the circular shape needed to make a spiral (transition) curve from a circular curve.

shim, pack [mech.] A thin steel plate inserted between two surfaces to fill a gap. A shim is thinner than a pack.

shin [mech.] (1) A replaceable edge of a *mouldboard*.
(2) A railway *fishplate*.

shingle beach A beach of stones that are more or less of the same size, enabling the beach to be self-draining. The large holes between the stones allow water through, so the beach is steeper than a sand beach.

Shinkansen The Japanese bullet train.

ship caisson, sliding c. [hyd.] A tall, floating box which is used for closing the entrance to a lock, dry dock or wet dock. It fits on to grooves in the floor and sides of the entrance when this is closed and into a recess (camber) in the dock wall when the entrance is open. Caissons have been made of riveted or welded steel and of reinforced concrete. See **step**.

shiplap boards See (*B*).

shoe A *pile shoe*. Also a *cutting curb*. See also (*B*).

shop weld [mech.] A weld made at the workshop. As with riveting, shop welding is often cheaper and stronger than site welding, but some joints must be made at the site if the pieces transported are to be of easily manageable size.

shore, shoring See (*B*).

shore protection [hyd.] Prevention of scour by *breakwaters*, *graded filters*, *groynes* and every sort of *revetment*.

short bored piles [stru.] On a clay soil, piles of about 25 cm dia. or even less, and 2 m long, which carry light foundation loads such as those of a house to below that part of a clay that cracks in summer.

short column [stru.] A column which is so short that if overloaded it will fail not by *crippling* but by crushing. For columns of this sort the *slenderness ratio* does not need to be calculated except as a check that it is low enough. See **lateral support**.

short ton, net t. The US ton of 2000 lb. Cf. **long ton**.

shot blasting machine A self-contained unit which keeps its dust and throws steel shot at steel to be cleaned before painting, or concrete before a road repair, etc. See **sand blasting**, **wheelabrating**.

shotcrete, sprayed concrete *Gunite* with aggregate larger than 10 mm, a development of the late 1950s from the *NATM* (new Austrian tunnelling method) for lining tunnels in two stages, using shotcrete immediately after *blasting*, with aggregate up to 30 mm and an accelerating *admixture*. Steel mesh has been used to strengthen the shotcrete, and sometimes *roof bolts* also, but some users have replaced these by steel-fibre reinforcement. The mix can be made to set in as little as 15 minutes, providing safe cover for workers in large excavations such as the Milan–Rome motorway, 24 m (79 ft) wide, driven in the 1960s. See **dry mix**, **wet spraying**.

shot firing See **blasting**.

shot hole UK miner's term for a hole drilled for *blasting*.

shovel (1) A hand tool for moving coal, stone, concrete, etc., or for mixing concrete, plaster or mortar.

(2) A mechanical *excavator*.

shrinkage (1) During setting, concrete can shrink by as much as 1% (initial shrinkage) but its subsequent 'hardening shrinkage' is less. It can amount to 0.0004 of its length at one year or to half this value at one month. Shrinkage is tripled if a test-piece is kept at the normal 70% relative humidity rather than at 90%. Much of the shrinkage comes from the *cement paste*, less from the aggregate. The higher the *W/C ratio*, the higher is the shrinkage. See **moisture movement**; cf. **creep**.

(2) [s.m.] The shrinkage of earth banks after construction. Since earth fill measured before excavation may occupy a smaller space when finally compacted in the embankment (but a

larger space after excavation and before compaction) there is considerable confusion about the meaning of this term. The usual comparison is between the volume of material in the compacted bank and that of the excavation from which it came. US railway construction figures show 10% initial shrinkage and 15% after complete settlement. An exception is dry clay from deep pits, which swells on contact with air. See **bulking**.

shrinkage-compensating cement *ACI*'s name for one type of *expanding cement*.

shrinkage joint See **contraction joint**.

shrinkage limit [s.m.] The highest water content for a given soil at which a reduction in water content decreases the volume of the sample. It is at the limit between the solid and the plastic states of a clay, and is usually shown by a colour change, the clay becoming much paler at water contents below the shrinkage limit. See **consistency limits**.

shrink-mixed concrete *Ready-mixed concrete* partly mixed at the batching plant to reduce its volume and enable a *truck mixer* to carry more than an *agitating truck*.

shuttering That part of *formwork* which either is in contact with the concrete or has the *form lining* attached to it.

side board A board used for timbering the sides of a heading, usually held by *side trees*.

sideboom cat, s. tractor, boom c. (see diagram, p. 400) A *crawler* tractor used in pipelaying. It lifts pipe and lowers it into a trench with its *jib* at right angles to its crawlers.

side-dump bucket A hydraulically

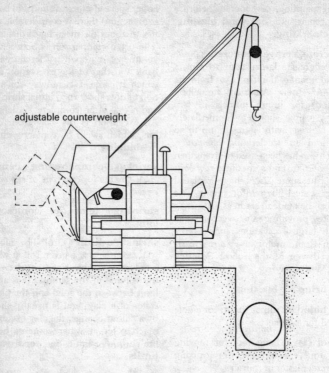

adjustable counterweight

Sideboom tractor (dozer) laying pipe. (F. C. Harris, *Construction Plant*, Granada, 1981.)

controlled digging bucket used on a loader, which can dump to either side.

side-entrance manhole A *deep manhole* in which the access shaft is built not over but to the side of the *inspection chamber*. A passage leads to the inspection chamber from the foot of the shaft, or the inspection chamber merges with the shaft. Large sewers must have side entry for the safety of the workers supervising them.

side forms *Road forms.*

side-jacking test [s.m.] A load test on a soil formation in a trial pit made by jacking apart two vertical bearing plates on opposite faces of the pit, a method developed from the **plate bearing test**.

sidelong ground A hillside slope along which the ground must be cut to expose the formation for a road or other structure. Cuts on slopes often interfere with the stability of the soil uphill and may cause landslips.

side pond [hyd.] A storage pond beside a lock chamber which reduces considerably the water loss in lockage.

side rail See **check rail**.

sidesway [stru.] Slight sideways movement of a *frame* in its own plane caused by wind or other horizontal forces or unsymmetrical loading.

sidetracking In *directional drilling*, the drilling of a branch hole at any point.

side tree Posts 7–15 cm thick holding the *head trees* and *side boards* in a heading.

sidewalk US term for the *footway* of a road.

sidewall core A small soil sample obtained in an oil well or other borehole by using explosive to shoot a tube into the side of the hole, withdrawing it and hoisting it to the surface.

siding machine A machine which cuts back the edge of a grass verge to a required line.

sieve analysis [s.m.] *Screen analysis*.

SIFCON *Slurry-infiltrated fibre-reinforced concrete*.

sight distance The *visibility distance*.

sight rail [sur.] A horizontal board set at a certain height above a required level, e.g. 1.5 m above the invert level of a drain. With a *boning rod* 1.5 m long and two sight rails it is possible to check the levels set out for the drain in the trench for the full distance between the rails and for a short distance beyond them.

sight rule [sur.] An *alidade*.

signal man, spotter US term for a *banksman*.

signal processor See **converter**.

silane See **isobutyl silane**.

silent pile-drivers *Hydraulic pile-drivers* and *vibrating pile-drivers*.

silica SiO_2. Silicon dioxide, which occurs as crystalline quartz and non-crystalline chalcedony, agate, flint, sardonyx and many other varieties. The greater part of *sand*, *sandstone* and *quartzite* is silica. Although some of its varieties are semi-precious gems, silica is the commonest known solid material. Geologists regard free silica as acid. See **acid rock**, **silicon**.

silica brick *Refractory* brick which contains over 90% silica, and being bonded with lime will stand temperatures from 1650° to 1750°C before it softens.

silica fume See **microsilica**.

silicate injection See **Joosten process**.

siliclastic rocks Clays, shales, silts, sands, gravels, etc.

silicon One of the commonest elements on the earth, present in sands, clays and other building materials. It is indispensable for making *integrated circuits* and is a *semiconductor* with the advantage of possessing an oxide, *silica*, which is a first-class electrical insulator.

silicon carbide, carbon silicide SiC. An important constituent of the abrasive *Carborundum*.

sill (1) A timber laid across the foot of a trench or heading under the *side trees*. A *liner* nailed to the sill keeps the side trees apart.

(2) [hyd.] The horizontal overflow line of a measuring notch, dam spillway or other weir structure. See also **lock sill**.

silo A tall, often cylindrical, reinforced-concrete tower (occasionally of steel or timber) used for storing grain, cement or similar materials. The smaller silos are sometimes built of precast segments which are reinforced by spiral wires tightly wrapped round the outside and protected by a cement mortar or *gunite* cover. Tall silos may be built with *slipforms*.

silt (1) [s.m.] Granular material finer than sand but coarser than *clay*, i.e. from 0.002 to 0.06 mm (see **classification of soils**). It feels gritty between the fingers but the grains are difficult to see. It can be distinguished from clay by the *shaking test* or by rolling it into a thread. A thread of silt crumbles on drying, a clay thread does not. *Rock flour* and *loess* are materials of silt size. See **shaking test**.

  (2) [hyd.] See **sediment**, **siltation**.

siltation, **silting** [hyd.] Deposition of *sediment* from water. Silt can be removed by dredging or by building *groynes* to scour it away. Most harbours have to be dredged. Siltation of a dam may be reduced by diverting heavily silted water away from it, or by preliminary settlement of the silt outside the dam.

silt box A loose iron box at the bottom of a gulley for collecting grit. It is pulled out and emptied occasionally.

silt displacement A method of using a *shield* for tunnel driving in nearly fluid silts in the USA. As the shield is driven forward the silt is forced into the tunnel through two rectangular openings, like toothpaste from a tube. The miners cut it with wires and load it out. See **blind shield**.

silt ejector A *hydraulic ejector*.

silt grade material [s.m.] Material of *silt* size.

silt pond, **tailings p.** A settling pond at a gravel pit, mine or mineral beneficiation plant, where the water used for washing the mineral must pass slowly before flowing off as clear water to a stream. Silt ponds must be carefully fenced to prevent public access since the settled silt is extremely fine, like *quicksand*, and could be dangerous to people walking on it. See **tailings dam**.

silt tests To make sure that concreting sand does not contain excessive silt, at least two simple tests can be made on site:

  (1) The *shaking test*.

  (2) About 50 mm of sand are poured into a jam jar. The sand is then submerged in salt water, shaken and allowed to stand for three hours. Any silt visible on top of the sand should be not more than 3 mm deep.

similarity, **dimensional similarity** See **dimensional analysis**.

simple beam, **simply supported b.** [stru.] A beam subject to *simple bending*.

simple bending [stru.] The bending of a beam which is freely supported and has no *fixed end*.

simple curve [sur.] An arc of a circle joining two straights with no *transition*.

simple engine [mech.] Usually a reciprocating engine from which the steam or compressed air is passed to the atmosphere after expanding in one cylinder only.

simple framework [stru.] A *perfect frame*.

simply supported beam See **simple beam**.

Simpson's rule [d.o.] A rule for estimating the area of an irregular figure. The figure is first divided into an even number n of parallel strips of width d. The lengths of the boundary ordinates (lines separating the strips) are measured. There will then be $n + 1$ boundary ordinates, of lengths h_0, h_1, h_2 and so on to h_{n-2}, h_{n-1}, and h_n. The area of the figure is

$$\frac{d}{3}[h_0 + h_n + 2(h_2 + h_4 + \ldots + h_{n-2})$$
$$+ 4(h_1 + h_3 + \ldots + h_{n-1})].$$

Cf. **prismoidal formula, trapezoidal rule**.

single-acting [mech.] A description of *reciprocating pumps* or compressed air or steam engines in which only one side of the piston works and every second stroke of an engine is a power stroke. Every second stroke of such a pump delivers fluid. A single-acting pile hammer drops its weight under gravity alone. Cf. **double-acting**.

single-cut file [mech.] A file with *cuts* (4) in one direction only, used for filing soft material.

single-pass soil stabilizer [s.m.] A powerful machine with four rapidly rotating toothed milling wheels in contact with the soil. These rotors pulverize the soil to a measured depth, mix it with a liquid binder (water, bitumen, etc.) and intermingle the soil with cement or other solid binder which has been spread ahead of the machine. The original soil stabilizers (1944) were single-rotor, multiple-pass machines, but the multiple-rotor, single-pass stabilizers cover the ground more quickly. After the soil has thus been treated and has had liquid added to bring it to its *optimum moisture content*, the road has to be graded, shaped, rolled with sheepsfoot rollers, wobble-wheel rollers and steel rollers, and finally given a wearing surface. Per centimetre of depth, soil cement costs about half as much as reinforced concrete, and it is economical as a base in almost any large road programme, provided that the soil type is right.

single-ply membrane A roof waterproofing sheet intended to supersede the 3-plies needed with bitumen felt roofs. It is usually of PVC or synthetic rubber, occasionally of polymer-modified bitumen with glass reinforcement. O'Hare Airport, Chicago, had one over its restaurant in 1961 which was found to be in good condition in 1983. In warm climates, even if protected by gravel, such roofs may have an automatic sprinkler which at 29°C (85°F) or more cools it with a water spray. The sheet may be held down by *solvent welding*.

single-sized aggregate For concretes, an aggregate of a size between two adjacent sieves in the series listed under *fineness modulus*. Thus aggregates from 38 to 19 mm or from 19 to 9.5 mm are single-sized. But roadstones may need much closer sizing, so this definition may not be accepted for them and is in any case too wide. *Macadam* may be from 19 to 13 mm and this would be a half-size material.

single sling A *sling* with an iron or steel ring at one end and a hook at the other end. See **two-leg sling**.

single-stage compressor [mech.] A machine which compresses air to its full pressure in one cylinder (or one stage of a centrifugal compressor). These are little used for pressures above 4 bars and only for small capacities, since compression in stages is more economical of power.

single-stage pump [mech.] A centrifugal pump with one *impeller*. It could be built to lift water 180 m, but generally a lift of 30 m per stage is normal. For 180 m, therefore, a six-stage pump would ordinarily be used.

sinker drill [min.] A large hand-held *rock drill* used for shaft sinking and other down holes. It may weigh 30 kg or more.

sinking bucket [min.] A *kibble*.

sinking pump [min.] A pump built for keeping a shaft dry during sinking. It may be driven by electricity, compressed air or steam, but is always robust, to withstand falling stones, and built so that its height is about ten times its greatest width. In this way it occupies very little space in the crowded shaft bottom. It is often a *submersible pump* and may be a *borehole pump* with a protective frame round it.

sintered carbides [mech.] A term sometimes preferred for *cemented carbides*. See **sintering**.

sintered clay *Expanded clay* aggregate (*B*). See **lightweight aggregate**.

sintered fly-ash A *lightweight aggregate* made by pelletizing *fly-ash* and then sintering it – available in the UK since 1962.

sintering Heating a powder until it begins to melt. This gives it some mechanical strength while maintaining its porosity and is used in the manufacture of *lightweight aggregates* such as Aglite, Leca or Lytag. Cemented carbides are made by sintering hard carbides and metal powders of various melting points with cobalt, which has a low melting point and fuses them together.

sinuous flow [hyd.] *Turbulent flow*.

siphon [hyd.] (1) A closed pipe, part of which rises above the *hydraulic gradient* of the pipe, but not more than the head due to the atmospheric pressure (maximum 10 m). Water may thus be forced through it by atmospheric pressure.

(2) See **sag pipe**.

siphon spillway [hyd.] A dam spillway which is built as a siphon passing over the crest of the dam. The water must rise to the crest of the siphon before the siphon primes itself and begins to flow, but the siphon will go on flowing until the water falls below its inlet (which is below the crest). There can thus be a difference of more than 1 m between the two levels at which flow starts automatically and stops automatically.

SI system The Standard International system of units, the current version of the *metric system*, suitable for scientific or technical work. It is not yet completely accepted, since the SI pressure unit, the pascal, is not the only unit in use. $1 \text{ MPa} = 145 \text{ psi} = 1 \text{ N/mm}^2 = 10 \text{ bar} = 10 \text{ atm}$.

site (lot USA) Land for building on.

site engineer Someone who works not on *structural design* in a drawing office but on site, ensuring that work is built according to the drawings and specification. The site engineer helps the *agent* in managing the site, but usually his or her first activity is to do any setting out that is too difficult for the foremen and requires familiarity with land-surveying instruments. Many site engineers become *agents* or *resident engineers*.

site investigation An activity which includes *ground investigation* but is much wider. The investigation of a site should expose all its previous uses. A consulting engineer who undertook a site investigation, believing it to be his task to find what sort and depth of piles should be used, was found guilty of negligence because he did not find the seriously contaminating chemical waste on the site. The judge must have sympathized with him because he had to pay only £2 damages. If a consulting engineer's brief (instructions) are unclear or ambiguous he or she should re-write them clearly and submit them to the client for confirmation.

site weld A weld made at the site, unlike a *shop weld*.

size analysis, s. distribution, s. grading See **grading curve**.

skeleton [stru.] A building-frame of steel or concrete.

sketch See **drawing**.

skew bridge A bridge which is oblique to and therefore longer than the gap.

skids Lengths of round or rough timber, steel rail or pipe placed under a heavy object when it is being moved, to prevent it sinking into the earth.

skid-steer Steering a tracked vehicle by locking one track, and running the other so as to turn sharply. The locked track skids. It can also be applied to wheeled loaders.

skilled man A leading hand in a modern trade which has expanded too quickly for the apprenticeship system to become general (general *riggers, bar benders, steel erectors*). They are generally not called tradesmen though they may have more skill and adaptability than many tradesmen. Most of them work to drawings. Since there is no apprenticeship system, most skilled men have risen from labourer.

skimmer [sewage] A device for removing floating *pollutants*.

skimmer equipment, skimmer A digging bucket mounted to slide along an *excavator* jib. The jib is held horizontally and the skimmer takes a slice of soil as it moves away from the machine. Skimmers do not dig below the level of the crawler tracks.

skimming (1) Removing the irregularities of the surface of the soil.

(2) [hyd.] Diverting surface water by a shallow overflow to obtain the cleanest water.

skin friction (1) The resistance of the ground around a pile or caisson to its movement. It is proportional to the area of contact and for caissons in motion is from 10 to 25 kN/m^2. (1.5 to 3.6 psi), but the friction on starting to move is several times as much, even with the bottom edge completely undercut. See **negative skin friction, bentonite mud**.

(2) [hyd.] That part of the resistance to flow that is caused by *roughness* rather than by eddying.

skip A steel concreting bucket, *kibble* or carrier on an *aerial ropeway* or refuse container. All of these different types are of different shapes.

skirt A dropped edge of a footing, which confines the foundation soil and prevents *scour*, e.g. at the two ends of a *culvert*. The skirt of the Condeep *gravity platform* is of steel plate 2–4 m deep, but drainage of the area inside it is possible.

slab (1) **suspended s.** The thin part of a reinforced-concrete floor between beams or supporting walls.

(2) Any large, thin area of concrete such as a wall, a road or a roof. See also (*B*).

slab and beam floor See **beam and slab floor**.

slab track [rly] Rail track with rails fixed directly to precast concrete or with sleepers cast in. Slipform-paved, continuously reinforced concrete track has been used abroad and by British Rail.

slack-line cableway, slack-line excavator (see diagram, p. 406) Digging equipment resembling a stationary *dragline*, though it may be built to have some mobility. Its track rope, hanging from a high mast, is anchored on the far side of the excavation either to a low tower or to an anchor block, and the

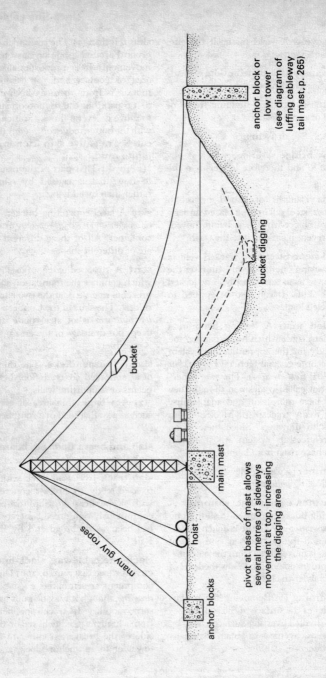

anchor block or low tower
(see diagram of luffing cableway tail mast, p. 265)

bucket digging

bucket

main mast

pivot at base of mast allows several metres of sideways movement at top, increasing the digging area

hoist

many guy ropes

anchor blocks

Slack-line excavator.

digging bucket travels along it hauled by another rope. Its range is far beyond that of a dragline, particularly over a riverbed. The span may be up to 300 m, digging as much as 24 m below water. The far tower may travel on wheels widely separated to provide stability.

slack-water navigation, still-water n. [hyd.] A *navigation* with no current or only a slight one. This is made possible in a river by raising the water level with *weirs* creating *reaches* joined (or separated) by *locks*. The water is thus deepened and the current made slower, allowing large vessels to pass.

slag The waste glass-like product from a metallurgical furnace, which flows off above the metal. The slags most used in building and civil engineering in Britain are *blast-furnace slags* (*B*).

slag cements Many different, excellent *hydraulic cements* made by grinding *blast-furnace slag* and mixing it with lime or Portland cement. See **Trief process**.

slant bar US term for a *bent-up bar*.

sledge hammer, sledge A heavy two-handed double-faced hammer used for breaking stone and for timbering.

sleeper (1) [rly] (**rail tie** USA) A steel, precast concrete or wooden beam passing under the rails of the *permanent way* and holding them at the right spacing. Wooden sleepers usually are pressure-creosoted if of softwood and measure 2.6 × 0.13 × 0.25 m, but much longer, thicker ones called timbers are used under *crossings*, *turnouts*, etc.

(2) A *foot block*.

(3) The horizontal leg of a Scotch *derrick*.

sleeve A steel tube slightly larger in bore than the reinforcing bar it must contain. Two bars can be joined to the sleeve by threading, welding or other means and thus effectively extended. A sleeve congests the section less than lapped bars. See also **crimped coupler** (*B*), **steel sleeve jacking**.

sleeve grouting [s.m.] The use of *packers*, *tube à manchette*, etc.

slender beam A beam which would tend if overloaded to fail by buckling in the compression flange. For concrete it applies to beams which are longer than 20 times their width. The compressive stress in such a beam must be reduced in proportion to its *slenderness ratio*. The greatest slenderness ratio allowed for concrete beams is 60. For timber and steel the allowable slenderness ratios are higher because the tensile strengths are also higher.

slenderness ratio [stru.] The effective height of a column divided by its *radius of gyration*, the value $\dfrac{l}{k}$ in the formula for the *Euler crippling stress*. This is simplified for concrete, in which columns are usually rectangular and calculations are therefore based on the effective height divided by the least width. If this exceeds 15 the allowable stress is considerably reduced until, at 45, the allowable stress is 0. For steel a slenderness ratio of 200 is often allowed, and for timber about 150. For brickwork, stone or mass-concrete walls, no greater ratio of effective height to breadth than 24 is allowed, and usually 18 is the maximum. At ratio 1 for masonry the stress *reduction factor* is 1, at ratio 18 it is 0.30, and at 24 it is 0.20.

slewing (1) [mech.] Rotation of a crane jib. It may be simultaneous with *derricking*.

(2) [rly] Sideways shifting of rail track for realignment.

sliced blockwork *Blockwork* for breakwater construction built in sloping, nearly vertical courses so that the placing of the submerged blocks is much easier than in *coursed blockwork*. Each block being lowered by the crane must rest on two of its neighbours and slides naturally into position.

slick See **outfall**.

slickensided clay [s.m.] A description of *stiff-fissured* clay.

slickensides Polished, grooved surfaces on the faces of cracks in rocks, formed by movement along them. They are usually seen in fault planes. The scratches show the direction of movement but do not necessarily show which side has been upthrown and which downthrown.

slide [s.m.] See **landslip**.

slide rail [mech.] A steel or cast-iron mounting for a belt-driven machine which enables it to be moved as the belt stretches, so as to tighten the belt. Generally the electric motor is bolted to the slide rails and it is the motor which is adjusted.

sliding caisson [hyd.] A *ship caisson*.

sliding forms *Slipforms*.

sliding gate [hyd.] A *crest gate* which has a high frictional resistance to opening and can therefore be used only in small sizes. For this reason the *roller gate* has been developed.

sliding joint In concrete work, a gap in the steel and the concrete, fitted with sliding surfaces to ease relative movement in the joint plane. See **movement joints**.

sliding-panel weir [hyd.] A *frame weir* with wooden panels which slide between grooved uprights.

sliding-wedge method [s.m.] The *wedge theory* for determining graphically or by calculation the passive or *active earth pressure* on a *retaining wall*. When the properties of the soil are known from laboratory tests on undisturbed samples this method is preferable, particularly for clays, to the use of *Rankine's theory*.

sling A length of rope or chain for hanging an object from a crane hook. A steel wire rope with a loop at one end is a special type of sling called a *bond*. Slings may be single, two-leg, three-leg or four-leg, with *reeving* or non-reeving end links. The latest rope slings are braided, not twisted, and may be of synthetic fibre.

slip (1) [s.m.] A small *landslip*.
(2) **slipway** A concrete or stone slab which slopes down to the water's edge. It supports a vessel being built or repaired. See **cradle**, **traversing slipway**.

slip circle [s.m.] (see diagram) The assumed *circular arc* of failure of a clay bank.

slip coat In *pipe jacking*, a coating of *epoxy resin* sometimes painted outside concrete pipe to reduce the soil friction.

slip dock A dock with a sloping bottom and a gate to keep out water.

slip factor The *coefficient of friction* between a *high-strength friction-grip bolt*, or group of bolts, and the steel members which are gripped. The slip load is found by testing assemblies and the slip factor deduced from the tests. From it can be calculated the safe working load at 70% of the slip load.

slipform (1) **continuously moving forms** *Formwork*, for building a reinforced-concrete silo, lift shaft or other wall of smooth outline and uniform cross-section, which is raised con-

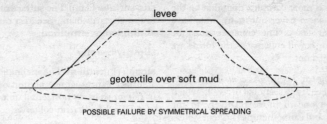

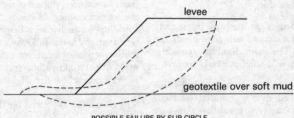

Use of geotextile spread over soft mud to strengthen it in tension and enable it to carry a levee.

tinuously or at short intervals at about 0.3 m/h (faster in hot weather, slower in cold weather). This rapid way of casting high walls has the advantage of no *construction joints*, but faultless site organization and supplies are needed as well as a high capital investment in jacks, formwork, the crane and the working platforms to carry the workers and the formwork. For continuity of supply it may be essential to mix the concrete on site. Since slipforming can enable a 20 storey high lift shaft of a tower block to be built in ten days it can be a great boost to the morale of all on the site. It enables the permanent lifts to be installed quickly.

Slipforming can enable a concrete building to be put up much more quickly than a steel-framed one unless many months have been available for designing and fabricating the steel.

It is possible to stop slipforming at weekends but this is not usual in the UK.

(2) A narrow section of *formwork* in slab or wall shuttering that can easily be removed, and is designed to be struck first, thus making it easy to strike the remaining larger panels. It may also be called a wrecking piece or wrecking strip.

(3) A loose temporary vertical plate in a wall mould to separate the expensive *face mix* (*B*) from the cheaper backing mix during casting – withdrawn as soon as the mould is full.

slipform paving train, s. paver A succession of roadbuilding machines which differs from the *fixed-form paving train* in travelling on its own caterpillar tracks and guiding itself by sensors on tensioned wires. Its equipment is more complex than that on the fixed-form paver but many units have similar tasks. The concrete must be stiff or it will collapse as the forms are dragged on.

slip joint [stru.] A *contraction joint* between two sections of wall, which allows one to move relatively to the other and can sometimes be kept waterproof. It usually consists of a vertical tongue at one side fitting into a groove at the other side.

slip lining *Pipe renewal* by merely inserting a smooth new pipe inside the old one. The main disadvantage is the reduction in diameter. See **U-liner**.

slip membrane A sheet of *polythene* film placed over a blinded (sanded) *sub-base* or *base course* to allow *shrinkage* without cracking of a concrete slab cast on it. With *long-strip construction* and prestress, there may even be two sheets. The Concrete Society estimates that even so one third of the prestress (applied at the mid-depth of the slab) is lost in friction.

Below the polythene, to ensure that slip is possible, any stones must be covered with sand or other fine material carefully levelled to within 10 mm ($\frac{3}{8}$ in.) See **prestressed concrete pavement**.

slip plane [s.m.] A hazardous sloping surface such as a clay bed or fault plane, along which a *landslip* may be likely.

slip ramp, s. road, ramp A road connecting a *motorway* to a road at another level.

slip scraper [min.] A scraper used for moving stocks of material above ground. It consists of a hoist with two ropes, one fitted to each side of the scraper bucket, and two *return sheaves* which may be on movable poles held by guy ropes.

slip surface [s.m.] The surface of failure of an earth bank. See **slip circle**, **sliding-wedge method**.

slipway See **slip**.

slope (1) **gradient** The inclination of a surface expressed as one unit of rise or fall for so many horizontal units.
(2) [min.] In British collieries, a *drift* (4).

slope correction [sur.] A deduction from a measured slope length to derive the corresponding horizontal length. For slopes which rise a height h in a taped length l, the following approximation is within 1 mm for a rise or fall of 10 m per 100 m slope and the error is very much less for slighter slopes.

The correction is $\dfrac{h^2}{2l}$, and for this slope works out at 0.500 instead of 0.5013, which is more exact. See **tape corrections**.

slope gauge [hyd.] A depth indicator consisting of a staff laid at a slope and graduated correspondingly to give a more precise reading than a vertical rod.

slope staking [sur.] The marking of the ground surface by pegs or stakes at points where earthworks in cut or fill meet it.

slotted insert A metal slotted fixing built into a concrete wall, column, etc. by lightly nailing it to *formwork*, which releases it when the formwork is stripped. The slot allows slight adjustment of the fixing bolt or tie slipped into it.

slough (1) Of an earth slope, to slide or break off.
(2) A secondary river channel in which the flow is sluggish (USA).

slow bend, easy b. A bend in a pipe or duct of larger radius than usual to reduce resistance to flow, or to enable stiff cables to be inserted, etc.

slow bend test See **longitudinal bead test**.

slow sand filter A *graded filter* for purifying drinking-water. See **sand filter**.

slow test [s.m.] A *drained shear test*.

sludge Solids settled out from sewage, or in smaller quantity from *raw water*. It is often about 99% water and very difficult to dewater. Sludge must be separated from *effluent* because it contains at least half the *pollutants*. Sludges from the various *sewage* and water treatments are of many different types.

sludge digestion [sewage] *Digestion*.

sludge disposal [sewage] Sludge may be disposed of by burning (the most expensive method), by dumping at sea (the cheapest) or on land. The last is not safe for industrial sludges, which often contain poisonous metals. Many British cities dispose of their sludges in the North Sea – 5 million tons/year in all, some of which is poisonous industrial sludge incinerated at sea – but Continental countries have strongly objected to this polluting practice. The USA forbids dumping at sea. Untreated sludge, at over 99% water, is easily pumped, but *sludge thickening* is often essential.

On the farm, in dry weather, preferably in spring, sludge can be delivered by tanker lorry into furrows ploughed for it, for burial soon afterwards. The disadvantage of any delivery by tanker lorry to a field is the damage done to the soil structure by the wheels crushing the soil. It can be partly overcome by fitting the tanker with large, low-pressure tyres – flotation or balloon tyres. See also **digestion**.

sludge gas *Sewage gas*.

sludge gulper A *gulley sucker*.

sludger (1) A *sand pump*.
(2) A tool for scraping rock chips from a drilled hole before insertion of explosive.
(3) A pump of any sort for *sludge*.

sludge thickening, s. concentration, s. de-watering [sewage] Reducing the water content of *sludge* from the 96% or more which it starts at – a procedure which dramatically reduces its bulk and the expense of handling it. *Activated sludge*, at 99.2% water and 0.8% solids, when thickened to 96% water and 4% solids loses four fifths of its original volume. Several methods of thickening exist: gravity settlement in *sedimentation tanks*, flotation, adding chemicals to help subsequent *filtration*, sedimentation or *centrifuging*, and finally the most expensive, drying with heat. Thickening reduces the bulk and cost of any plant needed for subsequent sludge treatment as well as the amount of heat needed for drying or incineration. See also **digestion**.

slug [mech.] A relatively small quantity of fluid in a pipeline which does not mix with the main fluid in the pipe. For instance, the *slurry* of *oil-well cement* for cementing an oil-well *casing* is driven through the gap at the foot of the casing and up between the casing and the rock outside it by the pressure from the drilling fluid pumped down the drill rods. The cement slurry is a slug sandwiched between two masses of drilling fluid.

sluice [hyd.] (1) A gate to hold back water. Small ones slide vertically, large ones are of many different types.
(2) A channel for taking a rapidly flowing stream of water.

slump loss See **superplasticizer**.

slump test A test for the stiffness of

wet concrete. A conical mould is filled with concrete, well rammed, and then carefully inverted and emptied over a flat plate. The amount by which the concrete cone drops below the top of the mould is measured and is called the slump. This test is valuable only when the same aggregates are used all the time and in the same proportions. It then gives a rough idea of the water content of the mix. This otherwise most useful test cannot be applied to stiff concretes with slump of less than about 20 mm (1 in.) See **compacting factor test**, **flow-table test**, **nuclear density gauge**, **V.-B. consistometer**.

slurry Any fluid mixture of fine solids and water, particularly one which contains cement or *bentonite*.

slurry erosion See **pipeline transport**.

slurry-infiltrated fibre-reinforced concrete (SIFCON) Concrete containing steel fibre, grouted with fine-grained cement slurry after placing. With ordinary placing, 2% fibre is the absolute maximum but S I F C O N can accept 20%, though 5% might be more usual. Thin slabs can be made with a very high resistance to wear and spalling.

slurry shield, bentonite shield, hydroshield A *closed-face shield* for a *slurry tunneller* excavating a face into which water or a clay or *bentonite* slurry is pumped under controlled pressure to restrain inrushes of water or mud. The slurry lubricates the cutters and its return pipe removes the cuttings from the face. An air-lock may have to be provided for workers to go into the face and remove or break boulders. The slurry shield is used in Japan for driving through any weak soil from soft clay to water-bearing gravel, above or below water, but the stones must be smaller than 50 mm (2 in.) One of the world's largest tunnels, driven in this way, of 11.2 m (37 ft) dia. is an *aqueduct*. The tunnel is kept on line by a laser beam aimed at a target on the shield. Using water instead of bentonite reduces the difficulties of settling solids from a bentonite slurry, which may be considerable.

slurry transport *Pipeline transport*.

slurry trench, guide t. When a *diaphragm wall* is begun, a guide trench is dug 1 m (3 ft) deep for the full length of the wall to guide the digging *grab*, prevent damage to the trench sides, and exclude unwanted soil from the trench. It is lined with concrete on both faces so must be dug at least 0.4 m (16 in.) wider than the grab. Apart from guiding the grab, it stores *bentonite* slurry. See **slurry wall**.

slurry tunneller A tunnelling machine or microtunneller (MT) for soft waterlogged ground which makes *compressed air* (2) unnecessary. The machine works in a *slurry shield*. Slurry may also drive the cutting wheel if it is not electrical. The electric power cable, apart from supplying the driving power, may send back information about the direction and position of the cutting head. One Japanese MT of 40 cm (16 in.) dia. uses a temporary series of steel linings with integral slurry feed and return pipes. When the hole has been completed, these 'cans' are pushed into the *receiving pit* and the new, permanent pipe is pushed in with much less force than normal *pipe jacking* demands.

slurry wall, s. trench A type of *diaphragm wall* built for watertightness rather than strength, e.g. to create a *groundwater dam*. The slurry may include soil, *bentonite*, *fly-ash*, *cement*, other materials or a mixture.

slusher See **scraper loader**.

slush ice *Frazil ice*.

smithing [mech.] The *forging* of iron or steel when hot.

smith welding [mech.] *Forge welding*.

smooth [mech.] A description of the *cut* of a file.

smoothing iron A hot iron tool for smoothing asphalt and sealing joints in it.

snatch block A block or sheave with an eye through which it can be fixed by lashing to a scaffolding or pole. It is possible to put a rope over the groove on the wheel without threading the end through. For this purpose one of the side straps of the wheel is hinged. It can be lifted, the rope passing under it and over the pulley, and the strap then refixed. It is much used by *riggers*.

SN curve See **stress/number curve**.

snore A pump working 'on snore' has very little water to suck up. Some pumps are damaged by working on snore.

snots Leakages of *cement paste* through *formwork* which create unsightly blobs on the face of the concrete.

snow course A line marked on the area drained by a stream, along which *snow samples* are taken to estimate snow depth and density for the spring melting.

snow density The water content of snow expressed as the ratio of the depths of snow before melting and as water.

snow load [stru.] A *live load* for which a flat roof in temperate or cold climates may be designed. In the south of England it is taken as 75 kg/m^2.

snow plough A blade for pushing snow from the front of a locomotive. On roads, a rotary snow plough is a self-propelled vehicle with an air blast or a rotating blade that throws the snow aside.

snow sample A core of snow taken by a sampling pipe on a *snow course*, from which the *snow density* can be measured and compared with previous years.

snow sampler A set of light jointed tubes for taking *snow samples* and a spring balance which reads directly the depth of water corresponding to a given weight of snow.

soakaway (dry well USA, **rummel** Scotland) A pit which may be either empty or filled with large stones and is lined (if at all) with stones or bricks laid without mortar. Now made of precast concrete with *weepholes*. Surface water drains into it to soak away into the ground.

sock lining, flexible l., hose l., resin l., reverse l. (Comecop Mexico, **Copeflex** France, **Inliner** Germany, **Insituform** UK, **Paltem** Japan. Some of these compete with each other) (see diagram, p. 414). *Pipe renewal* by insertion of an inside-out polyester felt and polythene 'stocking' into the old pipe after it has been cleaned or reamed as required. The felt is soaked with thermosetting resin usually applied on site but sometimes in the factory. It is then inverted into the old pipe by water or compressed-air pressure and sometimes cured by hot water pumped through it. The water pressure forces the lining into contact with the old pipe and ensures good bond. The method is cheap compared with most alternatives and is extremely quick once the access shafts have been dug. A 450 mm (18 in.) brick sewer 294 m long was lined in two hours. In Japan the method renews

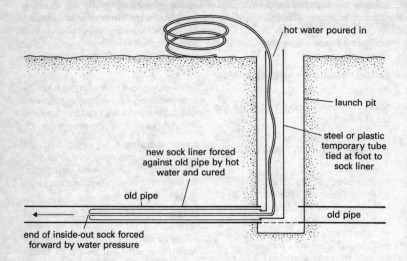

hot water poured in

launch pit

steel or plastic
temporary tube
tied at foot to
sock liner

new sock liner forced
against old pipe by hot
water and cured

old pipe

old pipe

end of inside-out sock forced
forward by water pressure

Sock lining. The length, L metres, to the next manhole is measured, and L metres of sock are provided. The tail end is tied up and not untied until the far manhole is reached. (Insituform.)

25 mm (1 in.) gas connections to houses. It has been used by British Gas.

In Belgium, in 1987, curing by ultraviolet light was tried by Insituform. It was successful and believed to be 20% cheaper than curing by heat. There is no temperature *shrinkage*, which improves bonding with the old pipe. There is also no need to cool the resin-wetted sock laboriously with carbon dioxide snow to prevent premature hardening of the resin. The cured liner is translucent, enabling branches to be located easily. In the tropics, the shelf life of factory-resin-wetted liners is indefinite even without refrigeration, provided they are kept dark (in black bags).

soffit See **crown**, also (*B*).

soft clay [s.m.] A soft *clay* is one which can easily be moulded in the hand and dug with a spade.

soft sand Fine sand for plastering, bricklaying, etc., unsuitable for concreting. Cf. **sharp sand**.

soft-suspension theory [stru.] The theory that foundations carrying vibrating machinery would automatically not be *resonant* if the foundation were large enough has been proved false. Foundations must be designed for a ground pressure and area which are not dangerous for the frequency of the machine which they carry.

software Another word for computer *programs*.

soil (1) [s.m.] In the engineering sense, soil is *gravels*, *sands*, *silts*, *clays*, *peats* and all other loose materials including topsoil, down to bedrock. In agriculture only the *A-*, *B-*, and *C-horizons* are soil.

(2) *Sewage* as opposed to surface water or waste water.

soil analysis [s.m.] See **mechanical analysis**.

soil auger [s.m.] See **shell-and-auger boring**, **auger**.

soil-cement [s.m.] *Soil stabilization* by the use of cement.

soil consolidation [s.m.] See **artificial cementing**, also **consolidation**.

soil dowelling [s.m.] *Soil nailing* through a *slip surface* with *micropiles*.

soiling See **resoiling**.

soil mechanics The investigation of the composition of soils, their *classification*, *consolidation* and strength, the flow of water through them, and the *active* and *passive earth pressures* in them. The science was so named at the first International Conference of Soil Mechanics and Foundation Engineering at Cambridge, Massachusetts, in 1936. *Ground engineering* includes much of soil mechanics. The instruments and methods include many that are the same as in *rock mechanics*. See also **critical voids ratio of sands**, **frictional soil**, **soil stabilization**.

soil mixer [s.m.] A machine used for pulverizing soil in *soil stabilization*. The term includes *plant mixers* as well as travel mixers.

soil nailing Driving or grouting into holes members such as 50 × 50 × 5 mm steel angles, 5.5 m long, at 0.7 m centres both ways into a steep slope so as to create a *reinforced-earth* retaining wall. The exposed earth surface is covered with steel mesh, which in turn is usually *shotcreted*. Sometimes the shotcrete is used as the back *formwork* for a concrete retaining wall. In the USA soil nailing has used 115 mm (4½ in.) dia. holes 7–9 m (25–30 ft) long at 1.5 m (5 ft) centres, with rebars grouted into them.

soil profile A vertical section showing the succession of *soils* at a site.

soil reinforcement See **reinforced earth**, **soil nailing**.

soil sample [s.m.] Any specimen of soil, e.g. an *undisturbed sample*.

soil sampler, clay s., sampling spoon [s.m.] A tube which is driven into the ground with the object of obtaining an *undisturbed sample*. They are used mainly for clays, since the technique of getting undisturbed samples of clean sands is very much more complicated and difficult. See **core catcher**, **Raymond standard test**.

soil shredder [s.m.] A machine used in *soil stabilization* consisting of two half drums which just do not touch, rotate in opposite directions, and break up soil.

soil solidification *Soil stabilization*.

soil stabilization, s. solidification [s.m.] Any artificial method of strengthening a soil to reduce its shrinkage and ensure that it will not move. Common methods are mixing the soil with cement, waste oil or imported soil, *compaction* or merely covering with a *primer*. It is a cheap method of making roads which carry little traffic, and is therefore used in Africa and America where populations and traffic are less dense than in Europe. Soil stabilization was developed in the southern states of the USA from 1935 onwards, and that country now has over 100 million m² of soil-cement pavement. Other common binders are lime (specially for clays), fly-ash and bitumens. See **bulk spreader, processing, single-pass soil stabilizer, soil mechanics, soil mixer, soil shredder, spotting, wet-sand process**.

soil survey [s.m.] A thorough examination of the soils at a site, recorded in a report on their strengths, etc. See **soil mechanics**.

soldier (1) In trench timbering, an upright which is held in place by struts to the soldier on the other side. On their outer face the soldiers hold the horizontal sheeting against the earth. In this timbering, *walings* are not used. (Cf. **puncheon**.) Hence any heavy upright timber.

soldier beam A steel *rolled section* driven into the ground to carry the force from a horizontal sheeted earth bank (USA).

solenoid [elec.] A common *actuator* consisting of soft-iron strips encircled by a coil of wire through which a current passes to strengthen or reverse the magnetism. The *electromagnet* thus created has many uses in automatic controls or *protective equipment* such as *circuit breakers*.

solid map A *geological map* showing no *drift* but only those formations below the drift, that is the solid, usually called bedrock by engineers.

solid web [stru.] A web of a beam consisting of a plate or other rolled section but not a *lattice*. A box girder is usually regarded as a solid web girder.

soling Large stones. See **pitching**.

solum The ground below the lowest floor of a building. See (*B*).

solution cavity An underground cavern created artificially or naturally by circulating water which dissolves salt or other soluble rock.

solution injection [s.m.] See **artificial cementing**.

solvent-cemented joint, solvent welding A *spigot-and-socket joint* (*B*) between pipes made of unplasticized PVC and some other plastics. The spigot is coated with a solvent that dissolves its surface and joins it to the socket permanently. The method can be used on pipes to carry water under pressure.

sonar/SONAR (sound navigation and ranging), echo sounding A technique akin to *radar*, often used for locating submerged objects or water depths.

sonde, probe A long tube used in *well logging*, containing *geophysical* equipment; it is lowered into a hole and has an electrical cable connecting it to the surface, that provides a record of the rock properties as the sonde passes through each rock.

Sondes provide either a continuous record of the hole properties for the length being logged, or 'snapshots' at intervals. More than 50 types of sonde exist, but all are cylinders of about 100 mm (4 in.) dia. to go easily into a 150 mm (6 in.) dia. hole. Several together could make a string 30 m (100 ft) long. *Well logging* usually takes place as the sonde is slowly hoisted to the surface, sometimes inside the drill string.

sonic pile-driver US term for **a vibrating pile-driver**.

sorted sand [s.m.] Civil engineers and geologists disagree fundamentally about the meaning of the terms 'well sorted' or 'badly sorted' or '*graded sand*'.

sounding (1) [sur.] Determining the depth of the ocean or river bed by an echo-sounder or sounding line. Sounding with a measuring staff is possible but not for depths above 5 m.

(2) [s.m.] Driving a steel rod into the soil to measure the depth of bedrock. See **penetration tests**.

sounding lead [sur.] The *hand lead* on a *sounding line*.

sounding line [sur.] A *lead line*.

southing [sur.] A distance measured southwards from an east–west axis. See **latitude**.

space frame, s. structure [stru.] *Plane frames*, including *trusses*, repeat themselves at regular intervals and need to be held or stiffened against wind loads by bracing or walls. Space frames on the other hand are three-dimensional and so may be stable against wind by themselves. But their main advantage from the architect's viewpoint is that they span large gaps with few or no intervening columns. They usually cover such large spans that it is impossible for them to compete in cost with plane frames and their much smaller spans. Space structures include domes, *double-layer grids*, *folded-plate roofs*, *lamella domes*, suspended cable structures, tent structures and many others. Shells, being very specialized, are usually not considered as space structures.

Any structure outside earth's gravity, in interstellar space, is a space structure.

space grid, s. deck [stru.] A flat, *double-layer grid*.

space lattice [stru.] A *space frame* which is built of *lattice* girders, or any other open framework.

spacer (see diagram, p. 418) A small piece of specially shaped precast concrete, clip-on plastic or bent steel which keeps the correct *cover* over the *reinforcement*. See also **bar chair**.

spad [sur.] US term for a *spud*, a surveyor's nail.

spader See **grafting tool**.

spall See **pitching**, also (*B*).

spall drain See **rubble drain**.

spalling Loss of chips or lumps from a concrete surface.

span [stru.] The distance between the supports of a bridge, truss, arch, girder, floor, beam, etc. See **clear span**, **effective span**.

spandrel wall A wall carried on the extrados of an arch, filling the space below the *deck*. See **spandrel** (*B*).

sparrow peck A rough texture made by striking the surface with a pointed chisel or hammer. See (*B*).

SPAT (self-piercing and tapping) screw A pointed, hard-steel screw forced into metal sheet by a special 'gun drill' before it is turned to form the thread. Cf. **huckbolt**, **pop rivet**.

spatterdash See **dashbond coat**.

special steel [mech.] *Alloy steel*.

special structural concrete According to BS, special concretes are those that contain *admixtures*, *lightweight aggregates*, extra-dense aggregates or *cements* other than *Portland*, *blast-furnace*, *low-heat* or *sulphate-resisting*, or that involve any special surface finish or structural requirement.

specific adhesion The chemical bond between glued or cemented substances as opposed to the *mechanical bond*. Strong specific adhesion can exist between two quite smooth surfaces, mechanical bond cannot.

specification A collection of work descriptions which are too long and complicated to go on drawings or into a *bill of quantities*. Written in the order chosen for the *sequence of trades*, its function differs with the type of *contract*. In contracts without a bill of quantities it is an important part of the

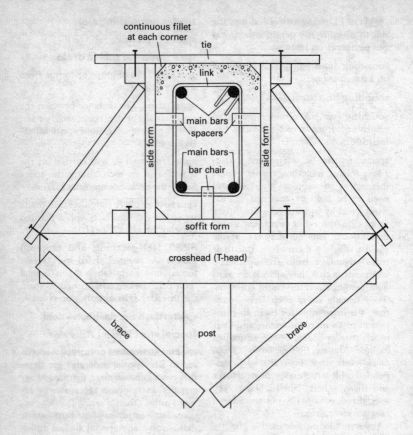

continuous fillet
at each corner

tie

link

main bars
spacers

main bars

bar chair

side form

side form

soffit form

crosshead (T-head)

brace

post

brace

Wooden formwork for a typical isolated reinforced-concrete beam. (James Maclean, builder, Tasmania.)

contract documents and describes the work, but in those with a detailed bill it may not even be part of the contract, though it is provided. In the event of conflict between a drawing and the specification, the drawing is usually accepted as correct.

specific gravity See **relative density**.

specific retention [hyd.] The moisture content by weight held in a soil against gravity by capillary attraction, after being saturated and allowed to drain for a week. It is only about 5% in uniform sand such as dune sand but in a clay may be three times this amount. Consequently rain will pass more easily through the dune sand than through the clay, and the groundwater beneath it will be recharged more quickly.

specific speed [mech.] (1) The specific speed of a centrifugal pump of known performance is:

$$\frac{3.65 \times \text{rpm} \times \sqrt{\text{m}^3/\text{min.}}}{\text{head at maximum efficiency (metres)}^{0.75}}.$$

The specific speed describes the performance of a certain design of impeller and is fixed for any given design. It is the speed in rpm at which a geometrically similar impeller of suitable diameter would turn to deliver 1 m³/min. at 1 m head.

(2) The specific speed of a water turbine of known performance is:

$$\frac{\text{rpm} \times \sqrt{\text{bhp}}}{\text{ft (m) head}^{1.25}}$$

The specific speed, like that for a pump, indicates the design of the *runner*. It is that speed at which a geometrically similar runner of suitable diameter would turn to develop 1 bhp under 1 ft (m) head. The conversion factor to UK hp and feet from metric hp and metres is: metric specific speed = 4.446 × UK specific speed.

For heads above 450 m only the Pelton wheel is suitable; between 450 and 60 m there is a choice between Francis and Pelton turbines. Other choices exist for the lowest heads. Pelton wheels have specific speeds from 10 to 20 rpm, Francis turbines from 20 to 120 rpm, and Kaplan turbines from 120 to 150 rpm.

specific surface The ratio of the exposed area of a powder to its weight or volume, usually expressed in m²/kg. It is a measure of the fineness of a powder and is used for describing cements. The greater the fineness of a cement, the more quickly will it harden. See **ASTM Portland cement, fineness modulus**.

specific yield [hyd.] The specific yield of an aquifer is the amount of water it yields when it drains by gravity. The value can occasionally reach 45%, but a good gravel yields 25% and a clay rarely more than 3%. High porosity is essential for a high specific yield but does not guarantee it. Clays are as porous as the best aquifers but their yield is minute.

spelter [mech.] Zinc of less than 99.6% purity used for galvanizing. In the USA *hard solder* (*B*) is also called brazing spelter.

sphericity Roundness. It is measured by the fraction: (surface area of sphere of same volume) divided by (surface area of particle under consideration).

spider line, s. web See **cross hair**.

spike See **screw spike, track spike**.

spile (1) A wood pile.
(2) [min.] **spill** A sharp-edged thick board: a *forepole*.

spiling *Forepoling*.

spillway, wasteway, waste weir [hyd.] An overflow channel, particularly one over a dam.

spillway gate See **crest gate**.

spiral curve, spiral [sur.] In joining a circular arc to a straight on a road or railway, a spiral provides a *transition* which is often used. It is a curve whose radius gradually increases from the point where it joins the circular curve to the point where it becomes straight. It is more used in railways than in roads.

spiral pipe Steel pipe can be made by helically winding a steel ribbon to the appropriate diameter. One of the largest diameters, 2 m (6½ ft), was made at the site of the Galata Bridge across the Golden Horn in Turkey in 1988. Imported from Scotland, the mill rolled the pipe from coils of steel strip for the piles 90 m (295 ft) long to carry the bridge piers. The piles were test loaded to above 2000 tons after being filled with concrete.

spiral reinforcement, helical r. Steel bar wound continuously round the main bars of a cylindrical concrete column or pile instead of *links*.

spiral winding of pipe lining *Pipe renewal* by winding PVC or steel ribbon round the inside of an old pipe to make a narrower-diameter new pipe inside the old one. Spirally wound steel pipe can later be expanded into contact with the old pipe. Spiral winding is unsuitable for very rough or very distorted pipes, whether brick or cast iron.

splash zone The levels between tide marks, exposed to both air and sea-water, where steel needs the heaviest and most careful protection against corrosion, usually concrete encasement, painting or *wrapping* as well as *cathodic protection*.

splice [stru.] A joint, often between members of the same cross-section, so as to extend their length, e.g. stanchions, concrete piles or reinforcing bars. Stanchions can be extended by bolted *fishplates* (like rails) or by welding; piles by welding or lapping their reinforcement and filling in new concrete made of *rapid hardening cement*. Steel *rebars* can be spliced by mechanical clamps or screwed *couplers*. See also **grip length**, **crimped coupler**, **sleeve**.

splice bar (1) **s. piece** [rly] A *fishplate*.

(2) Where two adjoining *reinforcement* bars do not have enough *lap*, a third bar of the same diameter may be inserted at their junction to lap them both.

split bubble [sur.] A *prismatic coincidence bubble*.

split-set stabilizer A simplified *roof bolt* consisting of a split, tapered tube, driven into a hole drilled for it. The visible end carries a steel plate (Ingersoll Rand).

splitting test, point-load strength t. A test of a concrete cylinder which gives a reasonable value of the tensile strength. It has the advantage for a site engineer that the small inexpensive test unit is portable and that it can be loaded up by a hand jack.

spoil See **waste**.

spontaneous liquefaction [s.m.] See **liquefaction**.

spool A cast-iron *distance piece* between timbers.

spot level [sur.] The elevation of a point.

spotter, signalman US terms for a *banksman*.

spotting [s.m.] In *soil stabilization*, laying bags of stabilizer in position on the ground to be stabilized, at regular intervals.

spot welding Usually *resistance spot welding*, but *MIG* and *TIG* spot welding also exist.

spout-delivery pump [mech.] A pump like the contractor's *diaphragm pump* which delivers water no higher than itself. Cf. **force pump**.

Sprague and Henwood core barrel [min.] An American *core barrel* mounted inside a diamond drilling bit on ball bearings. It can take rock cores down to 25 mm dia., but is generally not used for taking undisturbed samples of *soils*, since the movement of the core barrel disturbs the core.

spray bar The pipe with jets spraying binder on to a road from a *pressure tank*.

sprayed concrete See **gunite**, **shotcrete**.

sprayed mineral insulation See (*B*).

sprayer A *tank sprayer*.

spray lance The pipe of a *hand sprayer* carrying the jets for spreading the binder on to the road.

spray lining, spray-on l. *Pipe renewal* by passing a spraying machine through an old pipe to coat it with cement mortar or polymer mortar (epoxy, polyester or polyurethane). A real advantage of spray lining is that there is less closing and reopening of connections to houses. This labour is unavoidable with many other pipe renewal methods. The best linings are smooth and resilient, but *infiltration* must be stopped or the lining will not stick. See **pipe lining**.

spread, rate of s. The area covered by a given quantity of material such as chippings or road binder. **See spreading rate** (*B*).

spreader (1) A machine which travels on railway track and spreads dumped material with its 4.5 m wide blades.
(2) See **concrete spreader**.
(3) A strut in tunnel or trench timbering.

spreader beam, strongback, yoke A stiff beam hanging from a crane hook, and with ropes or chains hanging from different points along it. It is used for lifting a long reinforced-concrete pile, a large glass sheet or any other long fragile object to prevent damage during lifting.

spread footing A *footing* which is wide and therefore usually of reinforced concrete.

spreading box An appliance which receives road materials and spreads them in a uniform layer without compacting them.

spread recorder In bridge testing, an instrument which measures the outward spread of an *abutment* during loading. Cf. **rotation recorder**.

springing [min.] Enlarging a blasting hole by *chambering*.

spring points [rly] *Points* which are held closed by springs, except when they are *trailing points*, and the flanges of the train wheels open them.

spring washer [mech.] A *washer* which consists of a steel ring cut through and bent to a slow helical curve. It prevents a nut from unscrewing and may be used instead of a *lock nut*.

sprinkler [sewage] A distributor of *effluent* on a *trickling filter* or in *land treatment*. On a trickling filter it is a rotating perforated pipe which releases a spray of water containing few *suspended solids*. For land treatment the effluent may contain more suspended solids and may be released at low pressure from relatively large holes in a pipe that is only slightly above ground. See also (*B*).

spud (1) [sur.] (**spad** USA) A nail used by mine surveyors for hanging a plumb bob as a mark for a survey station, generally with a hole or notch in it for the plumb bob string.
(2) **anchoring s.** A steel post on a *dredger* which may be lowered by a toothed rack or by ropes until it is fixed in the bottom to serve as an anchorage. Two or more may be needed fore and aft.
(3) A heavy, hard-pointed metal pipe, used in *spudding* for a *driven pile*.

spudcan The foundation plate rigidly fixed to the bottom of a leg of a *jack-up rig*.

spudding Before a pile is driven through hard ground, a *spud* (3) may be dropped in the hole several times to penetrate the rock. When the spud has penetrated the rock it is withdrawn and normal pile-driving continues.

spud vibrator US term for an *internal vibrator*, a *poker vibrator*.

spun concrete Pipes and hollow poles have been made of centrifugally cast concrete since 1933, and their almost glazed outer surface may be preferred in *pipe jacking* to the slightly rougher surface of vibrated pipes. See also **centrifugally cast GRP**.

spur, s. dyke (wing dam USA) [hyd.] A *groyne* built out from a bank of a river, having a head so armoured that it cannot be removed by scour. It diverts the flow from a scoured part and may encourage silting elsewhere.

square-mile foot [hyd.] A volume of water 1 ft deep over 1 mile².

square thread [mech.] A robust screw thread used for transmitting a thrust, but, unlike the *buttress screw thread*, capable of transmitting it in either direction.

squibbing See **chambering**.

squirrel-cage motor [elec.] A robust alternating-current motor used for many purposes. The rotor consists of a number of stout, parallel copper or aluminium bars on the perimeter, joined to end rings of the same metal. It has the advantage that there need be no electrical connection between the rotor and the outside of the motor. It can therefore without difficulty be given a flameproof enclosure.

SRC *Steel-reinforced concrete.*

SSCV (semi-submersible crane vessel) A *floating crane*.

stability (1) [stru.] The resistance of a structure to buckling, sliding, overturning, or collapsing. A structure can be tested (on paper) for stability by verifying that it tends to return to its original state after being disturbed.

(2) The stability of an emulsion is its tendency to remain an emulsion and to resist *breakdown*.

(3) See **dimensional stability** (*B*).

stabilization (1) See **stability**.

(2) [sewage] The treatment of an *effluent*, *sludge* or *sewage* to a state when it is no longer likely to stink. *Biological treatment* is the commonest method. Most methods supply air or oxygen, sometimes by *photosynthesis* in a stabilization pond (oxidation pond or lagoon).

(3) In *raw water* treatment, a stabilized water is one that does not deposit solids whether from solution as calcium carbonate or from *suspended solids*.

stabilized soil [s.m.] See **soil stabilization**.

stabilizer [s.m.] Any material added to a soil in *soil stabilization*. It may be a chemical which absorbs water, a waterproofer, like bitumen, a cement or a resin. Resins are required in smaller quantities than other stabilizers, only 2–5 kg/m² per 15 cm thickness being needed.

stadia hairs [sur.] Two horizontal lines in the *reticule* of the telescope of a *level* or *theodolite*, symmetrically above and below the line of sight. They are set at a distance apart such that they subtend a particular known angle at the eye, usually equal to 0.01 radian. This means that any *intercept* of *L* m on a remote staff proves that the staff is at a distance of 100 *L* m plus or minus the *additive constant*. In this case 100 is the multiplying constant.

stadia rod [sur.] A special levelling staff with bold graduations for stadia work.

stadia work, tacheometric surveying, tacheometry [sur.] The use of the *stadia hairs* and a levelling staff for determining the distances of visible points. It is an extremely quick way of mapping, particularly when used with the *plane table*. An accuracy of 1 in 500 is obtainable, which is often more accurate than the plotting of the results. The only complication arises with inclined sights, for which a correction must be applied. The rod is held vertical (with the help of a spirit level or plumb bob) and will therefore not be perpendicular to the line of sight of the telescope except when rod and telescope are on the same level. For all ordinary purposes the true horizontal distance is $\cos^2 d$ times the apparent horizontal distance, $100 \times$ staff *intercept + additive constant*, where d is the angle above or below the horizontal. The use of a horizontal staff, though slower, can be much more accurate than the vertical staff method. Stadia work, because of *EDM*, is obsolescent. See **anallactic telescope**.

staff (rod USA) [sur.] A rod of wood or light alloy with easily read graduations painted on it, used in levelling or *stadia work*.

staff gauge [hyd.] A graduated scale on a rod or metal plate or the masonry of a pier on which the level of the water may be read.

staff man, s. woman (rod person USA) [sur.] Someone who carries a *levelling staff* for a surveyor doing *stadia* or levelling work.

stage [hyd.] The water level measured from any chosen reference line; in the USA, **gage height**.

stage grouting *Grouting* in stages, at increasing pressures and/or increasing depths through the same holes, re-drilled while the grout is soft. See **artificial cementing**.

staging A working platform on to which earth is shovelled in pit sinking. It may be so called because the shovelling is done in stages, each staging being about 1.2 m above the next below.

stairs See (*B*).

stalk [stru.] (1) **stem** The vertical part of a reinforced-concrete *retaining wall*.
(2) The central outstanding part of a tee-section. Cf. **table**.

stanchion, stauncheon, etc. A vertical steel *strut*. Cast-iron stanchions were used until about 1870 before cheap steel was available, but steel is now preferred because it is less brittle. A concrete strut is usually called a column.

standard (1) [sur.] A U-shaped metal casting fixed on the upper plate of a theodolite, carrying the telescope trunnions.
(2) See **standardization**.

standard deviation [stat.] The square root of the average of the squares of the *deviations* of all the observations. It measures the dispersion or spread of the observations and is the square root of the *variance*. See **coefficient of variation, least squares adjustment, probable error**.

standard diving gear, standard diving dress, hard hat diving gear, helmet diving gear The old type of diving dress with lead-soled boots and weights and a heavy tinned copper helmet. It protects the *diver* from the cold and from dirty water, but in other circumstances *SCUBA* is often preferred. In standard diving dress the diver can receive the air supply for breathing by a hose from the surface.

standard error [stat.] The *standard deviation* of many samples of the mean. It shows the amount of inconsistency between the sample and the consignment mean.

standard gauge [rly] A width of 1.435 m between inner faces of the rails of a track, slightly increased on sharp curves and reduced to about 1.432 m on continuously welded straight track. See **broad gauge**, **narrow gauge**.

standardization Agreement between producer and consumer under the authority (in Britain) of the *British Standards Institution* (BSI) on certain tests, dimensions, tolerances and qualities of a product for its purposes. When agreement is reached it is published as a *British Standard* (BS).

It is becoming increasingly urgent to use European standards (prefixed EN). Many BS in fact are also EN standards. EN are published by CEN, the European Committee for Standardization. CEN shares a building in Brussels with its sister organization for electrical standards, CENELEC. The London office of CEN/CENELEC is BSI. The aim of CEN/CENELEC is *harmonization* of the standards of their members – the west European members of ISO, the International Standards Organization. CECC is the electronics committee of CENELEC. See **classification society**.

standardization correction [sur.] A *tape correction* for a tape which is not of the correct length when pulled at the correct tension at the *standardization temperature*, or a correction for an *EDM* instrument which is not operating at the standardized wave frequency.

It is an added or subtracted correction of so much per 100 m or so many parts per million.

standardization length See **comparator base**.

standardization temperature [sur.] The temperature at which a tape is compared with a standard tape of known length.

standard pile A *guide pile*.

standard section A *rolled-steel section*, or a light alloy *extruded* section.

standard specification See **standardization**.

Standard wire gauge See (*B*).

standing derrick See **derrick** (1).

standing pier A bridge pier with spans each side of it, as opposed to an abutment pier.

standing-water level [s.m.] The level at which the groundwater finally stands in a hole or pit which is left open for some days. All the pores of the ground are filled with water below this level, but above it any water in the pores is held there against gravity by *capillary* attraction and the larger pores may not be filled. See **water table**.

stand pipe [hyd.] A pipe rising vertically from a soil mass or water main under pressure. It may be closed at the top by a valve for drawing off water, or it may be open-topped and rise to a height above the hydraulic gradient to show the water pressure in the pipe or soil. If discharging to an open tank at the top it forms a *surge pipe*. The open top ensures that the pressure near the foot of the stand pipe cannot be higher than that due to the head of water at the open top, and high pressures due to *water hammer* are thus released.

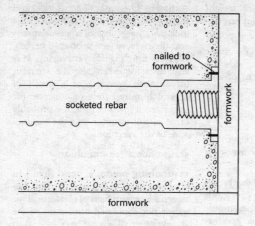

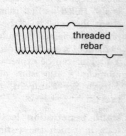

Threaded socket bar: improvement on starter bars. (Halfen Fixing Systems.)

stank (1) A small timber *cofferdam* made watertight with clay.

(2) To make watertight; to seal off.

stanking off Damming the flow of a *sewer*, e.g. for *pipe renewal*.

stapling gun A machine which drives powerful staples, e.g. to hold hardboard to a ceiling or wall. It may be spring-operated or power-operated.

starling Piles (usually timber) driven into the river bed upstream of a bridge pier (and downstream in tidal water) to protect it from floating rubbish, ice and so on.

starter, motor s. [elec.] *Protective equipment* which ensures that a motor shall not receive too high a current when starting up. It may be automatic or hand operated.

starter bar, starter (see diagram) A bar projecting through a *construction joint* so as to knit the adjoining masses of concrete effectively together. A starter bar should be embedded each side of the joint by a length which is usually at least 45 times the bar diameter, so that the new bar has a full *grip length*.

starter frame, stub A *kicker*.

statically determinate frame, isostatic f., perfect f. [stru.] A frame in which the bending moments and reactions can be determined by the laws of statics alone.

statically indeterminate frame, hyperstatic f., redundant f. [stru.] A frame in which the bending moments and reactions cannot be calculated from the equations of statics because the frame has more members or more fixity than a *perfect frame*. In such a frame additional equations must be introduced to determine the forces in each member. Often these equations make use of *strain energy* or *deflection*.

static head [hyd.] The difference in level between two points, such as between the reservoir level and the tailwater level of a water turbine.

static load The amount of weight placed on a structure without any increase for *impact*.

static moment [stru.] The static moment of a section about an axis YY is also called its first moment of area about the axis. It is the sum of the products obtained by multiplying each element of area *a* by its distance *x* from YY. This can be expressed thus: first moment = the sum of $(x \times a)$ or $\Sigma (xa)$. See **centroid, moment of inertia**.

static penetration test [s.m.] Soil tests in which the testing device is pushed into the soil with a measurable force, as opposed to *dynamic penetration tests* in which the testing device is driven in by blows from a standard hammer. *Plate bearing tests*, the *penetrometer* and the *cone penetration test* are examples, possibly also the *vane test* and the *Proctor plasticity needle*.

statics [stru.] The study of forces and bodies at rest, a branch of mechanics which is the basis of structural engineering.

static suction lift See **suction head**.

station [sur.] A *survey point*.

stationary dredger A *bucket-ladder dredger* which is not self-propelled and discharges its dredged material into a hopper barge or a pipeline. Mining dredgers are also in the same sense not self-propelled, but this is never mentioned since it is taken for granted.

statistical uniformity [stat.] A description of the quality of materials or manufactured goods which is stable in the sense that statistical methods can define the relationships between samples, consignments and batches. The *level of control* is a measure of the statistically uniform variation of a product.

statistics The study of *populations* and their variations from the *mean*. This includes the study of measurements and their errors (see **probable error**), which is much used in surveying. It is also used for estimating the strength (or mineral content) of a batch of several hundred tons of concrete (or ore) from a few samples, the probable error of the estimate and the significance of the variations.

stauncheon See **stanchion**.

staunching piece, s. bead [hyd.] In concrete dams a vertical gap left between successive bays of concrete. This gap is not concreted until most of the *shrinkage* in adjoining bays has taken place. It is shaped like a concrete column and may be reinforced. See **stank**.

staunching rod [hyd.] A strong rubber rod in contact with a *crest gate* which is compressed between the gate and the structure to form a watertight joint.

stay [mech.] (1) A tie bar or diagonal brace to prevent movement.
(2) A *guy* rope.

stay pile A pile driven or cast in the ground as an anchorage for a *land tie* holding back a *sheet-pile* wall, etc.

steady flow [hyd.] *Streamline flow*.

steam boiler See **boiler**.

steam curing Maturing of precast concrete members in a steam oven to accelerate the chemical action of *curing*. After two hours in a steam oven, highly stressed members of a precast, prestressed concrete bridge have been stripped from their moulds. See **autoclaving**.

steam engine [mech.] A piston-driven engine, a *prime mover* worked by the force of the steam on the piston.

It is a typical *reciprocating* engine. Cf. **steam turbine**.

steam roller A road roller driven by a *steam engine*. These are now obsolete in Britain although they are much simpler mechanically than the more economical *Diesel engines* which are replacing them.

steam shovel The earliest *excavators* were driven by *steam engines* with their own coal-fired boiler on board. Electrical or *internal combustion engine* drive is now usual.

steam turbine [mech.] A machine with vaned wheels driven round by the force of high-pressure steam blown on to the vanes. It is useful for power generation and large stationary engines generally, since it is much more efficient than the *steam engine*.

steel (1) [mech.] **carbon s.** An alloy of iron and carbon with less than 1% of all other components. *Dead-mild steel* may have 0.1% carbon or less, *high-carbon steel* up to $1\frac{1}{2}$% carbon. The hardness of high-carbon steel is increased and its ductility lowered by heating and *quenching*. Cf. **alloy steel**.

(2) [min.] A *drill steel*. Its end is either threaded to take a *detachable bit* or forged with a bit by the mine smith.

steel band [sur.] A *band chain*.

steel bender, s. fixer See **bar bender**.

steel cored conveyor A *belt conveyor* containing embedded steel strands allowing heavy loads to be carried long distances. For its two 15 km long coal conveyors in the Selby coalfield, raising 2000 tons of coal/hour each, British Coal use one steel cored conveyor and one *cable belt conveyor*.

steel erector, constructional fitter and e., iron fighter A *skilled man*, one of a team of two who climb on to a *steel frame* and fix each end of a steel beam as it is lowered to them by a crane. See guy **derrick** (diagram).

steel-fibre reinforced-shotcrete (SFRS) *Shotcrete* can produce very thin-walled products if it is steel-fibre reinforced, e.g. tunnel linings and repairs, furnace linings, prefabricated bathrooms or box-section harbour gates only 50 mm (2 in.) thick. Partitions for pigsties 25 mm (1 in.) thick are glued to the concrete floor slab. Urine drain slabs are only 20–30 mm thick. In Sweden spun shotcrete poles for street lighting or power transmission have been made 8–13 m (26–43 ft) long, with walls from 30 to 60 mm ($1\frac{1}{4}$ to $2\frac{1}{2}$ in.) thick, containing 1.1–2% fibre. The fibres are longer than usual – 500 mm (20 in.) long, 1.2 mm dia. high-tensile steel.

Norway, because of its geology, has 30% of the world's underground power stations. One contractor working on them between 1975 and 1980 changed over from hand-operated dry shotcrete to remote-controlled wet shotcrete because of better productivity, less loss by rebound and safer working conditions. He uses a *water reducer* and *microsilica* to replace 8% of the cement. Steel fibres eliminate the time-consuming labour of fixing wire mesh. The shotcrete is easily pumped and impermeable. Short fibres are used because they do not ball up and are easier to compact than long fibres.

steel-fibre reinforcement (SFR) From 0.5 to 2% by volume of steel fibres some 50 mm (2 in.) long and of 0.5 mm dia. (10–40 kg/m³) can greatly improve the first-crack bending strength of concrete, its direct tensile strength, ductility and thermal conductivity, but not its compressive strength. In *refractory* concretes the resistances to thermal shock and spalling are improved. Small percentages of fibres can be added easily though even 2% are difficult to mix (cf. **Sifcon**). Fibres with good bond and consequently high pull-out strength can be used more sparingly than smooth chopped round wire. Good bond is obtained from pitted fibres or those with rectangular or crescent-shaped cross-section. *Admixtures* containing chlorides, fluorides, sulphates, sulphites or nitrates must be avoided because of corrosion hazard.

Steel fibres 0.6 × 40 mm, used in the 30 cm (12 in.) thick concrete lining to the tunnel under the Saône and Rhône at Lyon in 1988 raised the tensile strength of the concrete to 8 N/mm^2 (1160 psi).

steel fixer A *skilled man* who cuts and bends steel *reinforcement* and wires it into position ready for the concrete to be poured round it, and attends during the casting of concrete to correct any bars which are out of place.

steel frame A load-carrying building skeleton formed mainly of *rolled-steel sections* meeting each other at right angles. Nowadays in cities most of the site connections are made by bolts. Riveted connections are no longer made, because of advances in *welding* techniques and the noise of riveting. The brick infilling of a steel building frame was in the past considered to contribute nothing to the strength of the frame, but research has shown that it is an important load-carrier.

steel grades See **grades of steel and concrete**.

steel-grit blasting [mech.] A form of *shot blasting*.

steel-reinforced concrete A Japanese technique of inserting a stanchion or joist in a reinforced-concrete column or beam so as to improve the earthquake resistance of the building, sometimes with *shear studs* welded to the rolled section.

steel ring [min.] A *ring* of pressed steel 10 mm thick, used in the USA for lining a tunnel of circular section. Shaft rings are usually of cast iron and known as *tubbing*.

steel sheet In the UK formerly sheet was thinner than 5 mm, plate was 5 mm or thicker. Now 3 mm is the limit between steel sheet and plate.

steel sheet piling *Sheet piling* of interlocking rolled-steel sections driven vertically into the ground along the edge of a guide waling before excavation is begun. When the sheet piling is completed the excavation can be begun in safety. It keeps out flowing ground and often also water, but requires heavy strutting against either of these, unless its penetration depth below the lowest dig level is relied upon to support it. In this case the penetration depth is about equal to (or more than) the length of piling above dig level; also a very strong piling section must be used. Strutting or tying back saves steel, and is therefore used where possible.

steel-sleeve jacking In soft waterlogged ground, steel sleeve jacking has succeeded where other methods of *pipe jacking* or *microtunnelling* have failed, especially where other pipes or cables must be inserted inside the steel tube. The first steel tube is jacked into the ground for its full length. Another length is then welded on to the rear end and this again is jacked either for its full length or until the jacking force becomes excessive. Excavation inside the sleeve begins only when it is clear that there will be no inrush of mud or water. If water or mud begins to flow in, excavation is stopped, the face can be boarded up and jacking recommences either with the original sleeve or with a smaller one. See also **buried shield**.

steel tape [sur.] See **tape**.

steel-wheeled roller A *tamping* or smooth-wheeled *roller* or a *loader* on which the four rubber-tyred wheels have been changed for *sheepsfoot rollers*. They are used for roadbuilding, compacting refuse tips, etc.

steel-wire rope [mech.] See **rope**.

steening, steining, etc. The lining of a well or *soakaway* with stones or bricks laid usually dry, sometimes with mortar.

stell, stull [min.] An *outrigger* on a loader or *boomheader* to jack it against the ground or the side of the tunnel to steady it.

Stellite [mech.] *Hard-facing* alloy of cobalt, chromium and tungsten with a little carbon, manganese, silicon and iron. It is very much harder than steel, and is applied by welding to the wearing parts of drilling tools such as *fishtail bits* or dredger bucket teeth. Cf. **cemented carbides**.

stem See **stalk**.

stemming The material used for filling the rest of a *blasting* hole after the explosive has been inserted. Modern methods use sand in plastic bags of the right diameter. Formerly loose clay was used, rolled into sausage shapes; loose sand can be poured into downward holes.

step To step a lock gate is to place it in a vertical position. *Ship caissons* are often too large to travel by any method except towing by sea. They therefore arrive at the port of use in a horizontal position in which they are stable. The process of stepping is slow and difficult. The term may be derived from the stepping of a mast – placing it in a vertical position on the strengthening timber or step over the keel.

step iron, foot i. A galvanized heavy *malleable-iron* staple built into the wall of an *inspection chamber* at about 0.2 m (8 in.) vertical spacing to enable someone to climb in and out.

stepped foundation A *benched foundation*.

stereometric map [sur.] A map showing valleys and hills which can be fully visualized by merely looking at the map.

stereoplotter [air sur.] A *plotting instrument*.

stereoscope [air sur.] An instrument for looking at a pair of photos; it enables a three-dimensional view to be clearly seen.

stereoscopic pair [air sur.] Two photographs overlapping and taken on the same survey, which provide *stereoscopic vision* when viewed with a *plotting instrument* or *stereoscope*. One pair can be oriented with three levelled points and two points of known coordinates, but this is the minimum *ground control*.

stereoscopic vision Ordinary human vision, in which depth or distance can be estimated. See above.

Stevenson's formula A formula for the height of waves, developed by gales travelling over F sea miles of water, the *fetch*. It states that the height in metres is equal to $0.46 \sqrt{F}$. It applies only to unobstructed deep water, since even submerged sandbanks can cause the waves to break. Its values are too high for fetches of 300 sea miles or more. One sea mile = 1852 m.

stick welding US term for *manual metal-arc welding*.

sticky cement *Pack-set cement*.

stiff clay [s.m.] See **clay**.

stiffened suspension bridge A *suspension bridge* with *stiffening girders*.

stiffener, web s. [stru.] A small member, such as an angle, welded to a slender beam or column *web* to prevent buckling of the web. These are often placed under concentrated loads. See **joggle**.

stiffening girder A *girder* built into a *suspension bridge* to distribute the loads uniformly among the *suspenders* and thus to reduce the local deflections under concentrated loads. A suspension bridge does not need stiffening if the maximum deflection is less than one three-hundredth of the span. See **self-anchored suspension bridge**, **Severn Bridge**.

stiff-fissured clay, slickensided c. [s.m.] *Clay* which is stiff at depth when kept dry but is filled with a network of cracks through which water can pass easily. The London clay, like others of this type, is subject to slips in hilly areas. See **artificial cementing**.

stiff frame [stru.] A *redundant frame*. A frame described as just stiff is a *perfect frame*.

stiff-leg derrick [mech.] A US term for the Scotch *derrick*.

stiffness [stru.] (1) Of a material, the stiffness is its resistance in GN/m^2, etc., to deflection, more often called its *modulus of elasticity* or E value.

(2) Of a structural member such as a beam or column, its resistance to bending or buckling. It is proportional to the E of the material, to the I of the section, and to the reciprocal of its span. For beams of the same material the stiffness is thus proportional to $\dfrac{I}{l}$.

stilling pool, water cushion [hyd.] An enlargement and deepening of the river at the foot of a dam spillway to lower the speed of flow and reduce *scour*.

stilling well, gauge w. [hyd.] A chamber which communicates with the main body of the water by a small inlet. It may contain a *recording gauge*.

still-water navigation [hyd.] A *slack-water navigation*.

stimulation [hyd.] Artificially increasing the output from a well by methods such as applying acid to an *aquifer* containing limestone or dolomite, or fracturing it with explosives or with high-pressure water, or by combinations of these methods.

stinger, stern ramp, pontoon A support in the stern of a *lay barge*, which prevents the pipe buckling as it is lowered to the seabed. In 120 m of water, about 370 m of pipe will be hanging, with 46 m of stinger at the forward end of the pipe.

stirrup A *link*.

stock rail [rly] The fixed outer rail at *points*, against which the *switch tongue* is held.

Stokes's law, S. formula (1850) [s.m.] An expression for the settling velocity of spherical particles in fluid. It is utilized for determining the effective diameters of those parts of a soil sample which cannot conveniently be sieved, being smaller than 0.07 mm. The *terminal velocity* in water is

$$\frac{(S - S_w)gd^2}{30n} \text{ cm/min,}$$

where

S, S_w = relative densities of particle and water

g = acceleration of gravity (981 cm/s each second)

d = particle diameter, mm

n = viscosity of water gm/cm/s.

See **elutriator**, **vel**.

stone See **coarse aggregate**.

stone-block paving *Sett* paving in which the stones are accurately cut to a rectangular shape to reduce the joint thickness to the minimum.

stone blower [rly] Rail-borne or hand-held machines for packing *ballast* below *sleepers* by compressed-air blast. The rail-borne version was developed by British Rail with Plasser.

stone drain A *rubble drain*.

stone pile, s. column A vertical hole in soft ground made by *vibroflotation* or other methods, and filled with stone in the same way as a *sand pile*. A *bearing pile* is thus built more cheaply than of concrete.

stop end A *stunt end*.

stop-end tube A clean steel tube, dropped into the end of a length of *diaphragm wall* before concreting, can be removed as soon as the concrete has set and before it has hardened, forming the smooth, curved surface of a joint to the next lift of wall.

stop log, s. plank, flashboard [hyd.] A baulk, plank, precast concrete beam or steel joist which fits between vertical grooves in walls or piers to close up a spillway or other water channel. The baulks are laid horizontally on top of each other to form a *cofferdam*. CF. **needle weir**.

stop valve [mech.] A valve used for turning on or shutting off completely a supply of fluid. See **discharge/gate valve**.

Storaebelt Danish for the *Great Belt*.

storm pavement, A gently sloping bank to a *breakwater*.

storm sewer, stormwater s. A sewer which normally carries no flow but after heavy rain carries *stormwater* overflow directly to a river.

storm surge A rise in water level caused by storm winds. It can reach 1 m (3.28 ft) around the UK. Yet another reason for rise or fall in sea level is variation in atmospheric pressure. This may cause as much as 10 mm change in sea level for 1 mbar change in atmospheric pressure (more in the Baltic). BS 6235:1982 *Fixed offshore structures* recommends adding 0.6 m (2 ft) to wind-induced storm surge for this reason. See **wind shear**.

stormwater (1) Water discharged from a *catchment area* after heavy rain.
(2) In *combined* or *partially separate* drainage systems, a stated excess of the drainage water above the dry-weather flow at which the sewage is allowed to overflow into a river after passing through *stormwater tanks*. Usually the total flow must be three times the average dry-weather flow before the sewage overflows to the river.

stormwater overflow, storm o., separation weir [sewage] In a *combined* or *partially separate* sewerage system, a weir within a *sewer*, which allows excess water in it to overflow after heavy rain, discharging it into a storm sewer. The overflow is often designed to start when the flow reaches six times the *dry-weather flow*. Normally only sewers of 500 mm (20 in.) or larger dia. have storm overflows.

stormwater tank A tank in which *stormwater* waits until capacity becomes available in the *sewage* works to treat it.

straight-run bitumen A bitumen obtained as residue after distilling a suitable petroleum.

strain [stru.] A change in length caused usually by a force applied to a piece, the change being expressed as a ratio, the increase or decrease divided by the original length. *Elastic strain* is wholly recoverable, but *permanent set* is not. The instantaneous strain of good concrete loaded to 14 N/mm² can be 0.06% or more, and this increases with time because of *creep*. So great is the creep that the final strain of similar concrete stressed only to 3.5 N/mm² is also 0.06% after a year of creep. The strain of steel stressed to 100 MPa (14,500 psi) is also about 0.06% and so is that due to its thermal expansion or contraction over 50°C.

strain ageing [stru.] Increase of strength and hardness with time after cold working or other overstressing. It is seen in steel and iron, but also occurs in *light alloys*.

strain energy [stru.] (1) The energy stored in an elastic body under load.

(2) A method of structural analysis based on the amount of work (energy) stored in a loaded frame. It is a method of wide application, but many simultaneous equations must be solved. See **resilience**.

strain gauge [stru.] A sensitive instrument for measuring small deflections in machines or structures. From these deflections the strains can be calculated. *Acoustic* and *electrical resistance strain gauges* are well known, but mechanical and optical strain gauges also exist.

strain hardening [stru.] *Hardening* due to cold working.

strake [mech.] A metal lug or cleat fixed to a pneumatic-tyred wheel or conveyor belt to improve its grip.

strand (1) [mech.] A number of wires grouped together by twisting. With other strands, usually laid helically round an oil-filled hemp core, it makes a rope. Most ropes contain six strands round a hemp core and most strands are built of six wires round one central

wire of the same diameter, so that the pattern of 6 round one goes from the strand to the rope. Other strands contain 19 wires (12 round 6 round 1) or 37 wires (18 round 12 round 6 round 1). See **lay**, **locked-coil rope**, **rope**.

(2) Compared with the strand in a hoisting or haulage rope, strand for *prestressed concrete* has the same structure except that its core is invariably of steel. As in ropes, the commonest strand has seven wires. Compared with wire *tendons*, strand allows more force to be put into a given concrete section, and *die-formed strand* is a further improvement.

Because rope or strand tends to unwind under load, it stretches much more than straight steel wire. The *modulus of elasticity* is less than one third that of straight steel wire, so the elastic deflection is tripled. But there is an even larger difference in the once-only 'permanent constructional extension' when rope or strand is first loaded, which does not occur with straight steel. According to *Kempe* this amounts to 1% for a heavily loaded rope or 2% for one with many bends. But Blake's *Civil Engineer's Reference Book*, 1975, quotes 0.5%. This too is large.

stranded caisson A *box caisson*.

Stran-steel US *cold-rolled* steel sections fixed together in pairs by spot welding to provide a nailing slot between them.

strap [mech.] A metal plate or thin bar used for fixing butt-jointed timbers or other members. The *fishplate* is developed from it.

strata control [min.] *Roof control*.

stratified flow [hyd.] See **density current**.

stratified soils [s.m., hyd.] Soils or rocks that have been laid down by deposition from the water of rivers, lakes or seas. Because they have pronounced horizontal layering (stratification) their permeability is usually greater in this direction than vertically.

stratigraphy The study of soils or rocks, their properties and geological time sequence. Such information is useful to engineers who have to make excavations or find *groundwater*.

streamline flow, or **laminar/ steady/viscous f.** [hyd.] Fluid flow in which the movement is continuous and usually parallel, the movement at each point being constant. The head loss due to friction is only proportional to the first power of the velocity. The upper limit of speed of streamline flow is Reynolds's *critical velocity*. See also **turbulent flow**.

street A ribbon of land used for a public *highway* in a town. In the USA the term is often reserved for east–west highways. Cf. **avenue**, **boulevard**, **freeway**, **parkway**.

street refuge See **refuge**.

strength [stru.] (1) The strength of a material is measured by its greatest safe working stress. This is equal to the *yield point*, the *ultimate strength* or the *proof stress* divided by an appropriate *factor of safety*. See **modulus of elasticity**.
(2) The strength of a structural part is its ability to resist the loads which fall on it.

strength of materials The calculation of stresses due to tension, compression, shear force, bending, torsion and any of these stresses combined, including the study of failures of materials and of deflections. This is an important subject for structural engineers and is an essential preliminary to the study of *structural design*. See **theory of structures**.

stress [mech.] The force on a member divided by the area which carries the force, formerly expressed in psi, now in N/mm^2, MPa, etc. See **compression**, **effective pressure**, **fatigue**, **principal stress**, **shear**, **tension**.

stress analysis [stru.] The determination of the stresses in the different parts of a loaded structure. See **photoelasticity**.

stress circle [stru.] *Mohr's circle*.

stress concentration [stru.] Sudden changes of cross-section in a structural member such as notches, holes or screw threads cause stress concentrations which can be detected by *photoelastic* analysis. Such sudden changes of section are called stress raisers, and can be avoided by curving the surface of the bar at each change of section. The change of shape is thus made gradual instead of rapid, and the stress also changes gradually, avoiding *notch effect*.

stress-corrosion cracking [stru.] Cracking of metal under stress in corrosive surroundings. *Brittle fractures* in *austenitic* stainless steel wires stressed in the presence of chlorides have been disastrous.

stressed-ribbon bridge (Spannbandbrücke in German) A bridge that may be of *prestressed concrete*, but has an uncommon, *catenary*-shaped deck. Several examples exist in Europe, over the Rhône at Lignon near Geneva, another at Freiburg and a conveyorbelt bridge in Switzerland that passes through a tunnel. Heavy abutments or anchorages are needed to carry the tension at each end but big differences in level are possible.

stressed-skin construction, geodetic c. [stru.] Terms borrowed from the aircraft industry, in which *sandwich construction* (B) of frames or fuselages is usual. As the name suggests, the surface material is structural and not, like slate, a mere weather-resisting cladding. It has been used by the makers of plywood houses, in which the plywood wall cladding is glued to the load-bearing members. *Shells* are special examples of it.

In buildings clad or roofed with steel troughing, the troughing can carry forces in its own plane (not at right angles to it) provided it is securely fixed by *self-tapping screws* (*B*), bolts or pins fired through the troughing into the frame. Hook bolts are unsuitable. Seam fasteners must not work loose. *Pop rivets*, bolts or deep self-tapping screws are adequate. Purlin–rafter connections must be designed to resist the forces transmitted by the skin. Ideally sheeting should be considered to add strength in stressed-skin construction only when it forms part of a building mainly intended to exclude the weather (snow, wind). Removal of sheeting then implies removal of load.

stress-free temperature [rly] The temperature, 27°C. (81°F), at which British Rail's *continuously welded rail* becomes free of stress after *destressing*.

stress/number curve, SN c. [mech.] A curve obtained in *fatigue testing*. It shows the range of stress in a material plotted against the number of cycles to failure.

stress probe A British type of push-in *pressuremeter*.

stress raiser See **stress concentration**.

stress relieving [mech.] Heating steel to a temperature below the *critical point*, followed by slow cooling to relieve internal stress. Cf. **normalizing**.

stress/strain curve A curve, showing the result of a (usually tensile) test on a metal test piece, in which the *strains* are plotted against *stresses*. The stresses are calculated as the load divided by the original area of the piece, in spite of its *contraction in area* above the *yield point*. Cf. **load/extension curve**.

stretcher A *liner*. See also (*B*).

striding level [sur.] A sensitive *level tube* used for cross-levelling a *theodolite* telescope. It is fitted at each end with a leg projecting downwards from the tube. These legs stand on the telescope trunnions.

strike [min.] The horizontal line in the plane of a fault or a sedimentary rock, at right angles to the hade or *dip*.

strike off To draw a *screed board* across fresh concrete between edge forms and remove the excess.

striker [mech.] A helper to a *black-smith* or *angle-iron smith*.

striking (1) Removing *formwork* or other temporary supports from a structure.

(2) See **struck capacity**.

striking plate A piece of waste metal kept close to the workpiece in *arc welding* before the *electrode* and arc are transferred to the work.

stringer, s. beam [stru.] A long horizontal member which ties together the heads of trestles in wooden trestle bridges or provides support under a *rail*, parallel to it, in a steel railway bridge.

strip (1) To strike *formwork*.

(2) [air sur.] See **vertical photograph**, also (*B*).

strip footing, s. foundation A foundation for a wall or for a close succession of piers. It is ribbon-shaped, projects slightly each side of the wall, and may be of brickwork, timber (if permanently submerged), mass concrete or reinforced concrete.

stripping (1) Clearing a site of turf, brushwood, topsoil or the first layer of soil.

(2) *Striking* formwork.

(3) Loss of binder or aggregate from a road surface. See **ravelling**, **scabbing**, also (*B*).

(4) **air stripping** [sewage] An *advanced treatment*, blowing air through sewage *effluent* to remove carbon dioxide, hydrogen sulphide or (after the addition of lime) ammonia from it. A disadvantage is that the ammonia pollutes surrounding air, water or snow. Air stripping may be used also to remove nitrogen and carbon dioxide after the *deep-shaft process*, but air pollution laws limit its use in the UK. Cf. **breakpoint chlorination**.

strongback A *spreader beam* or a proprietary steel or aluminium *soldier* in *formwork*.

struck capacity The capacity of a bucket, mine car, *kibble*, skip or other container, calculated as if it were full of water, so that all material above the edges is imagined to be struck off with a straight edge. The rated capacity of an *excavator* bucket is its struck capacity.

structural analysis The early part of a *structural design*, which consists of determining what forces are carried by all the parts of a structure and what proportions the forces bear to the loads on it.

structural design [d.o.] Calculating the sizes of members to carry loads in a structure in the most economical way. A structural design is carried out in two main parts, the first being a rough estimate of the loads with the *structural analysis*. When the structural analysis is checked and found satisfactory the

loads are more accurately estimated. The second half of the design consists of the proportioning of the members according to the calculations of the first half, together with the adjustment of the original calculations to any final altered sizes of parts. This is the main work of a *structural engineer*, since most conventional building types have been sufficiently analysed and need no further analysis in the drawing office. See **theory of structures**.

structural designer [d.o.] One who makes *structural designs*. He or she is an experienced *draughter* and usually though not necessarily a *chartered engineer*.

structural designer-draughter [d.o.] A designer who draws his or her own designs and generally has no draughtsman working for him or her.

structural draughter [d.o.] One who makes structural drawings, i.e. drawings of those parts which carry load in buildings and structures, usually in metal or concrete but sometimes in brickwork or masonry. He or she may be, but does not need to be, a qualified structural engineer. The main tools are a ruling pen, some scales and a set-square.

structural engineer A *chartered engineer*, in Britain a corporate member of the ICE or IStructE. His or her main abilities are in *structural design*.

structural steelwork *Rolled-steel joists* or built-up members fabricated as building frames, by riveting, welding or bolting, or all three.

structure (1) The load-bearing part of a building.

(2) Anything built by people, from a hydraulic fill dam built of earth or a pyramid of stone to a hydroelectric power station or an earth satellite. A

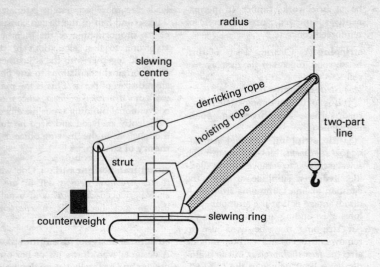

Strut-jib crane.

structure is not necessarily roofed, a building is.

structure gauge [rly] The outline of a tunnel which the tunnel builder must work to. It is bigger at all points than the *loading gauge*.

strut [stru.] A *long column*, usually of wood or metal, not necessarily vertical.

strut-jib crane (see diagram) A *crane* which has a strut projecting from the top of the cab, carrying the pulley for the *derricking* rope, as opposed to a *cantilever crane*.

Stub's iron wire gauge See **Birmingham wire gauge**.

stub wall A *kicker*.

stucco See (*B*).

stud [mech.] A rod threaded for its full length, a screw with no head. See (*B*); also **stud gun**, **stud welding** (*B*).

student member, s. engineer A man or woman studying civil engineering in a recognized course of study may join an *engineering institution* from the age of 17. This makes it easy for young people to join their chosen institution at low cost. After obtaining an engineering degree and full-time work, the student normally becomes a *graduate engineer*.

stuffing box [mech.] A recess, filled with packing, fitting tightly round a piston rod to prevent leakage of steam from an engine or water from a pump.

stumper A *tree-cutter*.

stunt end, form stop, stop end, day joint Vertical shuttering placed across a wall, slab or trench to form a *construction joint* which ends the day's concreting.

sub-base A bed of suitable material such as gravel, laid sometimes under a *base course* of a road to strengthen it, improve the drainage, etc.

subcontract A *contract* between a specialist firm and a *main contractor*.

subcontractor A firm doing specialist work for a *main contractor*. Subcontractors should be approved by the *client* or *consultant* before they are engaged. Many large British contractors began as railway subcontractors from 1840 onwards, including McAlpine, Monk, Mowlem and Wimpey. See **nominated subcontractor**.

subcritical flow [hyd.] In an *open channel*, flow is said to be slow, tranquil or subcritical when the *Froude number* is less than 1. See **hydraulic jump**.

sub-grade, formation The natural ground below a road, airstrip, or other structure or foundation.

sub-irrigation Irrigation by raising the water level near the roots of plants.

submerged-arc welding An *automatic welding* method in which a bare or covered consumable wire *electrode* travels along the weld under a mound of flux. The lower part of the molten flux protects the weld. The method is used for joining thick steel plates in a horizontal position.

submerged float [hyd.] (1) A *rod float*.
(2) A *subsurface float*.

submerged tunnel See *immersed tube*.

submerged weir [hyd.] A *drowned weir*.

submersible A submarine to carry one or more persons, designed for work in deep water where divers cannot work. It may have a mechanical hand (manipulator), television cameras or geophysical survey equipment for examining a pipeline or the seabed. A

2800 ton Vickers submersible surveyed 21 km of seabed pipeline in 13 days in 1972. It was claimed to work at 420 m depth.

submersible pump [mech.] A centrifugal pump which may be driven by compressed air or electricity and can be wholly submerged in water. A submerged pump can lift water much higher than one on the surface, which is always limited by the maximum *suction head*. See **borehole/deep-well/sinking pump, filter well**.

subsidence, settlement (see diagram, p. 438) Downward movement of the ground surface. South-east England is subsiding at 3 mm yearly, with the north and west of Britain rising at the same rate. Land drainage increases the sinking, especially on peat.

(1) Filling of a mined void reduces settlement, but in Britain most coal mining subsidence amounts to more than half the seam thickness over the years even though some filling is used.

(2) Bulk removal of fluid from a borehole causes settlement. By 1987, because of the removal of gas and oil from the chalk 3000 m below the seabed, the six Ekofisk oil and gas production platforms in the North Sea had sunk 6 m, being at the centre of a *cone of depression* 6 km across. They were then jacked up a total of 6 m (20 ft). Phillips Petroleum used 104 jacks, each able to lift 700 tons, and raised them in five days, cutting through their supports above water level and re-fixing them after the jacking, to make them 6 m longer. The high *porosity* of the reservoir rocks is believed to have caused both the rich yield and the settlement.

In Mexico City subsidence from the pumping of water is so serious that pumping in the city is prohibited.

At Terminal Island, California, oil extraction caused 6 m (20 ft) of subsidence and one volume of seawater was pumped in for each volume of oil removed, with complete success. Central London settled 20 cm (8 in.) between 1830 and 1970. See also **inherent/interference settlement, gross loading intensity, porewater pressure, rising groundwater under cities, settlement crater**.

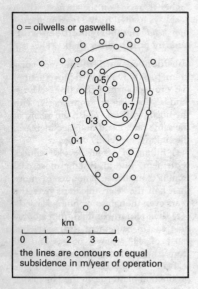

O = oilwells or gaswells

0·5
0·7
0·3
0·1

km

0   1   2   3   4

the lines are contours of equal subsidence in m/year of operation

Subsidence at the Ekofisk Oilfield, North Sea. (M. E. Jones, M. J. Leddra and M. A. Addis, *Reservoir Compaction and Surface Subsidence Due to Hydrocarbon Extraction*, HMSO, 1987.)

subsoil (1) [s.m.] The weathered *soil* immediately below the topsoil.

(2) The ground below *formation level* also called the subgrade or foundation.

subsoil drain A *field drain* laid just below the ground surface and covered with stones. If it connects to a foul drain the water must pass through a trap first. To protect a road foundation

from *frost heave*, subsoil drains are sometimes laid about 1.5 m down so that capillary water stays at least 0.3 m below the foundation.

substructure That part of any structure which is below road level, in particular the foundations and piers of a bridge. The substructure of a suspension bridge can include its towers which are 150 m high.

subsurface erosion [s.m.] *Piping*.

subsurface float [hyd.] An underwater *float* tied by a line to a surface float which indicates its movement.

subsurface flow [hyd.] See **density current**.

subtense bar [sur.] A bar of accurately known length (usually 2 m), held horizontally on a tripod at a point whose distance is required. The surveyor notes the readings of the horizontal circle of his *theodolite* at each end of the bar and, by subtraction, deduces the angle it subtends. From this angle the distance from theodolite to bar can be calculated, or read from tables. It is slower but much more accurate than *stadia work* with a vertical staff. An accuracy of 1 in 7000 has been claimed.

subway (1) In the UK an underground footway.

(2) In the USA an underground railway.

suction-cutter dredger [hyd.] A *suction dredger* provided with a rotating *clay cutter* at the end of the suction pipe. It can dig stiff clay or gravel. See **draghead dredger**.

suction dredger [hyd.] A *dredger* without digging buckets, which digs by powerful suction pumps. The mud-and-water mixture is either pumped away along a *floating pipeline* to land which is being reclaimed, or it is dumped in a hopper barge, the clear water being allowed to overflow until the mud in the barge is thick enough to be removed. This dredger is often used for providing the fill for *hydraulic-fill dams*.

The reservoirs of Taiwan have suffered severe silting like many others, and it has been found profitable to use a suction dredger to clean them down to 80 m (260 ft) depth, extracting water with a solids content of 30% and pumping it as far as 3 km at 150 m³/min.

suction head, static s. lift The height to which a pump can lift water on its suction side, measured from the free water level in the sump. Theoretically this is about 10 m (1 atm.) with water at freezing point, 0°C, and diminishes to 0 at boiling point, 100°C. In practice, good pumps at ordinary temperatures should not have to suck more than 4–6 m. Centrifugal pumps lift less than reciprocating pumps.

suction pad, vacuum p. In *vacuum lifting*, a device which clings by vacuum to the piece being lifted. It connects and disconnects quickly, so for concrete is faster than lifting by crane hook, and spreads the load better.

suction valve [mech.] A *check valve* on a suction pipe near its entry to the *sump*.

sudden drawdown [hyd.] Rapid drop in water level such as is caused by tidal variations. It produces a critical condition in the design of a dam or quay wall or any earth slope, since the support given by the water is removed at a time when the soil is saturated.

sudden outburst, gas o. [r.m.] An explosion, usually of coal at a coal face, caused by or at least accompanied by the release of a large volume of methane, sometimes with other gases, often kill-

ing the miners at the face. Outbursts of methane have also occurred in salt mines and potash mines. Unlike *rock bursts*, sudden outbursts can occur in shallow mines, and in Britain occur in certain anthracite mines, very much shallower than those where rock bursts happen.

sue load [hyd.] The heavy and increasing load on a *slipway*, transferred from the front wheels of the *cradle* as a ship is drawn out of the water.

sulphate-bearing soils [s.m.] If groundwater contains more than 0.1% SO_3, or if a clay contains more than 0.5% SO_3, *high-alumina cement* should be used for all concrete in the ground. *Portland pozzolan cement* may sometimes give enough protection at lower cost. No precautions need be taken with foundation concrete in water which contains less than 0.02% SO_3 or clay which contains less than 0.1% SO_3. See also **sulphate-resisting cement**.

sulphate-resisting cement, ASTM Portland c. type V A *cement* that is generally available in the UK. It resembles *ordinary Portland cement* but is less easily attacked by sulphates in the soil. It generates slightly less heat than *ordinary* but more than *low-heat cement*. It is sometimes used as a substitute for low-heat cement, which is not usually available in the UK. See **super-sulphated cement**.

sulpho-aluminate cement *Expanding cement*.

sulphur concrete The sulphur surplus in Canada has resulted in sulphur being used as a substitute for cement. At least two ways of mixing it exist but both involve heating above the melting point of sulphur, 119°C (240°F).

The powdered sulphur mixed with the aggregate is heated quickly to 140°C (284°F) and then cast.

Alternatively the aggregates alone are heated to 180°C (356°F) and fed into a tilting drum mixer with powdered sulphur and a workability agent such as silica flour. Moulds are overfilled to allow for contraction on cooling. Mixes by weight are about 19%S, 31% sand, 46% coarse aggregate, 4% silica flour. Full strength is normally reached in 6–8 h.

The concrete resists chemicals except alkalis, but has high *creep*, poor resistance both to freeze-thaw and fire, and corrodes steel in damp conditions. (A. M. Neville and J. J. Brook, *Concrete Technology*.)

sulphur dioxide See **flue-gas desulphurization**.

sulphur-infiltrated concrete Concrete can be impregnated with sulphur in a roughly similar way to the impregnation with polymer of *polymer-impregnated concrete*.

summary of reinforcement A *cutting list*.

summit canal A canal which rises through its locks to a summit and then falls through other locks. The first summit canal was probably the northern 1120 km (700 mile) length of the Chinese *Grand Canal* from Huaian to Beijing (Peking), opened about AD 1290. Cf. **lateral canal**.

sump A pit in which water collects before being baled or pumped out. The pump suction dips into a sump.

sumpers [min.] British term for *cut holes*, especially in shaft sinking.

supercritical flow [hyd.] In an *open channel*, flow is said to be fast, rapid or supercritical when the *Froude number* is more than 1. See **hydraulic jump**.

superelevation, cant, banking [rly] The tilt given to a pair of rails at a bend is 150 mm (6 in.) max. on British Rail's *continuously welded rails* and 110 mm elsewhere. For roads and conventional railways the outer rail is higher than the inner rail. But for rope-drawn trains the slope is reversed: the inner rail is higher.

superficial [d.o.] Relating to a surface or an area.

superficial compaction [s.m.] The *compaction* of soils by the *frog rammer*, hand punning, vibration, *pneumatic-tyred rollers*, *sheepsfoot rollers* or similar methods in layers usually not exceeding 15 cm. Cf. **sand piles**, **vibro-flotation**.

superimposed load Normally *live load*, but superimposed dead loads also exist, e.g. in bridge design, the possibility of spandrels filling with water (BS 5400).

superintendent US title for a contractor's *agent* or a *resident engineer* or resident architect.

superload [stru.] *Live load* or *superimposed load*.

superplasticizer, high-range water reducer An *admixture* in concrete which can reduce by 30% the requirement of mixing-water, compared with the maximum 15% achieved by a *water reducer*, both at normal dosages. It can produce a *flowing concrete* that does not segregate, and needs very little vibration, sometimes none. Superplasticizers and water reducers are generally *retarders* of set. Both usually entrain 2–3% of air.

There are several reasons why these admixtures should be added late in the mixing cycle. Some of them very slightly accelerate the *initial setting time*. One type, melamine-based superplasticizers, causes a 'slump loss', i.e. the concrete stiffens after 30–60 min. But the main reason is probably that superplasticized concrete is so fluid that it can easily spill.

Strengths of 40 MPa (5800 psi) in precast concrete can be achieved after 8–18 hours, and values of 100 MPa (14,500 psi) have been exceeded. Generally superplasticizers can produce impermeable, strong concrete which is workable and pumpable. But with such fluid concretes, *formwork* must be particularly solid. See **Seattle**.

superposition [stru.] The principle of superposition is a principle which simplifies structural calculations, and can be used for solving the forces in *redundant frames*. Briefly it means that the stresses in a member due to one system of loading can be added to the stresses in it due to another system of loading. The redundant frame is split up first into two or more *perfect frames* having some members in common. The forces in the perfect frames are determined by statics and those which occur in the same member are added algebraically, the algebraic sum being the total force in the member. The method is exact for symmetrical loading and only slightly inaccurate for unsymmetrical loading.

supersonic air knife Sound travels at about 340 m/s (1100 ft/s) in air. A supersonic air knife releasing an air jet travelling at twice this speed, made by Briggs Technology of Pittsburgh, helps the US gas industry with its *pipe renewal*. Used as a *microtunnelling* device it breaks up sandy soils (probably not clays) into fine particles that are removed by a suction pipe. In 1988 the knife cost £1000 and the suction tool £1500. The compressor needed is a standard unit of 3.5 m³ (125 ft³)/min output at 0.6–0.7 MPa (90–100 psi) pressure.

superstructure [stru.] The visible part of a structure; that part above the *substructure*.

supersulphated cement, metallurgical c. A *cement* made mainly from blast-furnace slag, which resists attack by sulphates even more than *sulphate-resisting cement*. It resists weak acids provided that the water/cement ratio of the concrete is 0.45 or less.

support The removal of coal or ore under a railway, canal or reservoir constitutes removal of support and may be a cause for legal action.

support moment [stru.] *Hogging moment*.

suppressed weir [hyd.] A *measuring weir* whose sides are flush with the sides of the channel. It has therefore no *end contractions*, and is thus the contrary of a *contracted weir*. The contractions may be suppressed at one or both sides, at the bottom or at all three positions.

surcharge Any load above the level of the top of a *retaining wall*. Surcharges may be temporary (live) loads such as lorries, locomotives, cranes or stacked goods; or permanent (dead) loads such as earth sloping up from the top of the wall or a building above the top of the wall. A surcharge increases very considerably the *active earth pressure* on the wall. This increase can be removed completely by a *relieving platform*. See **barge bed**.

surcharged sewer A *sewer* which in places is running full and under pressure, with its manholes filling. The resulting exfiltration followed by *infiltration* can remove fine soil around the pipe and cause it to collapse. Surcharging may flood the streets by the sewage rising and forcing the manhole covers off.

surcharged wall A retaining wall carrying a surcharge, such as an embankment, above its top.

surface-active agent A *surfactant*.

surface bonding A way of using *glassfibre-reinforced cement* (GRC) as a 'structural plaster', developed in the USA by the Glassfibre Reinforced Cement Association. A wall of blocks or bricks is laid dry, with no mortar, then wetted and plastered with GRC on both sides, either by spray machine or by trowel. The wall surfaces must be clean and must be wetted afterwards. The wall is quickly built, inexpensive and insulating. Repairs may also be done like this.

surface concrete The outermost layer of concrete is much more porous than the interior and was claimed, in research results announced in 1988, to have pores 150,000 times larger than those in the mass.

surface detention [hyd.] The thin sheet of water that covers the ground during rain. When this water reaches a stream it becomes *run-off*.

surface dressing A wearing surface consisting of a layer of chippings or gravel on a thin layer of fresh road tar or bitumen.

surface float [hyd.] A *float* on the surface of water.

surface tension The property of a fluid surface to behave as if it were covered with a tight skin (the meniscus), so that a needle or fine powder can be made to float on cold water (for example) without being wetted. The wetting power of water can be increased and its surface tension and *capillarity* reduced by mixing soap or other *surfactants* with it.

surface-tension depressant [min.] A *surfactant*.

surface vibrator Several types of *vibrator* commonly used on road or airfield concrete, such as vibrating *screeds*, plate or grid vibratory tampers, and vibratory roller screeds (ACI).

surface water The *run-off* from unpaved or paved land or buildings, as opposed to *soil* or waste water.

surface-water drain Any pipe for rainwater in the ground. It is usually of at least 10 cm dia.

surfacing In a road, the *wearing course* either alone or combined with a *base course*.

surfactant, surface-active agent A substance which emulsifies, disperses, dissolves or penetrates other substances, or makes them froth, in fact a *depressant* of *surface tension*, such as a soap or detergent or *air-entraining agent*. In waste water, all these substances are *pollutants*.

surge (1) [stru.] A horizontal force applied to a high level of a structure usually by a crane accelerating or braking. Crane surge is applied at crane rail level. It cannot exceed 0.25 of the greatest weight at each wheel, since this is the highest coefficient of friction between crane wheel and rail. See **wracking forces**.

(2) [hyd.] A change in the pressure in a pipe caused by a change in the speed of the fluid flowing in it, often caused by the closing or opening of a valve. These pressure changes modify the forces exerted by the pipe, e.g. at its bends or anchorages. See **surge pipe**, **water hammer**.

surge pipe [hyd.] An open-topped *stand pipe* for releasing surge pressures when water does not need to be saved. The water spills over the top. See **surge tank**.

surge tank [hyd.] An open tank connected to the top of a *surge pipe* to avoid loss of water during pressure *surges*. A surge tank is usually connected to the pressure pipes leading water to the turbines at a water-power station.

survey (1) A map or drawing showing, for a *topographical survey*, the layout of the ground; for a *geophysical* survey, the variations in gravity, resistivity, magnetic intensity, radioactivity, temperature; or, for a *geological map*, the outcropping strata. Drawings are increasingly being replaced by computer files which enable the survey to be viewed on a screen. Dimensions or quantities are then calculated by the computer instead of scaled from a dawing. See **soil survey**, **solid map**, **traverse**.

(2) To make a survey.

surveying, land s. [sur.] Measuring the earth's surface so as to produce a *survey*. Surveying is either *plane surveying* for small areas or *geodetic surveying* for large areas.

surveying errors See **error**, **tape corrections**.

surveyor A vague term which may mean a member of several very different professions which fall into two groups, first mining or *land surveyors*, second *quantity* or *building surveyors* (*B*). The Royal Institution of Chartered Surveyors includes all types of surveyor in its membership.

surveyor's draughter [d.o.] One who prepares maps or drawings for a *land* or mining *surveyor*. He or she needs no diploma but may be a qualified surveyor. (*Quantity surveyors* rarely prepare drawings.)

surveyor's transit See **engineer's transit**.

survey point, s. station, traverse p., change p. (**hub** USA) [sur.] A peg having a nail head flush with its surface, driven into the ground, or in a mine into the roof, carrying a *spud*. A *theodolite* centred over the nail or under

the spud measures angles, and if provided with an *EDM instrument* also distances. It may be a *traverse* point or a point in a *control network*. On a nation-wide survey network such points are marked on a brass plate set in concrete.

suspended floor [stru.] Any floor, whether of timber or concrete, not resting on the soil but spanning between end supports.

suspended-frame weir A *frame weir* in which during floods the frames are hung above water level from a bridge.

suspended solids Solids in *sewage* or *raw water* that do not settle quickly and are not dissolved. They can be removed by *sedimentation*, *filtration*, *centrifuges*, etc. In rivers, suspended solids may become *sediment*.

suspended span The middle, short, freely supported span of e.g. a *cantilever bridge*.

suspended structure, construction [stru.] Suspended structures, apart from the suspension bridge, are relatively new and of two main types. The first includes tents, canopies or awnings of textile, often of impressive appearance, difficult or impossible to calculate but splendid for exhibition purposes, and initiated by the German architect Otto Frei. Any tent has masts under pure compression, with ropes and tent material under pure tension. The other type of suspended structure is a modern multi-storey block. The columns around the edges of the building are replaced by vertical tie bars or cables that are much more slender and so do not obstruct vision. The main floor loads are carried up by the tie bars to beams cantilevered at roof level from one or more central piers or *core walls*. Because of the overhanging beams at roof level this type has been

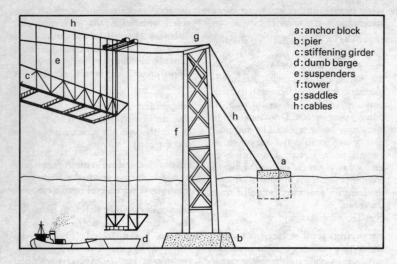

a : anchor block
b : pier
c : stiffening girder
d : dumb barge
e : suspenders
f : tower
g : saddles
h : cables

Suspension bridge.

called an umbrella structure. *Inflatable structures* are also suspended. See **tension structure**.

suspender A vertical hanger in a *suspension bridge*, by which the road is carried on the cables.

suspension bridge (see diagram) A bridge hung from ropes, chains, cables, or pinned steel or iron bars passing over towers at each bank. The cables or ropes are held by anchor blocks or solid rock behind the towers. The track or road is hung from the cables above by suspenders – rods or ropes each side carrying the weight. Suspenders are usually vertical but on the *Severn Bridge* for the first time they were made sloping so as to improve the stability of the bridge in wind. For modern bridges see **Akashi–Kaikyo Bridge**, **stiffening girder**, **Verrazano Narrows Bridge**.

Early Chinese decked suspension bridges had eight or more chains mostly below the deck, carrying it directly, perhaps with two more above the deck acting as railings. Consequently decks were curved, not flat. An ingenious rope suspension bridge built in AD 494 on the Huei River in eastern China had ten ropes which could be slackened by windlasses to submerge it. The bridge then blocked the river against large boats. It could not be burned and no enemy could cross the river. Another Chinese suspension bridge, built in AD 65, was the Lan Chin bridge over the Lantshang river in Yunnan, spanning 76 m (250 ft). Its iron chains carried a timber deck. Footbridges of rope with no deck have long been known and still exist, some of them of ropes made of spun bamboo fibres or even of growing vegetation (creepers).

suspension cable A steel-wire rope carrying a suspension bridge. Two are needed for each bridge. They are often spun on the site.

suspension-cable anchor A mass of

masonry in soft ground, or a fixing deep into rock, on the land side of a suspension bridge tower, to hold the ends of the suspension cables.

sustained yield, safe y. [hyd.] The highest annual rate of withdrawal from an *aquifer* or basin, which does not have undesirable results such as damage to the aquifer, or *encroachment* of nearby seawater. Normally the safe yield must not exceed the average annual *recharge*.

swabbing Removal of loose deposits, slimes and tiny animals from relatively smooth pipes. A plastic foam swab is driven by the water pressure along mains which may be as large as 1.37 m (4½ ft). In badly encrusted pipes the method does not work and the swab may even block the pipe. Cf. **air scouring**.

swaging A *pipe renewal* method devised by British Gas, in which the diameter of medium-density *polythene* tube is reduced by passing it through a hot die followed immediately by a chilling die to prevent re-expansion. Once in place the pipe reverts to its original diameter in 21 days, or more quickly if it is pressurized with air or nitrogen for 24 hours. 200 m lengths of previously 325 mm dia. pipe have been die-shrunk and inserted into a 305 mm bore pipe by Northwest Water Authority. *Rolldown* is slightly different.

swamp shooting See **peat blasting**.

swash height [hyd.] The height to which the water from waves reaches on a beach. For a shingle beach it is often taken as 1.7 times the height of deep-water waves.

sway [stru.] Sideways movement of a frame, particularly *sidesway*.

sway rod, s. brace [stru.] A diagonal brace to resist wind or other horizontal forces on a building.

Swedish cylindrical surface [s.m.] See **rotational slide**.

swelling pressure [s.m.] The pressure exerted by a contained clay when it absorbs water. It can amount to considerably more than the pressure of the overlying soil.

swelling soil, expansive s. [s.m.] Like a sponge, any clay shrinks when it dries and swells when it absorbs water. In the UK, subsoils dry out less than those in warm, arid countries, where subsoil expansion has caused severe troubles with crushing of cellars, lifting of foundations or cellar floors, and other damage to the building above.

But in the UK, as in warm countries, when a deep foundation is dug in clay it is usual to carry the building on piles going even deeper, which will be long enough to resist the uplift of the expanding clay if it should ever get wetter. The piles are anchored below the clay. Another reason for expansion of clays is unloading. If much load has been removed from above it by excavation, the clay can be expected to swell afterwards as it sucks in water. In such deep foundations, therefore, a gap of about 60 cm (2 ft) is commonly left below the lowest floor slab to allow the clay below it to expand upwards. Cf. **heave**; see **montmorillonite**.

SWG [mech.] *Standard wire gauge (B)*.

swing bridge, pivot b., turn b. [stru.] A *movable bridge* that swings on a vertical pivot at its centre, to allow vessels to pass.

swinger A pointed bar about 0.9 m long, used for moving *runners* in trench timbering.

swing-jib crane A crane with a horizontal *jib* and counter-jib to carry the counterweight. The jib and counter-jib pivot through 360°. The hook radius is changed by 'trolleying' in or out along the jib. See **tower crane**.

switch [elec.] A device which opens or closes an electric circuit.

switch blades [rly] *Points*.

switchgear [elec.] Plant, including switches or *circuit breakers*, for controlling or protecting a power circuit.

switch tongue [rly] See **points**.

SWL Safe working load. See **safe load indicator**.

Sydney Harbour Bridge (1932) A steel, two-hinged trussed arch bridge of 509 m clear span, erected by temporarily cantilevering the ends from each shore until they met in the middle. Cf. **Kill Van Kull Bridge**.

synclastic shell A *shell* with double curvature, e.g. a dome. Cf. **anticlastic shell**.

systematic errors, cumulative e. [sur.] Those *errors* which are either always positive or always negative, as opposed to *compensating errors*.

system de-watering *Groundwater lowering*.

T

table [stru.] The horizontal part of the 'T' of a *tee-beam*. Cf. **stalk**.

tacheometer [sur.] A *theodolite* telescope which measures distance by its *stadia hairs*.

tacheometric surveying, tacheometry See **stadia work**.

tachometer A revolution counter, an indicator of rotational speed.

tack coat A thin coat of bitumen, road tar or emulsion laid on a road to improve the adhesion of a course above it.

tack rivet A rivet which does not carry load, but is put in for convenience of construction or to comply with regulations.

tack weld A temporary weld that holds steel parts together, but is remelted during fabrication.

Tacoma Narrows Bridge A *suspension bridge* of 853 m central span and 335 m end spans which collapsed in November 1940 a few months after its completion. It failed during a stiff breeze which caused a fluttering movement of oscillation and twisting (*aerodynamic instability*), but was rebuilt in 1950 to withstand winds of 200 kph. See **vortex shedding**.

tail bay, t. gate, t. race, tailwater [hyd.] A downstream bay, lock gate, channel, etc.

tailings dam [min.] An embankment of fine refuse from mineral beneficiation built up to make a *silt pond*. Tailings dams can fail catastrophically.

tail rope [min.] A haulage rope which passes round a *return sheave*, as opposed to a direct (or main) rope.

tail seal In a shield-driven tunnel in water-bearing ground the *shield* is unusually long, to enable at least two *segmental rings* of lining to be built on the *tailskin*. The tunnel has to be sealed against the entry of water or mud from the ground, as well as against the grout pumped behind the rings. The tailskin of the shield is therefore fitted with thick rubber strips or steel brushes which press on the lining to seal off any inflow.

tailskin A *shield* rear end forming a *tail seal*, and used for building the *segmental rings*.

Tainter gate [hyd.] A *radial gate*, named after Burnham Tainter.

tally [sur.] A distinctive plastics (formerly brass) label, marking every metre on a surveyor's 20 m chain. The tallies at 5, 10 and 15 m are numbered and red, the others are yellow, of different shape and not numbered (BS 4484).

tamp To *pun*.

tamped finish, off-tamp f. The ridged surface of a concrete slab as left by the *tamper*, also suitable as a base for a *topping*.

tamper (1) A *screed* board, possibly carrying a *vibrator* to make a *tamped finish*.
(2) A rail-borne machine which uses vibrating tines to pack *ballast* beneath *sleepers* and bring the rails to the required line and level.

tamping roller, steel-wheeled r. A tractor for *compaction* of soils, having two or more large steel wheels with tapered steel projections. Although resembling the *sheepsfoot roller* (apart from the shape of the projections) it travels faster and is more suitable for soils without clay.

tandem roller A road roller having rolls of about the same diameter behind each other on the same track. Ordinarily there are two rolls but see **three-axle tandem roller**.

tangent distance The distance from an *intersection point* to a *tangent point* in the setting out of a road or railway curve.

tangent point A point on a curve at which it joins a tangent or changes its curvature (in the setting out of roads or railways).

tangent screw [sur.] A fine-adjustment screw which moves the line of sight a short distance. One is usually fitted on both the horizontal and vertical circles of a *theodolite*.

tank farm Several oil tanks with their pipes, pumps and controls.

tank sewer See **detention tank**.

tank sprayer A *pressure tank* on wheels.

tape [sur.] A ribbon of steel, *invar* or fibreglass coated with pvc, etc., graduated usually in metres and millimetres and coiled in a steel or plastic case when not in use. Used by surveyors, engineers and builders, the most accurate tapes are of invar, but the steel tape is accurate enough for nearly all purposes in engineering. See **band chain**, **tape corrections**.

tape corrections [sur.] The following corrections are applied as a matter of course to all lengths measured accurately by steel or invar tape: *slope, temperature, sag, standardization, gravity* and *sea-level corrections*. Standard tension for steel or invar tapes is usually 5 kg at 20°C.

tapered-flange beam A *rolled-steel joist* in which the two surfaces of the flanges are not parallel. The inner surfaces are at 95° to the web, not at 90°. Cf. **universal stanchion**.

tapered washer A *bevelled washer*.

taper file [mech.] A triangular *file* with fine teeth for sharpening saws.

tare The weight of an empty lorry, wagon or other container.

tar emulsion See **bituminous emulsion**.

target, traverse t. [sur.] An object centred over a traverse point or other *survey point* which can be easily and accurately sighted by a *theodolite*. The special targets in use in *three-tripod traversing* often have their own built-in bubble tube and *optical plummet*, enabling the theodolite to use the same *set-up*.

target cost contract A type of *cost reimbursement contract* in which a target cost is estimated beforehand. On completion the difference between the target cost and the actual cost is shared in proportions agreed between *client* and *contractor*.

target rod, t. staff [sur.] A *levelling staff* with a sliding target which is raised or lowered by the *staff man* until the instrument man signals to him that it is on his line of sight. The staff man then reads and notes the staff reading. A *laser sensor* is conveniently used in this way. Cf. **self-reading staff**.

tarmacadam, coated macadam A road material consisting of stone coated with tar or a tar–bitumen mixture. It has very little fine aggregate and a high proportion of voids.

tarmacadam plant A plant for making *tarmacadam*.

TARP The Tunnel and Reservoir Program of Chicago, begun about 1979, was more than half completed by 1990. Its 168 km of tunnels, 90–100 m underground, of from 3 to 11 m dia., with capacities of up to 1 million m³ each, act as *stormwater tanks*. Explosives are forbidden, so many *fullface tunnelling machines* have been used in TARP. MSD (Metropolitan District of Greater Chicago), which operates TARP, was created as the Sanitary District of Greater Chicago in 1889 but changed its name to MSD in 1955. It has always been known for bold, determined action. Early in this century it even reversed the flow of the Chicago River, which then was flowing into Lake Michigan with much of the *sewage* effluent from the city. The Sanitary District cut three trenches through the drainage watershed of the river and made it flow into the Mississippi River instead of into the lake. Milwaukee's similar project has to use ground *freezing*.

tar paving Tarmacadam surfacing laid in one or two courses for very light traffic or playgrounds.

taut-line cableway A *cableway* with the same general construction as the *slack-line cableway* but with a taut suspension rope. It may have both towers stationary, both mobile on parallel tracks, or one stationary and the other travelling radially.

technician engineer (since 1988 **incorporated e.**) A person registered as an engineer by the *Engineering Council* above the level of *engineering technician* but below that of *chartered engineer*. He or she is normally an associate member of an *engineering institution*.

tectonic plate See **earthquake**.

tee-beam (1) **tee-iron**, **tee-section**

A *rolled-steel section* shaped like a T, composed of a *stalk* and a *table*.

(2) Part of a reinforced-concrete floor in which the beam projects below the slab. In *hollow-tile floors* the stalk cannot be seen, since the soffit of the hollow tile coincides with the soffit of the tee-beam.

tee-iron, tee-section A *tee-beam*.

telecontrolled power station [elec.] A hydroelectric power station, e.g. in the Alps, which is not manned but wholly controlled by radio. See **remote control**.

telemetry Sending data by telephone or radio. The UK *water service plcs* are expected to spend £100 million on telemetry before 1992. In 1987 a single scheme for the Yorkshire water authority cost £20 million. A different sort of telemetry devised at Queen's University, Belfast, comes from a 24 kg radar-operated 'sentry' with a 12 V car battery, emitting microwaves at a frequency of about 10 GHz. On detecting intruders it radios its master receiver, whch actuates an alarm. See also **mains-borne signalling**.

telfer, telpher, monorail An electric hoist hanging from a wheeled cab moving on a single overhead rail, occasionally from a steel rope. It is used in factories, hung from roof girders and over dams being built. An overhead gantry may be built to carry the rail. The difference between an *aerial ropeway* and a rope-borne telfer is that telfers are driven by a motor in the cab, ropeways are pulled by a rope driven by a stationary engine.

teller [sur.] A *tally*.

telltale [stru.] A piece of thin glass or other indicator, cemented firmly across a crack in a structure that is settling, to indicate by its cracking whether the crack continues to widen. The date is marked on it.

Tellurometer [sur.] *EDM* instruments made by Tellurometer Ltd, which use very high frequency radio. They are used in identical pairs, a 'master' and a 'remote', both active in distance measurement simultaneously. Other instruments, which use infra-red light, are lighter and more compact, and may be mounted on a *theodolite*. They use a prism at the remote station to reflect the light.

telpher See **telfer**.

temper [mech.] (1) To temper steel which has been hardened by *quenching* is to reheat it to a temperature below the *critical* and then to cool it. The *hardness* of the steel is thus reduced (tempered) by an amount which varies with the temperature from which it is cooled. The higher the tempering temperature, the softer is the steel. See **hardenability**.

(2) To toughen non-ferrous metal by *annealing* and sometimes rolling.

(3) The state of hardness of metal. Thus a metal at low temper is relatively soft and *ductile*. A metal at high temper is hard and brittle.

temperature coefficient The coefficient of change of electrical resistance per °C. A positive temperature coefficient (usual for metals) is one which raises the electrical resistance as the metal heats up. With a negative temperature coefficient, the resistance falls as the material (usually a *semiconductor*) heats up. See **resistance temperature detector**, **thermistor**.

temperature correction [sur.] The temperature correction of a steel tape used at a temperature different from its *standardization temperature* is equal to 0.000011 m per metre of measured length, L metres, per °C of difference. This amount, in millimetres, 0.011 × L × T°C, must be deducted if the temperature is below the standardization temperature, added if it is above the standardization temperature. See **invar**, **tape corrections**.

temperature gradient A change in temperature per unit length. In building, the temperature gradient through an outside wall is carefully studied before the position for the *vapour barrier* (*B*) is decided.

temperature-matched curing Calculation of *concrete curing* times so as to determine minimum times for striking *formwork* or for the stressing up of *prestressed concrete*. It is discussed in BS draft for development 92:1984, and in British Patent 1300099. Cf. **maturity of concrete**.

temperature sensor A *sensor* for measuring temperature, usually with electrical output. Only a few types are mentioned in this book: *resistance temperature detectors*, *semiconductor sensors*, *pyrometers*, *thermistors*, *thermocouples*.

temperature steel [stru.] Reinforcement which is inserted in a slab or other concrete member to prevent cracks due to *shrinkage* or *temperature stresses* from becoming too large. It generally amounts to a minimum of about 0.1% of the cross-section in any direction, the requirement for a slab, which is two-dimensional, being therefore about 0.2% altogether. If, as usual, more steel than this is inserted to take bending or other stress, this requirement is disregarded.

temperature stress [stru.] A stress caused by temperature change. If the expansion or contraction due to temperature is restrained, the member concerned suffers compression during rising temperature or tension during falling temperature. See **strain**.

tempering [mech.] See **temper**.

template See (*B*).

temporary adjustment [sur.] An *adjustment* to a surveying instrument made at each *set-up*, such as levelling or centring. Cf. **permanent adjustment**.

temporary way (**construction way** USA) The temporary rail track used for building the *permanent way*.

tenacity [mech.] *Tensile strength*.

tender (**bid** USA) A written statement by a *contractor* of the price or rates for which he proposes to do certain work.

tendering The *consultant*'s work of preparing drawings and a *bill of quantities* and the contractor's work of pricing the *items* of the bill. Tendering may be open, meaning it is advertised and any contractor who wishes may tender, or selective, when only a few contractors with a reputation for the type of work to be done are invited to tender.

tender total The total of the *priced bill*. The *consultants* and the *client* compare the bills from the contractors and choose either the lowest *tender* or the most reputable contractor. Cf. **contract sum**.

tendon [stru.] A *prestressing* bar, *cable*, rope, *strand* or wire. See **Polystal**.

tensile force [stru.] See **tension**.

tensile strength [mech.] A loose but convenient term often used as an abbreviation for *ultimate tensile stress*. It is much higher than the greatest safe *stress*. See **tension**, **working stress**.

tensile strength of concrete See **concrete tensile strength**.

tensile test [mech.] A test in which a piece of standard shape, made of metal, mortar, concrete, wood, etc., is pulled in a testing machine until it breaks. Often the load is recorded for every measured extension so that a graph called the *load/extension curve* of the test can be drawn. See **beam test**, **necking**.

tension [stru.] A pulling force or *stress*. Metals and wood take tension well, but masonry, including concrete, is generally not allowed to take any tension except in the dispersal of concentrated loads. In this case bricks with good adhesion (not pressed bricks), *frogs* (*B*) and joints filled solid may be allowed a tension of 0.1 MPa (14 psi). Unreinforced concrete in foundations may be allowed 0.3–0.7 MPa (40–100 pi). See **proof stress**, **ultimate tensile strength**, **yield point**.

tension carriage [mech.] A frame carrying a pulley round which an endless belt or rope passes to be tensioned. The frame can move, e.g. for tensioning the traction rope of an *aerial ropeway*. See **conveyor loop**.

tension correction [sur.] The *tape correction* to be applied to a tape used at a tension which is not that at which it was standardized. If the difference in tensions is *p*, *l* the measured length, *a* the cross-sectional area of the tape and *E* its *modulus of elasticity* the correction is $\dfrac{l \times p}{a \times E}$. This must be added to *l* when the standardization pull is lower, and subtracted from *l* when the standardization pull is higher than that at which *l* is measured.

tension drum, return d., r. pulley

tension flange

(see diagram) At the far end of a *conveyor belt* from the drive motor, the return drum pulls (tensions) the belt so as to increase the friction on the drive drum between the belt and the drum at the other end. Without friction the belt will slip, possibly causing a fire, and in any case will not move. If the space at the tension drum is congested, tensioning at mid-length is possible. The method shown in the diagram can be developed to make a *conveyor loop* take-up.

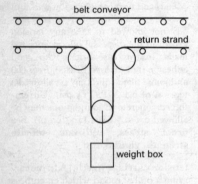

Tensioning of belt conveyor in the event of congestion at the return drum.

tension flange [stru.] The side of a beam which is in tension, usually the lower side, particularly in the middle of a beam on two supports. It is the upper side of a cantilever beam.

tension-leg platform A floating *offshore platform* (see diagram, p. 295) which is held to the seabed by vertical 'tethers'. Conoco's Hutton platform in the North Sea is in 147 m of water and anchored by 16 tubes, four at each corner.

tension pier (see diagram) In very deep water where bridge piers would be too expensive, a tension pier might be feasible.

tension pile [stru.] An *anchor pile*.

tension sleeve [stru.] A *screw shackle*.

tension structure [stru.] Possibly the only structures that involve no stresses other than tension are pressurized *inflatable structures* (*air houses* (*B*)), often called balloon structures, or radomes. All contain a fluid, air or water. *Fabridams* contain water. All stresses are tensile and are carried on the 'balloon skin'. Ordinary structures with more than the usual amount of tension are often called *suspended structures*.

tensor, hydraulic t., hydrostressor Equipment used in laying *continuously welded rails* that pulls a rail at about 70 tons maximum pull, so as to bring it (in cold weather in the UK) to the length that it would have at about 27°C. The pull needed to lengthen a rail weighing 54 kg/m is about 1.6 tonnes per °C.

terminal (1) The shore end of an undersea pipeline.
(2) A telephone link to a *mainframe computer* can enable a microcomputer to send or receive its data and thus to act as its terminal. To do this the micro needs both a suitable *modem* and a *communications program*.
(3) A *VDU*.

terminal velocity, free-falling v., settling v. [s.m.] The maximum velocity which a body can attain when falling freely in a fluid (usually air or water). It can be determined for spherical particles by *Stokes's law*.

terotechnology [mech.] The design, installation, commissioning, operation, maintenance and replacement of machinery, plant or equipment.

terracing See **contour ploughing**.

terrazzo See (*B*).

tertiary treatment, advanced t., polishing Any *sewage treatment* that

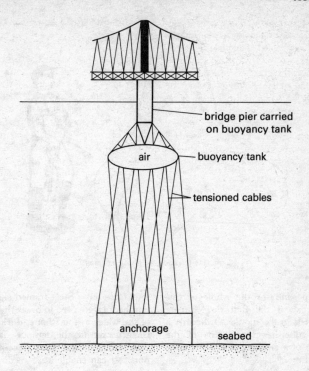

bridge pier carried on buoyancy tank

buoyancy tank

air

tensioned cables

anchorage

seabed

Proposed tension pier for bridge in deep water. (J. G. A. Croll, Oleg Kerensky Conference on Tension Structures, London, 1988.)

immediately follows *secondary treatment*. A typical one would involve *microstrainers* or *land treatment*. It should produce an *effluent* with less than 30 mg of *suspended solids* per litre and less than 20 mg/l of *biochemical oxygen demand*.

Terzaghi [s.m.] The founder of modern *soil mechanics*. See **consolidation, graded filter**.

test core A solid cylinder cut from rock or hard concrete by *diamond drilling*. Standard cores are drilled to either 100 or 150 mm dia., aiming at a length of twice the diameter, but cores never break off to the precise length. Test

cores provide proof of the concrete's *compressive strength*, its density, the cover to the steel, etc. Non-standard cores as small as 50 mm are also useful and much cheaper and quicker to drill. The most valuable information provided by any core is obtained by examining it. This is one reliable way of identifying the damaging alkali–silica gel which weakens mature concrete in the *alkali–aggregate reaction*. See **splitting test**.

test cube A 15 cm or 10 cm cube of concrete used in Britain to show the concrete strength. It is crushed usually at an age of seven or 28 days. It is kept

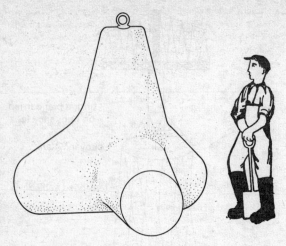

Concrete tetrapod weighing 15 tons.

in damp sand for the whole of this period except the first day, during which it is in the mould. Ordinarily a 7 day cube has two thirds of the strength of a 28 day cube. See **cylinder test**.

testing See **type testing, destructive tests, non-destructive testing, semi-destructive testing**.

testing machine [mech.] A machine used for loading test pieces or structural members to determine their deformations at given loads and their breaking strengths. The commonest machines are for tension, compression, *impact* and *fatigue*.

test piece [mech.] A piece of mortar moulded to shape or a piece of metal or wood turned (or cast) or cut to shape, for testing in a testing machine. See **coupon**.

test pit Term used in the USA for *trial pit*.

tetrahedron A solid figure having

four faces only. Steel-framed skeleton tetrahedra were used to break the force of the water in the Rhône, during the construction of the famous dam at Génissiat, France, 1945. See **tetrapod**.

tetrapod (see diagram) An equiangular figure of the same general type as a tripod, with a fourth leg rising from the intersection of the other three. Since it has four legs, one leg is always vertical when the other three are on a flat base and it is extremely stable. In *breakwaters* (Casablanca) large reinforced-concrete tetrapods weighing about 15 tons were (1950) piled up to break the force of the waves and reduce their scour. See **tetrahedron**.

Thaulow drop table A test unit which estimates the flowability of very stiff concretes, comparable with the *V.-B* and *compacting-factor tests*.

thefts from sites Thefts from sites have been reduced by 34% by stamping all stealable objects with numbers and recording them on a computer database. Another precaution is to paint everything with a unique-coloured paint. Second-hand plant should not be bought unless the maker's certificate of origin is provided – a fairly sure indication the unit is not stolen. 90% of thefts take place in the 60 hours between 6 pm on Friday and 6 am on Monday.

theodolite [sur.] (**transit** USA) A *land surveyor*'s instrument for measuring horizontal and usually vertical angles. Its telescope rotates on a horizontal axis, the *trunnion axis*, carried on a forked standard fixed to the circular upper graduated plate. The lower circular plate or limb is carried on the *tribrach* which has three levelling screws and a device to centre the instrument over the *survey point*. A vertical graduated circle is usually also provided, though in the first linkage between the French and English triangulations in 1785 the theodolite used had no vertical circle. The horizontal circle, however, was large, of 36 in. dia. (90 cm). See **electronic distance measurement, optical plummet, totalstation instrument**.

theory of structures The study of structures and their stability and strength. It does not properly include *strength of materials*, but in this book 'stru.' covers these subjects and *structural design*.

thermal boring *Thermic boring*.

thermal cracking Cracking of a concrete mass as it cools after *hydration*.

thermal mapping See **infrared thermography**.

thermal runaway See **semiconductors**.

thermel [elec.] Any instrument which includes a *thermocouple*.

thermic boring, thermal b., oxythermic lance A method of boring holes into concrete by the high temperature of a burning steel tube (lance) packed with steel wool through which an oxy-acetylene or similar fuel gas mixture is passed to ignite the end of the lance. After ignition, oxygen alone is fed, to keep it burning. If concrete is heavily reinforced it can be cut more easily than if unreinforced, since the reinforcement burns in the oxygen jet. See **jet drilling**.

thermistor A *temperature sensor* based on a semiconductor bead or rod, usually with a range from -30 to $+200°C$. It responds more quickly than the *resistance temperature detector*, being smaller. But its variation is not *linear*, though a microcomputer, if connected to it, can correct for nonlinearity.

thermit, thermite A mixture of aluminium powder and iron oxide used in the *alumino-thermic* welding of steel rails and other steel parts. A temperature above the melting point of steel is generated, so that the parts to be welded must be enclosed in a mould of clay or other *refractory* material. A generally troublefree electrical and mechanical bond is thus formed. See **ignition powder**.

thermocouple Two wires of different metals connected at both ends, which are maintained at different temperatures, one hot the other cold, will generate a voltage and a small current will flow. One junction is the measurement junction, the other is the remote junction. The voltage generated depends only on the two junction temperatures, not on any intermediate temperatures. Many *pyrometers* depend on thermocouples. Their measurement junction is usually enclosed in a protective probe. See **thermel**.

thermo-osmosis [hyd.] Water and

thermoplastics

water vapour move, other things being equal, in the direction of heat flow, i.e. towards cooler ground, because this is also the direction of increasing surface tension. Thermo-osmosis is important in icy regions because it can cause accumulations of ice underground. Cf. **electro-osmosis**.

thermoplastics Plastics which soften when heated. They include a large proportion of the plastics in use – most if not all of the plastic pipes: ABS (acrylonitrile butadiene styrene), acetal, acrylics, cellulosics, nylon, polybutylene, *polycarbonate, polypropylene, polythene, PVC (polyvinyl chloride)*, polystyrene, and polytetrafluorethylene or *ptfe (B)*. Cf. **thermosets**.

thermosets *Plastics (B)* that are probably less widespread in industry than the *thermoplastics*, though the following resins are structurally important: alkyd, *epoxy*, furane, melamine formaldehyde, *phenolic formaldehyde, polyester, polyurethane* and urea formaldehyde. Unlike thermoplastics they do not soften on heating. See **fibre-reinforced plastics**.

thickening agent, suspension a. An *admixture* which should reduce or eliminate *segregation* in concrete by improving its *cohesiveness*. It also reduces *bleeding* and is especially useful for concrete that must be pumped. Cellulose ethers and polythene oxide are typical thickeners. They may also be used in *glassfibre-reinforced concrete* to hold the sand in suspension.

thickening-time test A test of cement used by oilwell engineers to find out for how long it may be possible to pump a hot *grout* made of the cement. The grout is held in a rotating cup containing a fixed paddle connected to a *torque* meter. Surrounded by hot oil or water, the cup rotates until the torque reaches a forecast value beyond which pumping is thought to be impracticable.

thimble [mech.] A pear-shaped ring formed of steel sheet built into the end of a *fibre* or steel rope by splicing or doubling the rope on itself to form a convenient loop. Often used with *bulldog grips*. See **reeving thimble**.

thin section [min.] A slice of rock cut to about 0.02 mm thickness by grinding and polishing, for examination under the *petrographic microscope*. The rock is fixed to a glass slide by Canada balsam or other glue.

thin shell [stru.] See **shell**.

thin surfacing A *bituminous carpet*.

third-order, tertiary triangulation, trilateration [sur.] A *triangulation* or *trilateration* with sides from 1 to 10 km long. See **second-order**.

thixotropy A property first intentionally developed with oilwell *drilling fluids* in the USA early this century. They were made of *bentonite* slurry so that they would hold the chippings from the drilling bit which would otherwise lock the drill pipe and prevent drilling. When a thixotropic fluid is allowed to stand it gels and holds even large rock chips. Such fluids are used in *slurry trenches*, etc.

three-axle tandem roller A *tandem roller* with three rolls in sequence covering the same ground.

three-dimensional vision See **stereoscopic vision**.

three-hinged arch, three-pinned a. An arch which is hinged at each support and at the crown. It has the advantage that each half can sink relatively to the other without damaging the arch. For this reason it is often used where *differential settlement* is likely. It is statically determinate.

three-legged derrick See **derrick** (4).

three-leg sling A *sling* made of three chains or ropes with a hook at the end of each. The chains are hung from one *thimble* or ring.

three-light display A *laser sensor*, attached to a unit of earthmoving plant (bulldozer, grader, etc.) so as to show the driver whether he or she is high, low or dead right. A green light shows that he or she is dead right. Above it is a yellow light which means too high, and below it another yellow light which indicates too low.

three-point problem [sur.] (1) The problem in *plane-tabling* of locating the *set-up* on the plan when only three points on the plan can be seen from the set-up. See **resection**.
(2) Determining the dip and strike of a seam or vein or fault when only the elevation and position of three points on it are known. (These are obtained by cores from boreholes, usually with a core barrel which is not *oriented*.)

three-tripod traversing [sur.] *Traversing* with a *theodolite* that can be detached from its *tribrach* without disturbing the tribrach centring. Three similar tripods are in use and each carries in turn the theodolite, the *target*, and sometimes a *subtense bar* or a staff for distance measurement. As soon as the observer has finished taking readings at one point, he or she unclamps the theodolite from the tribrach and fixes the sighting target on to it. He then moves on to the *foresight* station and clamps on the theodolite while his assistant brings up the rear tripod (his previous *backsight*) and sets it up at the next foresight station. Much of the labour of set-ups is thus avoided. The tripods may each have their own optical plummet and centring head. See **electronic distance measurement**.

throat In a fillet weld, the greatest thickness of weld metal.

through bridge A bridge in which the lower *chord* carries the deck and traffic. Cf. **deck bridge**.

thrust [stru.] (1) A horizontal force; particularly the horizontal force exerted by retained earth.
(2) An inclined force such as that from a *raking shore* (*B*), or the total force in an arch.

thrust bearing See **end thrust**.

thrust block A heavy concrete *anchorage* cast to prevent movement at a bend or expansion joint in a pipe.

thrust boring *Pipe jacking*.

thrust pit, launch p. access shaft A working-shaft at the beginning of a *pipe-jacking* or *microtunnelling* job, from which the pipes are inserted and the excavated muck is hoisted. Cf. **reception pit**.

thrust ring A steel ring bearing on a concrete pipe to spread the *pipe-jacking* load uniformly round it, with wooden slips between it and the concrete to prevent crushing of the concrete.

thrust wall A wall at the rear of the *thrust pit*, spreading the thrust from the pipe jacks. It is usually reinforced concrete.

tidal currents Currents caused by the tides around the UK are highest in straits such as the Strait of Dover and those between the mainland and Ireland or the Hebrides, where they may reach a maximum of 1.75 m/s (6 ft/s) at spring tides.

tidal dock A dock which has no gates. The water inside is therefore always at the level of the water outside it.

tidal lag [s.m.] The delay between high tide (or low tide) in an estuary and the highest (or lowest) resulting level of the neighbouring groundwater.

tidal power Electrical generation by low-head turbine-pumps or pump-turbines. It has been practised at the estuary of the River Rance near St-Malo since 1968 with 24 pump-turbines of 10 MW each, generating a nett annual output of 540 GW-h, mostly on the ebb tide, usually for about two hours in 12. *Pumped storage* is used.

tidal range The greatest difference between levels of high and low tides when the tides are most extreme. It varies from 1.3 m in the middle of the North Sea (and probably other seas) but rises to 11 m (36 ft) in the funnel-shaped inlets of the Bristol Channel and Brittany. See **storm surge, tidal power**.

tide gauge, t. predictor An instrument with which the tides in any known set of channels can be predicted.

tie (1) [stru.] A member carrying tension.

(2) A *link* around the *main steel* in a concrete column, beam or pile.

(3) [rly] US term for a railway *sleeper*.

tieback An *anchorage* or the tie rod connected to it.

tied retaining wall [stru.] A *retaining wall* anchored to a *dead man*; the contrary of a *cantilever wall*.

tie line [sur.] A line which joins opposite corners of a four-sided figure and thus enables its sides and angles to be checked by *triangulation*.

tie rod [stru.] A *tie*, generally a steel rod, often threaded.

TIG welding, tungsten-electrode inert-gas shielded-arc w. (gas-tungsten arc (GTA) w. USA) Welding of most commercial metals with a non-consumable tungsten electrode and a shield of argon or other gas provided through a hose, either pure or mixed with others. The shielding gas and the filler rod are varied to suit the metal being welded. For *automatic welding* the torch is usually water cooled. Manual welding is also possible.

tight-fitting bolt See **turned bolt**.

till, tillite [min.] Mainly US terms for *boulder clay*.

tilt [air sur.] The angle between the vertical and the optical axis of a camera. It is rarely more than 3° and can in practice be kept within 1°, which is usually satisfactory in *vertical photographs*.

tilting gate [hyd.] A *crest gate* for dam spillways, so designed that the water pressure opens it at a certain level. It closes automatically when the water level has dropped to normal.

tilting level [sur.] A *level* with a *bubble* fitted on the telescope so that the axis of rotation does not need to be vertical. The telescope must, however, be levelled at each sight.

tilting mixer A *concrete mixer* which discharges its contents by tilting the rotating drum. It is a common small *batch mixer*, standardized in Britain at sizes from 200 litres upwards.

tilt-up construction, tilt slab Precasting walls horizontally, lying on top of each other separated by building paper or plastic film. When the concrete is mature the wall units can be lifted by crane and tilted up into position. One saving is the cost of wall forms.

timbering The support of the ground in excavations whether with wood, steel, concrete or light-alloy units, but usually with wood.

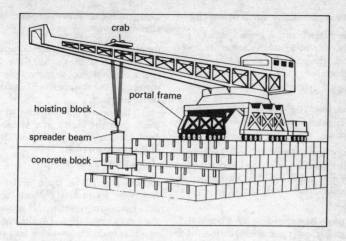

crab

portal frame

hoisting block

spreader beam

concrete block

Titan crane setting 30-ton blocks in breakwater of coursed blockwork.

time and motion study See **motion study**.

timekeeper A clerk who calculates workers' wages from their hours worked, or from their tonnage or other output.

timing [air sur.] The time interval between successive exposures of an air camera determines the distance between them, the *air base*, which varies with the aircraft speed and height. The interval is calculated to give about 60% lap between successive *vertical photographs* so that all points are generally photographed twice and some three times. The size of photograph, focal length of camera, speed and altitude of aeroplane all enter into the timing calculation. See **photogrammetry**.

tine (1) A prong of a harrow or rake.
(2) An excavating point or tooth in the mouth of a *dragline* bucket, *excavator* bucket, *grab*, *scraper loader*, etc.

tip grade US term for the *toe line* of piles.

tipping lorry A lorry which can discharge its contents backwards, or occasionally sideways.

tipping wagon, jubilee w. A small wagon on narrow-gauge track, pivoted for side-tipping or end-tipping.

Titan crane (see diagram) A *swing-jib crane* usually of at least 50 tons' capacity, and able to command a distance outside its legs, unlike the *Goliath crane*. It is used for building *blockwork* piers and for ship-building.

TML *Transmanche Link.*

TNT [min.] *Trinitrotoluene*, a powerful *explosive*.

to-and-fro ropeway A *jig back.*

toe (1) The part of the base of a dam or *retaining wall* which is on its free side, away from the retained material. Cf. **heel**.
(2) [min.] The toe of a blasting hole is the inmost end of it, where the explo-

sive is placed. See **burden**, also (*B*).

toe filter [s.m.] A *graded filter* on the free side of an earth dam at its lower end, designed to protect it against *piping*.

toe level A *toe line*.

toe line (tip grade USA), **t. level** The level to which the feet of piles are driven.

toggle mechanism [mech.] A mechanism used for applying heavy pressure from a small available force, used in the *jaw breaker* and many other applications.

Tokyo Bay crossing Two bored tunnels 4.9 km (3 miles) long connected to two *artificial islands* and to a 4.4 km (2.75 mile) bridge, due for completion in the year 2000, to cost rather more than the *Channel Tunnel*. The tunnels, of 13.9 m (49 ft) outside dia. are to be driven in poor silt and sandy clay with their crowns 15 m (49 ft) below the seabed. The Kawasaki artificial island of 200 m dia. will carry the tunnel ventilation machinery, and is at the tunnel midpoint. The other artificial island, Kisarazu, provides the way out from the tunnels to the bridge. See also **Kansai international airport**.

ton A unit of weight. In Britain the ton is now usually 1000 kg, the metric tonne, but was formerly 2240 lb (the long ton). In the USA more usually 2000 lb (the short ton). The long ton is equivalent to 1016 kg, i.e. 1.016 tonne. See **hundredweight**.

tool See (*B*).

top frame, ground f. A *frame* in timbering, set at or just below ground level to guide the first *setting* of *runners*.

topographical surveying [sur.] The work of the *land surveyor*, i.e. measuring or mapping land.

topographical surveyor See **land surveyor**.

topping (1) A layer of mortar, usually not less than 50 mm (2 in.) thick, laid over a concrete floor to make a smooth surface for the floor finish. See **monolithic topping**.

(2) The concrete, also at least 50 mm thick, over the blocks in a *hollow-tile floor* forming the compression flange of the tee-beam ribs.

(3) A *wearing course*.

topsoil [s.m.] The layer of the soil which by its humus content (up to 20%) supports vegetation. It is usually the top 15 cm of the soil in Britain, but in wet tropical climates may be 1 m or more deep. It is the upper part of the *A-horizon*, occasionally ending at the *B-horizon*.

torpedo [min.] A paper or cardboard tube containing explosive, used after *chambering* a hole which cannot be cooled sufficiently to insert unprotected explosive safely. The torpedo has 1 cm thick walls, may be of 12 cm dia., and is plugged at the foot by a wooden plug to which a line is attached, passing through the torpedo and carrying its weight. The torpedo can thus be quickly and safely lowered into the hole.

torque, torsion, twist [mech.] The twisting effect of a force on a shaft applied tangentially, like the twist on a haulage drum which winds rope on to its circumference.

torque converter On heavy equipment, a hydraulic coupling that makes use of slippage to multiply torque. This improvement on the (cheaper) mechanical clutch demands a less skilled driver. It also eliminates shocks to the machine from unskilled gear-changing and provides full torque all the time, keeping the engine working at full power, with its speed varying only about 50 rpm. Cf. **hydrostatic drive**.

torque wrench See **impact spanner**.

Torshear bolt A development (1960) of the *high-strength friction-grip bolt*, which makes correct tensioning of the bolt certain and inspection easy. The bolt thread is longer than usual, and the extension to the thread is separated from the main part of the thread by a groove cut round it to a precisely controlled depth. A special compressed-air-driven *impact spanner* grips the extension to the thread and rotates it, simultaneously holding the nut against rotation. (Access is thus only needed on one side, and one person can always tighten any bolt alone.) When the torque applied by the tool exceeds the strength at the groove, the bolt shears, and the tension is correct.

torsion *Torque*.

tortuous flow *Turbulent flow*.

total pressure [s.m.] The pressure on a horizontal plane in a mass of soil due to the weight of the material above it plus any applied loads. See **effective pressure**.

total-station instrument, electronic tacheometer [sur.] A *theodolite* with digital read-out which automatically averages the two readings of both the horizontal and vertical circles. It includes an *electronic distance measurement* device, a controlling *microcomputer*, and a *data logger* which makes manual booking obsolete.

toughness [mech.] The resistance of a material to repeated bending and twisting, measured by the amount of work in kilojoules needed to break it in an *impact test*. Toughness implies tensile strength and ductility.

toughness index [s.m.] The ratio

$$\frac{index\ of\ plasticity}{flow\ index}$$

tough way Quarrymen's word for a direction in rock which is opposite to the *rift*, and along which the rock is tough to break.

tower crane A *swing jib* or other crane mounted on top of a tower, usually so placed to command a congested city site where it occupies a small space. Tower cranes are much used in London where some contractors have found it worth while to incorporate the steel frame of the tower in the building so as to enable them to keep the crane in the same position throughout construction (possibly in part for its advertisement value). Occasionally the feet of the tower are placed on wheels which travel on rails.

tower excavator A *slack-line excavator*.

tower gantry See **derrick tower gantry**.

towers [stru.] Possibly the world's highest tower is the Warsaw radio and TV mast, 645 m (2117 ft) high (1974), of tubular galvanized steel, stayed with 15 guy ropes. But the highest free-standing tower may be the 625 m high Toronto radio-TV mast of prestressed concrete, with prestressing cables 450 m long made by Losinger Systems. Slightly shorter, at 537 m (1762 ft), is the Ostankino TV tower at Moscow. Its designer calculated that its top would sway about 1.4 m (4.6 ft) in wind.

Several dozen American guyed TV towers are 610 m (2000 ft) high. This is the greatest height allowed by the Federal Communications Commission. One of these, near Kirksville, north-east Missouri, only eight months old and stayed with nine guy ropes at three different levels, collapsed in June 1988, killing three steel erectors repairing it. This lattice tower had faults in the heat-affected zone of the welds of 1130 diagonal braces, of which the steel erectors had replaced 800. It was not heavily loaded at the time.

The Eiffel Tower (1889) was originally 300 m high but its antennae raise it to 319.5 m (1050 ft). It was built of wrought iron (puddled iron). See also **guyed tower**.

township [sur.] In the USA, the 36 sq. miles enclosed between adjacent range lines (which run north–south 6 miles apart) and adjacent township lines (which run east–west 6 miles apart).

tracer (1) To detect a flow of water or determine its source or the origin of pollution, an easily detected 'tracer' is injected, e.g. a fluorescent dye. But in a dirty effluent this is not easily seen. So other methods include lithium salts, detectable at extreme dilution using flame photometry. Bacterial counts also are detectable at high dilutions but the analysis may have to wait 24 hours for the bacterial culture. Radioactive tracers are less favoured because of hazards to staff and to drinking-water. A rubber ball containing a radio bleeper can show the flow in a buried sewer.

(2) Bright-coloured plastic netting laid over buried cables or pipes as a warning to excavator drivers.

tracing cloth [d.o.] See **tracing linen**.

tracing linen, t. paper [d.o.] Transparent linen or paper for making drawings or tracings from which *contact prints* can be taken. See **drawing paper**.

track [rly] The rails, sleepers and bal-

lasted formation, which carry a train or crane, or the single rail of a *telfer* or monorail.

track cable A steel-wire rope, often *locked-coil rope*, on which the wheels of the carriers of a *cableway* or *aerial ropeway* travel. In a *monocable ropeway* the track cable is also the traction rope.

tracking Ruts in a road surface, caused by vehicles following in each other's tracks.

track-laying tractor [mech.] A *tractor* which travels entirely on crawler tracks. Cf. **half-track/wheeled tractor.**

track man [rly] A maintenance worker on rail track, who supersedes the former lengthmen and platelayers.

track mat To enable vehicles to travel easily over mud, a series of hinged steel platforms or a *corduroy road* may be laid on the ground.

track spike [rly] A square-section, heavy steel nail driven into a wooden sleeper to hold a flat-bottomed rail, usually through a baseplate.

track stringer A timber about 10 × 25 cm laid under each rail instead of using *sleepers* where the ground is soft. See **stringer**.

traction rope The haulage rope of an *aerial ropeway* or cableway, as opposed to its *track cable*.

tractive force [rly] The amount of pull available at the drawbar of a tractor, locomotive, etc. It is equal to the locomotive weight × the coefficient of friction on level track. The coefficient of friction is a maximum of 0.25 on sanded rails, usually much less. Therefore the maximum pull is 0.25 × the weight of the locomotive. On upgrades the tractive force is reduced by the loco weight × the gradient, on downgrades

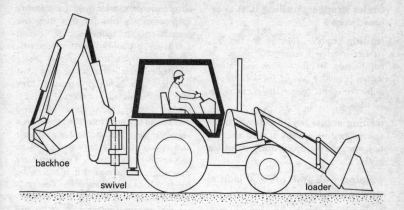

Tractor-loader-backhoe or backhoe-loader, etc. (JCB Sales Ltd, Rocester, Staffs.)

the tractive force is increased by the same amount.

tractive resistance, coefficient of traction [rly] The frictional resistance to motion per ton hauled. It is the ratio of the *tractive force* to the weight of the train. At starting it is 2–3 times the running resistance. Running resistances on level track vary from 1.8 kg/ton with roller bearings on clean smooth track to 23 kg/ton with plain open bearings on bad track. It is made up of rolling resistance between rails and wheels, bearing friction, air resistance and additional resistance due to curves and slopes.

tractor [mech.] A self-propelled vehicle which may be *track-laying*, *half-track* or *wheeled*, generally used for towing a *bowl scraper*, *rooter*, *grader* or *plough*, but often used as a mount for a *dozer*, winch or other implement. Normal tractor horsepowers vary from 25 to 115, *drawbar pulls* from 2700–12000 kg in first gear (2.5 kph) to 700–

4500 kg in fifth gear (5–8 kph). These values are for tracked vehicles. Wheeled vehicles are much faster. A tractor fitted with a *power take-off* can be used, when properly loaded with sidebooms and counterweights, as a crane, dragline or pile driver.

tractor-loader-backhoe (TLB) (see diagram) A *loader* with a *backacter* behind.

tractor-scraper See **bowl scraper**.

tradesman, craftsman See (*B*).

traffic engineering (transportation e. USA) Road markings, signs, signals, the geometry of roads, parking, loading, unloading, traffic flow theory, environmental aspects of roads.

traffic restraint Singapore charges motorists a fee, enforced by police, for entering congested areas between 7.30 and 10.15 a.m., reducing the morning peak by 60–75%. See also **infrastructure, road pricing**.

trailer dredger, trailing d. A *draghead dredger*.

trailing cable [elec.] A flexible, rubber-insulated conductor or set of conductors in the same insulation, which provides power for a *conveyor*, *crane*, *dragline*, *loader*, *telfer*, or other plant.

trailing points [rly] *Points* which are approached by a train in such a direction that it first meets their pivot (or heel).

training wall [hyd.] A wall built to contain a river.

training works [hyd.] Any structure designed to influence the flow, scouring or silting capacity of a river. See **dyke**, **groyne**, **levée**, **spur**.

trammel [d.o.] A board containing two grooves which intersect at right angles. In these grooves the two ends of a beam compass can slide and describe ellipses.

Trammel drain A perforated *field drain* laid in a trench in the usual way except that a porous, synthetic sheet material is wrapped round it, leading water to the drain from a 'filter slot' in the trench, containing sand. First installed in 1975, it is said to be highly effective.

tram rail [rly] A rail with its top surface slotted for the wheel *flange*.

tramway See **aerial/twin-cable ropeway**.

Träneberg Bridge A reinforced-concrete-arch bridge near Stockholm with a clear span of 181 m. The arches are rectangular, hollow box-girders with about 0.75% total steel. See **Parramatta Bridge**.

transducer An electronic device usually for measurement. People who operate chemical or other partly or wholly automatic processes need to measure their variables and for this they use instrument transducers, of which *EDM* are probably the most sophisticated. All transducers have a *sensor* which detects a signal whether of temperature, gas or water pressure, flow rate, water level, concrete density, sludge density, radio wave, etc., and this signal passed from the sensor to a *converter* becomes usable – at a current and voltage usually between 4 and 20 mA and less than 20 V. The converter may digitize the signal if it has to be input into a *microcomputer*. The converted signal may be sent to a transmitter which may first amplify it for use on a display panel or to operate an *actuator*, switch or recorder.

Non-instrument transducers include actuators, loudspeakers, etc.

transfer [stru.] For prestressed concrete, transfer is the moment when the concrete is stressed and some of the load is transferred from the steel to the concrete.

transformer [elec.] A device for converting one voltage of alternating current to another voltage. The ease with which power can be transformed (stepped up or down) in this way is the main reason for the supremacy of alternating current over direct current wherever power is consumed far from the generator.

transient A wave of air pressure (sound) or water pressure or voltage (surge) that occurs only for a moment. They are usually sudden increases in pressure and are important in hydraulics as well as in the design of electrical networks, since they can rise to high values that may cause destruction in pipes, open channels or harbours.

transistor An electronic device discovered in 1948 that superseded the thermionic valve. It is widely used in radio as an amplifier and in computing or telecoms as a switch. *Integrated circuits* contain thousands of them.

transit [sur.] (1) US term for a *theodolite*.

(2) The *culmination* of a star.

(3) To *change face* of a theodolite.

transition curve, easement c. A curve which eases the change between a straight and a circular curve. It is often a spiral.

transition length [rly] The length of a *transition curve*.

transit man [sur.] In the USA, an instrument man who is working at a transit instrument.

transit-mixed concrete *Truck-mixed concrete.*

Translink Short for *Transmanche Link*.

translocation [sur.] In satellite surveying, a method of increasing the accuracy of position fixing. Receivers are placed at two or more points, the position of one point being known. Simultaneous readings from the satellites are recorded at all receivers and the positions of the new point(s) are calculated. Translocation can increase the accuracy of the position fix by a factor of 10 compared with the single-position fixing for which the systems were orginally designed. See **satellite-Doppler**, **global positioning system**.

Transmanche Link (TML) The joint venture of the ten contractors who have united to drive the *Channel Tunnel*; on the British side Balfour Beatty, Costain, Tarmac, Taylor Woodrow, Wimpey; on the French side Bouygues, Dumez, SAE (Société Auxiliaire d'Entreprises), SGE (Société Générale d'Entreprises), Spie Batignolles. Their *client* is *Eurotunnel*.

transmission length [stru.] The *grip length* of a *prestressing* tendon in concrete. For 9 mm dia. strand it is about 2 m; for 18 mm dia. strand about 5 m, but varies with the *bond* strength of the concrete and the force in the tendon.

transpiration [hyd.] Discharge of water from plants into the air as vapour. It increases with increasing wind, sunshine and temperature, and varies from plant to plant, but between the soil moisture contents of *specific retention* and *wilting coefficient* the amount is more or less constant for each plant of a given size and species. Transpiration reduces the daytime flow of springs in the summer but barely affects it at all in winter when the leaves have fallen.

transportation speeds in water

	diameter, mm	moved by a current of (m/s):
Fine sand	0.4	0.15
medium sand	1.1	0.22
coarse sand	2.5	0.3
gravel	2.5–25	0.75
pebbles	25–75	1.2

These figures show what a stream can do. But to carry sand in a pipeline without blockages is a different matter. In 1989 2.5 million m³ of sand from the Goodwin Sands, 4.5 km out at sea, were pumped at 4 m/s (13 ft/s) through a 800 mm pipeline to create fill for the Channel tunnel junction at Folkestone.

transporter bridge A bridge consisting of a *lattice* girder spanning between the tops of two towers at each side of a gap. It carries vehicles across the gap in a container slung at road level by ropes under a crane *crab* on the girder. See below.

transporter crane, cantilever c. A long *lattice* girder carried on two lattice towers which may be fixed or travel on rails at right angles to the girder. A *crab* travels along the girder and a grab or hoist hangs from the crab. It can thus be used for excavating or stockpiling material, or for loading or unloading ships or trains. When called a cantilever crane, usually at least one end of the girder overhangs its tower. See **cableway transporter, materials handling**.

transverse loading [stru.] *Beam* loading.

trapezoidal rule [d.o.] An estimate of the area of an irregular figure divided into *n* strips of equal width *d*, each strip being a trapezium. If the length of the first strip is l_1, that of the last l_n, and the sum of the intermediate strips is Σl, the area is equal to $\dfrac{d}{2}(l_1 + 2\Sigma l + l_n)$. *Simpson's rule* is more exact.

trap points, catch p. [rly] *Points* built into a running track either to prevent conflicting movement of trains or to catch (and harmlessly derail) an unbraked disconnected carriage running away downhill.

trash rack Parallel bars or a *screen* across a stream to catch floating rubbish, always provided at intakes for a turbine or for *raw water*.

trass A volcanic ash from the Eifel mountains near Coblenz, resembling *pozzolan*. See **Portland pozzolan cement**.

traveller (1) [mech.] A pair of beams (or a single beam) carrying the moving hoist (*crab*) of an *overhead travelling crane*. They are mounted on wheels at each end and run on the crane rails.

(2) The central *boning rod* which is moved along the ground to check the ground levels between two *sight rails*.

travelling forms Usually large, built-up *formwork* for casting walls or the linings of tunnels or culverts, built on a carriage on rollers or wheels, enabling the forms to be moved without dismantling them. Movement may be intermittent or continuous. If continuous, the system can be regarded as *slipforms* and in fact some authorities do not distinguish between the two. To avoid confusion it might be better to regard travelling forms as moving horizontally and slipfoms as moving vertically.

travelling ganger A *walking ganger*.

travelling gantry A wheeled *gantry* which travels on rails and is provided with a hoisting *crab*.

travelling screen [hyd.] (1) A canvas diaphragm in a frame which fits closely across the section of a uniform channel and travels with the water. It is a *float* which gives a measure of the average velocity.

(2) A movable *trash rack*.

travel-mixer [s.m.] A self-propelled *soil mixer* which takes in soil at its front end from a *windrow*, and discharges it after mixing it to the *optimum moisture content* with *stabilizer*. See **mix-in-place, pulverizing mixer, single-pass soil stabilizer**.

traverse [sur.] A *survey* consisting of several connected lines of known length which meet each other at measured angles or *bearings*. See **closed/ compass/open traverse, closing error**.

traverse tables [d.o.] Tables of the differences of *latitude* and departure for different angles. They are generally more used by navigators as a quick check on their position than by surveyors.

traversing [sur.] See **progression, traverse**.

traversing bridge: retractable b.

[stru.] A *movable bridge* that retreats from the waterway to allow a ship to pass.

traversing slipway A slipway for ships which weigh less than about 500 tons, on which the ship after being hauled up the slipway can be moved sideways (traversed) to another berth for repair, leaving the *slip* free.

traxcavator A *loader* or *excavator*.

treamie See **trémie**.

tree cutter, stumper, treedozer A horizontal toothed blade placed ahead of the mouldboard of a bulldozer, or a tractor equipped with this blade.

treenail See (*B*).

trémie (also **tremmie** USA) A sheet-metal hopper with a pipe leading out of the bottom of it, used for placing concrete under water. The foot of the pipe is so arranged as to be always below the level of the concrete, with the top of the concrete in the pipe above water level. In this way the concrete in the pipe is always under a higher pressure than the water outside it and is unlikely to be diluted by water. Placing by a trémie prevents *segregation* of the concrete, and closely resembles the placing of the grout in grouted aggregate concrete.

If a large volume has to be concreted, this must be done by several trémies simultaneously, about 2.5 m apart, carried on staging or barges. Flow out of the trémie is reduced by lowering the hopper and increased by raising it. The concrete should rise uniformly at about 0.4 m/hour. The rate of concreting must be related to the size of the panel so that no panel takes more than four hours. This ensures that the first concrete to pass through does not stiffen before the panel is completely concreted. The concrete level is checked by sounding with a pole. At the start of concreting the trémie pipe is full of water and this must be expelled by a suitable plug placed in the hopper to separate the water from the concrete. If nothing else is available a ball made from old cement bags will do. The weight of the concrete pushes it through the pipe and mixing of water with concrete is prevented. But if the pipe outlet is lifted above the concrete, it will again fill with water and the process of expulsion of water will have to be repeated. If a concrete pump can reach the same distance as the trémie, it may be able to replace the trémie completely. Quite apart from under-water concreting, a trémie is useful to prevent *segregation* where there is a large drop between the point where concrete is delivered and that where it is placed. See **diaphragm wall**.

trémie seal The depth of submergence, below the concrete, of the bottom end of a *trémie* pipe.

trenail See **treenail**.

trench A narrow, long excavation with timbered or bare sides, vertical or battered.

trench box Equipment that provides quick support in a trench and is dropped into place with a crane. It consists of two parallel steel-faced sheets held apart by struts, which may be adjustable. One type, made of composite steel joists and glassfibre with steel cladding, weighs 2 tons but measures 2.44 × 2.88 m. The boxes may be stacked on top of each other in a deep trench, but must be removed while the excavator is deepening the trench.

trench compactor A *frog rammer*.

trench drain A *French drain*.

trench excavator, trencher, ditcher A wheeled or tracked machine that digs trenches. For soft ground one equipped with a *bucket-ladder excavator* is suitable. For hard ground a *rotary excavator* (*bucket-wheel excavator* or *rock wheel*) must be used. Neat, clean-sided trenches from 0.18 m wide × 1 m deep to 1.5 m wide × 5 m deep are dug quickly. A *backhoe* can also dig trenches, though less fast, but adequately for any small site. Submarine trenchers exist for sea pipelines, controlled by the *bury barge*.

trench hoe A *backacter* or *backhoe*.

trenching machine See **trench excavator**.

trenchless pipelaying, t. moling Apart from the old way of driving a timbered heading, there are at least six trenchless ways of placing a buried horizontal pipe or cable: by a *mole plough*, *pneumatic impact mole*, *pipejacking*, *steel-sleeve jacking*, *hydraulic impact mole* or a *microtunnelling machine*.

trench-type tunnel An *immersed tube*.

trepan An obsolescent tool weighing from 2 to 30 tons and of from 2.4 to 5.5 m dia. dropped on a string of boring rods down a shaft being excavated in rock under water. At each stroke it was dropped about 0.3 m and turned slightly, in the *forced drop shaft*, *Honigmann* and *Kind Chaudron shaft sinking* methods. See **Betws**.

trestle A timber, reinforced-concrete, or steel structure, generally a *bent* connected to similar parallel bents each side. It usually supports a temporary or permanent bridge or an *aerial ropeway*. The supports may be *raking piles*.

trestle bridge A bridge resting on trestle *bents*, which are of timber in a temporary structure, of steel or reinforced concrete in a permanent structure. It is often used over weak ground

since this sort of structure can be very light in weight.

trial erection The joining of steelwork in the *erecting shop* before delivery to the site.

trial mix A small quantity of a *designed mix* made for *concrete testing*.

trial pit, t. hole (**test p.** USA) A pit dug to determine the type of ground, or to prospect for mineral.

triangle [d.o.] US term for *set square*.

triangle of error [sur.] The triangle formed in the graphical solution of the *three-point problem* in plane tabling, when the three lines drawn from the three points do not meet owing to inaccuracies.

triangular notch, t. weir, vee n. [hyd.] A *measuring weir* of V-shape used for measuring small discharges. A useful mnemonic formula for discharge over a 90° notch is in British units: Q cusecs = $2.5h^{2.5}$, h being the depth of the water at the apex in feet.

triangulation (1) [sur.] The measurement of a large area of land by covering it with a network of triangles of which all the angles are accurately known. One side of a triangle (the *base line*) is also measured with great accuracy, and from it the lengths of all the other sides of the triangles are calculated by trigonometry. Cf. **trilateration**.

(2) [stru.] Designing a *truss* so that every shape enclosed by its members is triangular, no quadrilateral without a diagonal being allowed. If the joints are assumed hinged, the truss is a *perfect frame*.

(3) In *computer graphics* complex polygons often have to be reduced to triangles so as to simplify computer procedures – another sort of triangulation.

triaxial compression test, con-

fined c. t.. [s.m.] A test of a soil sample contained in a rubber bag surrounded by liquid under pressure. A load is applied by a piston to one end of the rubber bag and the deformations, loads and pressures are recorded. *Undrained tests* are rapid, but *drained shear tests* (often called slow tests) require much more time and are correspondingly more expensive. Cf. **unconfined compression test**, **vacuum method of testing sands**.

triboelectricity Frictional, static electricity, an electrical charge acquired by rubbing different materials together. Triboelectric noise is that part of *electronic noise* (or interference) in a circuit which comes from mechanical stress in cables – bending, stretching, twisting. A triboelectric axle *sensor* has been devised by the TRRL. It counts vehicle axles.

tribology The science of friction and lubrication.

tribrach [sur.] The frame under a *theodolite* which carries the three *foot screws* below it.

trickling filter, bacteria bed, biological f., contact bed, continuous f., percolating f. [sewage] A bed of filter medium such as rock, clinker, slag or specially made plastics units which exposes *sewage* effluent to the air and thus to the action of micro organisms that oxidize it. In types with a *dosing tank* the flow is intermittent. It is not a filter in the sense of a *graded filter* or a *sand filter*.

Trief process A method of making *Portland blast-furnace cement* by grinding the slag wet at the site and there mixing it with the cement.

trigonometrical station, trig s. [sur.] A *survey point* in a *triangulation* or *trilateration*.

trigonometrical survey A survey based on *triangulation* or *trilateration*.

trigonometric levelling [sur.] Calculating the difference in level between two points from the slope length between them and the angle of elevation from one to the other.

trilateration [sur.] Measurement of all the sides of a network of triangles, usually by *electronic distance-measuring instruments*, a method of establishing accurate survey points that has superseded *triangulation*, at least for large work. *First-order trilaterations* have longer sides than second-order or third-order ones.

trimmers [min.] See **rib holes**.

trimming (1) The final tidying up of an earthwork surface.
(2) Framing round or otherwise strengthening an opening through a floor, roof or wall, whether of timber or other material. See (*B*).

trinitrotoluene (TNT) A high explosive which produces considerable quantities of carbon monoxide, but it can be used for dry holes in civil engineering, in large quarry blasts in headings, or in *chambering*. See **detonating fuse**.

trip coil [elec.] A *solenoid*-operated device for operating a *circuit breaker* or other *protective equipment*.

tripod [sur.] A three-legged support (with telescopic legs if used underground) for a surveying instrument.

trommel A *rotary screen*.

troughed belt (see diagram, p. 470) A *conveyor belt* with *idlers* at an angle to each other.

trough floor A *one-way slab* cast in reusable, fibre-glass, U-shaped moulds about 500 mm (20 in.) deep, forming grooves on the underside of the finished slab.

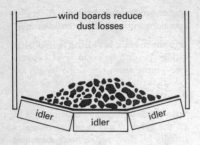

wind boards reduce dust losses

idler idler idler

return strand

idler

Cross-section of troughed belt conveyor. See also diagrams on pp. 32, 452.

troughing, trough sections *Rolled-steel sections* shaped like a broad U and connected together in *bridge decks* or other heavy floors with the U alternately upwards and downwards. Troughing is often covered with a concrete slab, for which it acts as the permanent formwork in *composite flooring*.

trough sections See **troughing**.

trowelling machine A *power float*.

truck mixer, transit m. A powerful lorry carrying a mixer in which concrete batched at the *batching plant* is mixed while travelling to the site. Because the concrete is not pre-mixed, the mixer contains less concrete than a structurally similar agitating truck. Mixing has to be at a speed between 7 and 16 rpm. The advantage of truck mixing is that traffic jams between batching plant and site are less worrying than with ready-mixed concrete.

truck-mounted concrete pump A *mobile concrete pump*.

truck-mounted crane A *mobile crane*.

truck-mounted drill, earth borer A powered *auger* mounted on a lorry. It can drill holes for *bored piles* or small shafts of 0.76 m dia. to 6 m depth in 10 min, and sometimes of larger diameter. This is many times faster than *shell-and-auger boring* by two or three people with a three-pole *derrick*.

true bearing [sur.] The horizontal angle between any survey line and the true north. See **bearing**.

true meridian [sur.] The geographical north–south plane, not the magnetic meridian.

true section [d.o.] A cross-section drawn with the same vertical and horizontal scales.

true-to-scale print [d.o.] A *contact print* made with black ink lines on tracing or opaque paper or cloth. It is generally a print of the best quality but may need more than the few minutes required for *blueprints* or dyelines.

trunk sewer, trunk main A large *sewer* or *main*.

trunnion axis [sur.] The horizontal axis of rotation of a *theodolite* telescope.

truss [stru.] A frame, generally nowadays of steel (but also sometimes of timber, concrete or *light alloy*), to carry a roof or bridge, built up wholly from members in tension and compression. It is generally a *perfect frame* or nearly so and may be *pin jointed*. There is usually some fixity at the joints which is not taken into consideration in the calculations but adds to the stiffness of the frame. See (*B*).

trussed arch [stru.] A steel arch built of rolled-steel sections, like the *Sydney Harbour* or *Kill Van Kull bridges*.

trussed beam An old type of timber

or cast-iron beam stiffened to reduce its deflection by a wrought-iron tie-rod (camber rod) fixed below it and tensioned by one or more struts to the beam. Cast-iron trussed beams were abandoned after the failure of the Dee Bridge in 1847.

tsunami [hyd.] A seismically generated wave of great destructive force, known in the north Pacific around Japan.

tubbing [min.] A cast-iron lining for circular tunnels or shafts, built up from segments of a circle, flanged and drilled for bolting together. One set of segments (360°) is called a *segmental ring*.

tube (1) [stru.] See **tubular sections**.
(2) A concrete pipe of any diameter with joints within its thickness.
(3) See below.

tube à manchette [s.m.] A perforated tube, with rubber sleeves (manchettes) covering the holes, inserted into a borehole to be grouted. When the grout pressure in the pipe increases the sleeves expand and release the *grout*. The hole in the ground is about 10 cm dia. and the tube à manchette is rather smaller, about 8 cm. The first grout injected is of plastic consistency. A second tube, of 2 cm dia., is then introduced for the fluid grout injection. This tube has *packers* each side (above and below) of the chosen grout outlet, which by their position select the release hole in the tube à manchette. In passing through the sleeve the grout breaks through the plastic grout and into the surrounding ground. The plastic grout prevents vertical leakages and the exact point of injection can be chosen, leading to high economy in grouting materials and time of grouting.

tube jacking See **steel sleeve jacking**.

tube railway An underground electric railway running in a cylindrical tunnel, excavated by mining, not *cut-and-cover* methods. In London each tunnel of 3.66 m (12 ft) bore is occupied by one track. The tunnels are driven in pairs using *shields* and are lined with cast-iron or concrete *segmental rings*.

tuberculation, incrustation Pimpling (rusting) in a cast-iron pipe.

tubing The pipe installed inside the *casing* of a borehole to extract oil, gas or water.

tubular scaffolding Standard steel scaffolding is made of welded or seamless tube of 48 mm (2 in.) dia. and 4 mm thick. Steel tube weighs about 4.4 kg/m against about 1.7 kg/m for aluminium tube, but is half the price and much stronger, though it rusts.

tubular sections, hollow s. Apart from the expensive but non-rusting light-alloy *extruded sections*, there are two types of hollow steel sections – very light ones made by *cold rolling*, and the heavy, rectangular or circular tubes described below.

Steel tubes, rectangular, square, or circular, are strictly speaking not *rolled-steel sections* though they are manufactured hot and, for design purposes, are treated by the structural designer in the same way. The smallest sizes, up to about 125 × 50 mm, are made by the welding together of two plates bent to a channel shape. Larger sizes than this are made as seamless tubes, a process more laborious and expensive than rolling. These sections are obtainable in Britain up to 460 mm outside dia. and 19 mm wall thickness, or 500 × 300 mm (20 × 12 in.). Such massive sections can be extremely useful as struts or beams. They are easily painted to protect them from rust, and the inner surfaces do not rust at all if the ends are sealed.

tucking board A narrow horizontal board in a *tucking frame*, the thickness of the *poling boards*, placed on edge horizontally over the top of one *setting*, behind the waling to ensure that the upper setting of poling boards is held by the *waling*.

tucking frame, middling f., poling f. A *frame* in timbering, in which *walings* support *poling boards* at their top and bottom ends.

tungsten carbide [mech.] A cemented carbide, a very hard material which may be brazed with an oxyacetylene flame in a cobalt *matrix* on to the cutting edge of a *core barrel, fishtail bit* or other boring tool.

tungsten inert-gas welding See **TIG welding**.

tunnel An underground passage, open to daylight at both ends. If open at only one end it is a *drift* or *adit*. The London underground railways are probably the world's largest underground network and include tunnels 30 km long; but one of the world's longest tunnels in 1973, completed in that year, was the Orange–Fish tunnel, South Africa. Of 5.35 m finished dia., with a concrete lining 23 cm thick, it is 82 km long. It irrigates the semi-arid Great Fish River valley, in the eastern Cape Province, with water from the Orange River.

tunnelling Tunnelling methods vary with the hardness of the ground, the size of the tunnel and whether the ground is waterlogged. Most tunnels in soft ground are driven by *shields*, mechanically with *tunnelling machines* (TBM).

In medium-hard ground and even in the hardest rock, tunnels are now driven by *fullface TBMs*. Drilling and blasting, formerly the only method in hard rock, is giving way either to a

TBM or to a heavy *hydraulic hammer* of up to 3500 kg in weight, powered by the hydraulics of the *excavator* that carries it. But it is not possible to get every hydraulic excavator into a tunnel cross-section below 30 m² (300 ft²). In the largest tunnels where two excavators can work side by side, one of them will be loading while the other is breaking ground. See also **microtunnel, microtunnelling machine, multi-face tunneller**.

tunnelling machine, tunnel boring machine (TBM) A machine for excavating a circular tunnel. The first ones were designed for penetrating relatively soft rock but they are now able to drive fast through the hardest rock. A rotating cutting wheel breaks the ground, which drops through slots in the cutting wheel for removal. The cutting wheel diameter is larger than the bore of the lined tunnel by twice the thickness of the *tunnel lining*. In soft ground, instead of TBM a *backhoe* may be used. TBMs have been successful in soft ground but they are inflexible and generally suitable only for the type of rock they were designed for. In variable ground the TBM may have to be changed, a cause of long delay. In soft ground a *shield* must be used, normally a *closed-face shield*, with a *tail skin* at the rear end long enough for building at least two *segmental rings*.

Behind the cutting wheel in softground TBMs is a watertight steel wall, the bulkhead, through which the muck is removed, either by a screw conveyor or by *pipeline transport*. The bulkhead provides protection against an inrush of mud or water and enables the face to be pressurized if need be, either by compressed air or by injected slurry. TBMs do not weaken the surrounding ground as explosives do but they are a heavy investment. One TBM for the *Channel Tunnel* has a train of associated equipment 215 m long behind it. At large diameters the cutting wheel is often supported and driven from its outer drum next to the shield. In early TBMs it was driven by a central shaft which obstructed the removal of muck. Other tunnelling methods involve *boomheaders* or *hammer tunnelling*.

tunnel lining In soft ground, tunnel linings are normally *segmental rings* of wedged concrete in good ground or bolted cast iron in bad ground. They may be made waterproof sometimes by grouting behind after erection. In rock, protection against water is achieved by continuously welded plastic sheets next to the *shotcreting* over the rock, all round the upper part of the tunnel. The plastic sheets are designed for drainage by dimpling, channels or felt bonded to them. Concrete linings are usually not less than 30 cm (1 ft) thick, and are cast over the plastic waterproofing. See also **gunite**, **shotcrete**.

tunnel vault See **barrel vault**.

tunnel ventilation Road tunnels longer than 400 m (1300 ft) generally need mechanical ventilation. But if there are two tunnels, one for each direction of traffic, permanent fans may be avoided because the traffic itself forces the air through at as much as 3 m/s (10 ft/s). It is advisable to install booster fans in case of contrary wind, but the Dutch believe such one-way tunnels are feasible up to 2000 m, with booster fans for occasional use.

tup A *drop hammer*.

turbidity Lack of clarity, an undesirable property of drinking-water, caused by *suspended solids*. They can be removed by *water treatment*. Air pollution or fog can make air turbid.

turbidity current [hyd.] See **density current**.

turbine [mech.] A rotating *prime mover* driven by water, gas or high-pressure steam, often used for driving an electric generator. See **specific speed**, **steam/water turbine**.

turbo-drill An oilwell drill which overcomes some of the difficulties of *rotary drilling* at depths below 3 km. It eliminates drill rods and operates as a turbine driven by *drilling fluid* or water pumped down at high pressure. By the 1970s, however, they were branching out. Turbines of small diameter were being used for short directional holes near the surface, and in hard rock where the support of drilling fluid was not needed they were driven by compressed air. One 45 mm dia. drill makes holes of 76 or 90 mm (3 or $3\frac{1}{2}$ in.) dia.

turbulence [hyd.] A state of flow in which the liquid is disturbed by eddies. See below.

turbulent flow, eddy/sinuous/tortuous f. [hyd.] Flow at a speed above the *critical velocity* of Reynolds. Flow in this state is the opposite of *streamline flow* and is unsteady and eddying.

turfing The covering of an earth surface with growing grass cut from another site. It can also be a *revetment* to slopes which are usually covered by water, made by laying turves on the slope according to a technique like *sliced blockwork*.

Turin Exhibition Hall The first use of *ferrocement* in a public building. It spans 100 m (328 ft) and had to be built in the seven months between autumn 1947 and early summer 1948. Pier Luigi Nervi built the roof of curved, corrugated ferrocement panels 40 mm ($1\frac{1}{2}$ in.) thick, cast in winter under cover and joined up on the site with mortar. This success enabled Nervi to build many other striking ferrocement structures of large span.

turn bridge [stru.] A *swing bridge*.

turn buckle [stru.] A *screw shackle*.

turned bolt, tight-fitting b., bright b. A bolt used in steel-to-steel connections, turned on a lathe to reduce it to a truly circular shape of exact dimension. The thread (and the nut) are of the same nominal diameter, d, as for a *black* bolt but the unthreaded part (barrel or shank) is appreciably larger ($d + 1.5$ mm). The hole diameter as for a black bolt is $d + 1.6$ mm and the shank clearance is therefore only 0.1 mm. The shank is consequently a tight fit in the hole and for this reason these bolts are allowed to carry 20% more load than a black bolt or rivet. Because of the exorbitant labour expense in reaming each hole to exact size, they are obsolete and have been replaced by *high-strength friction-grip bolts*.

turning point [sur.] A *change point* or *survey point*.

turnkey contract See **design and build**.

turnout (1) [rly] A junction between one track and another leading off it including the *points* and any accessories.

(2) [hyd.] A junction from an irrigation canal to a subsidiary canal taking water from it. It may be a wooden box for a small turnout or a glazed stoneware or concrete channel for larger ones.

turntable [rly] A round platform pivoted at its centre. Locomotives are run on to it and sent off in the reverse direction or on to another line as required.

TÜV (Technischer Überwachungsverein) A large consulting engineering firm separately organized in each German state (*Land*), with very wide official responsibilities and a branch office usually in each county (*Kreis*). It checks whether cars are roadworthy and issues the appropriate certificate every two years. Without its certificate of approval of structural calculations for housing etc., no construction, or even digging, can begin. Among its other responsibilities are the possibilities of air or water pollution and it may verify that refuse disposal and recycling are suitably done without harm to the environment.

twin-block sleeper [rly] A rail *sleeper* built of two blocks of concrete, one under each rail, joined by a steel bar.

twin-cable ropeway, tramway An *aerial ropeway* with carriers running on parallel track cables in opposite directions, both rows of carriers being pulled by the same *traction rope*. The track cables are two heavy, stationary, often *locked-coil ropes*, anchored at one end and tensioned by weights at the other, usually carried on opposite sides of the same towers.

twin-twisted bars Two similar steel bars twisted together for use as *reinforcement*.

twist See **torque**.

twisted deformed bars Concrete *reinforcement* made by *hot rolling* with herringbone or other upstanding ridges to improve *bond strength*. They are then twisted cold, to raise the *yield point*.

two-hinged arch, two-pinned a. A *rigid frame* which may be arch-shaped or rectangular but is hinged at both supports.

two-leg sling A *sling* made of two chains or ropes hanging from one *thimble* or link, and each having one hook at its lower end.

two-part sealant See **polysulphide sealant**.

two-pinned arch A *two-hinged arch*.

two-stage compression [mech.] Air compression in two stages, usual for pressures above 4 bars or horsepowers above 100. Two-stage compression with an *intercooler* is usual when compressing air for rock drilling. Cf. **four-stage compression**.

two-stage tendering At an early stage of a project a *contractor* is chosen on the basis of *tenders* based on an approximate *bill of quantities*. This enables the contractor to advise the *consultants* at an early stage before the design becomes final.

two-way grid [stru.] A *space frame*, a *grid* (3). See **double-layer grid**.

two-way slab See **one-way slab**.

tying wire See **annealed wire**.

type testing Testing of a new material or product such as a pipe, brick or roof truss, to show that it is suitable for its purpose. If the type tests are successful they are done only once but they must be followed up by regular quality control tests to ensure consistency and quality of the products, with *representative sampling* and *non-destructive testing*.

tyres See **rubber tyres**.

U

U-liner A method of *pipe renewal* with *polythene* tube, installed by Pipe Liners Inc. in the USA. Polythene tube when hot can be squeezed to take up a 'very curved banana' cross-section, also described as a U-shape, thus

In this crushed form it is easy to insert into an old, leaky pipe that needs relining. The polythene tube can be made circular again by pumping hot water into it.

ultimate bearing capacity of a pile The ultimate bearing capacity of a pile is the load that causes it to settle one tenth of its diameter, unless some other definition be found from the load/settlement curve. This is sometimes determined by the constant rate of penetration test, conveniently by jacking down from a beam connected to two anchor piles. The pile, if in clay, is jacked down at 0.75 mm/min, or if in sand or gravel at double this rate. The maximum force on the graph (before it begins to decrease) is the ultimate bearing capacity. If no maximum is reached, the force at a penetration equal to one tenth of the pile diameter can be taken as its ultimate bearing capacity. For a single pile, a factor of safety of 2 or 3 should be applied to the ultimate bearing capacity, to give the safe load.

ultimate bearing pressure The pressure at which a foundation sinks without increase of load. In *plate bearing tests* the ultimate bearing pressure is taken as that pressure all over the plate at which the total settlement amounts to one fifth of the plate width.

ultimate compressive strength [stru.] The stress at which a material crushes, the usual way of defining the strength of a brick, stone or concrete.

ultimate limit state See **limit state**.

ultimate strength, breaking s. [stru.] The highest stress (of any sort) which a material can withstand before breaking.

ultimate tensile stress, breaking strength, u. strength [stru.] The tension at which a specimen breaks, divided by its original area before testing began.

ultra-filtration A low-pressure, low-energy *membrane process* which strains out 10-micron or even smaller particles, down to 0.002 micron. It uses extremely fine, easily clogged filters. The downstream flow is therefore divided into two streams, one dirty, one clean. To prevent the filter clogging, it is constantly washed by the incoming stream removing the dirt to the dirtier stream. What passes through the filter is almost pure water since even dissolved salts are removed.

ultra-high-early-strength cement, Speed c. Very finely ground, therefore expensive, non-standard *Portland cements*, formerly sold in the UK as 417 cement or Swiftcrete, or in Belgium as Speed. The *specific surface* of 417 is from 700 to 900 m²/kg, and of Speed 450–500 m²/kg. No *admixture* is needed to reach a cube strength of 28 MPa (4000 psi) at 24 hours, 48 MPa

(7000 psi) at three days and 68 MPa (9000 psi) at 28 days, so the cement is suitable for winter concreting or road repair. See also **rapid-hardening, jet cement**.

ultrasonic flow meter [hyd.] Equipment for measuring the flow in a river or stream, which, unlike *measuring weirs*, can be used in a navigable channel. There are two transmitter-receivers, on different banks, one appreciably downstream of the other, and located at about 0.6 of the depth from the surface, at the level of the average flow rate. Electronic circuitry determines the frequency of the pulses received, those downstream being at a lower frequency than those in the upstream receiver. From the difference in the frequencies the flow rate can be determined, since the ultrasonic pulse velocity is known. This use of the *Doppler shift* has been practised also in an analogous method, the laser Doppler anemometer for measuring air flow.

ultrasonic pulse attenuation *Non-destructive* measurement of the viscous rather than the elastic properties of concrete. Strong concretes have high pulse velocity and low pulse attenuation (damping) properties.

ultrasonic testing Two methods of non-destructive testing (ultrasonic pulse velocity and *ultrasonic pulse attenuation*) are used for determining the strength of concrete and other materials. In the velocity method a pulse of longitudinal vibrations, transmitted by an electro-acoustical *transducer* held on to the concrete, is received by a second transducer or probe, and its speed through the concrete is electronically measured. Ultrasonic frequencies are used so as to obtain a pulse with a sharp onset and to generate maximum energy in the direction of the pulse. The velocity increases as the concrete

quality improves. This method is often used in conjunction with the *rebound hammer*. It can also be used to determine the presence of defects. The relation between strength and velocity is very complex, and varies with the cement and aggregate content and type, the maturity and the curing of the concrete. Sound can be heard at frequencies up to about 20,000 cycles/s but it becomes inaudible at higher (ultrasonic) frequencies. Testing frequencies for concrete are between 20,000 and 150,000 cycles/s and for other materials may be up to 20 million cycles/s.

Ultrasonic *sensors* can penetrate the walls of a reactor to determine liquid levels, flow rates, sludge density, etc.

ultraviolet irradiation A method of *disinfection* of drinking-water which has been difficult to monitor and therefore less popular than it should have been. Provided that the water is crystal clear it effectively eliminates microorganisms, leaving no smell. An ultraviolet radiation dose of 16 mW-s/cm^2 reduces *Escherichia coli* bacillus by 99.999%. The desirable wavelength is 260 nm. A lamp has a life of about 3000 hours – four months continuously.

unbonded screed A *screed* like those made for a *floating floor* (*B*), therefore structurally separate from the concrete floor slab and liable to *curling*. It must be at least 75 mm (3 in.) thick and cast in small panels, not larger than 10 m^2 (100 ft^2). Large aggregate (20 mm) with greater thicknesses helps to prevent curling. Cf. **granolithic screed**.

unconfined compression test [s.m.] A crushing test on a soil sample which is carried out without lateral restraint. (See **triaxial compression test**.) Half the unconfined compressive strength of a clay is generally equal to its *cohesion* or shear strength. The test can be carried out on a clay sample 76 mm long, of 38 mm dia. at the borehole, with portable apparatus. It is of little use for clean sands or gravels, but extremely useful for cohesive soils.

unconfined water *Groundwater* that is overlain by air-filled, therefore unsaturated ground, unlike *confined water*.

underfloor heating See (*B*).

underflow (1) [hyd.] Movement of water in the soil, under ice or under a structure.

(2) [s.m.] The oversize material of a *classifier*.

underground railway A railway laid wholly or mostly below street level. See **cut-and-cover**, **tube railway**.

underpin To provide new, deeper support under a wall or column without removing the superstructure, so as to allow the load on the building to be increased, or to allow the ground inside or outside it to be lowered, or to prevent settlement of the foundation. It is the construction of foundations for a building which exists. See **pretesting**, **dry pack**.

underplanting A contract on which the machinery is overloaded is said to be underplanted. Similarly, one where the machinery has not enough work to do is overplanted.

under-reaming, belling The widening out of the foot of a *bored pile* or of a foundation pier such as the *Gow caisson* to increase its area or to give an anchorage against lifting by wind or in *permafrost regions*. See **reamer**.

undersize [s.m.] See **classifier**.

underwater excavator (see diagram) Bulldozers have been used under water, but two long vertical pipes are needed to bring in combustion air and take out exhaust gases. Also the bulldozer will weigh at least one seventh less under water than it does out of it, so it will have correspondingly less push, since steel is about seven times as dense as water. An air connection is needed from the top of the fuel tank to the air intake, to allow fuel to flow. *Draglines* also dig under water.

undisturbed sample [s.m.] A sample of cohesive soil from a borehole or trial pit which has been *remoulded* so little that it can be used for laboratory measurements of its strength without serious errors. Generally a *soil sampler* is driven into the ground, turned through 360° to shear the core at the foot of the tube, extracted from the borehole with the sample within it, filled with paraffin wax, covered with screw caps, and sent to the laboratory for testing. Cores may also be tested at the site with the *unconfined compression* apparatus.

undrained shear test [s.m.] A *quick test* or *consolidated quick test*.

unequal angle An *angle section* in which the two legs are of unequal length. Cf. **equal angle**.

uniform building code (UBC) Probably the most important of the many US sets of building regulations, and published by the US International Conference of Building Officials. Each state can enact and enforce its own building laws.

uniformity coefficient [s.m.] See **coefficient of uniformity**.

uniform sand [s.m.] Any *sand* with grains mainly of the same size, therefore with a steep *grading curve*. A coarse uniform sand has grains between 2 and 0.6 mm; a medium-sized uniform sand is from 0.6 to 0.2 mm, and a fine uniform sand is from 0.2 to 0.06 mm.

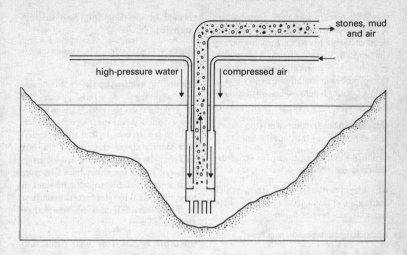

high-pressure water | compressed air

stones, mud and air

Underwater excavation by a combined air-lift pump and hydraulic ejector in loose soil. A crane, an air-compressor and a high-pressure water pump are needed. (Prof. Knaupe, *Erdbau: gewinnen, transportieren, einbauen*, Volkseigener Betrieb für Bauwesen, 1987.)

A *graded sand* contains a mixture of some or all of these sizes and is not uniform.

unit hydrograph, u. graph [hyd.] A method of deducing storm flows that has superseded the *Lloyd-Davies* or 'rational' formula with its unjustified assumptions that the rainfall is constant in time and space and that the *impermeability factor* is also constant. The unit graph shows, for a particular basin, the time-varying flows from 1 in. (25 mm) of rainfall over it. The time bases of floods from storms of equal duration are the same, therefore the unit graph can be deduced from another of the same duration but different rainfall on the basin. Unit graphs are useful for forecasting peak flows. If possible, one should be obtained for each duration of storm.

unit stress [mech.] Expression used in the USA for *stress*.

unit weight The weight of unit volume of material, its *density*.

universal beam [stru.] See **universal stanchion**.

universal motor [elec.] A motor which is usually of less than 1 hp and can work on alternating or direct current.

universal stanchion, u. column, broad-flanged beam A rolled-steel section in which the flanges, unlike a tapered flange beam, are of the same thickness throughout.

unplasticized polyvinyl chloride (uPVC) A stiff material of which corrugated roofing panels may be made, as well as pipes in lengths of 6 m (20 ft) or 9 m (30 ft). Their *spigot-and-socket joints* (*B*) are made either by a *solvent cement* or by a rubber ring. They are available in diameters up to 60 cm (2 ft). Wall thicknesses vary from 3 mm for small diameters at 60 m working head of water up to 22 mm for 60 cm dia. at 90 m working head. About 40% of all new UK drinking-water distribution pipe being installed in 1987 was of uPVC, but *polythene* is probably a good competitor.

unsoundness In cement testing, excessive expansion of hardened cement.

unstable A description of a structure which lacks *stability*.

unstable frame [stru.] A frame which contains too few members or too little fixity to be a *perfect frame*.

upending test See **dump test**.

uplift [s.m.] (1) An upward force due to *flotation*. Upward flow of water may cause fine sands to become *quicksands*. Such uplift can be prevented by draining the water or sealing the leaks.

(2) [stru.] Lifting of a structure caused by (1), by *frost heave* or, on the windward side, by wind force, or in a dry climate by swelling soil. **See underreaming**.

upper transit [sur.] The upper *culmination* of a star or the sun.

upsetting, setting up [mech.] Increasing the diameter of a red-hot steel bar by striking it on the end. This is sometimes done for that part of a bar which is threaded, to ensure that the threaded part stands above the rest of the bar and that the thread does not weaken it. It is not usual for short bars but may save an appreciable amount of steel in long tie-rods. Hot steel *rivets* are upset in the hole which they fill when they are hammered.

upstand A beam or wall projecting above a slab and monolithic with it.

up time *Committal rate*.

upward-flow filter An uncommon type of filter for *raw water*. It needs less backwashing than other filters.

U-value See (*B*).

V

vacuum concrete, v. dewatering
Concrete poured into formwork fitted
with a *vacuum mat*. The vacuum mat
sucks out air and surplus water from
the concrete and produces a dense,
well-shrunk concrete from which the
forms can be removed for walls and
other vertical faces in 30 min after the
beginning of processing. The concrete
reaches its normal 28 day strength in
10 days and has a 25% higher crushing
strength, although it can be placed in a
really workable, sometimes sloppy condition. Savings in cement and formwork can thus be appreciable. The
method is used in the 'spray-dewater-
ing process' for making *glassfibre-
reinforced cement* products. Flat
sheets can be dewatered from below.
After dewatering they may be shaped
to various curves while the concrete is
green.

vacuum lifting The raising by crane
hook and sling of concrete slabs cast on
the ground, through a suction attachment to the sling. The suction attachment works like a *vacuum mat* and
therefore has a lift equal to about 100
kN (10 tons) for each m² in contact
with the slab.

vacuum mat A stiff flat metal screen
faced by a linen filtering fabric, the back
of which can be kept under a partial
vacuum by a vacuum pump connected
to it through a hose. It is used in making
vacuum concrete. The linen filter makes
a flat smooth surface and can be replaced
by a patterned material when the surface
to be processed requires it. A 7 hp motor
is required for keeping up the vacuum
of a 3.66 m (12 ft) square panel of
vacuum mat. See also **vacuum lifting**.

vacuum method [s.m.] See **groundwater lowering**.

vacuum method of testing sands
[s.m.] A method of *triaxial* testing of a
sand sample by maintaining a partial
vacuum in the rubber tube containing
the sample. The outside of the tube is
under atmospheric pressure. The resultant pressure on the tube is thus
equal to the difference between the
two pressures.

vacuum pad A *suction pad*.

vacuum pump, air p. [mech.] A
pump which extracts steam or air from
a space so as to maintain it at a pressure
below atmospheric. The vacuum is usually measured in millimetres of mercury below atmospheric, the maximum
being about 740 mm. Such pumps are
needed for all the vacuum processes
listed above and for the condensers of
steam turbines.

vacuum sewerage The use of plastic
sewer pipes from 75 mm (3 in.) upwards under a vacuum between 175
and 500 mm of mercury has been
found both practical and economical in
the East Anglian fens, where normal
gravity drainage is difficult because
there is little fall in the ground surface
and trenches must therefore be deep
and, with water only 1 m or less below
ground, expensive.

A gravity system for 61 houses would
have needed three underground pumping stations pumping to a fourth main
pump. The vacuum system does not
need falling drains. They can go uphill
about 7 m (23 ft) but not more. Consequently the plastic pipe can be in a
shallow trench. The vacuum mains

start at 75 mm (3 in.) and increase in diameter as their load increases. But, unlike stoneware drains, they are laid to curves. The plastic vacuum mains collect from about 20 local sumps to which the drainage from each house passes by gravity. The single vacuum pump station is above ground. Consequently in 1985 the gravity system was estimated to cost £316,000 and the vacuum system £248,000, 20% less.

The longest main is about 600 m long. An 'interface valve' periodically allows the vacuum mains to collect sewage from each sump, when the level in the sump has risen high enough. Transportation speed is high at about 5 m/s (16 ft/s). Rather more care than usual is needed for the recording of the position of the buried mains since they are not straight. On the other hand illegal connections are immediately obvious.

vadose zone, unsaturated z. The *zone of aeration*.

value-cost contract A *cost-reimbursement contract* in which the contractor receives a higher fee by saving money for the client.

value engineering On large jobs an organized approach to reduce costs that do not help quality, durability or other needs of the client. It is separate from and later than the first design and ends with the value engineers trying to persuade the designers to accept their improvements. In this way it resembles *peer review* but its aim is to reduce cost. Peer review aims to improve quality.

valve [mech.] A device to open or close a flow completely (stop valve) or to regulate a flow (*discharge valve*).

valve tower [hyd.] A tower built up from the bed of a reservoir. From it the control valves of the pipes which draw off water at different levels are operated. It may be of cast iron, stone or concrete.

vanadium [mech.] The hardest known metal. It is used for making very strong alloy steels with chromium or manganese. These alloy steels contain from 0.2 to 1% of vanadium.

vane test [s.m.] A four-bladed vane is inserted into a soil at the foot of a *borehole*. It is rotated by a rod at the surface with a measured force until the soil shears. This *in-situ soil test* gives shear strengths that have been found to give consistent results down to 30 m depth, except in *stiff-fissured clays*.

vapour pressure, v. tension At any given temperature the saturated vapour pressure of a liquid is the pressure of the vapour above the liquid. When the vapour is not in contact with its liquid, the vapour pressure is usually less than the saturated vapour pressure unless condensation is occurring (dew is falling). The saturated vapour pressure increases as the temperature rises. See **saturated air**.

vapour tension *Vapour pressure*.

variance [stat.] The square of the *standard deviation*. It is the average of the squares of the deviations of all the observations.

variation [sur.] See **magnetic variation**.

V.-B. consistometer test, Vebe time A British standard alternative to the *compacting factor* test for the *workability* of concrete, as well as its *slump*. A non-corroding metal cylinder, 20 cm high, of 24 cm bore, contains a strong conical mould of 20 cm dia. at the foot and 10 cm at the top, 30 cm high, which is filled with concrete. The mould is lifted and the concrete subsides into the 24 cm dia. cylinder. A transparent, slightly weighted disc is lowered gently on to the concrete. A vibrator is switched on until the whole lower surface of the disc is seen to be covered with cement grout. Measured by stopwatch, this time is considered to be that of full compaction, and an indication of its workability in seconds of Vebe time.

o–3 Vebe seconds are equivalent to a slump of 60–180 mm; 3–6 Vebe seconds are equivalent to a slump of 30–60 mm; 6–12 Vebe seconds = 10–30 mm slump; more than 12 Vebe seconds = 0–10 mm slump.

V-cut See **wedge cut**.

VDU (visual display unit), VDT (v.d. terminal), monitor A cathode-ray tube (TV screen) which in business computing usually provides 24 lines of typing, each 80 characters wide.

vector (1) In calculations of forces, a vector is a line whose length is proportional to the size of the force, and whose direction represents its direction.

(2) A vector of disease is a fly, insect, mosquito, mouse, rat, louse, snail or other form of life that carries a bacterium, fungus, virus or other cause of disease from one host to another. See **vector control**.

vector control Elimination or reduction of *vectors* of disease. Drainage of stagnant water eliminates mosquitoes and the malaria they carry. Shredding of refuse and controlled tipping on municipal tips discourages rats, cockroaches, seagulls and flies. DDT has been used to kill insects but is now illegal in many countries because it is persistent and may survive in other forms of life to poison humans. Also, malaria vectors are becoming immune to DDT.

vee notch [hyd.] See **triangular notch**.

vehicle-detector pad A plastic-covered strip at right angles to a road centre line, built into the road surface in a metal rectangular frame. The weight of a vehicle passing over it switches on the traffic lights ahead. See **loop detector**.

vel A unit of size, abbreviated from 'velocity in an air elutriator'. The vel is a measure of the size of an air-transported particle. Thus a particle of x vel in size has a maximum free-falling velocity in air of x cm/s. At 20 cm/s free-falling velocity, (particles 20 vel in size) coal particles are below 76 microns (0.076 mm).

velocities in pipes [mech.] Normal allowable velocities in pipes for different fluids are as follows in m/s: air 9–15, compressed air 7–12, steam 50–75, water 1.5–3. See **transportation speeds**.

velocity head, kinetic h. [hyd.] The energy per unit weight of water due to its velocity (v). It is equal to $\dfrac{v^2}{2g}$ kg-m/kg, where v is in m/s and g is 9.81 m/s each second. The numerical result in kg-m/kg, if written in metres, is also equal to the vertical distance the fluid must fall freely under gravity to reach its velocity v.

velocity of approach [hyd.] The mean velocity in the channel of a *measuring weir* at the point where the depth over the weir is measured.

velocity of retreat [hyd.] The mean velocity immediately downstream from the *measuring weir*.

velocity ratio [mech.] The distance through which the force applied to a *machine* moves, divided by the distance moved by the load. See **mechanical advantage**.

vena contracta [hyd.] The narrowest

veneer

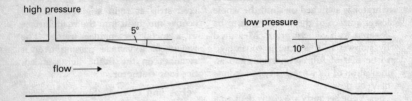

Venturi tube (meter). The great length may be a disadvantage but the head loss is small.

point in the cross-section of a *nappe* or jet beyond the plane of the weir or hole from which it issues. The stream lines converge towards this point and diverge away from it. At this point they are therefore parallel.

veneer See (*B*).

Venice See **groundwater**.

Venturi flume [hyd.] A *control* flume consisting of a short contraction followed by an expansion to normal width. The difference in level of the surface of the water gives a measure of the flow. See **Parshall measuring flume**.

Venturi meter, V. tube [hyd.] (see diagram) A *flow meter* for closed pipes in which there is a constriction (throat) followed by an expansion to normal width. The pressure is measured at the throat where the pressure is reduced and upstream where the width is normal. Tubes from these points lead to gauges. The quantity flowing is related to the pressure difference between these points. This meter can be made into a *water meter*. The head loss of a Venturi meter is usually less than 10%.

Venturi tube See **Venturi meter**.

verge, margin, reservation The unpaved ground next to a road or railway,

often of rough grass, but forming part of the legal highway. See also (*B*).

vermiculite See (*B*).

Verrazano Narrows Bridge The world's longest *suspension bridge* when it was built in 1964 in New York, and possibly still the heaviest. Its main span is 1298 m, with six lanes on each of its two decks. See **Akashi–Kaikyo Bridge, Humber Bridge**.

vertex See **crown**.

vertical alignment See **longitudinal profile**.

vertical circle [sur.] The metal or glass ring graduated in degrees and fractions of a degree, which shows the angle of slope of the telescope on a *theodolite*.

vertical control [sur.] Connecting a survey with *bench marks* that are well within the accuracy required for the survey. In air surveys, the levelling of points on the ground which can be identified on the *vertical photograph* to orient a *stereoscopic pair*.

vertical curve To provide a gradual change from one slope to another, a curve is inserted between two lengths of a road or railway at different slopes. At the foot of a hill the curve is concave upwards, and at a summit, convex upwards.

vertical interval See **contour interval**.

vertical photograph [air sur.] A photograph of the ground taken from the air with a camera pointing vertically down. This is the usual photograph from which *mosaics* are built up. Generally the photographs are taken with the pilot travelling along a strip at a *timing* calculated to give 60% lap between each photograph and its neighbours in the same strip. Neighbouring strips also overlap but only by 30% Cf. **oblique photograph**; see **collimation mark**, **orthophoto**, **plotting instrument**.

vertical sand drain [s.m.] A boring through a clay or silty soil, filled with sand or gravel to enable the soil to drain more easily. Its special purpose is to accelerate the *consolidation* of a loaded cohesive soil. Owing to the low *permeability* of clays, their consolidation is slow since it depends only on the removal of water. Sand drains accelerate the drainage of a clay loaded by an earth dam or other heavy weight if their spacing is appreciably less than the thickness of the clay. They do not need to reach a permeable soil below the clay, but it is preferable that they should do. Sandwicks, developed by Cementation Ground Engineering Ltd, ensure continuity of the sand plug and consequently of the drainage and enable the drainage holes to be small without risk. They are tubes woven from polypropylene, which is permeable to air and water, able to hold sand without loss or breakage, and filled with selected sand before insertion into the hole. See **Kansai International Airport**.

viaduct A bridge of many spans.

vial [sur.] A *level tube* or a circular bubble.

vibrated concrete Concrete compacted by vibration whether from a vibrating table for precast concrete or from an *internal vibrator*. It requires very much less water for effective placing than does concrete compacted by punning; therefore it is much stronger. The *formwork*, however, must also be stronger when the concrete is to be vibrated. Concrete in *hollow-tile floors* is not vibrated.

vibrating pile driver A device containing vibrating electric motors, which is mounted on top of sheet piles or bearing piles to be driven. It can be silent and effective in suitable soil. The principle of vibration can be used also in *pipe jacking*, and in the extraction of piles or pipe.

vibrating plate A unit handled by one person, which is suitable for *compaction* of soil in a confined space. The vibration of the horizontal plate, some 1×1.5 m or larger, fluidizes the soil beneath, enabling the operator to push the plate in any direction.

vibrating roller A self-propelled or towed *roller* with a mechanically vibrated roll. See **compaction**.

vibrating table A steel table with a *vibrator* which compacts a *precast concrete* unit in a mould resting on the table.

vibration of concrete, vibrators Stiff concrete is most easily placed with the help of a vibrator. Vibrators make very stiff concrete flow and enable it to achieve full strength by thorough compaction. Vibration drives out air and brings the concrete into close contact with the steel, the formwork and any inserts. Formwork must be strong and watertight. The time of vibration depends on the *workability* and the *admixtures*. Over-vibration can cause *segregation*. Very strong concrete was made without vibration at *Seattle*. See **Vibroflot**, **vibrating plate**.

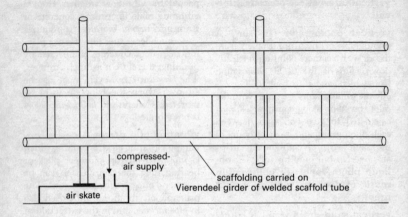

compressed-
air supply

air skate

scaffolding carried on
Vierendeel girder of welded scaffold tube

With four air skates, one at each corner, even a steel tubular scaffold becomes mobile. (Aero-Go Air and Roller Skates.)

vibration of foundations [stru.] Heavy or powerful machines permanently installed in a building should have foundations designed to be of such mass and shape that the machine frequency is $1\frac{1}{2}$–2 times the natural frequency of the combined system of foundation and machine. In any ground a smaller difference than this is undesirable, but in sand or gravel it is dangerous and may lead to rapid compaction of the subsoil and settlement of the machine foundation. See **resonance**.

vibrator *Contractors' plant* for compacting fresh concrete at 3000–10,000 vibrations/min. Internal vibrators (pokers) are used for compacting walls, beams or columns. Surface vibrators for slabs exist in many sizes; some of them are vibrating screed boards. Vibrating tables are used for precasting small units, such as hollow blocks.

vibrocompaction [s.m.] *Dynamic consolidation.*

Vibroflot, vibroflotation, vibro-replacement A *geotechnical* process for compacting clean sands or gravels. A vibrating cylinder of 38 cm dia., 2 m long with water or air jets at both ends is lowered into the ground by a crane. Its vibrations and the lower water jet allow it to sink easily and to compact the surrounding soil. The upper jets enable it to be withdrawn easily. The compaction obtained can be seen at the surface by the size of the crater produced; it may be of 2.4 m dia. The process works best with clean sand but soft silt or clay can also be compacted. The crater is filled with clean gravel or coarse sand to the original level and re-vibrated until a strong, thick column of gravel is formed to the required depth for use as a foundation. *Sand piles* treat the same sort of soil but by impact rather than vibration.

Vicat needle An apparatus for measuring the setting time of a cement by the pressure of a special needle against the cement surface.

Vickers hardness test [mech.] A hardness test used for testing very thin hard *cases* by the diamond pyramid, or for softer materials with a hard steel ball. The indentations are measured by a low-power microscope. Vickers hardness numbers are related to *Brinell hardness numbers*.

video camera An ordinary video camera has been developed for use with a *microcomputer* to record and analyse traffic.

Vierendeel girder, open-frame g. [stru.] (see diagram) A *Pratt truss* without the diagonal members and with rigid joints between top and bottom chords and the verticals. Many welded steel or reinforced-concrete trusses have been built in this way in Belgium and they are named after Prof. Vierendeel of that country. They are convenient to use when diagonal members would be obstructive, as e.g. in stiffening a *buoyant raft* between basement level and ground or first floor.

Vignoles rail [rly] A *flat-bottomed rail*.

virtual slope [hyd.] The *hydraulic gradient*. It shows the loss in pressure per unit length due to friction.

viscometer, viscosimeter An instrument for measuring viscosity which estimates the gel strength of a *drilling fluid* by measuring the viscosity during flow and after it has been standing for some minutes. See **thixotropy**.

viscosity [hyd.] The resistance of a fluid to flow. It is equal to the tangential force per unit of area required to maintain unit relative velocity between two parallel planes at unit spacing.

viscous flow [hyd.] Another description of *streamline flow*.

visibility distance, sight d. The distance for which an object is visible to an observer, both at the same height above the road, sometimes specified as 1.5 m.

visible horizon See **sensible horizon**.

vitrification Conversion to glass. One way of preparing radioactive waste for disposal is to vitrify it instead of pouring it into steel canisters coated with concrete.

voided slab, cored s. [stru.] A slab resembling a *hollow-tile floor*, in which the hollow clay blocks are replaced by blocks of foamed polystyrene or other *void formers*.

void former A block of expanded polystyrene foam or other lightweight material, or an empty box of concrete, clayware, cardboard, sheet metal or plastic placed at or below the neutral axis of a slab to reduce its weight, particularly in a bridge. Void formers must be firmly anchored to prevent them floating.

voids [s.m.] Spaces between separate grains of sand, gravel or soil, occupied by air or water or both.

voids ratio [s.m.] The ratio of the volume of voids to the volume of the solids in a sample of soil or aggregate. For clays the voids ratio may need to be defined at a stated pressure, since clay contracts with rise in pressure. See **critical voids ratio of sands**, **porosity**, **relative density of a sand**.

volumetric efficiency [mech.] The volume of water which leaves a pump cylinder for each piston stroke, divided by the volume swept by the piston (piston area × stroke).

volute [mech.] A spiral casing to a fan or a centrifugal pump, so shaped as to reduce gradually the speed of water or air leaving the *impeller*, transforming it into pressure without shock.

vortex shedding [stru.] A wind blowing on a chimney, tower or bridge will produce, in certain conditions, vortices (eddies) on alternate sides. If the frequency of this 'vortex shedding' approaches the natural frequency of vibration of the structure, the alternating buffeting will make it vibrate, perhaps dangerously. The *Tacoma Narrows Bridge* failed in this way. BS 8100, part 1, provides a flow chart for vortex shedding. On steel chimneys, vortex shedding can be reduced by welding a spiral ribbon outside, leading to the top.

voussoir arch An arrangement of wedge-shaped blocks set to form an arched bridge. After about 1910 this method was completely displaced by concrete or steel bridges, which were both cheaper and lighter in weight, until 1964, when the *Parramatta Bridge* was completed.

W

waffle floor [stru.] A *two-way* floor slab with deep square recesses below that eliminate structurally unnecessary concrete and consequently weight, making large spans possible. The moulds are reusable. See **diagrid floor**.

wagon drill A *rock drill* mounted vertically on a three-wheeled or four-wheeled wagon with steel or rubber tyres. It is used for drilling large or small dia. holes in quarrying and can drill 12 m deep or more.

wagon retarder [rly] See **retarder** (2).

wagon vault See **barrel vault**.

waist See (*B*).

wake [hyd.] Recurring eddies downstream of an obstacle.

waling (wale, waler, ranger USA) A heavy horizontal beam supporting *timbering* at the side of an excavation or the *formwork* for concrete. In trench timbering, pairs of walings are held apart by timbers called *struts* or *braces*.

walking dragline A very large *dragline* which stands on discs (feet) instead of on crawler tracks. The discs are of about 6 m (20 ft) dia, and are used for 'walking' (travel). The ground pressure under them is even less than under *crawler tracks*. Like other draglines, walking draglines can dig only towards themselves.

walking ganger A travelling *ganger*, one who supervises the work of several gangs and is thus not a working ganger.

wall chaser A *brickwork chaser* (*B*).

wall friction Friction between the back of a retaining wall and the retained soil. It generally improves the stability of the wall.

Wallingford procedure, WASSP Methods for *routing* channel flows.

Warren girder [stru.] A triangulated truss consisting of sloping members between horizontal top and bottom members and without verticals. Cf. **Pratt truss**.

wash boring Sinking a *casing* or *drive pipe* to bedrock by a jet of water within it. It is used for casing off loose ground before *diamond drilling* starts in the solid, but not for getting reliable soil samples.

washer [mech.] A steel ring placed under a bolt head or nut to distribute the pressure, particularly when the bolt passes through timber. The washer prevents the timber crushing. See **bevelled washer**.

washland [hyd.] Land enclosed between a river and a flood embankment which may be 1 or 2 km away and 40 km long. The washland both holds the water and allows it to flow.

washout valve (1) [hyd.] A *scouring sluice*.
   (2) A valve for draining a water main.

waste, spoil The excess of excavation over fill – building or mining rubbish.

wastewater, effluent Used water from industry, but also the US term for *sewage*.

wasteway, waste weir [hyd.] A *spillway*.

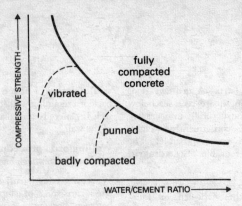

Effect of water/cement ratio on concrete strength. (A. M. Neville and J. J. Brook, *Concrete Technology*, Longman, 1987.)

water authority In 1974 the ten newly created water authorities of England and Wales took over from local authorities their responsibilities for *water supply, sewerage, sewage treatment* and land drainage, followed in Scotland by the *river purification boards* in 1975. They ceased to exist in September 1989, then becoming *water service plcs*. See **Camelford, director-general of water services, National Rivers Authority, privatization of the water industry**.

waterbar, water stop A jointing strip made of plastics (PVC) or other material, built into wet concrete or mortar so as to exclude water after the concrete has hardened. The waterbar may cross a joint and be cast into the pieces both sides of the joint, but this would not be done with rigid water-bars, only with rubbery ones.

water-bearing ground Ground below the *standing-water level*.

waterbound macadam A road surface of gravel or broken stone made by watering clay, sand or *hoggin* into the gaps between stones.

water/cementitious ratio The weight of the water in a mix divided by the weight of cement plus *cement replacement*. The distinction from *water/ cement ratio* is usual only for very meticulous work. See **Seattle**.

water/cement ratio (W/C r.) (see diagram) The weight of the water in a *mix* divided by the weight of the cement. The lower the ratio (the drier the mix), the stronger is the concrete, provided that it is properly compacted. The water absorbed by pores in the aggregate is not included in the W/C ratio. *Lightweight aggregates* absorb much water, hence the term 'free-water/cement ratio'. See **Abrams' law, porosity**.

water content [s.m.] See **moisture content**. (For timber see (*B*).)

water cushion [hyd.] A *stilling pool*.

water cycle The *hydrological cycle*.

water-filled structure [stru.] A steel

frame containing water in its hollow columns, and sometimes also in the beams, to prevent them weakening in case of fire and so to eliminate the need for clumsy-looking but fire-protective cover outside the steel. One spectacular example is the United States Steel Corporation building in Pittsburgh, containing nearly 2000 tons of water in its columns.

The US Steel tower is 256 m (841 ft) high and has 64 storeys. To limit the water pressure, water circulation in the frame is interrupted about every 16 storeys. The water contains additives and is replenished from tanks with ball valves. Water filling is claimed to be cheaper than the usual covering with insulation. European examples exist.

water gauge [min.] A vertical U-tube containing water, used for measuring the pressure difference caused by the mine fan or between the two sides of an air door. 60 mm of water gauge is equal to about 67 millibars pressure.

water hammer [hyd.] Any sudden very high pressure in a pipe caused by stopping the flow too rapidly. *Surge tanks*, *stand pipes* or relief valves are provided on large pressure-pipes to relieve them of this pressure which might burst them. The *hydraulic ram* works by water hammer.

water-jet assisted boomheader (WJAB) In 1987 more than 60 of these *boomheaders* were in use, developed jointly by manufacturers, the US Bureau of Mines and the British Coal Corporation. The water jet, applied at the point of cut, greatly reduces the dust and eases rock cutting. Pressures vary from 14 to 42 MPa (2000–6000 psi), with a maximum water consumption of about 50 litres/min. Possibly in order to reduce water nuisance, German researchers are investigating

much higher pressures of 140 MPa (20,000 psi).

water-jet drilling, water jetting See **hyperbaric, jetting**.

water lowering *Groundwater lowering*.

water meter [hyd.] An instrument which indicates the quantity of water which has flowed from a pipe. It is an integrating *flow meter*.

water metering Payment for water by quantity used, measured by a water meter in the house, usual in Continental countries, is likely to be common in the UK under the *water service plcs*. Another cause for metering is that the current basis for water-rates payment, the domestic rating system, was abolished in 1989–90. Industry and farms are already metered. Southern Water Authority together with three private firms has marketed a water meter which can be read through the phone by *mains-borne signalling*.

water of capillarity [s.m.] Water above the *standing-water level*, but in close contact with it.

water of hydration The water that combines chemically with cement.

water pollution control *Sewage treatment*.

waterproofing *Tanking* (B), applying a *damp-proof course* (B) or *cavity tanking*, etc.

waterproof membrane [s.m.] See **flexible membrane, tanking** (B).

water reducer, plasticizer, densifier An *admixture* which enables the water content of a concrete to be reduced by 10–15%, often with an improvement in *workability*. It should be added as late as possible in the mixing cycle. *Superplasticizers* are both much more powerful and more expensive.

water-repellent cement, hydrophobic c.. *Ordinary Portland cement* with stearates mixed in, often used in screeds and renderings. It is occasionally used also when a cement is needed that has to keep longer than usual in the bag without becoming lumpy. Water-repellent white cement keeps white longer than other cements used in concrete or mortar. *Pore fillers* work differently but they may also help to make a concrete impervious.

water requirement [hyd.] The amount of water from all sources required by a crop for its growth and maturity.

water-retentivity admixture An *admixture* to an oilwell *grout* which reduces loss of water to the surrounding ground.

water service plc Companies which must treat drinking-water and sewage. They were created in 1989 by the Stock Market flotation of the ten *water authorities*. Other water authority duties are taken over by the *National Rivers Authority*. Enormous investments were needed at the time of *privatization*, possibly £30,000 million by the year 2000, for replacing old pipes, improving water quality and sewage treatment to raise UK water and rivers to EEC standards. See **director general of water services**.

watershed [hyd.] (1) **water parting (divide** USA) The summit and dividing line between two *catchment areas*, from which water flows away in two directions.

(2) In the USA the term watershed may mean a catchment area.

water stop A *waterbar*.

water supply [hyd.] The *impounding*, treatment, pumping and piping of

drinking-water for customers by a public authority. In Britain *water service plcs* have existed only since 1989, but several private water companies happily coexist with them. The Water Act of 1973 reorganized the water supply and sewage disposal industries of England and Wales, as well as flood control and drainage of land generally. See **privatization of the water industry, waterworks**.

water table, saturation line [s.m.] The surface, which may be undulating, of the *standing-water level*. Cf. **perched water table**. It is important in *earth dams* that the saturation line does not reach the surface of the dam. For this reason a *graded filter* is often provided at the toe of the dam. The water table is the boundary between the zone of saturation below it and the zone of aeration above it.

For many centuries water was pumped from wells in London but this has stopped now that heavy industry has moved away. Consequently the water table which had been falling for centuries is now rising. The result may be that buildings in London will rise because of the swelling of the clay below them. See **rising groundwater under cities, subsidence**.

water test The *hydraulic test* for drains.

water treatment [hyd.] *Waterworks* processes that make water drinkable, including water softening to make it suitable for washing or for use in steam boilers. Iron also needs to be removed since it colours the water. See **preliminary treatment**.

water-tube boiler [mech.] A steam boiler in which the water and steam pass through tubes in the furnace. It is very much larger and more efficient than the *fire-tube boiler*.

water turbine [mech.] A wheel turned by the force of water, a development from the *water wheel* which was generally not of high power or very efficient. Turbines now made attain 500,000 hp. The main types used are the *Pelton wheel* for high heads, and the *Francis turbine* for low to medium heads, modified in the USA as the *Kaplan turbine*. The main difference between the Pelton wheel (impulse turbine) and the Francis type (reaction turbine) is in the very high speed of the Pelton jet. The Pelton jet gives its energy to the wheel and the wheel is not wholly surrounded by water but partly by air. *Reaction turbines* are not acted upon by a jet but by relatively slowly moving water at high pressure which completely surrounds the wheel and forces it round. See **specific speed**.

water wheel A wheel turned by the force of flowing water. It was the successor to the *treadmill* (*B*) and some water wheels were shaped like it. The term is now reserved for overshot and undershot wheels but it could logically be used for the much more powerful and efficient modern *water turbines*.

waterworks A site at which *raw water* is treated and sent out as a *water supply*. Treatment usually includes *screening*, *coagulation*, *flocculation*, *clarification*, *filtration* and *disinfection*, in that order. Although waterworks are now an essential part of life in the UK, it should be remembered that as their output increases, so does the volume of sewage. Therefore if sewage treatment plants do not increase their throughput more rapidly than waterworks, the amount of sewage pollution is bound to increase.

wattle work *Hurdle work*.

wave pressure The pressures on breakwaters from breaking waves may amount to 0.3 MPa (44 psi) on the Atlantic coast but to 0.1 MPa (14 psi)

in the more sheltered Great Lakes of North America.

waybeam [rly] A timber or steel beam parallel with and under each rail where a bridge is being repaired.

wearing course, carpet, road surface, topping The top, visible layer of a road.

weathering steel [stru.] Several low-alloy steels mentioned in BS 4360, 'Weldable Structural Steels', have various percentages of chromium (0.3–1.25%), manganese (0.6–1.5%) and copper (0.25–0.55%). Perhaps the best-known type is Cor-Ten in the UK and the USA. These steels rust appreciably in a wet climate.

web [stru.] The vertical plate joining the *flanges* of any *beam* or *rail*, of whatever material.

web stiffener [stru.] See **stiffener**.

wedge See **plug and feathers**.

wedge cut, V-cut, centre c. [min.] In tunnelling or shaft sinking, a layout for the *cut holes* in which they point towards each other and nearly meet.

wedge pile See **expanding pile**.

wedge theory [s.m.] Coulomb's analysis of the force tending to overturn a retaining wall (1776). He based it on the weight of the wedge of earth which would slide forward if the wall failed. See **sliding-wedge method**.

wedge-wire screen A screen made of wedge-shaped parallel wires with their wide edges uppermost. This sort of screen rarely clogs and is used for dewatering sands or grits over openings from 4 to 0.1 mm and with or without vibration.

weephole A hole to allow water to escape from behind a retaining wall and thus to reduce the pressure behind it. The water leaking through may make unsightly stains on the wall. Sometimes, therefore, weepholes are replaced by a *field drain* of ample capacity, carefully laid below coarse stone behind the wall, at the level where the weepholes would have been, to drain the water out to the end of the wall. This field drain may be called a back drain.

weigh batcher A *batching plant* for concrete in which all the ingredients of a mix are measured by weight (except water). A correction must be made for the weight of the water in aggregates, especially if they are stored out of doors as usual. A moisture meter can provide the necessary information.

weighbeam A steel plate measuring about 3×1 m which weighs any axle passing over it and may transmit the figures to a *transducer*. It is smaller than a *weighbridge*. See **Clifton Bridge**.

weighbridge A steel weighing platform at road level, large enough to take all the wheels of a large lorry. The weight of material loaded on to the lorry is found by subtracting its weight empty on entry from its weight full on coming out.

weighted average [stat.] An average obtained from a series of values after each value has received an appropriate 'weight'. The weight of each value is a factor by which it is multiplied, corresponding to its trustworthiness and importance. After the values have been multiplied by their weights they are added together, and then divided by the sum of the weights (not, as usual, by the sum of the values). The result, an average which takes account of the opinions of the observer about the values, should be more reliable than a simple average.

weighted filter See **loaded filter**.

weighting (1) [stat.] Applying weights to observations to obtain a weighted

value as for *degree days* (*B*), *weighted averages*, etc. The differences between partial *load factors* are a way of weighting different loadings.

(2) Loading a *pneumatic caisson* with cast iron, water, or other *kentledge* to make it sink and prevent it floating.

weir [hyd.] A wall built across the full width of a stream, with a horizontal crest over which the water flows. It is built to hold up the water, to help boats to pass through a lock or to reduce floods, or as a *measuring weir*.

weir head [hyd.] The depth of water in a *measuring weir* measured from the weir crest or the bottom of the notch to the water surface upstream of the weir. It does not include the *velocity of approach*.

Weisbach triangle [sur.] A *set-up* at the foot of a vertical shaft during *shaft plumbing* from which both of the plumb wires can be sighted, thus forming a triangle with an angle at the *theodolite* of up to half a degree. More accurate observations of the wires can be taken in this way than with the theodolite in line with the wires. The Weisbach set-up is used for orienting the underground workings.

weld [mech.] A joint between pieces of metal or plastics at faces which have been melted or made plastic by heat or pressure or both. *Filler rods* may be used. See **welding**, etc.

weldability The ability of a combination of parent metal and *filler rod* in a well-designed procedure to make a strong, corrosion-resistant weld. Weldability is sometimes wrongly regarded as the opposite of *hardenability*.

welded frame [stru.] A *bent* consisting (at its simplest) of two legs and a beam joined by welding. It is architecturally attractive, saves steel and sometimes may be cheaper than a frame of trusses and stanchions.

welded mesh Steel bars welded at their intersections into a rectangular grid as *reinforcement* for *slabs*. Many qualities exist.

welding [mech.] Joining two metal or plastic surfaces by fusion or pressure or both, heating them by electric arc, electric *resistance welding* or flame. Most methods of welding structural steel are *arc-welding* methods. See also **automatic welding**, **butt weld**, **carbon-arc/fusion/metal-arc/plastic welding**, **thermit**.

Welding Institute An organization unique in the UK in combining research with the functions of an *engineering institution*. Its membership includes some 5000 welding engineers and welding technicians. The research division, at Abington, near Cambridge, England, employs about 500 people and has research member companies in many countries.

welding on building sites, field w. Many specifications discourage the welding of steel on site, preferring *shop welds*, and forbid the welding of wet surfaces or any welding at freezing point or below.

welding thermit [mech.] See **thermit**.

well [hyd.] (1) A shaft or borehole sunk in the ground to obtain water, oil, etc.
(2) An *absorbing well*.
(3) A *well foundation*.

well borer, w. sinker A *skilled man* or tradesman who sinks a *shallow well* with pick and shovel, sometimes doing no timbering until reaching the bottom, and then steens the well with bricks laid dry (or with mortar) for the full height. The work is dangerous, but sinkers make their own estimates, do their own sinking, and often work only in their own district, where they understand the ground.

well-conditioned triangle [sur.] A triangle which is as nearly as possible equilateral. In such a triangle an error in the measurement of an angle has least effect in creating errors of length.

well curbing, pit boards US term for sheeting to keep earth out of a pit.

well drain An *absorbing well*.

well foundation A *foundation* used in India and the USA. It is excavated like the Gow or *Chicago caissons* by sinking a small-diameter pit.

well hole A large vertical borehole, of about 15 cm dia., used in quarries when blasting a heavy *burden*.

well logging, geophysical l., wireline l. Many sophisticated modern ways of discovering the properties of rocks around a borehole and the fluids they contain. Most have been developed from those of Schlumberger, a French mining engineer who first used them around 1930. His company flourishes in many countries. A 100 mm (4 in.) dia. *sonde* or probe, connected by electrical cables to recording devices at the surface, is dropped into a borehole on a wire rope. Measurements are made as the sonde (or series of sondes which may be 30 m long) is pulled slowly to the surface. The electrical resistivity of the soil is one variable usually measured, but electrical induction, electrical potential, gamma ray, acoustic and other methods exist. Resistivity surveys require an electrically conducting drilling mud. Cf. **borehole surveying**.

wellpoint A tube of between 38 and 63 mm dia., sunk usually by *jetting* into a water-bearing soil, fitted with a close-mesh screen at the foot and connected through a *header* pipe to a suction pump at the top. Usually a number of wellpoints are connected to one header. They are sunk outside an excavation to reduce the pumping required within it and to increase the strength of the ground by the flow of water towards the wellpoints, away from the excavation. An excavation in sand cannot become quick if effectively wellpointed. Filters are created around the feet of the wellpoints by mixing coarse sand with the jetting water as the wellpoint approaches its formation level at the end of the jetting. The sand must be coarser than that in the formation. The water flow is also reduced, enabling the coarse sand to settle around the pipe and form a filter around the tube screen. See **groundwater lowering**.

well sinker See **well borer**.

westing [sur.] A *coordinate* measured westwards from an *origin*; a westward *departure*.

wet analysis [s.m.] The *mechanical analysis* of soil particles smaller than 0.06 mm (the smallest convenient standard sieve). It is done by mixing the sample in a measured volume of water and checking its density after various periods with a sensitive *hydrometer*. Several days may be needed for a test. Even with a *centrifuge* to give the particles a settlement force of many hundreds of times gravity, the test is slow. See **dispersing agent, Stokes's law**.

wet cube strength Since the wet strength of any concrete is well below its dry strength, it is important in cube tests to specify whether the cube is tested wet or dry, and if wet, exactly how wet. For the sake of uniformity and fairness, many BS specify water saturation, with the surface water wiped off, but this is not always fair, particularly to aerated concretes, which have a low wet strength.

wet dock A *dock* in which the water is kept at high-tide level by dock gates which are opened only at high tide. A lock is usually provided to enable vessels to pass in or out at all states of the tide.

wet drilling [min.] Drilling in hard rock, with water injected down a hole in the centre of the drill to keep the dust down and reduce the danger of dust disease to the people drilling.

wet galvanizing [mech.] A coating of zinc laid on to steel by passing it through a bath of molten zinc on which floats a layer of flux. Before dipping, the steel is passed through pickling tanks. See **galvanize**.

wet mix Concrete with too much water. See **Abrams' law**.

wet-sand process [s.m.] A *soil stabilization* process for sand. Special road oil (SRO, a creosote with cut-back bitumen and a wetting agent) and hydrated lime are mixed with the sand in the proportions of about 6% SRO and 2% lime.

wet silt The form in which silt occurs in a river or pond, occupying about five times its volume dry.

wet spraying, slurry s., wet shotcreting *Shotcreting* or *guniting* with aggregate, cement and water premixed before the nozzle. The nozzle needs only two hoses, one for the mix and the other for the air blast; it is more difficult to place at a low water content than *dry mix*. Accelerator is usually added at the nozzle; see **steel-fibre-reinforced shotcrete**.

wetted perimeter [hyd.] The length of surface in contact with water in a channel, measured round its curves. See **hydraulic mean depth**.

wetting agent A *surfactant*.

wet welding *Welding* of steel under water; it requires great skill but is of limited use since the welds are porous and brittle. See **hyperbaric**.

wet well [mech.] The *sump* of a pumping station. Cf. **dry well**.

wharf A *berth* parallel to the waterfront. It may be of solid or open construction. In solid construction the wharf holds back all the earth behind it like a retaining wall. In open construction no earth is retained; the wharf is carried on piles of timber, steel or precast concrete driven into the bed. Open wharves can be used only for vessels of shallow draught. Cf. **jetty**.

wheelabrating [mech.] *Shot blasting* with steel grit or shot thrown from a fast-spinning wheel.

wheeled loader Normally a *loader*, though tracked loaders do exist.

wheel excavator [min.] See **rotary excavator** (2).

wheel gauge [rly] The distance from outside one wheel flange to outside the other. It is 13–19 mm less than the *gauge* of straight track.

wheel scraper A *bowl scraper*.

whip See **bond** (2).

Whipple-Murphy truss A bridge truss like the *Pratt truss*, but rather more complicated.

Whirley crane A large revolving crane used in the USA, mounted on wheels, rollers, skids or a gantry.

whiskers Tiny single crystals of much greater length than diameter (about 1 to 10 microns dia.) which have been proved to be at least 20, sometimes 50, times stronger than steel. The materials so far made as whiskers include alumina, (Al_2O_3), carbon (graphite) and silicon carbide (Carborundum). They might be developed to become a reinforcement for concrete. Glassfibres are strong for the same reason that whiskers are strong – their very small diameter reduces the possibilities for flaws.

white cement *Portland cement* made from clays and limes with no iron in them, and if burnt with coal, from coal with no pyrite (FeS_2) to discolour it.

white metal, anti-friction m. [mech.] A tin-based alloy, more than half tin, containing also lead, copper and antimony. It is used for lining bearings, capping ropes and making small castings.

Whitney stress diagram [stru.] The diagram of stress distribution in a reinforced-concrete beam according to the ultimate load theory (*plastic design*).

whole-circle bearing [sur.] A *bearing* which defines a direction by its horizontal angle measured clockwise from true north.

whole-tide cofferdam A *full-tide cofferdam*.

wicket A small gate or door forming part of a larger one, hence a sluice in a lock gate.

wicket dam [hyd.] A movable barrier made of *wickets* or shutters revolving about a central axis.

wide-gauge track [rly] *Broad gauge*. *Portal cranes* may have a gauge as broad as 3 m (10 ft).

wilting coefficient [hyd.] The water content of a soil, below which plants wilt and die even in damp air. It varies from 0.9% in coarse sand to 16% in clay.

winch [mech.] A hand-operated winding drum or a small engine-driven haulage or hoist.

wind beam, windbrace [stru.] A beam inserted into a structure solely to resist *wind force*.

wind energy

wind energy See **Burger Hill**.

wind force, w. load [stru.] The force on a structure due to *wind pressure* multiplied by the area of the structure at right angles to the pressure.

windmill A wheel turned by the wind to generate power, pump water or grind corn, etc. See **prime mover**.

wind portal [stru.] A portal designed to resist *wind force*.

wind pressure [stru.] The pressure in kN/m² of wall or roof area due to wind. It increases with the wind velocity roughly according to the formula stated under the *Beaufort scale*. On sloping roofs a suction on the lee side occurs which may be nearly as much as the pressure on the windward side.

windrow [s.m.] A ridge of soil consisting of spill from a *grader* or railway *ballast cleaner*. See **blading back**, **travel-mixer**.

wind shear, w. surge [hyd.] The effect of wind in raising the water level on a lee shore. In 1953 wind raised the sea level by 2.7 m in East Anglia and 3 m in the Netherlands. Wind-induced oscillations of water level occur in enclosed waters such as Windermere, Loch Ness, Lough Neagh and the Baltic Sea – in Lake Geneva as 'seiche'. In the Baltic the period of oscillation is about 15 hours, in the lakes from 20 to 45 min according to the size and *fetch* See **storm surge**.

windshield [stru.] Usually a tubular concrete wall that protects the flues within it from wind forces. The tallest in the UK in 1973 was at Drax power station, 259 m high. The second tallest, at Grain power station, tapers from 40 m outside dia. at the base to 24 m dia. at the top, 244 m above ground. The wall thickness at ground level is 1.5 m but only 380 mm thick for the top 100

m. There are five circular brick flues inside the windshield, each of 7 m dia. See **Dartford Bridge**.

Windsor probe See **penetration resistance test**.

wind tunnel A usually conical or cylindrical structure through which air is blown at measured speeds to test its effect on models of suspension bridges, towers and other shapes which may be affected by wind.

windy Scots term for pneumatic. Thus a windy drill is a *rock drill*.

wing dam [hyd.] A *spur*.

wing wall, abutment w. A wall at the abutment of a bridge which extends beyond the bridge to retain the earth behind the *abutment*.

winter gritter See **gritter**.

wire drawing (1) [mech.] *Cold drawing*.
(2) [hyd.] The pressure drop in a flowing fluid as it passes through a small opening, for example in a *Venturi meter*.

wire for prestressing *Prestressing* wire may be *as-drawn* or prestraightened.

wire frame A style of semi-perspective diagram made by *computer graphics*, in which the edges of solid objects are drawn, including those that in reality are hidden. Developing the wire frame to a picture that corresponds more with the seen image requires the use of a computer program for 'hidden-line removal'.

wire gauge [mech.] See (*B*).

wire lath US term for *expanded-metal* lathing.

wire-line core barrel A *core barrel* contained within the drill pipe of a diamond drill, which can be withdrawn by steel rope, and is suitable for vertical or steeply sloping holes. It avoids the delay of pulling out the drilling pipes when a core is being withdrawn and so allows fast coring progress.

wire-line logging A method of *well logging* in which one or more *sondes* are pulled up inside the drill pipe.

wire-mesh reinforcement Chicken-wire, *expanded-metal* or welded intersecting fabric of light rods, used for reinforcing concrete or mortar.

wire-rope [mech.] Steel-wire rope. See **lay, rope, strand**.

wobble-wheel roller A *pneumatic-tyred roller* with wheels which are suspended freely on springs to follow the irregularities of the road, used in *soil stabilization*.

wood-block paving Softwood blocks used as a road surface by laying them with the grain vertical on a smooth concrete base in hot tar which glues them to the concrete. It was much used in British cities before 1940, but the cost of wood blocks has now made it uneconomical. See also (*B*).

wood-stave pipe Pipes built of wood boards tied together with steel straps. They have only a short life when they are alternately wet and dry but if continuously wet they last longer. They are sometimes used in the USA for water supply and for transporting hydraulic filling. Melbourne, Australia, still had some in use in 1991.

work [mech.] The product of a force and the distance through which it moves, thus the work done in raising 5 kg through 2 m (or 2 kg through 5 m) is 10 kg-m. It is to be distinguished from *energy* and from power, which is a rate of doing work. Energy can, however, be expressed in the same units as work, and often is. See conversion factors and **efficiency, horsepower**.

workability, pumpability The ease with which concrete can be placed or pumped, especially through congested reinforcement. Wet concretes are workable but usually weak. Workability can be measured by the common *slump test* as well as by the *compacting factor*, *V-B consistometer* and other tests. It improves with more *cement paste* and with rounded rather than angular aggregate. See also **Abrams' law, consistency, water reducer**.

work-hardening, strain h. [mech.] See **hardening**.

working chamber The chamber at the foot of a *pneumatic caisson* in which the workers stand on the ground and work in *compressed air* at excavation or construction. It is connected to the outside world by a *man-lock*, sometimes also by a *materials lock* or *hydraulic ejector*.

working drawing [d.o.] A *detail*.

working shaft A shaft sunk to excavate a sewer or other tunnel and filled in after the sewer is built.

working stress See **permissible stress**.

world population See **pollutants**.

worm conveyor [mech.] A *helical conveyor*.

wracking forces [stru.] Horizontal forces in the plane of a *bent*, e.g. wind or crane *surge*, which tend to distort a rectangular shape into a parallelogram.

wrapping See **pipe wrapping**.

wrecking strip A *crush plate*.

wrought aluminium alloy [stru.] *Light alloy* which has been cold-rolled, forged, pressed, drawn or *extruded*, and is therefore not described as 'cast'.

wrought iron [mech.] Very malleable,

fibrous pure iron, with such a low carbon content (less than *dead-mild steel*) that it cannot be *hardened* by quenching. It is soft, does not rust so easily as steel but is more expensive, and therefore has been replaced by dead-mild steel or manganese steel even for such uses as small domestic water pipe or chain-making.

It has not been made in the UK since 1974, but blacksmiths, the main users, find old anchor chain suitable or the internal bracing of gasholders, the tie-rods from *trussed beams*, etc. Generally any iron that is not cast, and is older than about 1860–70, is wrought iron, not steel.

wythe US term for a *withe* (B), a leaf of a *cavity wall* (B).

X

XLPE, cross-linked polyethylene An unusually tough type of *polythene*, used as a cable insulation.

XPM *Expanded metal* (wire lath in the USA).

Xylonite [stru.] A *thermoplastic* material like celluloid, used for making models in *photoelasticity*.

Y

yard [sur.] A unit of length, 36 in. or 3 ft or 914.4 mm long.

yard trap A *gulley trap*.

yield (1) [mech.] The permanent deformation (*permanent set*) which a metal piece takes when it is stressed beyond the elastic limit. A piece which yields is in lay terms bent, stretched or *buckled*.

(2) A slight horizontal and slighter vertical movement of a loaded retaining wall. This movement allows the earth thrust to fall to the value known as the *active earth pressure*. See also **volume yield** (*B*).

yield point, y. stress [mech.] The stress at which noticeable, suddenly increased deformation occurs under slowly increasing load. This occurs for mild steel at a stress slightly above the *elastic limit*. For *light alloys*, and for cold-drawn or high-tensile steels, many of which do not have such a pronounced yield point as mild steel, the 0.1 or 0.2% *proof stress* is taken as the yield point for estimating safe stresses.

yoke (1) A *spreader beam*.

(2) Stiff steel angle-sections on timbers bolted in a horizontal frame round the *formwork* for a rectangular column. The formwork is wedged tightly to the yoke during the casting of the concrete.

Young's modulus [mech.] The *modulus of elasticity* of a material.

Ytong An *aerated concrete* used in the UK and elsewhere, sometimes reinforced with steel that is protected against corrosion. Consequently, unlike most aerated concretes, it can carry load.

Z
___

Z Abbreviation for *modulus of section*.

zee US spelling of Z, pronounced zed in Britain.

zeolite Many types of mineral used in *base-exchange* water softening. *Ion exchange* resins are 'synthetic zeolites'.

zirconia [min.] ZiO_2, zirconium oxide, a *refractory* which can be used at very high temperatures.

zoned construction of earthen dams Where there is not enough clay or where conditions are unsuitable in an earthen dam to place large volumes of clay, zoned construction makes good use of local materials. The clay *cut-off wall* has *graded filter* zones each side of it, transition zones beyond these and rock fill outside.

zone of aeration [hyd., s.m.] The ground above the *water table*. In deserts it may be 600 m (2000 ft) thick.

zone of saturation [s.m.] The ground below the *water table*.

APPENDIX 1: DATES

Civil engineering probably began with irrigation but some interesting dates before then are:

BC

30,000	(Palaeolithic period) Glaciers retreat from northern Europe.
9000	(Mesolithic period) Gold and silver jewellery, copper, pottery.
8500 –7000	First farming by slash and burn.
8000	People living in both Americas.
6000 –5000	Copper is smelted from its ores.
4500	Bronze is smelted from ores of tin and copper.

Engineering dates BC (*in the Middle East unless otherwise stated*)

5000 –4000	Irrigation of the river valleys of southern Iraq (Sumeria), Egypt, Indus Valley, China.
3500	Cart with solid wheels (see 1800 BC).
3000	(Neolithic period) First Egyptian pyramids. Cuneiform writings in cities of Sumeria imply the existence of banking. Field cultivation exists in monsoon Asia. Animals pull ploughs.
3000 –1700	Megalithic civilization (Stonehenge, etc.).
2800	Khanats.
2750 –2500	Sadd el Kafara dam near Helwan, Egypt, 11 m (37 ft) high.
2500	Bronze weapons in Kuban, south Russia.
2500 –1500	Indus civilization (Mohenjo Daro, Harappa) has water supplies and waterborne waste disposal.
2280	The Great Yu, successful hydraulic engineer, becomes emperor of China after taming the Yellow River.
2000	China produces bronze. Brick ziggurats built in Iraq.
1900	Minoan stone palace in Crete with waterborne waste disposal, but servants carry water upstairs in buckets.
1800	Cart with spoked wheels (see 3500 BC).
1700	Hammurabi's cuneiform text mentions the standardization of bricks.
1300	Alphabetic writing in Palestine and Syria (not cuneiform, hieroglyphs or pictographs).
1200	Iron age supplants bronze, possibly even earlier.
850	Bridge built over River Meles (Izmir, Turkey), still standing.

700 A stone-lined aqueduct 20 m (66 ft) wide and 80 km (50 miles) long was
–600 built by Nebuchadnezzar in 15 months to bring water to Nineveh,
Assyria.

600 First stone bridge in Rome.

612 Mobile horse-borne archers from south Russia conquer Assyria, one of
the most powerful kingdoms of the Middle East.

530 Eupalinos digs the first water tunnel on the island of Samos, 1 km long,
starting at both ends simultaneously.

512 Mandrokles of Samos builds a pontoon bridge over the Bosporus for
Darius I of Persia's army.

500 Lathe in common use for turning wood.
Steel swords made in China.

490 Hippodamus of Miletus plans the Athenian port of Piraeus, probably the
first town planning venture.

440 The Parthenon built by Iktinos in Athens.

350 Iron tools and weapons in China.

332 Alexandria built to a town plan designed probably by Deinokrates.

312 Appian Way, the first paved Roman road.
Aqua Appia, the first aqueduct, brings clean water to Rome, but see AD
226.

300 Iron smelted in Chinese blast furnaces, 1000 years earlier than in
Europe.

285 Ptolemy Philadelphus rebuilds the Egyptian canal to the Red Sea from
the Nile Delta.

265 Suspension bridges in China hang from spun-cane-fibre ropes.

260 Sostratos builds the Pharos (lighthouse) at Alexandria, containing 300
rooms on lower floors to house the garrison.

206 Iron-chain suspension bridges erected in China.

145 Aqua Marcia, the first high-level Roman aqueduct.

27 Pantheon in Rome built by Agrippa: dome of 43 m (142 ft) span.

18 Pont du Gard built by Agrippa to supply water to Nîmes.

Engineering dates AD

14 Water mills used for grinding corn, and later for other purposes.

99 Trajan's bridge over Danube near Orsova (the Iron Gate at Turnu
Severin), with 20 stone piers joined by wooden arches.

200 Wheelbarrows used in China.

226 The last of the 11 aqueducts was built to bring water to Rome, but by AD
1000 the only Roman water supply was the Tiber (see 312 BC).

260 Water-powered trip hammer used in China.

494 Submersible suspension bridge built in China over the Huai River.

537 Istanbul's Saint Sophia (Cathedral of Divine Wisdom) completed by
Anthemios of Tralles and Isidorus of Miletus, with brick arches spanning
30 m (100 ft).

500 Hollow clay pots used in dome of San Vitale Church, Ravenna, to reduce
–600 weight of dome.

600   Chinese Grand Canal joins the capital Chang An to Hangchow in the south.
      Windmills in use in Persia.
860   Jhelum River flooding controlled by Suyya.
1060 Locks built in Chinese canals; possibly even 100 years earlier.
1140 The first Gothic building, the abbey church of St-Denis, France, begun by William of Sens.
1174 Choir of Canterbury Cathedral, also Gothic.
      Building sites use the capstan, the treadmill and the horse-gin (and at bridges the water wheel) for hoisting materials until the steam engine is developed.
1176 London Bridge built; not replaced until 1831.
−1209
1212 Fire regulations forbid the use of thatch in London.
1283 Chinese and Mongol engineers complete the modern Grand Canal from Hangchow to Peking, a *summit* canal.
1300 Bricks come into use again in England, at first imported from the Continent.
1377 Duke of Milan bridges the River Adda with a 72 m (236 ft) span arch.
1400 First European canals with locks, in the Low Countries.
1402 Hammer-beam roof of Westminster Hall, London, completed; 20.5 m span.
1485 Leonardo da Vinci's mitre gate on canal locks near Milan, superseding overhead gates, allows tall ships to pass.
1495 First English dry dock at Portsmouth.
1580 Gravity dam, 42 m (138 ft) high, built at Alicante, Spain.
1596 First WC used in England, the beginning of English river pollution.
1597 Horse-drawn colliery railways with wooden rails.
1613 Hugh Myddelton's New River supplies water to London, probably through spigot-and-socket elmwood water pipes.
1642–9 English Civil War.
1653 Vermuyden completes the drainage of 302,000 acres of English fens.
1600 Many small navigations dug in England and in Continental Europe.
−1700
1654 Guericke in Magdeburg demonstrates the existence of atmospheric pressure.
1660 Royal Society founded by Charles II.
1663 London–Edinburgh road first made into a turnpike.
1666 Great Fire of London.
1668 Isaac Newton's first reflecting telescope.
1681 Languedoc canal completed; first tunnel driven by blasting.
      Leather water pipes introduced from Holland enable fire engines to pump at a distance from the fire.
1712 First Newcomen pumping engine installed near Dudley Castle, Worcs.
1716 Corps des Ponts et Chaussées formed by the French monarchy.
1743 Collapsing dome of St Peter's, Rome, investigated mathematically – one of the earliest instances of structural analysis.

1747   École des Ponts et Chaussées formed by the king of France, the first training school for civil engineers in Europe, giving France a century's lead in European structural theory and analysis.

1748   Smelting of iron with coke by Abraham Darby.

1750   Mapping of France by triangulation begins.

1750   Westminster Bridge piers built using cofferdams of dovetailed squared timbers, and wooden foundations 9 × 24 m (30 × 80 ft).

1759   Eddystone lighthouse completed by Smeaton, preceded by careful research into the making of hydraulic cements.

1760   Russian canals are well developed; 3000 80-ton barges a year travel by canal between the rivers Neva and Volga.

1762   Brindley completes the Duke of Bridgwater's canal, starting the 'canal fever'.

1769   James Watt first patents his steam engine.

1771   Smeatonian Club (Society of Engineers) formed in London.

1774   Perronet's Pont de Neuilly (not demolished until 1956).
John Wilkinson builds the first accurate boring mill for making steam engine cylinders.

1775   Watt and Boulton become partners to build steam engines.

1776   American Declaration of Independence.
John Wilkinson's 0.96 m (38 in.) dia. cylinder steam engine provides the blast for his iron furnace.

1779   Abraham Darby's cast-iron bridge spans the Severn at Ironbridge.

1780   Parys Mountain, Anglesey, is the world's largest copper mine.

1784   Henry Cort's puddling process for making wrought iron from pig iron.

1787   William Roy in Kent links the triangulations of England and France by theodolite.

1789   French Revolution begins.

1794   Completion of Charolais canal connecting the Mediterranean with the English Channel.

1795   École Polytechnique established to prepare French boys for a civil or military engineering career. Girls were admitted nearly 200 years later, when President Mitterrand's first tenure made it co-educational.

1796   Roman cement patented, a hydraulic cement whose constituents are found naturally in the London Clay at Harwich and Sheppey.

1799   Maudslay's all-metal screwcutting lathe invented.

1801   James Finley's iron-link suspension bridge built over Jacob's Creek, Pennsylvania, with two spans of 21 m (70 ft), patented in 1808, followed by similar but longer bridges.

1804   Telford begins the Caledonian Canal.
Marc Brunel, on the Admiralty's behalf, commissions Henry Maudslay to build machines for making wooden pulley blocks, completed in 1809.

1805   William Jessop's Thames River Dock, the first secure dock and storage.

1807   Fulton's steamship *Clermont*, the first commercially successful passenger steamship, regularly plies on the Hudson River.

1808   Maudslay and Marc Brunel's pulley-block machines completed.

1809   Cayley's glider flies.

1810   Stockholm–Göteborg canal, through the central lakes, 575 km (360
-32   miles) long, of which only 87 km had to be dug but largely in rock, with
58 locks, under Telford's direction.

1818   ICE founded.
L. J. Vicat's paper on hydraulic cement published.

1819   USS *Savannah* sails from New York to Liverpool in 31 days.
J. L. McAdam's *Practical Essay on Roads*.

1820   Canvass White in the USA patents hydraulic cement.
John Birkinshaw patents the rolling of wrought-iron railway rails.

1817   Erie Canal links the Great Lakes with New York; 580 km (364 miles)
-24   long.

1824   Aspdin in Leeds patents Portland cement.

1825   George Stephenson's Stockton–Darlington Railway opened, with some
lengths leased for horse-drawn trains.
Erie Canal completed in the USA.

1826   Neilson's hot blast for smelting iron greatly reduces cost of iron.
Telford's Menai suspension bridge (see 1849).

1827   Slow sand filters used for purifying London water.

1828   Marc Brunel uses concrete made of hydraulic cement for infilling the
bed of the Thames over his tunnel.
Corrugated iron made by J. Walter of Rotherhithe (see 1836).

1830   Liverpool and Manchester Railway opened. The first to use steam loco-
motives for all trains, to do away with leasings and to run the system
itself. It pioneered season tickets, excursion tickets and much else.
Telford's London–Holyhead road completed, including the Menai sus-
pension bridge.
Tarmac first used (Nottinghamshire), with coal tar.
In the USA steam-powered excavators build railways.
Earth-drilling derrick developed from ships' derricks.

1831   W. Bickford invents miner's safety fuze.
Furnace gases from blast furnaces used for heating the blast first at
Würtemberg.
Michael Faraday invents the electric generator.

1832   First horse trams in New York.

1834   Grand Pont Suspendu over the Sarine at Fribourg, Switzerland, built by
Joseph Chaley, spanning 273 m (895 ft), a world record until 1849. This
was the first European wire-rope suspension bridge.

1836   French chemist, M. Sorel, galvanizes sheet metal (see 1828).

1837   *Great Western*, the steamship designed by Isambard Kingdom Brunel,
with engines by Joshua Field, crosses the Atlantic under its own power.

1838   London streets paved experimentally with wooden blocks from Russia.

1840   Glasgow University, first chair of engineering.

1841   University College London, first chair of engineering.

1842   First shield-driven tunnel under the Thames completed by Marc Isambard Brunel and his son Isambard Kingdom Brunel.
Steam hammer invented by Nasmyth.

1845   Allegheny Aqueduct at Pittsburgh completed by John Roebling, the first US suspension bridge to be built by spinning in place.

1848   Lambot in France builds two dinghies of ferrocement.

1849   In France, Joseph Monier's reinforced-concrete tubs for orange trees; he patents RC beams in 1877, later RC bridges.

Britannia Bridge, tubular wrought-iron railway bridge over the Menai Strait, built by Robert Stephenson, helped by Fairbairn and Hodgkinson (see 1826).

Ohio Bridge at Wheeling, West Virginia, built by Charles Ellet, a French American, to 308 m (1020 ft) span, beating Chaley's suspension-bridge record of 1834.

1850   Russian navigable waterways amount to 80,000 km (50,000 miles).

Crystal Palace, London, built in six months, the first industrialized building (cast and wrought iron).

Zola Dam near Aix-en-Provence, one of the first arch dams; 36.5 m (120 ft) high.

Basse Chaine Suspension Bridge over the Maine at Angers, France, collapsed when 478 soldiers marched over it; 226 died. This bridge was carried on two steel wire ropes and spanned 102 m (335 ft). Virtually no French rope suspension bridges were built then until after 1870.

1851   Pneumatic caissons used for building the piers of bridges at Rochester and Chepstow.

Cast-iron frame buildings in cities.

Dover–Calais telegraph cable.

William Kelly's first steelmaking converters.

1854   Cholera in London proved by Dr John Snow to be caused by polluted water.

Zürich Polytechnic founded.

1856   Bessemer's first patent for steelmaking.

1857   William Kelly in the USA patents a 'Bessemer' process but goes bankrupt.

Cézanne's bridge over the River Tisza at Szeged, Hungary, uses compressed-air caissons for building the bridge piers and an early steam-driven concrete mixer.

1858   First mechanical rock crusher built for roadbuilding in New York.

Electric arc light in the South Foreland lighthouse.

1859   First steam-powered road roller built in France.

Bridge over the Rhine at Kehl, near Strasbourg, built by F. Saint-Denis, using pneumatic caissons for the piers.

1860   Cowper stoves halve the cost of smelting iron.

Steel railway rails.

1863   First London underground railway links Paddington with Farringdon St; 4 km (2½ miles) long, powered by steam locomotives.

1864   Dale Dyke reservoir, Sheffield, collapses on first filling. 244 lives lost.

1865   First English roads of Portland cement concrete.

1866   First transatlantic telegraph cable.

	Furens arch dam built to supply St-Etienne with water, then the world's tallest at 56 m (184 ft) high.
1867	Alfred Nobel patents dynamite and detonating caps.
1869	Suez Canal opened, designed by De Lesseps.
	Peter Barlow drives a 2.13 m (7 ft) dia. tunnel under the Thames from the Tower to Bermondsey, lined with iron rings, later to become the first tube railway, with one car containing 12 people hauled by steel wire rope.
1870	UK Tramways Act authorizes steam trams.
1857 –71	First Alpine tunnel, Mont Cenis, 12 km (7½ miles) long, begun with hand-drilling, finished with compressed-air drills.
1875	Hollow clay pots made in the USA to lighten concrete floors.
1876	Municipal refuse incinerator at Nottingham.
1872 –82	St Gotthard tunnel driven with the loss of 300 lives.
1877	UK Board of Trade allows the first use of steel in buildings.
	Monier patents reinforced-concrete beams.
1879	Tay Bridge, the world's longest bridge, collapses; 73 drowned.
1879	Edison's electric light bulb invented.
1880	Bacteria proved to be the purifiers of sewage.
1881	House of Commons lit by electric bulbs, dwellings follow.
1883	Brooklyn Bridge completed; suspension bridge by Roebling, 486 m (1596 ft) span.
1884	Parsons's first steam turbine patent.
	Electric trams in Germany.
1885	First Daimler petrol engine.
1883 –90	Forth railway bridge built to the design of Fowler and Baker.
1886	Niagara Falls hydroelectric scheme begins; 200,000 hp.
	Severn railway tunnel opened.
1888	John Dunlop's inflated bicycle tyre; his patent is allowed.
	Tesla in the USA invents alternating-current motor.
1889	Eiffel Tower completed.
1890	First electric tube railway opened, using two 120 hp electric motors (City and South London Railway from the City to Stockwell).
	First wholly steel-framed building, Chicago.
1890 –1903	Trans-Siberian railway construction.
1892	Hennebique uses steel links in concrete beams to resist shear force.
1893	New Croton aqueduct tunnel completed, New York.
	Immersed tube sewer tunnels sunk in Boston, USA (see 1910).
1894	Kiel Canal.
1897	Parson's *Turbinia* steams at 7 knots faster than any ship in the Royal Navy, proving the superiority of steam turbines.
1901	Russian oil production the world's largest, nearly 12 million tons.
1902	Engineering Standards Committee formed, later becoming the BESA

and eventually the BSI. It reduced the number of tram rail cross-sections from 70 to nine.

1903 The Wright brothers fly.
Henry Ford's first cars.
Trans-Siberian Railway completed.

1905 Inner Circle underground railway in London electrified. Until 1905 newspaper readers had to bring their own candles.
Victoria Falls Bridge over the Zambesi River; 150 m (500 ft) span, designed by Ralph Freeman.

1906 First broadcast of music and speech in the USA on Christmas Eve.

1910 Railway tunnels under the Detroit River (USA–Canada) sunk as immersed tubes (see 1893).

1928 Concrete placed by vibration in the USA.

1930 Plougastel Bridge near Brest built by Freyssinet with three arches of 187 m (612 ft) span; destroyed in 1945 and then rebuilt.
Concrete placed by vibration in France.

1930 Robert Maillart's tied-arch bridge for the Rhaetian Railway over the River Landquart at Klosters.

1940 Tacoma Narrows suspension bridge fails because of flutter in wind.

1946 P. L. Nervi rediscovers ferrocement and builds his 165 ton yacht.

1946 Freyssinet's five bridges over the Marne; prestressed, steam-cured, all of
–50 74 m (240 ft) span.

1957 Sputnik, the first artificial satellite.

1964 Chesapeake Bay bridge-tunnel, 28 km (17.5 miles) long, with four artificial islands, completed.

1982 Two bridges built of fibre-reinforced plastics, one at Beijing, China, of 20 m (65 ft) span, the other at Ginzi, Bulgaria, of 10 m (33 ft) span.

1987 Channel Tunnel between England and France begun; trains due to run in 1993.

APPENDIX 2: GREAT ENGINEERS

Adams, William Bridges (1797–1862) Inventor of the railway fishplate.

Agrippa, Marcus Vipsanius (63 – 12 BC) From 34 BC he was in charge of Roman aqueducts. He also built the Roman Pantheon in 27 BC and the Pont du Gard, a high-level aqueduct which brought water to Nîmes, in 18 BC.

Apollodorus of Damascus (active AD 98–117) The Emperor Trajan's engineer, responsible for important Roman buildings as well as the 20 span wooden bridge over the Danube at the Iron Gate (Orsova or Turnu Severin), on stone piers rising 46 m (150 ft) above the water.

Archimedes (287–212 BC) The last great Greek engineer and mathematician. Syracuse (now Siracusa), his home town, was a Greek settlement in Sicily where his relative the king allowed him to spend his life thinking. But he had many activities; as army engineer he defended Syracuse for years against the Roman besiegers; he worked on geometry and mathematics, and discovered hydrostatics, the use of levers, and probably also gear trains.

Armstrong, William George, Lord (1810–1900) A specialist in power hydraulics, who built possibly the first hydraulic *accumulator* in 1851, founder of the firm of Sir W. G. Armstrong Whitworth & Co.

Arnodin, Ferdinand-Joseph (1845–1924) French civil engineer, pupil of Marc Seguin and heir to his firm. He designed many bridges carried on steel-wire ropes, including probably the first *transporter bridges*. His first was in northern Spain at Portugalete, near Bilbao over the River Nervion, of 160 m (525 ft) span. He built others at Bizerta (Tunis), Rouen (the first in France), Bordeaux, Marseille (1905) and Rochefort. His ropes with alternate layers twisting in opposite directions may have been the origin of locked-coil ropes.

Arup, Ove Nyquist (1895–1988) Founder of Ove Arup & Partners and of Arup Associates, probably the largest consulting practice in the Western world he was responsible for Sydney Opera House and much else.

Aspdin, Joseph (1799–1855) Patented Portland cement in 1824, the year when *Vicat* in France was already building his first bridge with a hydraulic cement.

Baker, Sir Benjamin (1840–1907) Junior partner of Fowler with whom he designed the Forth Railway Bridge and other railway structures.

Bazalgette, Sir Joseph William (1819–91) Built the first 1330 km (833 miles) of intercepting sewers in London from 1859 to 1875; also bridges and Thames embankments. ICE president 1884.

Beardmore, Nathaniel (1816–72) Collaborator and partner of J. M. Rendel, well known for his *Manual of Hydrology*.

Belidor, Bernard Forest de (1693–1761) French mathematician and army officer whose engineering textbooks, used throughout Europe, were later updated by *Navier*.

Bell, Henry (1767–1830) A *millwright* who met *Fulton*, and in 1812 installed his

3 hp engine in the *Comet* of 30 tons, which regularly covered the 112 km (70 miles) from Glasgow to Greenock until 1820. It was the first regular European steamship, but in the USA *Fulton*'s ships were earlier.

Bentham, General Sir Samuel (1757–1831) Younger brother of the philosopher Jeremy Bentham, and inventor of the Panopticon from which Jeremy supervised his 1000 French prisoners. Samuel was apprenticed at 14 to the navy's master shipwright. In Russia he built barges for the Russian navy, fitting them with howitzers and the first naval guns to fire shells. His barges wrought havoc with the Turkish navy, and he was promoted to general and knighted. On his return he also did good work for the British navy, arranging with Marc *Brunel* for *Maudslay* to build machines to make 100,000 pulley blocks a year in an early mass production system. He invented steam dredgers and the caisson method of closing a dock gate. But the Admiralty disliked his exposure of corruption that reached the top.

Berkley, James John (1819–62) After leaving King's College, London, Berkley worked for G. P. *Bidder*, and from 1839 for Robert *Stephenson*. He went to India in 1850 as resident engineer for the first Indian railway, 1980 km (1237 miles) long, eastward from Bombay. But his health was ruined by the climate. He was awarded the Telford Medal in 1860.

Bernoulli, Daniel (1700–1782) One of an astoundingly brilliant Swiss mathematical family. His father, two uncles, two brothers and a nephew were all at one time professors of astronomy or maths at Basel. Daniel was the founder of mathematical physics but first studied medicine. His first papers in 1724, on fluid flow, probability, calculus and geometry, so attracted the St Petersburg Academy that he was invited there in 1725. His book *Hydrodynamica* was published in Russia. See **Euler**.

Bessemer, Sir Henry, FRS (1813–98) The son of a mechanical engineer who fled from the French revolution, Bessemer was an extraordinarily versatile inventor who showed how to make liquid steel cheaply. British annual steel output rose a hundredfold from 50,000 tons between 1855, the year of his first patent, and his death, when it was 5 million tons. He first worked for a printer and developed his method of 'gold-plating' to finance his other experiments.

Bidder, George Parker (1806–70) A child of astonishingly rapid mental arithmetic powers, taught at the age of six to count but only to 100. Wealthy people who had witnessed his calculations sent him to school, and to Edinburgh University, where he made a lifelong friend of Robert *Stephenson*, worked for him and helped him in his parliamentary battles by his fast calculations. He designed the Victoria Docks, London, a railway swing bridge and other engineering work. His son, a QC, and his grandchildren had similar gifts.

Boulton, Matthew, FRS (1728–1809) Successful Birmingham manufacturer, for many years the partner and wholehearted supporter of James *Watt*.

Bramah, Joseph (1749–1814) A cabinet maker who became a famous mechanical engineer and locksmith. Henry *Maudslay* worked for him between the ages of 18 and 26, and helped him invent the hydraulic press without which Brunel's ships and Stephenson's tubular bridge across the Menai Strait would have been impossible.

Brassey, Thomas (1805–70) A land surveyor with a genius for organizing big contracts. He worked with Joseph *Locke* on the London–Southampton and the Paris–Rouen railways and built in all some 7000 km (4500 miles) of railway.

Brindley, James (1716–72) A *millwright*, the greatest early canal builder, creator of the first English aqueduct, he wrote with difficulty and could not spell, yet the tunnels he drove for his canals were straight and level. He gained fame with his aqueduct over the River Irwell for the Duke of Bridgwater's canal, which he completed in 1761. The Duke was deeply in debt but the canal halved the price of coal in Manchester and by 1769 his £25,000 debts were repaid. Brindley drove many other canals.

Brunel, Isambard Kingdom (1806–59) Educated privately and at the Lycée Henri IV in Paris, he helped his father, Marc Brunel, at the first Thames Tunnel from the age of 17. At the age of 25 his design for the Clifton Suspension bridge was accepted, but was not built until 1863 because of money shortages. At 27 he was appointed chief engineer of the Great Western Railway. His mistake of setting the gauge at 2.13 m (7 ft) was eventually corrected. The GWR's successful steamship service from Bristol to New York in a ship of his design inspired him to build an even bigger ship that was an expensive failure.

Brunel, Sir Marc Isambard, FRS (1769–1849) Royalist French refugee, he became chief engineer of New York, returned to Europe; awarded Telford Silver Medal 1839 for the first tunnel under the Thames, which also used the first *shield*, but took 18 years from 1824 to 1842. Father of I. K. *Brunel*, he collaborated with *Maudslay* to achieve the Admiralty's aim of mechanizing the manufacture of 100,000 pulley blocks annually. He was jailed for debt but friends obtained money from Parliament to get him out.

Castigliano, C. A. (1847–84) Great Italian structural theorist whose strain energy theory was an elegant development of Menabrea's (1809–96) principle of least work.

Cayley, Sir George (1773–1857) English designer who built the first man-carrying glider. It flew for 500 m. He also invented crawler tracks for travel over soft mud.

Cézanne, L. J. Ernest (1830–76) French civil engineer who in Hungary, on his bridge over the River Tisza at Szeged (1857), used pneumatic caissons for building the piers, and a steam-driven concrete mixer. It was an inclined drum, 4 m (13 ft) long and of 1.3 m (4 ft) dia., which mixed 100 m³ (3500 ft³) of concrete in 10 hours.

Chézy, Antoine (1718–98) French civil engineer and researcher into fluid flow. Perronet's assistant for years.

Chia Ku Shan Shou (active AD 1315) Chinese woman civil engineer whose best-known road is in the mountains of Fukien.

Clapeyron, Benoît Paul Émile (1799–1864) French engineer who went to St Petersburg on graduation to teach pure and applied maths in the Russian capital. On his return to Paris in 1830 the Liverpool–Manchester Railway was in full swing and Clapeyron promoted railways with enthusiasm, at first in northern France, designing railways and their bridges. He was one of the first people to use steam expansively.

Clement, Joseph (1779–1844) An enterprising Westmorland boy who taught himself mechanics and became the best draughtsman in the country, at a time when mechanical draughtsmen were artists. By the end of 1813, working in Carlisle, Glasgow and Aberdeen, he had saved £100 and set off for London to work for himself.

Bramah took him on a month's trial but he then signed a five-year agreement because he loved Bramah's work. Bramah died inside a year and Clement left to be *Maudslay* Sons & *Field*'s chief draughtsman, working on marine engines. After Maudslay died in 1831 he started on his own, and employed only the best men, including Joseph *Whitworth*.

Coode, Sir John (1816–92) Articled as a civil engineering pupil to J. M. *Rendel*, Coode worked on Portland and other harbour works and the Great Western Railway, designed harbours at Colombo, Waterford, Portland (Australia), Table Bay, Cape Town and Dover. ICE President 1889.

Cort, Henry (1740–1800) In 1784 he patented puddling, a way of making *wrought iron* that was not superseded until *Bessemer* showed how to make steel cheaply in 1856. Before 1784 English iron was so bad that it was not allowed in the navy. Swedish or Russian iron had to be used.

Coulomb, Charles Augustin (1736–1806) French army engineer, the founder of *structural analysis* who also did original work on electrostatics, human fatigue (ergonomics) and *soil mechanics (Coulomb's equation)*. In 1783, as a consultant on the harbours and canals of Brittany, he was jailed for writing a critical report. But this did him no harm: he was soon put in charge of the royal water supply in Paris.

Cowper, Edward Alfred (1819–93) Invented the Cowper stove in 1860 for preheating the air to the blast furnace, reducing coke consumption to below 1 ton per ton of iron, thus further improving on *Neilson*'s hot blast. Cowper stoves are built in threes about 18 m (60 ft) high and of 8.5 m (28 ft) dia. Two are heated by waste gases while the third heats the blast, taking turns.

Coyne, André (1891–1960) A great designer of arch dams for French hydroelectric schemes, founder of the consulting firm of Coyne & Bellier. Kariba Dam is essentially a Coyne dam.

Daimler, Gottlieb (1834–1900) Apprenticed to a gunmaker, Daimler loved drawing plants and animals. From Stuttgart Polytechnic, 1857–9, he went to an engineering works where he met Wilhelm *Maybach*, 12 years his junior. They worked together also at his next job as manager of a works in Karlsruhe, and for *Langen* and *Otto* at Deutz, near Cologne. In 1890 they formed their own car company at Cannstatt near Stuttgart. In the first international car race in 1894, from Paris to Rouen, out of 102 starters only 15 finished but his Daimler won at 24 kph (15 mph). The prize was 80,000 francs. In 1900 they built the first Mercedes. Cannstatt is still Daimler's centre.

Darby, Abraham (1676–1717) A Quaker whose father, a farmer, made nails and locks. The son went to Holland to study Dutch methods of casting iron. He was the first Abraham Darby; there were two more. His foundry at

Coalbrookdale was near to water power and to coal and iron deposits. In 1779 it cast the iron bridge of 30 m (100 ft) span over the Severn near by (still standing), and in 1802 made the first railway locomotive with a high-pressure boiler for *Trevithick*. From 1718 it used coke for smelting iron.

Diesel, Rudolf (1858–1913) In 1893 he patented his idea and published his book on a 'rational heat engine'. His engine was installed in French submarines and elsewhere but *Stuart*'s engine probably preceded it.

Dudley, Dud (1599–1684) Smelted iron with coal (1619) in a blast furnace.

Duke of Bridgwater, Francis Egerton (1736–1803) *Brindley*'s sponsor. He deserves to be remembered for his single-minded determination to finish his canal in spite of heavy debts.

Dunlop, John (1840–1921) A Scottish veterinary surgeon in Belfast who invented the pneumatic tyre because his nine-year-old son complained about his bumpy tricycle. A similar invention 30 years earlier in Scotland had not been followed up and Dunlop's patent was allowed.

Edwards, William (1719–89) A stonemason, builder of many stone bridges in Wales including one with a span of 52 m (170 ft) over the River Taff at Pontypridd.

Eiffel, Gustave (1832–1923) French engineer from Dijon who worked for the Chemins de Fer du Midi until 1866, then set up on his own as a consultant. But in 1868, understanding where the money was, he became a contractor, building many steel bridges in France, South America and Portugal and selling bridge kits to be erected by unskilled people. Eiffel was connected with De Lesseps in the Panama Canal bankruptcy and was arrested in 1892 with the Lesseps, father and son, so his triumph with the Eiffel Tower in 1889 was short-lived, but he was released on appeal.

Foreseeing possible disaster he had sold his company, in 1889, to his assistants, Koechlin and Adolphe Salles, who had designed the tower.

He was the world's expert not only on steel structures but also on compressed-air caissons for building bridge piers as well as on the *incremental launching* of bridges. From 1893 until 1920 he worked most of the day in his laboratory investigating wind forces and, later, aircraft.

Ericsson, John (1803–89) A versatile Swedish inventor and army engineer, whose steam fire engine in London was so successful that jealous firemen working hand pumps took care to extinguish his boiler fire before they did their duty. In 1829 his lightly built locomotive, the *Novelty*, lost against the *Stephensons' Rocket* in the Rainhill trial.

In 1839 he crossed the Atlantic in 45 days in a tiny 30 ton vessel fitted with his screw propeller and stayed for the rest of his life in the USA. His armoured ships were successful in the American Civil War in 1862.

Euler, Leonhard (1707–83) A friend of Daniel *Bernoulli*, also from Basel, Euler was an extraordinarily prolific mathematical writer whose work on the *buckling* of long columns is still used by civil engineers. He studied maths under Daniel's father, Jean (Johannes), and followed Daniel to St Petersburg. In his lifetime he published 560 books and papers and invented sciences that were previously unknown. In his old age he went blind but his literary output increased because he persuaded his children to read proofs for him.

Eupalinos of Megara In 550 BC, on the island of Samos, Greece, the home of Pythagoras, Eupalinos drove an aqueduct 1000 metres (3300 ft) long through Mount Kastro, simultaneously from each end to meet in the middle.

Eytelwein, J. A. (1764–1848) Hydraulic engineer who was founder and first director of the Berliner Bauakademie (1799), forerunner of the university, and edited the first German building journal (1797).

Faber, Oscar (1886–1956) Danish civil engineer who used reinforced concrete with confidence before British engineers, and wrote a standard work on it. He refurbished the heating, ventilation and air conditioning of the Bank of England and the House of Commons. Before 1939 he was in such demand that he was reputed to demand a fee of £100 for writing a letter.

Fairbairn, Sir William (1787–1874) A Scottish *millwright*, from 1817 the partner of James Lillie in Manchester, building wrought-iron ships. In 1835 he set up a shipyard at Millwall, London, and from 1845, with *Hodgkinson*, designed Robert *Stephenson*'s Britannia railway bridge across the Menai Strait. In 1837 faced with a strike by riveters he developed a successful steam-powered riveting machine. He invented machines for testing iron and steel and put them to good use with the *Britannia Bridge*. Apart from his many iron ships he also designed the Lancashire boiler, with two fire tubes instead of one.

Field, Joshua (1787–1863) ICE founder member who was *Maudslay*'s partner from 1804 in Maudslay, Sons & Field, mainly building ship's engines, though it also made precision instruments. In 1838 *Brunel*'s paddle steamer *Great Western*, with Field's engines, steamed to New York in 13 days, confounding the critics. Her steam cylinders were of 1.87 m (73.5 in.) bore, with a stroke of 2.1 m (7 ft). The engines occupied nearly half the hull and the 500 tons of coal took up most of the rest. Smaller engines and screw propellers, though desperately needed, were not installed in the UK for many years.

Finley, James (1762–1828) A justice of the peace in western Pennsylvania who in 1801 persuaded the authorities to allow him to build a bridge with a deck carried by iron chains from two towers. This, his first suspension bridge, at Jacob's Creek, was 21 m (70 ft) long, 3.8 m (12½ ft) wide, with a stiffened wooden deck. By 1810 another 39 bridges to his design were built but he did not secure a patent until 1808. The first one failed in 1825 under the weight of a six-horse team, but his longest, at Newburyport (1810), of 74 m (244 ft) span and 12·2 m (40 ft) wide, lasted until it was replaced in 1909. It was carried by ten chains each side.

Fontana, Domenico (1543–1607) Italian architect-engineer best known for his water supply to Rome, his Roman churches and the moving of the 300 ton Egyptian obelisk to the front of the new church of St Peter's with the help of 907 men, 75 horses, a trumpeter to signal when to start work, a bell to signal when to stop, an executioner to instil discipline, and the wholehearted support of Sixtus V, Pope from 1585 to 1592. At a critical moment, when the ropes could be tightened no more, a man, in spite of the command for silence, shouted '*aqua*', to hose the ropes and tighten them further. The executioner moved towards him, but the ropes tightened, the work was successful and the Pope pardoned him.

Fowler, Sir John (1817–98) The son of a land surveyor near Sheffield, he set up on his own as a civil engineer in 1844, taking on *Baker* as his partner in 1875.

He designed the first underground railway, built by cut-and-cover along the main roads, opened in 1863 from Paddington to Farringdon St and in 1865 extended to Moorgate. At that time anyone tunnelling under a house was obliged to buy it first. Consequently digging up streets and relaying them was the only possibility. At first the Great Western Railway operated the underground but later quarrelled with the owners and withdrew with their 22 steam locomotives. The owners had to find another operator and hire other engines. The smoke-filled tunnels were particularly unpleasant in summer. ICE president 1865.

Fox, Sir Charles (1810–74) Fox, at first a contractor, built the London–Birmingham Railway under Robert *Stephenson* in 1837, and many other railways in the UK and abroad. His firm, Fox Henderson, knew how to erect the Crystal Palace because they had built large roofs for railway stations. The Crystal Palace was completed in seven months and later re-erected with enlargements on Sydenham Hill. Paxton, Fox and William Cubitt were knighted.

Fox's son, Sir Douglas Fox (1840–1921), was senior partner from 1874 in Sir Douglas Fox & Partners, consulting engineers. Later, with Ralph *Freeman*, it became Freeman Fox & Partners and from 1988 Acer Freeman Fox.

Freeman, Sir Ralph (1880–1950) In 1900 he passed brilliantly out of the Central Technical College (now Imperial College) and joined Sir Douglas Fox & Partners, remaining with them until his death, becoming a partner in 1912, and senior partner in 1921. In 1905 this great steel designer built one of his most spectacular bridges, of 150 m (500 ft) span, the Victoria Falls Bridge over the Zambesi River. From 1924 to 1932 he designed the Sydney Harbour Bridge with its four railway tracks, a road and two footways. Later he was concerned with the Severn and Humber road bridges. See **Fox**.

Freyssinet, Eugène (1879–1962) Founder of *prestressed concrete*. He began to think seriously about this material in 1904, while working for the French Department of Transport. In 1930 he built the greatest reinforced-concrete bridge of its time, at Plougastel over the River Elorn near Brest. The upper deck carries a road, the lower deck a railway over its three equal arch spans of 187 m (612 ft). It was here that he first used his *flat jacks*. From then on his main interest became prestressed concrete and he aroused interest in the postwar bridge-reconstruction of France by several more or less identical two-hinged, prestressed concrete arch bridges over the Marne, all of 74 m (240 ft) span, at Luzancy (1946), Esbly (1950) and elsewhere. Precast in a factory on site, the bridge units were steam cured.

Frontinus, Julius Sextus (AD ?30–?103) Administrator of 400 km of Roman aqueducts above and below ground, with a staff of 700 people. He wrote a book on the subject, *De Aquis*.

Fulton, Robert (1765–1815) An American who studied painting in London and in 1786 earned a meagre living there by painting pictures. He then turned to digging canals by the dredger he had invented. In 1797 he was in France, hawking his canal dredger and designing a submarine to destroy the British navy, powered by three men. In 1804, back in England, he offered help in the war against France. When the war ended Parliament gave him £15,000 for his pains and he returned to the USA. The steam engine he had ordered from England in 1806 arrived in 1807 and was built into his paddle steamer.

This, the first regular steamship, regularly covered the 240 km (150 miles) from New York to Albany in only four hours. It was the first of Fulton's many successful ships, but was not the first in the USA. J. Fitch's (1743–98) financially unsuccessful steamships sailed 1786–98.

Gauthey, Emiland Marie (1732–1806) French civil engineer, theorist in strength of materials and inventor of testing machines who designed an early *summit canal*, the Canal du Centre (Canal Charollais) connecting the English Channel to the Mediterranean. As *Navier*'s uncle, Gauthey cared for him from 1799 when his father died.

Gooch, Sir Daniel (1816–89) Apprenticed to Stepehenson & Pease locomotive works, Newcastle, Gooch became *I. K. Brunel*'s deputy for mechanical engineering in the Great Western Railway, but left in 1864 to lay the first transatlantic telegraph cable from Brunel's large ship the *Great Eastern*. In 1865 he rejoined GWR as chairman of the board of directors and saw the completion of their Severn Tunnel in 1886.

Greathead, James Henry (1844–96) A South African who came to Britain to study engineering and developed the Greathead tunnelling *shield* for soft ground, for many years the only type used. He drove the Tower subway under the Thames with it in 1869, using screw jacks to push it forward, reacting on the cast-iron lining behind. In 1896 he drove the two 3 m (10½ ft) dia. tunnels of the City and South London Railway near London Bridge, with probably the first shields containing *compressed air* to keep out water and mud.

Guericke, Otto von (1602–86) Guericke lived in Magdeburg most of his life and in 1654 he proved the existence of atmospheric pressure by his evacuated globe made of two copper 'Magdeburg hemispheres', which even 16 strong horses could not pull apart. Before this he had invented the air pump. Others suspected the existence of air pressure, but Guericke proved it by his showmanship.

Hallidie, Andrew Smith (1836–1900) Built the California Wire Works in 1857 for making wire rope. He created the first ropeways and in 1867 the San Francisco rope-drawn street car. His father, Andrew Smith, an English immigrant, is said to have invented steel wire rope, but in France, *Seguin* was probably much earlier.

Hawkshaw, Sir John, FRS (1811–91) A railway engineer who also designed fen drainage, water supplies and canals, and in 1838 recommended the Great Western Railway to discontinue its *broad gauge*. He designed the East London Line, which uses Sir M. I *Brunel*'s tunnel under the Thames. With Sir John Wolfe-Barry in 1850 he was on the committee to unite the District and Metropolitan underground lines in London, to create the Inner Circle. ICE president 1862–3.

Hawksley, Thomas, FRS (1807–93) At first articled to the Architect and

Surveyor of Nottingham he eventually designed many water supply, gas supply and sewerage schemes, especially in Nottingham, Liverpool, Leicester and Sheffield. He reported on the Dale Dike reservoir disaster of 1864 near Sheffield. ICE president 1872–3.

Hennebique, François (1842–1921) Son of a farmer in the north of France, he studied science in his spare time and eventually found a building contractor in Arras willing to take him on as an apprentice mason.

By 1867 he was a builder on his own, repairing churches, but began work on railway bridges, and soon had more work than he could handle. He was a pioneer of reinforced concrete and in 1880 patented his reinforced-concrete floors. In 1892 he gave up contracting and concentrated on RC design under the name of 'Maison Hennebique'. From 1890 to 1900 he built 3000 RC structures, and in 1900 moved to Paris, rue Danton, to build an eight-storey house for his office. In England his representative from 1892 was Louis Mouchel, who opened an office in Victoria Street, London, manned by French engineers.

Hodgkinson, Eaton (1789–1861) Structural theorist, experimenter on strength of materials, and originator of the wrought-iron I-beam, who collaborated with William *Fairbairn* in the design of *Stephenson*'s tubular railway bridges.

Hooke, Robert (1635–1703) Secretary of the Royal Society from 1677, a contemporary of Christopher Wren who sat with him on the committee to rebuild London after the Great Fire of 1666; well known for *Hooke's Law*.

Imhotep (3000 BC) Built the first Egyptian pyramid, the great Step Pyramid of Sakarra, and was made a god after death.

Jenkins, Ronald (1907–75) Brilliant mathematician who introduced elegance and rigour into the design of shell roofs; worked for *Ove Arup* for many years from 1943. His early work included the design of the penguin pool at the London zoo.

Jenney, William Le Baron (1832–1907) Built the Home Insurance Office, Chicago, in 1884, a ten-storey frame of Bessemer steel beams and cast-iron columns, one of the earliest metal skeleton buildings.

Jessop, William (1745–1814) Jessop's father died in 1761 shortly after completing the Eddystone lighthouse with *Smeaton*, so Smeaton adopted the boy and took him as his civil engineering pupil. He designed canals and the West India Docks in London (1800–1802), and the horse-drawn Croydon & Merstham Railroad (1803).

Kerensky, Oleg, FRS (1905–84) Son of the leader of the Russian Provisional Government of February 1917, Oleg with his brother Gleb and mother escaped from semi-starvation in Russia to London, where Mrs Kerensky supported them by working as a typist.

Oleg studied at the Northampton Engineering College (now the City University) and then entered Dorman Long's bridge drawing office, to work on Sydney Harbour Bridge under Ralph *Freeman*. In 1946 he joined Freeman, Fox & Partners as Principal Bridge Designer. Only in that year did he become a British subject. He designed or worked on many steel bridges, including the *Severn, Forth Road*, Auckland Harbour and Milford Haven. He contributed to or chaired many BS committees. IStructE president 1970.

Krupp, Alfred (1812–87) His father's death in 1825 caused the boy to leave school and take charge of the family works, casting steel dies for royal mints. But he built up the business to such an extent that he made a good show in the 1851 Exhibition in Hyde Park, London. A great designer, he believed in making things only of the best quality.

Langen, Eugen (1833–95) After studying in the Karlsruhe Polytechnic and travelling in Spain, England and France, Langen met *N. A. Otto* in Cologne in 1864. Otto had just started his gas-engine factory (petrol or gasoline did not exist) and Langen bought 10,000 thalers worth of shares. At the 1867 Paris Exhibition, Otto and Langen's engine had only one third of the fuel consumption of Lenoir's engine, and won the gold medal.

Orders poured in but money troubles were serious, as they always are with an expanding, successful business. They were overcome by the formation of their company 'Gas-motoren-Fabrik Deutz AG' at Deutz, near Cologne. By 1875 the first Deutz engines were running on liquid fuel – benzene – a real incentive to increase the power above the maximum then of 3 hp. Though their patents were quashed in 1886 the company still prospers.

Latrobe, Benjamin Henry (1765–1820) The foremost civil engineer (docks and river navigation) and probably the only trained architect in the USA in the early 1800s, he emigrated from England in 1796. He designed part of the Capitol. Baltimore Cathedral was his finest work.

LeTourneau, R. G. (1888–1969) An American who built the first *bulldozer*, the first *rippers*, *bowl scrapers*, rubber-tyred earthmoving units, the first *jack-up rig* for offshore drilling, and much other original equipment. This deeply religious man said of himself: 'I am just a mechanic that the Lord has blessed.'

Locke, Joseph (1805–60) Designer of many railways from 1835 onwards in Britain and abroad, including the Paris–Rouen. His routes avoided tunnels and so were not expensive, but his steep gradients caused trouble in frost. ICE president 1858. See **Brassey**.

MacAdam, John Loudon (1756–1836) An Ayrshire boy, brought up from the age of 14 in New York, where his uncle was a prosperous businessman. He returned to Ayrshire in 1782, having amassed some money, and took an interest in roadbuilding. In 1798, appointed agent for revictualling the Royal Navy in the western ports, he settled in Falmouth. Still interested in roadbuilding, in 1815 as surveyor-general of Bristol roads he tried out his theories, successfully because in 1827 he was made general surveyor of roads. Parliament allowed him £10,000 for his past expenses but he refused a knighthood on account of his age and passed it to his third son, James Nicoll MacAdam (1786–1852), who became chief trustee and surveyor of turnpike roads. J. L. MacAdam's grandson, William (1803–61), was also for many years surveyor-general of roads.

Maillart, Robert (1872–1940) A Swiss engineer whose striking, curved bridges appeal to all, even those who dislike engineering structures. He was a consultant on his own from 1902 and was one of the first to use *mushroom construction* (Zurich warehouse, 1910).

Mandrokles of Samos In 512 BC he built the first pontoon bridge over the Bosporus for the advancing army of Darius of Persia.

Maudslay, Henry (1771–1831) From age 12 to 18 he worked in Woolwich Arsenal, at first filling cartridges with powder. He became a blacksmith and was strong and big enough to do it well, being tall (1.88 m, 6 ft 2 in.). *Bramah*'s works in Pimlico attracted him when he was 18, and Bramah liked his work, but in 1797 would not raise his pay above 30 shillings (£1.50) per week though he had been foreman for many years, so he started his own small shop near Oxford Steet. He was given a good start by Marc *Brunel*'s Admiralty contract to build machines to make 100,000 pulley blocks a year. It took 5½ years but the machines produced 130,000 blocks a year and some of the machines worked for over a century.

At its prime, Maudslay, Sons & Field in Lambeth was Europe's leading engineering shop, employing several hundred men building powerful marine engines as well as doing precision engineering, an unusual combination. A perfectionist, the 'father of the machine-tool industry', he was the first to make standard screw threads. *Nasmyth* and *Whitworth* worked for his firm.

Maxwell, James Clerk (1831–79) A brilliant Scots mathematician, the first professor of experimental physics at Cambridge University, he converted Faraday's concepts into equations by his electromagnetic theory of light. Maxwell's lectures at King's College, London, included his mathematical treatment of *Rankine*'s graphical solutions of structural problems (trusses). His book *Theory of Heat*, 1870, went through several editions.

Maybach, Wilhelm (1846–1929) Son of a carpenter in Heilbronn, Maybach worked with *Daimler* in the Württemberg engineering works and followed him to Karlsruhe. They also worked together at Deutz for *Otto* and *Langen* and later at Cannstatt in Daimler's own firm. He left only in 1907.

Meikle, Andrew (1719–1811) A *millwright*, teacher of John *Rennie* and inventor of the threshing machine, descendant of a long line of excellent but little known millwrights.

Metcalf, John (Blind Jack) (1717–1810) A Knaresborough, Yorkshire, boy who was blinded by smallpox at the age of six but rode, swam, kept a dog, coursed hares, and played the fiddle and was paid for it. In the Scots rebellion of 1745 he was a musician in the company of Col. Thornton of Thornville Royal, Yorkshire. On his return to Knaresborough he started a carter's business, often going as far as Aberdeen. But he sold his carter's business to build roads, first from Knaresborough to Harrogate and eventually nearly 300 km (180 miles) of turnpike roads in Yorkshire, Lancashire and Derbyshire. He retired when he was 75. At his death he had 90 great-grand-children.

Mohr, Christian Otto (1835–1918) German civil engineer who invented *Mohr's circle* and other graphical solutions of structural problems.

Müller-Breslau, Heinrich, F. B. (1851–1925) The founder of modern German structural engineering. Like *Mohr*, he solved problems by graphical methods, but also designed Zeppelins and their very large hangars.

Murdoch (or **Murdock**), William (1754–1839) A *millwright* from Ayrshire, the right-hand man in Cornwall for many years of Boulton and *Watt*, taken on by Boulton in 1777 probably because of his painted wooden hat. Anyone who could turn an oval was a real craftsman. Murdoch was sent to Cornwall in 1779 and stayed there until 1798, servicing and erecting steam engines. His house in Redruth may have been lit by coal gas, because in 1794 he proposed gas lighting for the Soho works, but nothing was done until 1802 when, lit by gas, the firm began to make gasworks plant. In 1807 Murdoch was awarded the Rumford Gold Medal of the Royal Society for his paper on gas lighting.

Myddelton, Sir Hugh (?1560–1631) A London goldsmith of Basinghall Street, but born at Galch Hill, Denbigh, North Wales, and several times MP for Denbigh. Goldsmiths at that time were bankers, moneylenders and often also, like Myddelton, merchants.

London's water shortages caused disease and in 1609 he was empowered by the City Corporation to bring water to London from the Chadwell spring near Ware, 32 km (20 miles) away. The New River is nearly twice this length because it follows contours closely at its slope of only 1 in 31,680.

The work was completed in 1613 at the New River Head reservoir in Pentonville, but only after Myddelton had appealed to King James I for help against the violent opposition of farmers and landowners. Some 400 miles of wooden pipes (probably drilled elm logs with spigot-and-socket joints) were laid for distribution. No dividends were paid until 1633, but Myddelton also had profitable activities. His lead and silver mines near Aberystwyth worked a rich ore yielding 100 oz (2.2 kg) of silver per ton. King James I made him a baronet in 1622, waiving the usual fee of £1095 for a baronetcy in view of all the royalties the Aberystwyth mines had brought him.

Nasmyth, James (1808–90) Son of a fashionable Edinburgh portrait painter with a taste for science, Nasmyth inherited his father's scientific interests and talent for painting. With the brass foundry he set up in his bedroom he made a small steam engine to grind his father's pigments and a little steam carriage.

In 1829 he took ship for London and asked Maudslay to give him work, showing his drawings and model steam engine. He was taken on at 10 shillings (50 p) a week. In September 1830 he took three weeks' holiday to see the famous Rainhill trial in Lancashire, won by *Stephenson's Rocket*. After Maudslay's death in 1831 he worked on the firm's 200 hp engines for waterworks and then returned to Edinburgh to do any work available and to build his own machine tools to equip his works.

He set up as Nasmyth and Gaskell in Manchester and first built 20 locomotives for the Great Western Railway. In 1839 he invented the steam hammer – soon used for pile driving. He was financially so successful that he retired at the age of 48 to concentrate on astronomy.

Navier, Louis Marie (1785–1836) Brilliant discoverer of the *bending formula*, Navier was brought up from boyhood by his uncle, *Gauthey*, an eminent civil engineer who trained him so well that in 1806 he graduated first from the École des Ponts et Chaussées. The French government sent him to Wales in 1819 to see Thomas *Telford's* suspension bridge erected over the Menai Strait. The result in 1823 was his *Report & Memoir on Suspension Bridges*, the standard work for 50 years. He also updated *Belidor's* books *Science des Ingénieurs* and *Architecture Hydraulique*.

Neilson, James Beaumont, FRS (1792–1865) Although himself a Glasgow gasworks manager, Neilson invented the hot blast for an iron smelter friend, heating the blast to about 350°C with a coal fire (1826). It reduced the coke consumption per ton of iron to 5 tons from 8 tons with a cold blast. It also enabled previously unusable Scottish blackband ironstone to be successfully smelted, and anthracite to be used in place of coke. See **Cowper**.

Nervi, Pier Luigi (1891–1979) Italian engineer-architect who graduated in Bologna as a civil engineer in 1913 and set up on his own in Rome in 1920. His marvellous hangars for the Italian Air Force in 1935, now destroyed, were carried on four sloping columns, one at each corner, providing aircraft access along the full length of each side. The *Turin Exhibition Hall* and similar outstanding structures brought *ferrocement* to the notice of the world. British engineers have claimed that achievements like his are impossible in Britain with its strict separation between *contractor* and *consultant*. Nervi was both. In 1988 the UK Association of Consulting Engineers began to discuss changing the rules.

Newcomen, Thomas (1663–1729) An ironmonger in Dartmouth, a participant in *Savery's* work on the atmospheric engine. He lived 14 years later than Savery and set the first successful pumping engines to work in the mines. The first one, installed in 1712, dewatered a colliery near Dudley Castle, Staffordshire. But Newcomen's engines installed in Cornish mines used five times as much coal as *Watt's* and were quickly superseded from 1777 onwards. Coal brought by pack horse was expensive.

Otto, Frei (1925–) An architect whose tent-structures are devoid of ornament and look as if they were designed by a structural engineer, being built only of poles, cables and plastic sheets. They protect tennis courts, plants in deserts, Olympic stadia or building sites. He has even designed cooling towers of plastic sheet.

Otto, Nikolaus August (1832–91) The designer of the Otto four-stroke cycle for internal combustion engines, he first heard of the gas engine invented by LeNoir, a Frenchman, in 1861. He became absorbed by this machine which could supersede the steam engine. By 1862 he had built himself one and in 1864 he met *Langen*, who became his partner in the firm of Deutz.

Papin, Denis, FRS (1647–1712) Born in Blois, France, he studied natural science at the University of Angers at the age of 15. He spent years in England and worked with Boyle on the air pump, becoming an FRS in 1670. In 1688 he went to Marburg in Germany as professor of mathematics but returned to London to die in poverty though he had invented the pressure cooker, the centrifugal pump, the atmospheric engine, a submarine and a diving bell. He was a century ahead of his time.

Parsons, Sir Charles Algernon, FRS (1854–1931) Privately educated, using his father's foundry and workshop, at the age of 14 he built himself a steam car with differential gear. He studied maths and science at Cambridge. In 1877 he went to work for *Armstrong* in Northumberland but moved to Clarke Chapman in Gateshead, where in 1884 as a partner he patented his first steam turbine at an opportune time when industry was held back by the low power, low speed, vibration, great weight and inefficiency of the reciprocating steam engine. His first turbine was a 10 hp machine, very large for that time. In 1889 he left Clarke Chapman to build his own turbines at Heaton, Newcastle, but could not progress because his 1884 patent blocked him. Parsons came to an agreement with Clarke Chapman and developed his condenser. In 1897 his ships suffered *cavitation* but model tests in a glass tank solved this problem and he reached 34 knots in that year in his 30 m (100 ft) long ship, *Turbinia*. To show the Admiralty her speed he raced her past the Royal Navy in Spithead at 7 knots faster than any ship there. By 1905 the Admiralty decreed that every naval ship should be turbine-driven. He also worked on large reflectors for searchlights and invented non-skid car chains.

Paxton, Sir Joseph (1801–65) Manager of the Duke of Devonshire's estate at Chatsworth, and builder there of the largest glasshouse in Europe, Paxton was confident of his design for the Crystal Palace in 1849, though he was without formal education, having risen from simple gardener. The 7.5 hectare (18 acre) structure with three storeys was built in the six months between 26 July 1849 when Fox-Henderson's tender was accepted and 1 February 1850 – a month later than agreed. The 1060 cast-iron columns acted as rainwater downpipes as well as holding up the floors and roof.

Liberal MP for Coventry from 1854, Paxton was re-elected twice. He designed Mentmore Mansion, Bucks., and another large mansion in France, speculated in railway shares and was on the board of railway companies. He died wealthy but worn out by his many activities.

Perronet, Jean Rodolphe (1708–94) At 17 Perronet started work as an architect's assistant but entered the government service at 28. French civil engineering flourished because the king founded his roads and bridges department (Service des Ponts et Chaussées) in 1716 and the training school (École des Ponts et Chaussées) for it in 1747.

Perronet was quickly promoted and became the first director of the school in 1747. He greatly improved masonry bridges in France and his Pont de Neuilly near Paris was not replaced until 1956.

Perry, Captain John (1669–1732) A retired sea captain who had worked in Russia for Peter the Great on docks and canals. He closed the breach the Thames had made in its north bank at Dagenham between 1715 and 1720, after failures by many others. See **Vermuyden**.

Peter, chaplain of St Mary's Colechurch Designer of old London Bridge from 1176, completed in 1209 after his death. It had 20 small arches and a drawbridge in the middle to let occasional ships through. The houses on both sides provided useful rents for the bridge authority but left little room at street level except for pedestrians and pack horses. It survived until 1831 although the waterway was so constricted that there was a fall of 1.5 m (5 ft) between the water levels on one side and the other. This was exploited by installing water wheels for pumping water up.

Poisson, Denis (1781–1840) French mathematician, expert on celestial mechanics, probability, capillarity and elasticity (Poisson's ratio).

Prony, Gaspard François Clair Marie Riche (1755–1839) French civil engineer from the École des Ponts et Chaussées (1776–80) who was one of Perronet's assistants. After the revolution he had to survey the country's land ownership and then to draw up new trigonometrical tables to 14, 19 or 25 places of decimals helped by several hundred wig-powderers thrown out of work by changing fashion. In spite of such services to the revolution, Prony suffered, but Napoleon helped him, making him Inspector of the service of roads and bridges until his death. He is remembered for the Prony brake used in engine-testing.

Rankine, William John Macquorn, FRS (1820–72) Although not strong and therefore privately educated, Rankine entered Edinburgh University at the age of 16 but left two years later for an apprenticeship in Ireland with Thomas *Telford*'s former chief assistant, John McNeill. He returned to Scotland as a qualified civil engineer in 1842 after his Irish apprenticeship in railways, hydraulics and other work. He experimented and wrote much on the efficiency of heat engines, soil mechanics (then known as 'earth pressures'), cast-iron columns and graphic solutions of structural problems. His books were for many years standard civil engineering texts.

Rendel, James Meadows (1799–1856) After an apprenticeship to his *millwright* uncle, the lad from Devon worked in London for *Telford*, surveying for the proposed suspension bridge over the Mersey at Runcorn. He then settled in Plymouth, designing roads in north Devon. In 1824–7 he bridged the Catwater in Plymouth with the largest iron bridge then in existence and was awarded the ICE Telford Medal for it. He was a partner of Nathaniel Beardmore. In 1826 he designed Boscombe Swing Bridge, near Kingsbridge, Devon, the first to use hydraulic power for rotation. For other Devon bridges he received a second Telford Medal. He also designed harbour improvements at Portland, Grimsby, Leith, Littlehampton, Holyhead, Guernsey, Genoa and Rio de Janeiro, and railways at home and abroad. ICE president 1852–3.

Rennie, John (1761–1821) A friend and apprentice of the local *millwright*, Andrew *Meikle*. At 15 he was sent to Dunbar High School for two years, returned to Andrew Meikle, then went south to *Watt* in Birmingham. Watt sent him to supervise the millwork of Albion Steam Flour Mills in Southwark in 1784. Rennie used cast iron and wrought iron where wooden cogwheels had been used before. His water wheels had iron shafts that lasted longer than the wooden ones previously used.

Like *Telford* he built many roads, docks and bridges, including the first Waterloo Bridge over the Thames (1809–17), and Southwark Bridge (1817–19). His son, Sir John (1794–1874) was ICE president in 1845.

Reynolds, Osborne (1842–1912) FRS in 1877, professor of engineering at Owens College, Manchester, well known for his work on critical velocity of flow, *dilatancy* of silt, etc.

Riquet de Bonrepos, Pierre-Paul (1604–80) A Customs officer in the south of France who in 1662 wrote to Colbert, the Controller General of Finance, asking for money to build a canal from the Garonne at Toulouse to the Mediterranean at Cète, 250 km (158 miles) away. After many difficulties it was built from 1666 to 1681. Since the Garonne flows into the Atlantic, the canal joins the Mediterranean to that ocean. It climbs 63 m (206 ft) to the summit, which is 5 km long and fed by mountain streams from Riquet's reservoir. The canal falls 200 m (660 ft) from the summit to the Mediterranean. It cost 16 million livres (£1.3 million) but Riquet's heirs reaped no profit until 1724.

The Canal du Midi, also called the Grand Canal du Languedoc, was easily the greatest civil engineering work in Europe from Roman times until the nineteenth century. It has 100 locks, three major aqueducts, a tunnel (the first driven by blasting), many culverts and a large summit reservoir to provide water in summer.

Roberts, Sir Gilbert (1899–1978) Obtained his civil engineering degree at Imperial College and then worked for Ralph *Freeman* of Sir Douglas Fox & Partners on the design of Sydney Harbour Bridge. In 1936 he joined Sir William Arrol, eventually becoming director and chief engineer. He extended the use of welding and high-tensile steels and developed the *high-strength friction-grip bolt*. From 1949 a partner in Freeman Fox & Partners, he was in charge of the designs of the *Severn, Forth Road, Humber* and other steel bridges.

Roebling, John August (1806–69) With his son, Washington *Roebling*, he played an enormous part in US suspension bridge design and construction. The father, educated as a civil engineer in Germany, emigrated to the USA in 1831 with the intention of farming, but in 1841 was making the first American wire ropes. They replaced the thick hemp ropes that hauled canal boats over the inclined planes of the Allegheny mountains. His first wire rope suspension bridge carried an aqueduct of the Pennsylvania Canal over the Allegheny River. He built many other suspension bridges, including Brooklyn Bridge (486 m (1596 ft) main span) and Niagara Falls railway bridge (248 m (815 ft) main span), in reality a cable-stayed bridge. Like *Seguin* his main activity was rope making.

Roebling, Washington (1837–1926) Helped his father on many bridges, and when his father died, took over his father's wire-rope works and the Brooklyn Bridge, then the world's longest span, completed in 1883. He developed *pneumatic caissons* for building the piers.

St-Venant, Adhemar Barre de (1797–1886) A boy with a genius for maths who entered the École Polytechnique at the age of 16. Soon after, in 1814, the students were mobilized to defend Paris but St-Venant refused to fight for the 'usurper' Napoleon. He was expelled and had to take a minor job in the gunpowder service. He was not allowed to enter the École des Ponts et Chaussées until 1823 and even then his classmates would not speak to him. Nevertheless he completed his studies and entered the government service as an engineer on bridges, canals, roads and rivers, developing his theoretical interests in *elasticity*, especially *shear*.

Samuely, Felix James (1902–59) A partner in the consulting firm of Berger and Samuely in Berlin, he advised the Soviet government on welded steelwork, but started a British consultancy in 1933, working mainly on all-welded steel structures, then little known in the UK. From 1942 to 1944 he developed welded tubular construction for aircraft hangars, etc.

Savery, Thomas (1650–1715) An army engineer, friendly with *Newcomen*, who lived not far away, Savery was interested in mechanical propulsion and his 'vacuum-pumping' device was the predecessor of Newcomen's atmospheric engine. Steam was blown into a vessel connected with a sump to be drained. The vessel was then showered with cold water to condense the steam and cause the vacuum to suck water from the sump. More steam then blown in forced the water up the pipe. This, the first pumping engine to be used underground, was not very useful because it needed greater pressures than the boilers could raise.

Seguin, Marc (1786–1875) (or Séguin, often called Seguin the elder) The oldest of five sons of a French industrialist from Annonay, Marc, like several of his brothers, was sent by his uncle, Joseph Michel Montgolfier, who made the first balloon ascent and was Director of the Conservatoire des Arts et Métiers, to be educated in Paris. Marc had no formal engineering education but took a great interest in engineering and studied in Paris until he was 35 (1821). He then returned to Annonay to manage the firm and, in 20 years with his brothers, built 86 wire-rope suspension bridges. They made the wire for the ropes of Vicat's bridge at Argentat in 1828. They also built the St-Etienne–Lyon railway, completed in 1832, and Marc designed a multi-tubular fire-tube boiler for it.

Semenza, Carlo (1893–1961) One of the greatest Italian designers of arch dams, as *Coyne* was for France. He was chief engineer of the *Societa Adriatica di Elettricita*, now nationalized.

Siemens, Werner von (1816–92) Walked to Berlin in 1834 to join the Prussian Corps of Engineers but was told that they were full and was sent to Magdeburg to join the artillery. There he acquired a good scientific training. In 1846 his electric telegraph sent 40 characters a minute. In 1847 with his mechanic friend, Halske, he started the firm of Siemens & Halske to make electric telegraphs. As early as 1867 this inventive man was thinking of electric railways. Now the largest company in Germany, Siemens produce only electrical goods of every sort.

Smeaton, John (1724–92) Probably the first man to call himself a consulting civil engineer, Smeaton left school at 16 to work in his father's law office, but sent to London he apprenticed himself to an instrument maker and after 1750 became an instrument maker himself. In 1754–5 he visited Holland to study its canals and ports. In 1756 he started to design the first Eddystone lighthouse to be built of stone. The previous two had been wooden and failed. Calm weather was needed for inspecting the rock 22 km (14 miles) out of Plymouth, so he had to stay several weeks. Back in London he built a model of the proposed lighthouse. It was approved and he arranged for stones to be sent from the Portland quarry. His dovetailed stones could not be dislodged by the sea and their shaping was begun that winter. The first stone, weighing 2.25 tons, was placed in the summer of 1757 and the lighthouse was completed in 1759. Smeaton also worked on the drainage of the fens and the repair and construction of bridges and harbours. His bridge piers sometimes had to be sunk by cofferdams built of dovetailed square timbers. Like his successor *Rennie*, in his work on Ramsgate harbour he used a diving bell.

Smith, William (1769–1839) A surveyor from the west of England who by mining work and canals, especially the Somerset Coal Canal, became a civil engineer. From 1799 he was not only the country's foremost drainage and water supply engineer, but he was given the Wollaston Award by the Geological Society for his work on the geological succession. He was convinced that certain fossils always appear in strata of the same geological age and he collected fossils to support this theory. In 1815 he published the first large-scale geological map of England and Wales with part of Scotland, at a scale of 1 cm to 3.2 km (1 in. to 5 miles).

Stephenson, George (1781–1848) The first railway engineer had no formal education and no training except work at coal mines in north-east England. He taught himself to read and by 21 was engineman at Willington Colliery. (His son Robert, the civil engineer, was born in 1803.) By 1812 he was in charge of all the engines at several mines and by 1814 he had built some 39 stationary engines, including one of 200 hp.

In 1814 his *Blücher* locomotive was the first to haul coal on the surface. In 1815 George joined the Walker Ironworks part-time to build his inventions.

As originally conceived in 1821, the Stockton–Darlington Railway was to have been horse-drawn. George was made its engineer in 1823 and began surveying. In 1825 it was opened with George's engine *Locomotion* pulling the first people ever to travel by train. In 1825 he was appointed engineer of the Liverpool–Manchester Railway, with 73 bridges and the first deep railway cutting as well as a marsh to cross, but it opened in 1830. George was the first president of the Institution of Mechanical Engineers in 1847.

Stephenson, Robert (1803–59) Robert's father, George, paid for his schooling in Newcastle and for a year of university studies in Edinburgh. He helped his father on the Stockton–Darlington line and from 1827 was manager of Stephenson & Pease, his father's locomotive works in Newcastle. In 1824 he had gone to Colombia, originally to the silver mines, and travelled all over South America to find work but without success, returning home in 1827.

In 1833 he was appointed to his biggest job, the London–Birmingham Railway, the first to reach London from the north, opened in 1838. He also designed the Derby–Leeds and Manchester–Leeds lines and others but retired in 1840, though he continued as a consultant to visit Spain, Belgium, Sweden, Denmark and Switzerland on railway matters.

Stevin, Simon (1548–1620) Flemish hydraulic engineer, mathematician and physicist. Holland during his lifetime was struggling for freedom from its Spanish rulers, and trying to drain its inland lakes. Stevin was active in both these ways. His book, *A New Manner of Fortification by Sluices*, was the first book to describe how to stop invaders by flooding the land. His many books were written in Flemish (itself a political statement) but then translated into French, sometimes also into English and Latin.

Stuart, Herbert Akroyd (1864–1927) Studied engineering at Imperial College and then entered his father's works at Bletchley. In 1886 he began experimenting with a compression-ignition engine. Spark-ignition engines running on paraffin existed at the time. Like *Diesel* he thought of compressing the air before it came into contact with the fuel oil to be injected by pump into it, and thus ignited immediately. He patented the idea in 1890. By the end of the year his first 12 heavy oil engines had been built. In 1891 he licensed Richard Hornsby & Sons of Grantham, Lincs., to make his 'Hornsby–Akroyd engine'. Rudolf Diesel's engine, patented in 1893, was not truly practical until 1897. Even then, the oil was not injected by pump but by high-pressure compressed air which involved difficulties with multi-stage compression. When Stuart's patents expired in 1906, many engine builders adopted his 'airless injection'.

Suyya (active AD 860) Indian hydraulic engineer who found out how to control the Jhelum River in Kashmir, freeing former swamps for cultivation, thus raising food production and preventing floods.

Telford, Thomas (1757–1834) A mason from Eskdale, a remote district of south Scotland, Telford worked on the building of the New Town in Edinburgh, and in 1782 came to London to work on Somerset House, studying construction in his spare time. By 1784 he was in Portsmouth Dockyard building a house and a chapel.

The conscientiousness, intelligence and charm of 'Laughing Tam' soon gave him a good reputation, and after some work on Shrewsbury Castle he was appointed Surveyor of Public Works for Shropshire, where he designed his first bridges. This was the time of the 'canal boom'; in 1793 he was appointed Engineer to the Ellesmere Canal and in 1795 to the Shrewsbury Canal. Until then canals had been made watertight only by puddling – giving them a lining of 1 m of *clay puddle*. Telford was the first to use cast iron instead of clay on his aqueducts, which must have saved space, weight, time and money. He also built roads and bridges. ICE founder member and first president, Laughing Tam never married, though he was settled in a London house for many years.

Terzaghi, Karl Anton von (1883–1961) An Austrian engineer, the founder of *soil mechanics* as a science. He spent much of his life in the USA, luckily for engineers who do not read German. He received his doctorate of engineering from Graz University in 1911 and visited the USA before 1914. His *Erdbaumechanik (Soil Mechanics)* was published in 1925 and in that year he joined the staff of Massachusetts Institute of Technology. In 1929 he accepted the newly created chair of soil mechanics at Vienna University but in 1938 returned to the USA for the chair of civil engineering at Harvard University from 1946 to 1956. His consulting practice covered the world and he was chairman of Egypt's Aswan High Dam project until 1959.

Than-Ston-Rgyal-Po (1385–1464) A Tibetan Buddhist whose chain suspension bridge for pedestrians over the Brahmaputra River at Chak-Sam-Chö-Ri (1420) spanned 135 m (450 ft), and was seen by Westerners in 1903. Unlike Chinese suspension bridges of his period, the deck was level.

Thomas, Sydney Gilchrist (1850–85) With his cousin, Percy Gilchrist, an ironworks chemist in Wales, Thomas took out his first patent for a dolomite lining to the Bessemer converter that would remove phosphoric impurities from the iron ore. Further patents followed in 1878 and 1879. He received the Bessemer Gold Medal from Henry Bessemer himself. By 1929 world steel output had reached 104 million tons, 90% by his process.

Torroja y Miret, Eduardo (1891–1961) Spanish architect and engineer famous for his concrete shell structures, in particular his Zarzuela racecourse grandstand in Madrid with a shell cantilever roof projecting out some 13 m (42 ft). In 1951 he formed the Technical Institute of Construction and Cement and was its director until his death. *The Structures of Eduardo Torroja* is his autobiography (1958).

Tredgold, Thomas (1788–1829) Durham carpenter who worked in Scotland until 1813 and then came to London to work for an architect, William Atkinson. Author of several books, he was one of the 50 founder members of the ICE and wrote the ICE's definition of *civil engineering* in its Royal Charter of 1828.

Trésaguet, Pierre (1716–94) French civil engineer in the Corps des Ponts et Chaussées, Trésaguet used his roadbuilding method first in the road from Paris to Toulouse and the Spanish frontier. Inspector-General of the Corps from 1775, his method by 1788 had been used to build 19,300 km (12,000 miles) of road. He shaped the formation soil to a camber, laid large stones upright on it and filled their spaces with small stones. He emphasized that the wearing course had to be the hardest possible stones, even if imported from another district.

Trevithick, Richard (1771–1833) Cornishman from Carn Brea, where his father managed the Dolcoath mines. To avoid litigation with *Watt*, he designed his steam engines to work at high steam pressures. They were consequently much lighter than Watt's, but he had less success than he deserved.

Vermuyden, Sir Cornelius (1595–1683) A Dutch engineer invited to England to stem a breach in the north bank of the Thames near Dagenham. He was so successful that the king employed him to drain his land near Hatfield, Yorks. Vermuyden had the misfortune to live at the time of the Civil War and to be a friend of the king so he died in poverty, though his drainage of the English fens was a great achievement – over 120,000 hectares (300,000 acres) drained. See **Perry**.

Vicat, Louis Joseph (1786–1861) An unusually capable engineer, both analytical and practical, from the École des Ponts et Chaussées. His first task on qualifying seemed impossible at the time – bridging the torrential River Dordogne at Souillac.

Money was scarce because of Napoleon's wars. Cements that would harden under water (hydraulic cements) did not exist, though it was known that the Romans had used *pozzolans*. Vicat decided to make his own hydraulic cement, did so, published his results in 1818, and completed the bridge in 1824 and then a wire-rope suspension bridge over the Dordogne at Argentat in 1828. For the next 20 years he travelled over France locating limestone and clay deposits that could make good cement. His work resulted in wide use of concrete in France by 1830. The bridge at Argentat had towers made of precast concrete blocks. No stone was quarried. He conceived the spinning of ropes on site, but never used it. His methods of testing cement were practised until recently.

Vignoles, Charles Blacker (1793–1875) Born in Ireland of a Huguenot family.

When his father was killed in 1794 and the infant was taken prisoner by the French, Sir Charles Grey gave him an officer's commission to arrange for his release. In 1816 he was in the USA, making a survey of South Carolina and nearby states, but returning to England, he worked on the Liverpool–Manchester Railway and other railways and canals, and was the chief engineer of the first Irish railway, the Dublin–Kingston line. Inventor of the flat-bottomed rail, then known as the 'railway bar', now the standard rail throughout the world, he was consulted on many French, German, Swiss and Brazilian railways. ICE president 1870.

Vinci, Leonardo da (1452–1519) Born out of wedlock at Vinci near Florence but acknowledged by his father, Leonardo was a great painter, poet, sculptor, goldsmith, and military and civil engineer. He used lifting tackle and toothed wheels, and experimented with steam. He lived at the royal courts of Borgia and Milan and spent his last four years at the court of King Francis I of France.

Vitruvius, (Marcus Vitruvius Pollio, first century BC) His ten books on building, entitled *De Architectura*, were the chief authority on Greek construction for many centuries, though he himself was a Roman. They describe military as well as civil engineering.

Watt, James, FRS (1736–1819) Born in Greenock on the Clyde, then a much bigger town than Glasgow. Watt was not strong and was apprenticed to an instrument maker. In 1763 he repaired Glasgow University's *Newcomen* engine. Watt discovered why it was inefficient: it needed a condenser and Watt invented one. To feed his wife and children, Watt worked from 1766 to 1774 as a land surveyor, setting out Scottish canals.

In 1773 Watt's wife died and his partner, Dr Roebuck, became bankrupt. Watt luckily found another supporter in Matthew Boulton whose factory at Soho, Birmingham, made profitable metal articles and was driven by a water wheel. Boulton settled Watt's debts to Dr Roebuck and Watt moved to Birmingham with his two children. The Soho works then built the most modern engines of the period, including in 1776 a blower for a blast furnace, with a cylinder dia. of nearly 1 m (39 in.) and a pumping engine with a cylinder dia. of 1.27 m (50 in.). By 1790 Watt was a wealthy man, having earned £76,000 in royalties in 11 years. Though he retired in 1800 he continued to invent things.

Westinghouse, George (1846–1914) Born in New York State, of German stock. His father built steam engines and other farm machines. In 1865 he made a re-railing device for derailed wagons. In 1867 he patented his railway brake worked by compressed air and built the first one himself. At that time the only way to brake the train was for the guard to climb over the wagon roofs and to put on each brake by hand. Westinghouse took out some 400 patents for railway brakes, signals, alternating current, natural gas, etc.

His electrical generators used alternating current at a time when direct current was fashionable, but he built the first generators for the Niagara water-power station. In 1893 for the Chicago World Exhibition he was awarded the lighting contract against other contractors who held Edison's patents for electric light bulbs. Undeterred, he designed his own electric bulb with a replaceable filament and filled it with nitrogen. He installed the 10,000 hp generators for the exhibition. Although he lost money on the contract he gained enormously by the publicity.

This far-sighted man in 1912 regarded the outbreak of the 1914 war as imminent because Germany was industrially so far ahead of the UK. He thought this was because compulsory military service enabled men to adapt easily to factory discipline.

Whitworth, Sir Joseph, FRS (1803–87) Almost as much prized as 'FRS' are the letters 'Wh.Sch.', meaning 'Whitworth Scholar', someone with a fine engineering training earned by technical ability alone. Sir Joseph left £100,000 to endow these scholarships for people without money. Whitworth's education finished at 14 when he was sent to an uncle to learn cotton spinning. But in 1821 he ran away to Manchester to work at an engineer's bench until 1825. Then he was in London working for *Maudslay*, the country's finest toolmaker. While at Maudslay's he worked out how to obtain a plane surface on a piece of steel plate by first preparing two plates, rubbing them together to find the high points, and then hand-scraping (not grinding) them off. The final true surface is obtained with three plates.

Williams, Sir E. Leader (1838–1910) Designer of the Manchester Ship Canal.

Yu, the Great Yu, (active 2280 BC) The first hydraulic engineer. He tamed the Yellow River, stopped the flooding and became Emperor of China.

READ MORE IN PENGUIN

In every corner of the world, on every subject under the sun, Penguin represents quality and variety – the very best in publishing today.

For complete information about books available from Penguin – including Puffins, Penguin Classics and Arkana – and how to order them, write to us at the appropriate address below. Please note that for copyright reasons the selection of books varies from country to country.

In the United Kingdom: Please write to *Dept. JC, Penguin Books Ltd, FREEPOST, West Drayton, Middlesex UB7 OBR*

If you have any difficulty in obtaining a title, please send your order with the correct money, plus ten per cent for postage and packaging, to *PO Box No. 11, West Drayton, Middlesex UB7 OBR*

In the United States: Please write to *Penguin USA Inc., 375 Hudson Street, New York, NY 10014*

In Canada: Please write to *Penguin Books Canada Ltd, 10 Alcorn Avenue, Suite 300, Toronto, Ontario M4V 3B2*

In Australia: Please write to *Penguin Books Australia Ltd, 487 Maroondah Highway, Ringwood, Victoria 3134*

In New Zealand: Please write to *Penguin Books (NZ) Ltd, 182–190 Wairau Road, Private Bag, Takapuna, Auckland 9*

In India: Please write to *Penguin Books India Pvt Ltd, 706 Eros Apartments, 56 Nehru Place, New Delhi 110 019*

In the Netherlands: Please write to *Penguin Books Netherlands B.V., Keizersgracht 231 NL–1016 DV Amsterdam*

In Germany: Please write to *Penguin Books Deutschland GmbH, Friedrichstrasse 10–12, W–6000 Frankfurt/Main 1*

In Spain: Please write to *Penguin Books S. A., C. San Bernardo 117–6° E–28015 Madrid*

In Italy: Please write to *Penguin Italia s.r.l., Via Felice Casati 20, I–20124 Milano*

In France: Please write to *Penguin France S. A., 17 rue Lejeune, F–31000 Toulouse*

In Japan: Please write to *Penguin Books Japan, Ishikiribashi Building, 2–5–4, Suido, Tokyo 112*

In Greece: Please write to *Penguin Hellas Ltd, Dimocritou 3, GR–106 71 Athens*

In South Africa: Please write to *Longman Penguin Southern Africa (Pty) Ltd, Private Bag X08, Bertsham 2013*

READ MORE IN PENGUIN

PENGUIN SCIENCE AND MATHEMATICS

QED Richard Feynman
The Strange Theory of Light and Matter

Quantum thermodynamics – or QED for short – is the 'strange theory' – that explains how light and electrons interact. 'Physics Nobelist Feynman simply cannot help being original. In this quirky, fascinating book, he explains to laymen the quantum theory of light – a theory to which he made decisive contributions' – *New Yorker*

God and the New Physics Paul Davies

Can science, now come of age, offer a surer path to God than religion? This 'very interesting' (*New Scientist*) book suggests it can.

Does God Play Dice? Ian Stewart
The New Mathematics of Chaos

To cope with the truth of a chaotic world, pioneering mathematicians have developed chaos theory. *Does God Play Dice?* makes accessible the basic principles and many practical applications of one of the most extraordinary – and mindbending – breakthroughs in recent years. 'Engaging, accurate and accessible to the uninitiated' – *Nature*

The Blind Watchmaker Richard Dawkins

'An enchantingly witty and persuasive neo-Darwinist attack on the anti-evolutionists, pleasurably intelligible to the scientifically illiterate' – Hermione Lee in the *Observer* Books of the Year

The Making of the Atomic Bomb Richard Rhodes

'Rhodes handles his rich trove of material with the skill of a master novelist ... his portraits of the leading figures are three-dimensional and penetrating ... the sheer momentum of the narrative is breathtaking ... a book to read and to read again' – Walter C. Patterson in the *Guardian*

Asimov's New Guide to Science Isaac Asimov

A classic work brought up to date – far and away the best one-volume survey of all the physical and biological sciences.

READ MORE IN PENGUIN

PENGUIN DICTIONARIES

Abbreviations
Archaeology
Architecture
Art and Artists
Biology
Botany
Building
Business
Chemistry
Civil Engineering
Computers
Curious and Interesting
  Words
Curious and Interesting
  Numbers
Design and Designers
Economics
Electronics
English and European
  History
English Idioms
French
Geography
German

Historical Slang
Human Geography
Literary Terms
Mathematics
Modern History 1789–1945
Modern Quotations
Music
Physical Geography
Physics
Politics
Proverbs
Psychology
Quotations
Religions
Rhyming Dictionary
Saints
Science
Sociology
Spanish
Surnames
Telecommunications
Troublesome Words
Twentieth-Century History